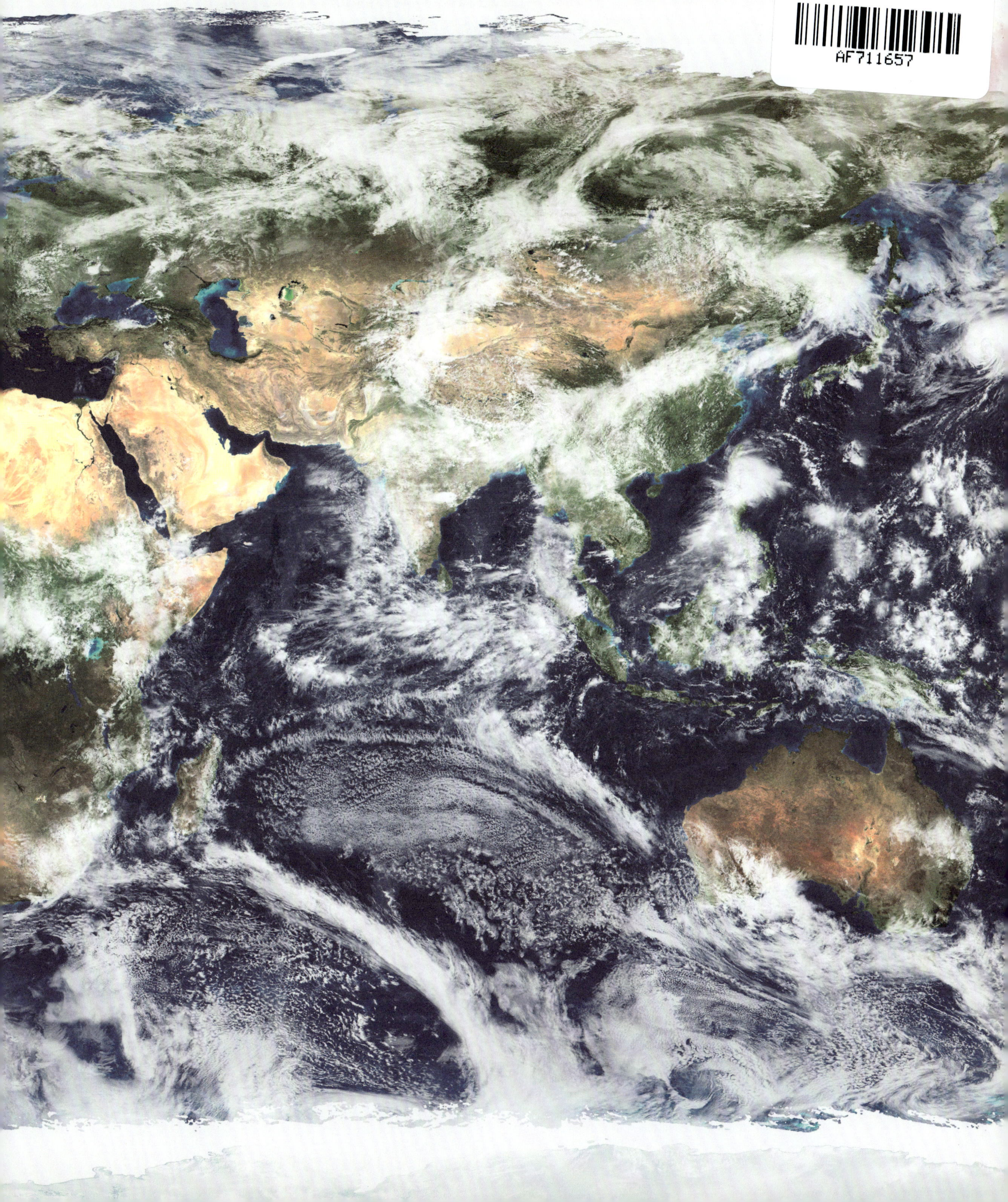

Unsere Erde – Satelliten erforschen unseren Planeten

Beautiful Earth – Our planet explored from above

Une Terre inconnue – Notre planète vue de l'espace

Nuestro planeta Tierra – Una maravilla vista desde el espacio

Dirk H. Lorenzen

UNSERE ERDE
Satelliten erforschen unseren Planeten

BEAUTIFUL EARTH
Our planet explored from above

UNE TERRE INCONNUE
Notre planète vue de l'espace

NUESTRO PLANETA TIERRA
Una maravilla vista desde el espacio

h.f.ullmann

In dankbarer Erinnerung an meinen wundervollen Lektor
Harro Schweizer (1949-2012)

In grateful memory of my wonderful teacher
Harro Schweizer (1949-2012)

À la gratitude et mémoire de mon merveilleux professeur
Harro Schweizer (1949-2012)

En memoria y agradecimiento a mi maravilloso profesor
Harro Schweizer (1949-2012)

Special edition

Book concept: Dirk H. Lorenzen, Harro Schweizer

Project management: Lars Pietzschmann

Book design: Prill Partners producing, Berlin
Cover design: Kati Klaeske
Cover photo: desert in Namibia (© USGS/NASA)

Repro: LVD GmbH, Berlin

Producer and editor: Harro Schweizer, Berlin

English translation: JMS Books llp (translation Malcolm Garrard, editor Sarah Chatwin)
French translation: Mariane Pontone
Spanish translation: Quality, Servicios Globales Editoriales, S.L., Barcelona (translation Pablo Díaz-Aller)

Updates:
Delivering iBooks & Design, Barcelona

Overall responsibility for production:
h.f.ullmann publishing GmbH, Potsdam, Germany

Printed in China, 2015

978-3-8480-0827-8 (D-Cover)
978-3-8480-0828-5 (GB-Cover)
978-3-8480-0888-9 (F-Cover)
978-3-8480-0889-6 (E-Cover)

10 9 8 7 6 5 4 3 2 1
X IX VIII VII VI V IV III II I

www.ullmann-publishing.com
newsletter@ullmann-publishing.com

Erste Seite: Die Erde, wenn überall Tag wäre (Montage)
Letzte Seite: Die Erde, wenn überall Nacht wäre (Montage)
Die Bilder auf den folgenden vier Doppelseiten:
6/7: Deutschland von oben: Inseln im nordfriesischen Wattenmeer, kombiniert aus drei Radaraufnahmen bei unterschiedlichem Wasserstand (TerraSAR-X)
8/9: Großbritannien von oben: Gut zu erkennen sind Wald- und Bergregionen, große Städte und unterschiedlich gefärbtes Wasser (Envisat)
10/11: Frankreich von oben: Radarbild von Paris, kombiniert aus zwei Aufnahmen. Unterschiedliche Farben zeigen unterschiedliche Höhen (TerraSAR-X)
12/13: Iberische Halbinsel von oben: die Hitze eines Hochsommertags – je stärker das Rot, desto heißer, hohe Wolken erscheinen kalt (Aqua)

First page: How the Earth would look if it could be daytime everywhere at once (montage)
Last page: How the Earth would look if it could be night everywhere at once (montage)
The images on the four following double page spreads are of:
6–7: Germany from above: islands in the Wadden Sea off North Frisia, a picture composed from three radar images taken during a succession of tides (TerraSAR-X)
8–9: Great Britain from above: forests, mountains, large cities, and differences in water coloration are all easy to make out (Envisat)
10–1: France from above: a radar image of Paris composed from two exposures. The variations in color reveal differences in elevation (TerraSAR-X)
12–3: The Iberian Peninsula from above: the heat of a day in high summer – the deeper the red, the higher the temperature; high cloud appears cold (Aqua)

Première page : la Terre, s'il faisait jour partout (photomontage)
Dernière page : la Terre, s'il faisait nuit partout (photomontage)
Illustrations présentées sur les doubles pages suivantes :
6/7 : L'Allemagne vue du ciel : îles frisonnes septentrionales dans la mer des Wadden, illustration combinant trois images radar acquises à des niveaux de marée différents (TerraSAR-X)
8/9 : La Grande-Bretagne vue du ciel : on distingue parfaitement les montagnes et forêts, les agglomérations et les variations de couleur de la mer (Envisat)
10/11 : La France vue du ciel : image radar de Paris, illustration combinant deux images. Les différentes couleurs représentent différentes altitudes (TerraSAR-X)
12/13 : La péninsule ibérique vue du ciel : représentation de la chaleur estivale. Le rouge foncé indique les secteurs plus chauds, les nuages hauts semblent froids (Aqua)

Primera página: La Tierra, si fuese de día en todas partes (montaje)
Última página: La Tierra, si fuese de noche en todas partes (montaje)
Ilustraciones de las siguientes cuatro dobles páginas:
6/7: Alemania desde el cielo: islas de las marismas de Frisia septentrional, combinación de tres tomas de radar con diferentes niveles de agua (TerraSAR-X)
8/9: Gran Bretaña desde el cielo: se pueden identificar muy bien las regiones boscosas y montañosas, las grandes ciudades y el agua de diferentes colores (Envisat)
10/11: Francia desde el cielo: imagen de radar de París, combinación de dos tomas distintas. Los diferentes colores muestran las diversas alturas (TerraSAR-X)
12/13: Península Ibérica desde el cielo: el calor de un día de canícula; cuanto más intenso es el rojo, más elevada la temperatura; las nubes altas aparecen frías (Aqua)

Inhalt
Contents
Sommaire
Contenidos

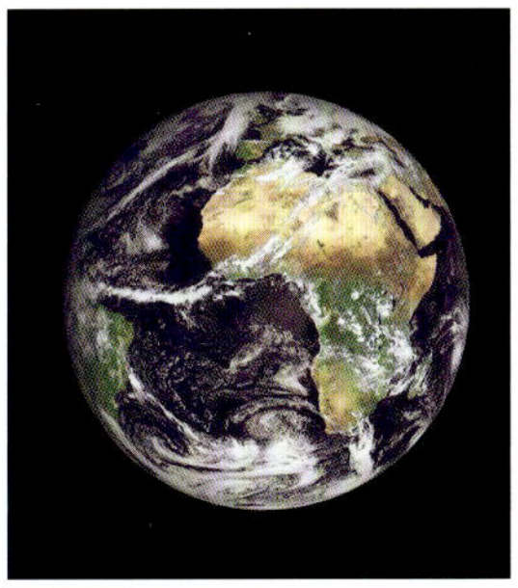
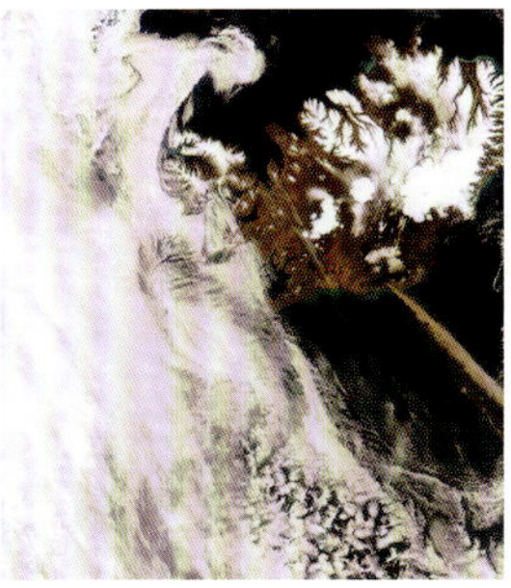
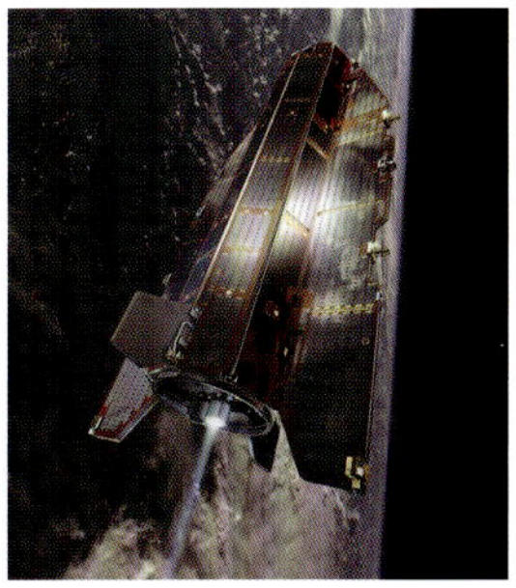
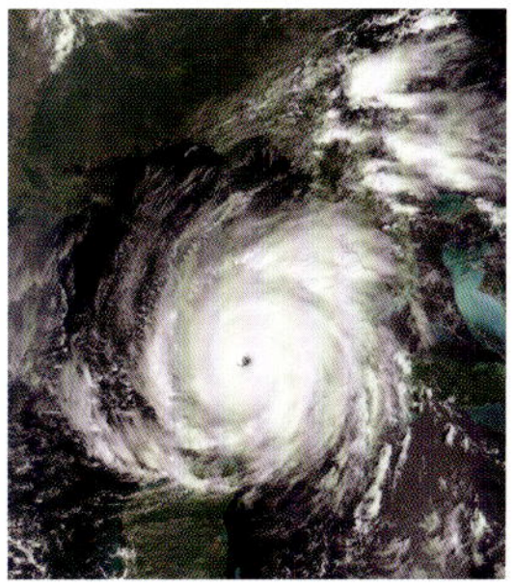
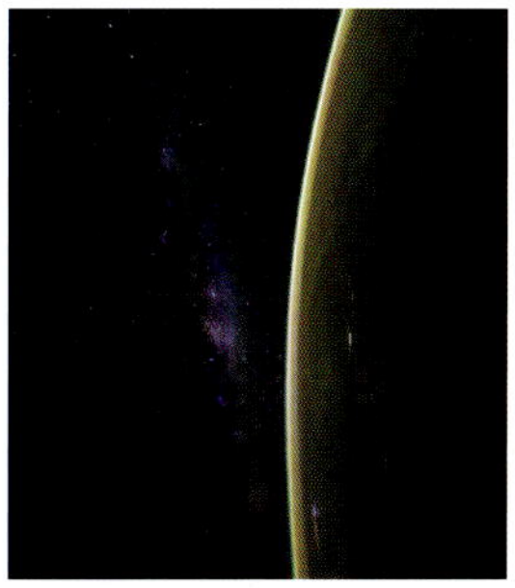

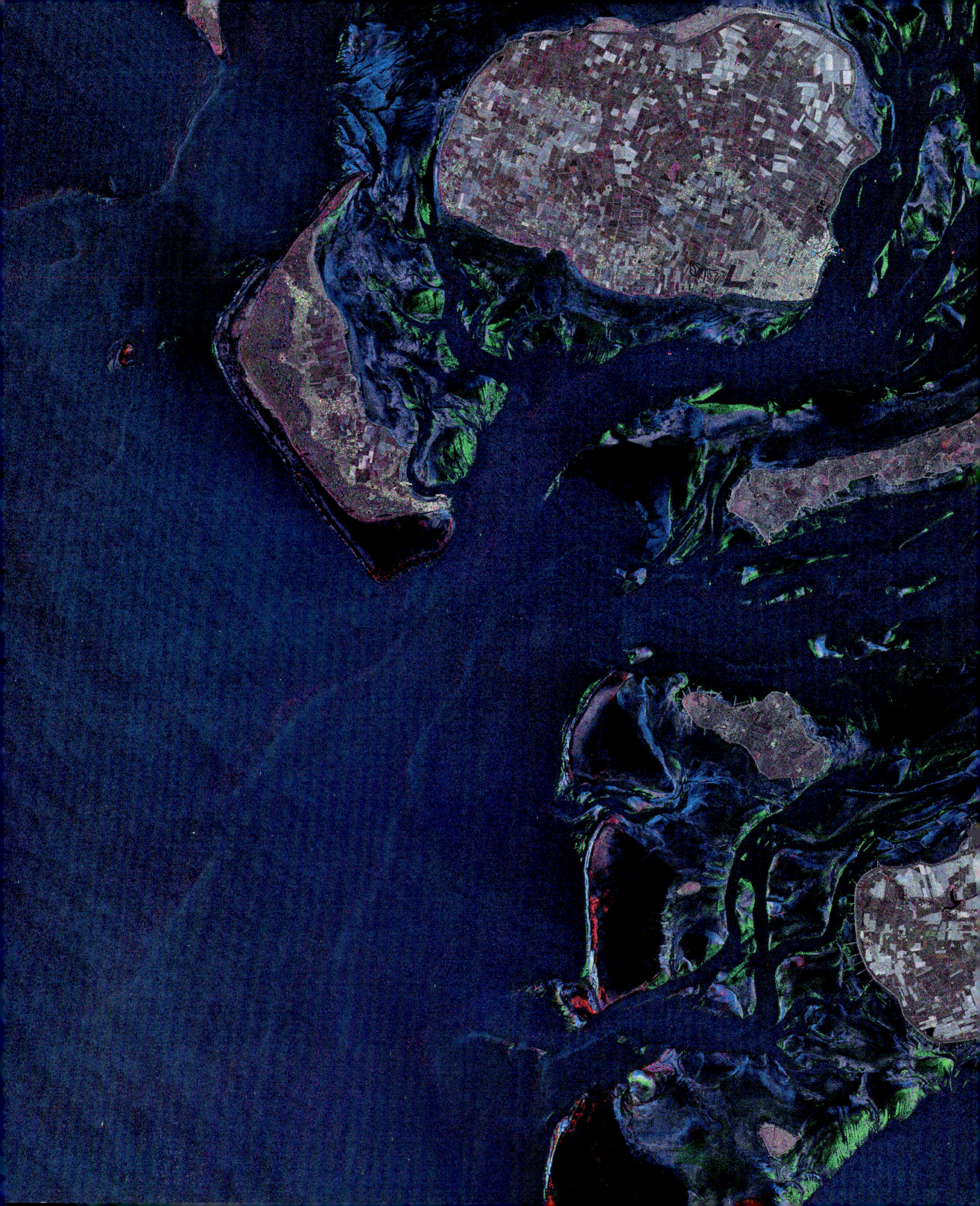

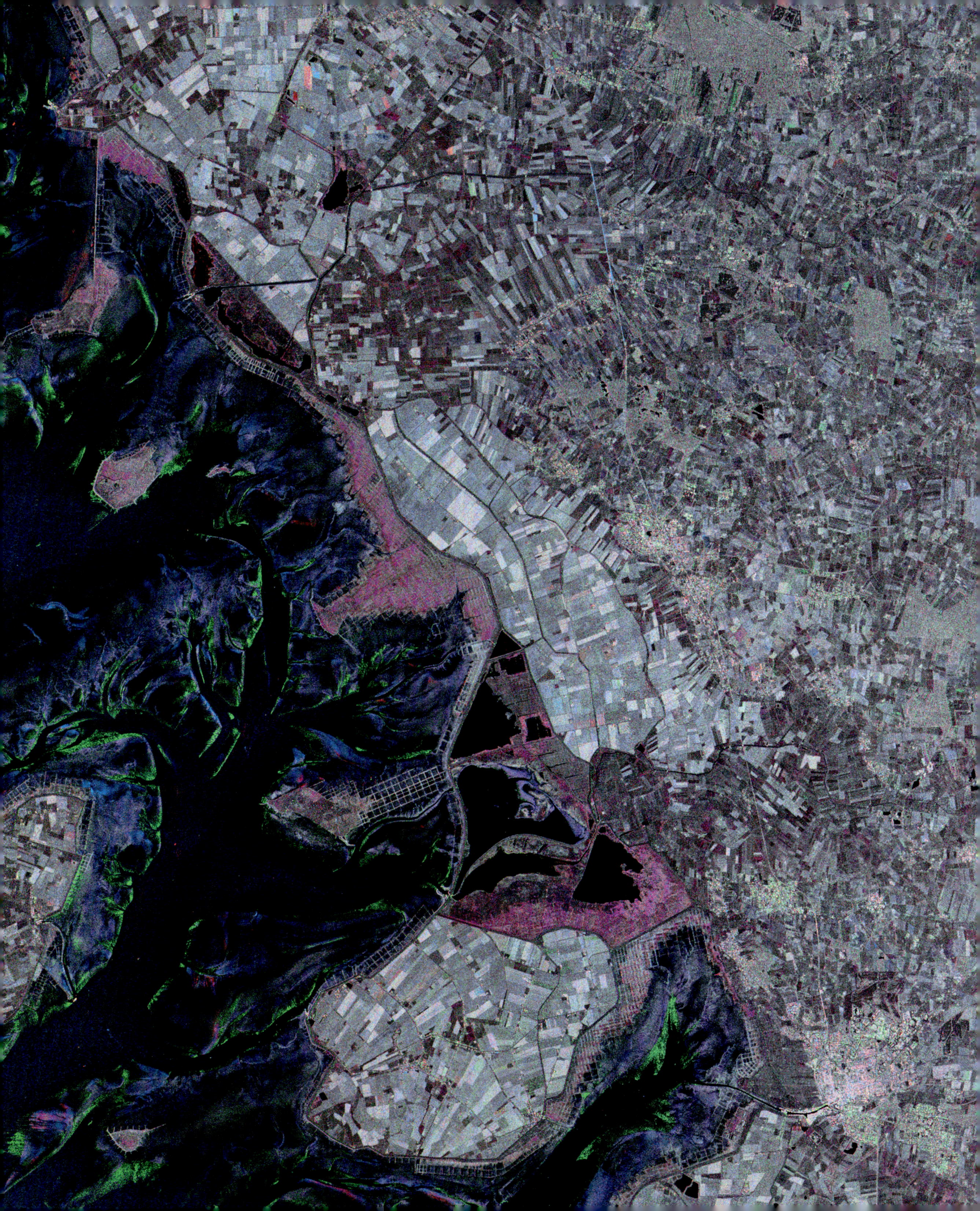

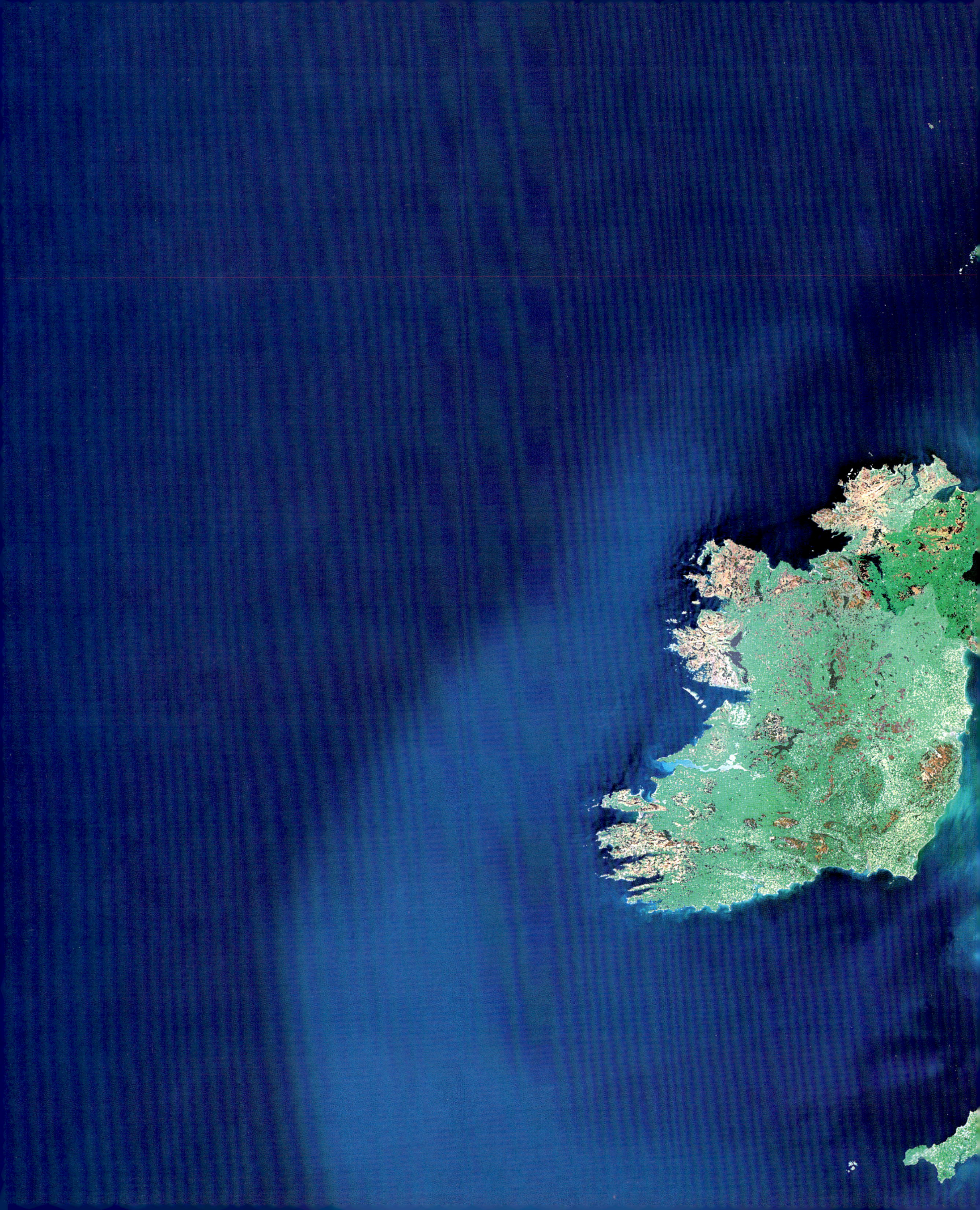

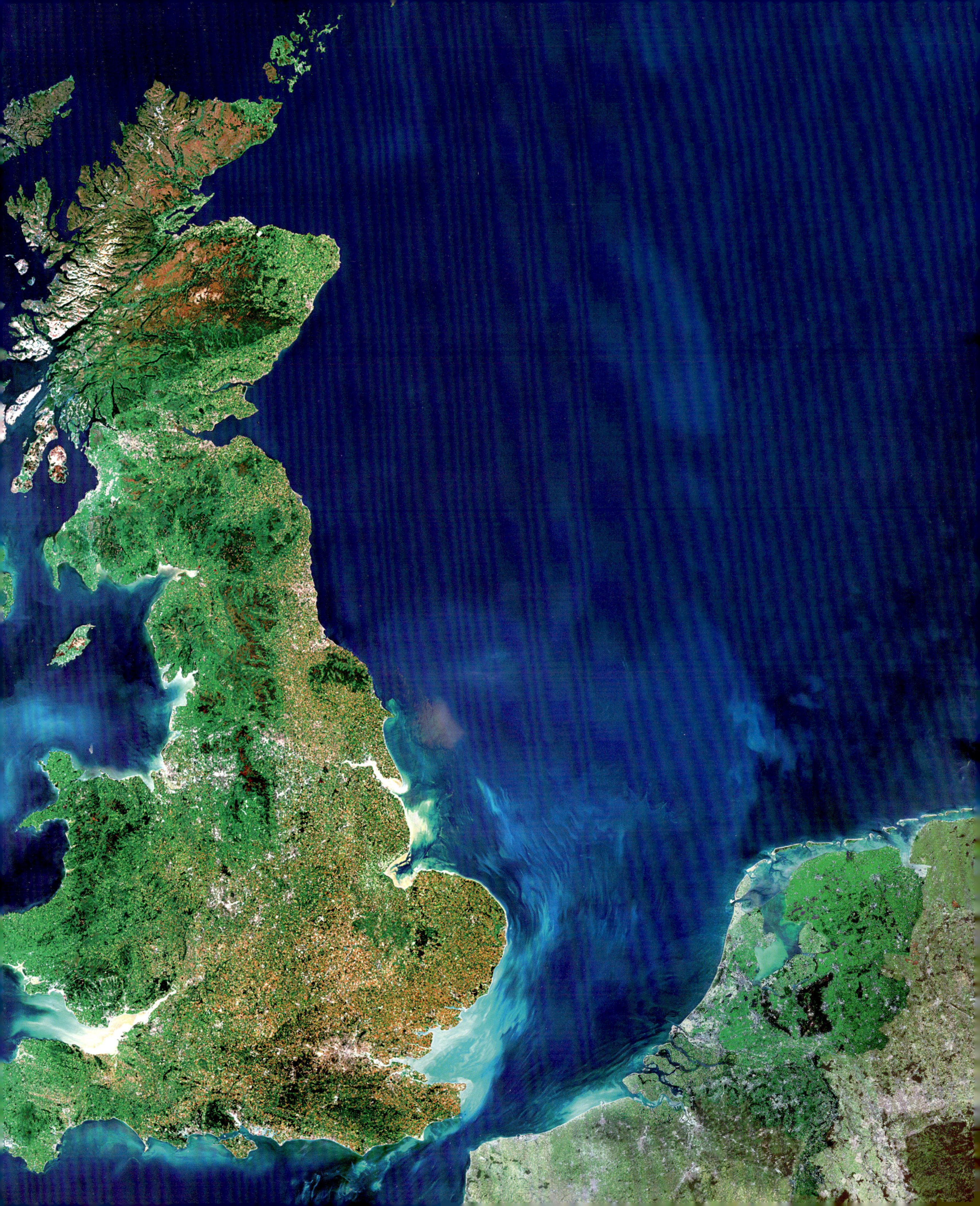

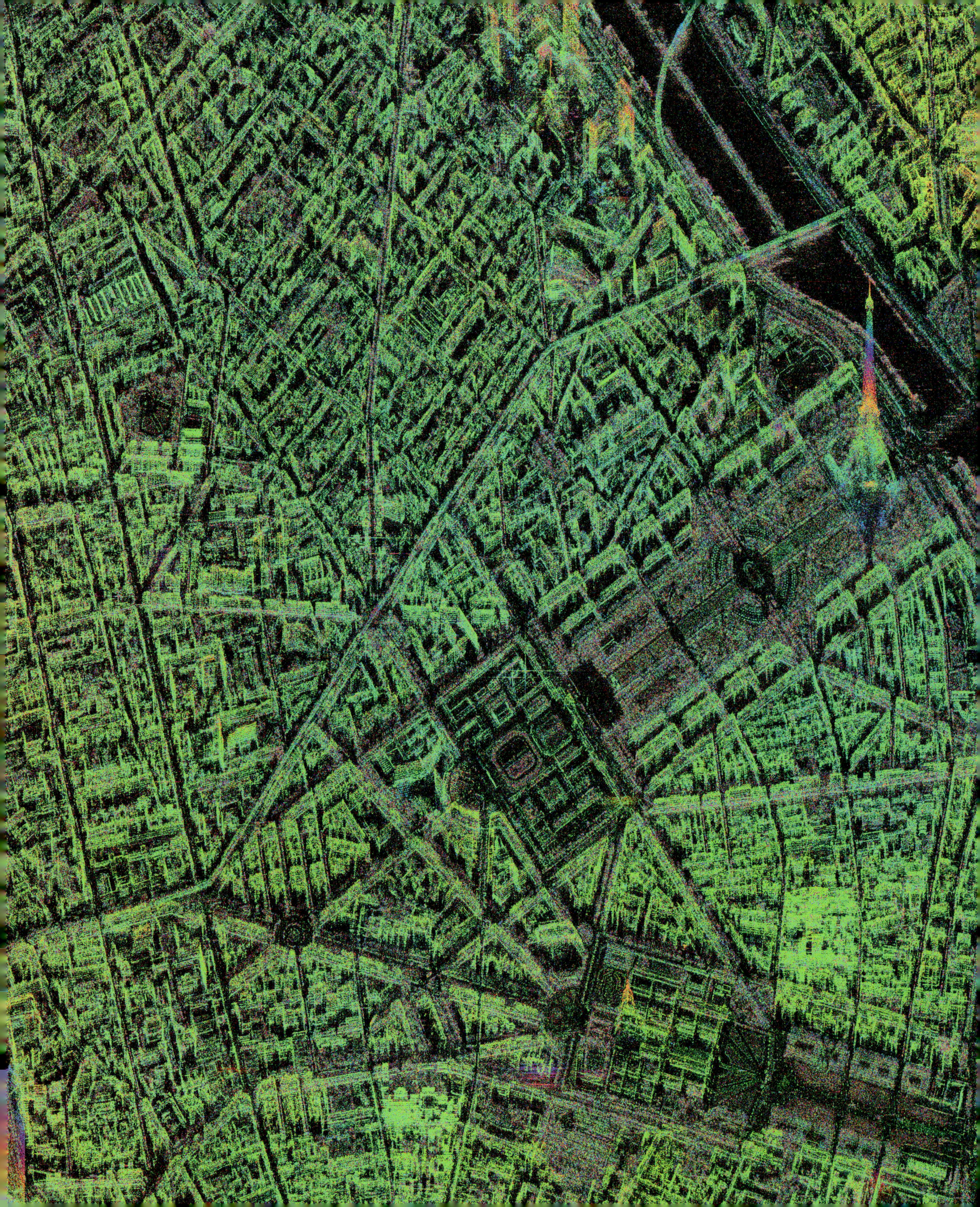

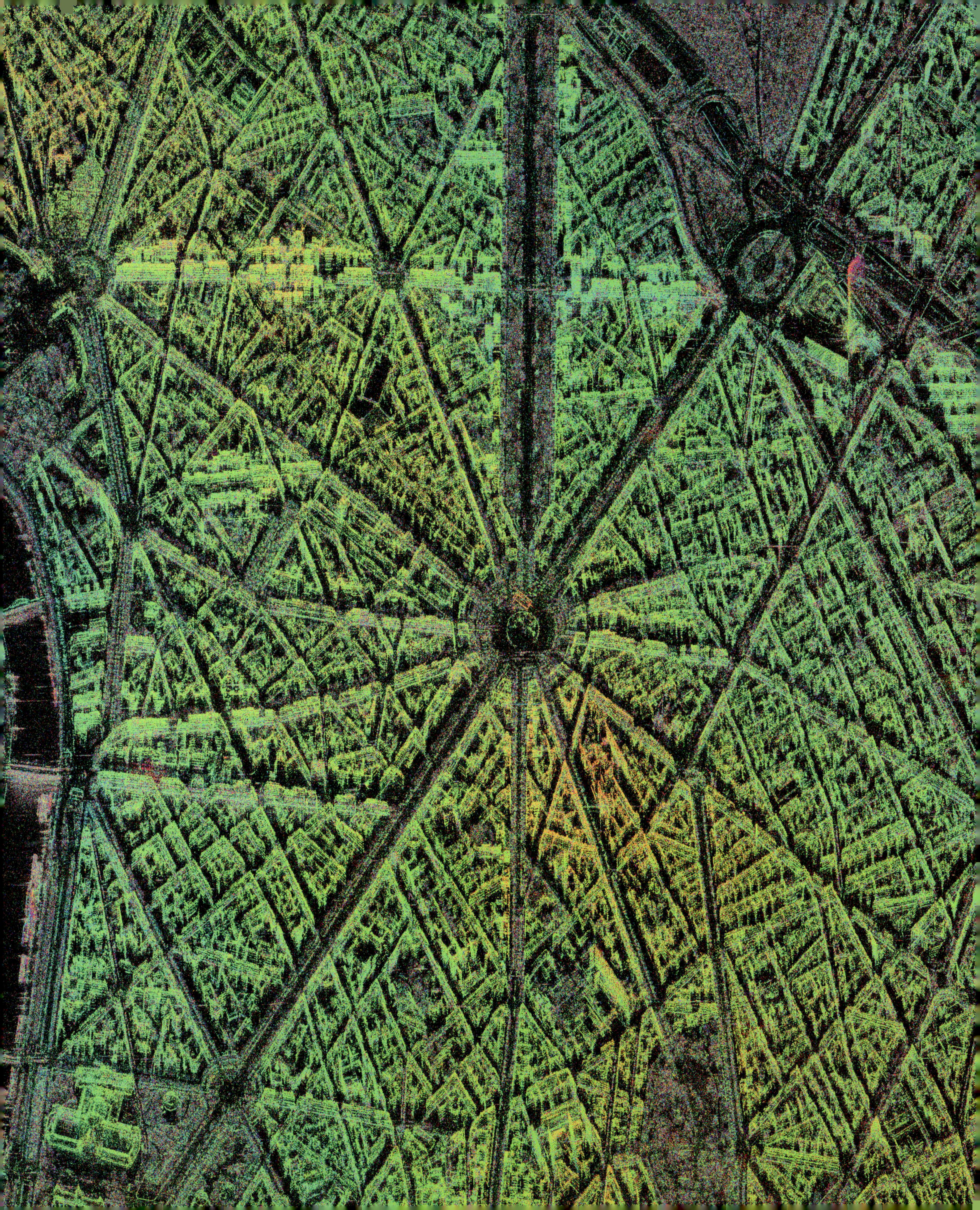

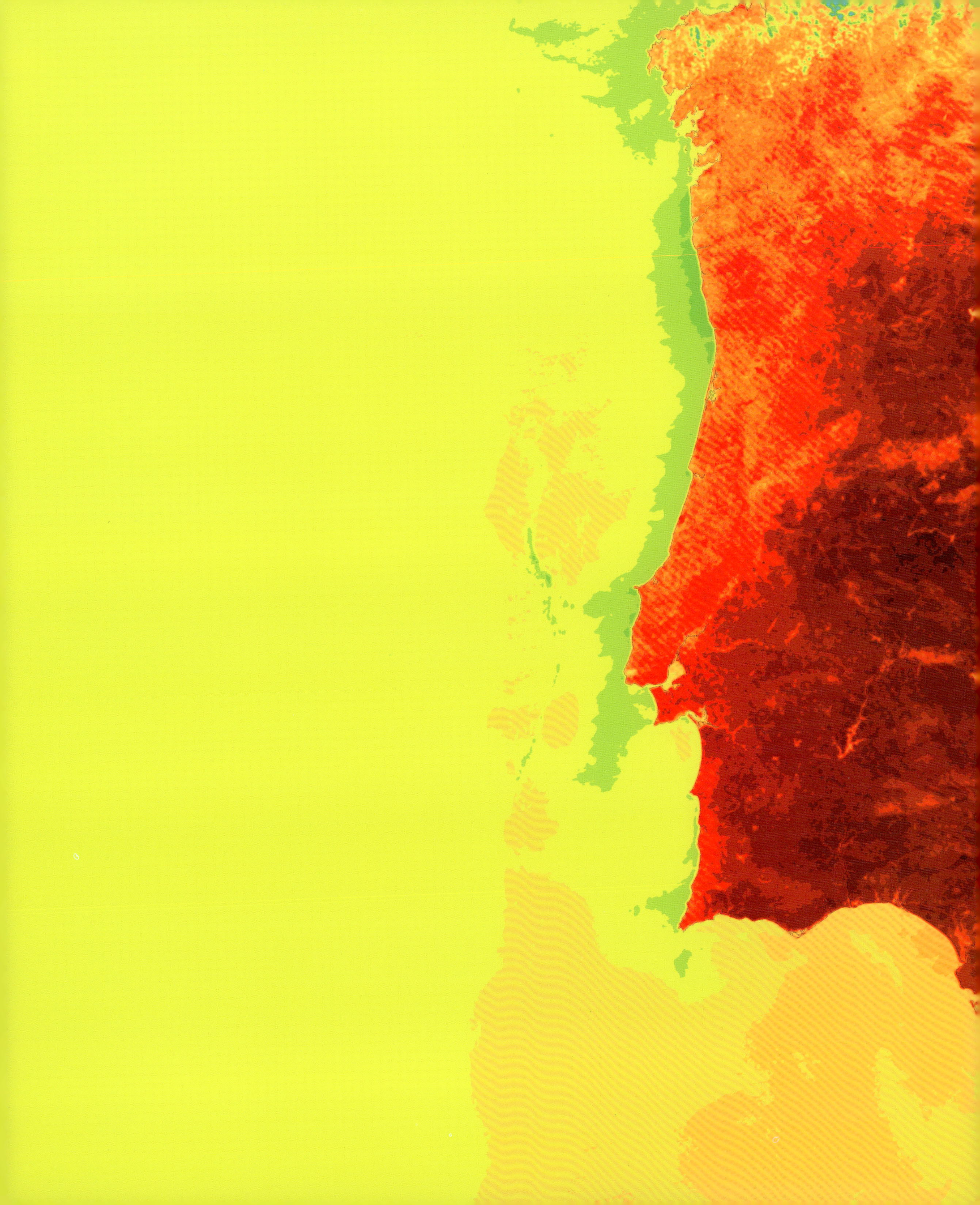

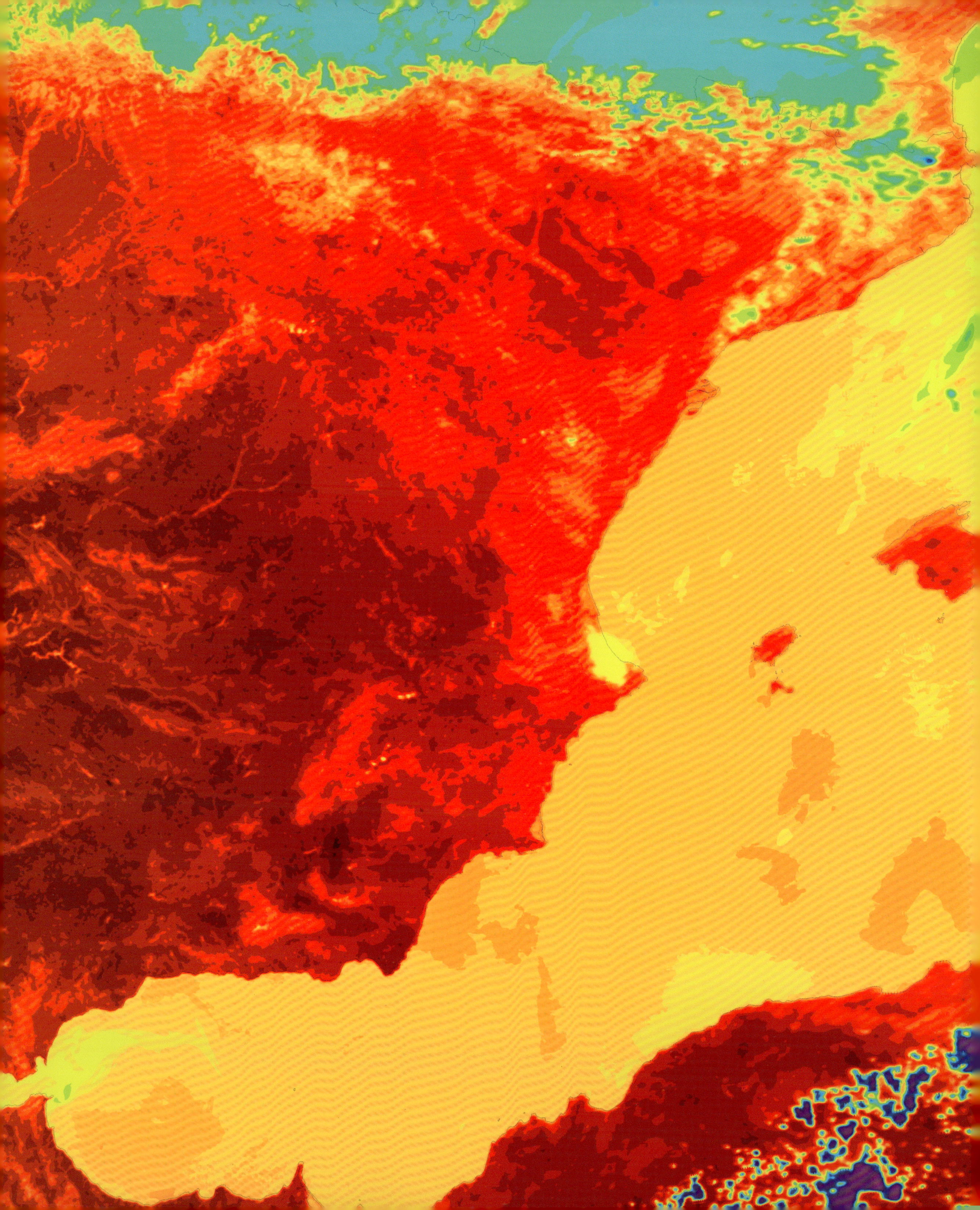

Eine neue Dimension der Verantwortung
A New Dimension of Responsibility
Une nouvelle dimension de la responsabilité
Una nueva dimensión de la responsabilidad

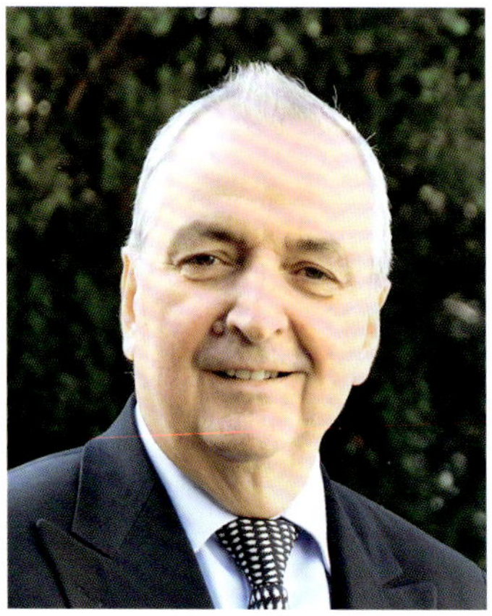

Klaus Töpfer

Bereits Mitte dieses Jahrhunderts werden etwa neun Milliarden Menschen unseren blauen Planeten Erde bewohnen. Besonders in Afrika und auf dem indischen Subkontinent wird es weiterhin einen steilen Anstieg der Bevölkerung geben. Es sind genau die Regionen der Welt, die bereits gegenwärtig durch Hunger, Armut und Perspektivlosigkeit gekennzeichnet sind. Es sind die Regionen, in denen es darum gehen muss, eine Entwicklung der Wirtschaft so zu gestalten, dass alle Menschen Zukunftsperspektiven erhalten, dass alle Menschen ein »Recht auf Entwicklung« verwirklichen können, wie es bereits 1992 auf dem »Erdgipfel« in Rio festgeschrieben und von allen Nationen einstimmig verabschiedet wurde.

Die massiven Entwicklungs- und Wohlstandsunterschiede, die gegenwärtig die Welt kennzeichnen, sind »keine Formel für den Frieden«, wie es Muhammad Yunus in seiner Dankesrede 2006 bei der Annahme des Friedensnobelpreises formuliert hat.

Diese Entwicklungsnotwendigkeit setzt voraus, dass wir mit den Leistungen in der Natur in ganz besonderer Weise intelligent und sparsam umgehen. Alle wirtschaftlichen Entwicklungen setzen voraus, dass Energie verfügbar gemacht wird, dass die Rohstoffe genutzt werden können,

By the middle of the century it is estimated that there will be some nine billion people living on our blue planet, the Earth. Africa and the Indian subcontinent in particular will continue to experience steep population growth, although it is precisely these regions of the world where hunger, poverty, and the despondency that results from such problems, are already widespread. These are the regions of the world where industrial development must be managed so that a new hope for the future may be available to all, and to provide a "right to development" for all—an idea first propounded and agreed unanimously by every nation that was represented at the Earth Summit held in Rio in 1992.

The massive disparity in levels of wealth and development that characterize the modern world "are no formula for peace," as Professor Muhammad Yunus reminded us in his acceptance speech for the 2006 Nobel Peace Prize.

This need for development presupposes that we exercise particular prudence and wisdom in our stewardship of the bounty that the natural world provides for us. All economic development is predicated on the assurance of

Notre planète comptera près de neuf milliards d'habitants dès le milieu de ce siècle. La population de l'Afrique et du sous-continent indien continuera notamment à croître rapidement. Il s'agit précisément des régions du monde qui souffrent, déjà à l'heure actuelle, de la faim, de la pauvreté et du manque de perspectives. Il s'agit des régions qui requièrent un développement économique capable de fournir à tous des perspectives d'avenir et de permettre à chaque individu jouir du « droit au développement », tel qu'il a été défini et adopté à l'unanimité par toutes les nations lors du « Sommet de la Terre » de Rio en 1992.

Comme l'a exprimé Muhammad Yunus dans son discours de remerciement prononcé à l'occasion de la remise du prix Nobel de la paix en 2006, les différences considérables en termes de développement et de prospérité qui caractérisent le monde actuel ne constituent « pas une formule pour la paix ».

Cet indispensable développement présuppose une utilisation intelligente et parcimonieuse des ressources naturelles. Tout développement économique repose sur la mise à disposition de l'énergie et des matières premières nécessaires qui constituent la base du processus de

A mediados de este siglo ya serán nueve mil millones de personas los que habiten el planeta Tierra. Especialmente en África y en el subcontinente indio se seguirá dando en el futuro un fuerte aumento de la población. Son éstas, precisamente, las regiones del mundo que ya están marcadas ahora por el hambre, la pobreza y la desesperanza. Son las regiones en las que es imprescindible organizar un desarrollo económico para que todas las personas tengan perspectivas de futuro, para que todos puedan hacer realidad su «derecho al desarrollo», tal como se consagró en la «Cumbre de la Tierra» de 1992, en Río de Janeiro, y fue aprobado unánimemente por todas las naciones.

Las enormes diferencias de desarrollo y bienestar que actualmente caracterizan al mundo, no son «una fórmula para la paz», como escribió Muhammad Yunus en su discurso de agradecimiento al recibir el Premio Nobel de la Paz en 2006.

Esta necesidad de desarrollo exige que manejemos los recursos naturales de un modo inteligente y ahorrador. El desarrollo económico exige que la energía esté disponible, que las materias primas que son la base indispensable del proceso de

die für diese wirtschaftlichen Entwicklungsprozesse unumgängliche Grundlage sind. Dazu gehören:

- die Leistungen der großen Ökosysteme, der Ozeane, der Atmosphäre, der Kryosphäre – die damit verbundene Bereitstellung und sinnvolle Nutzung von Wasser, die Sicherung von Böden und ihrer Produktivität;
- die Vielfalt der Tier- und Pflanzenarten, verbunden mit den darin eingebauten Genpotenzialen, diesem unerschöpflichen »Code der Natur« für die Lösung von Problemen;
- die Chancen, weitere erneuerbare Energiequellen zu erschließen – Solar- und Windenergie, Biomasse und Geothermie und vieles andere;
- die Stabilität der Atmosphäre zur Bewahrung des klimatischen Entwicklungsganges.

All dies und vieles mehr sind Herausforderungen, denen sich die Menschheit in Zukunft stellen muss. Die Antworten auf diese Herausforderungen, ob in technologischen Durchbrüchen oder in einer Transformation der Verhaltensmuster der Menschen, werden die Grundvoraussetzungen dafür bilden, dass es auf diesem Planeten eine friedliche Entwicklung, ein kooperatives Zusammenleben aller Menschen geben kann.

Über viele Jahrhunderte hinweg haben sich immer wieder Menschen darum bemüht, diesen Planeten Erde zu entschlüsseln, ihn zu erforschen, ja, ihn in Besitz zu nehmen. Dabei sind immer tiefere Einblicke in die Zusammenhänge der verschiedenen Ökosysteme wissenschaftlich erarbeitet worden. Die Position des Planeten Erde im gesamten System des Weltalls ist immer detaillierter entschlüsselt worden. Forscher und Entdecker haben die Welt, oft unter großen Gefahren für Leib und Leben, auf den verschiedensten Wegen durchstreift, haben schrittweise diesen Planeten aufgeschlüsselt und damit Hinweise darauf gegeben, wie der Mensch die darin enthaltenen Lösungen und Kräfte nutzbar machen kann. Viele Erfolge in den menschlichen Bemühungen, nachvollziehbare Erklärungen für die Komplexität der Wirkungszusammenhänge zu liefern, konnten erzielt werden.

ready access to energy resources and the practicability of exploiting the raw materials essential for such economic expansion processes. These include:

- the performance of the great ecosystems, the oceans, the atmosphere, and the cryosphere—and the availability and judicious use of water, along with the conservation of land and its productivity;
- the diversity of the Earth's flora and fauna, combined with its inherent genetic potential, that inexhaustible "code of Nature" that offers the solution to so many of the world's problems;
- the opportunity to access renewable energy sources: both solar and wind power, biomass reserves, geothermal energy, along with many others;
- atmospheric stability, to preserve the course of climate development.

All these and many more are the challenges that mankind must face in the future. The solutions to these challenges, whether they are to be found in technological breakthroughs or in the transformation of human behavior, will provide the basic platform for peaceful progress and mutually beneficial life for all on this planet.

Man has constantly striven for many centuries to unlock the secrets of the Earth, exploring it and attempting to gain mastery over even its farthest reaches. These efforts have provided science with ever deeper insights into the fundamental interconnectedness of our planet's various ecosystems. The position of planet Earth within larger cosmic structures has been narrowed down in ever greater detail, and explorers and researchers have traveled to the farthest reaches of the planet, often at very great personal danger, examining the world step by step and discovering ways in which its potential and power can be used for the good of mankind. Such human striving has resulted in many successes and helped us to unravel the convoluted knot of cause and effect. Explorers and researchers have never been

développement économique. Cela comprend :

- la performance des grands écosystèmes, des océans, de l'atmosphère et de la cryosphère, ainsi que la mise à disposition et l'utilisation judicieuse de l'eau, la protection des sols et de leur productivité,
- la diversité des espèces animales et végétales, ainsi que les potentiels génétiques qu'elles recèlent, cet inépuisable « code de la nature » pour la résolution des problèmes,
- la possibilité d'exploiter de nouvelles sources d'énergies renouvelables, telles que les énergies solaire et éolienne, la biomasse et la géothermie, et de nombreuses autres,
- la stabilité de l'atmosphère pour préserver le développement climatique.

Tous ces aspects comptent parmi les nombreux défis que l'humanité aura à relever dans les années à venir. Les réponses à ces défis, sous forme d'avancées technologiques ou de modification des comportements humains, constitueront les conditions de base nécessaires pour favoriser sur cette planète le développement pacifique et une cohabitation fondée sur la coopération de tous les êtres humains.

Au cours des siècles, l'être humain a sans cesse tenté de comprendre la planète Terre, d'en explorer les mystères, voire d'en prendre possession. Il a pour ce faire progressé dans l'étude scientifique des relations entre les différents écosystèmes. Il a expliqué, avec une précision croissante, la place de la planète Terre dans l'univers. Chercheurs et explorateurs ont emprunté les routes les plus diverses pour parcourir le monde, au mépris du danger. Ils ont progressivement cartographié la planète, fournissant ainsi à l'humanité de précieuses indications sur la façon d'utiliser les solutions et les forces qu'elle recèle. Les efforts humains, tendant vers une explication cohérente de la complexité des interrelations qui régissent la Terre, ont souvent été couronnés de succès.

Ces chercheurs et explorateurs ne se sont jamais contentés des limites du

desarrollo económico puedan ser utilizadas. Esto incluye:

- Los recursos de los grandes ecosistemas, de los océanos, de la atmósfera, de la criosfera y las existencias y el uso racional del agua con ellos relacionadas, la protección de los suelos y su productividad.
- La diversidad de especies animales y vegetales vinculada con la diversidad genética que atesoran, ese inagotable «código de la naturaleza» para la solución de problemas.
- Las oportunidades para desarrollar nuevas fuentes de energía renovable: energía solar y eólica, biomasa, energía geotérmica y muchas otras.
- La estabilidad de la atmósfera para proteger la evolución del clima.

Estos y muchos otros son los desafíos que la humanidad debe enfrentar en el futuro. Las respuestas a estos retos, ya sea en forma de avances tecnológicos o como transformación de los patrones de comportamiento de las personas, son requisitos básicos para que en este planeta se pueda dar un desarrollo pacífico, una convivencia serena de toda la humanidad.

A lo largo de los siglos han existido personas que han intentado descifrar este planeta Tierra, explorándolo, para de alguna forma tomar posesión de él. Al mismo tiempo, la ciencia fue desarrollando visiones más profundas de las relaciones entre los diferentes ecosistemas. La posición del planeta Tierra en todo el sistema del universo se ha ido descifrando cada vez con mayor detalle. Los investigadores y los exploradores han atravesado el mundo de las más diversas maneras, a menudo con gran riesgo para su integridad física, han desglosado poco a poco este planeta y han señalado de qué modo puede el hombre aprovechar las soluciones y energías que contiene. Estos esfuerzos de la humanidad por dar explicaciones comprensibles sobre la complejidad de los efectos, han obtenido grandes éxitos.

Estos exploradores nunca se han dado por satisfechos con esos límites del conocimiento y del crecimiento.

Diese Forscher und Entdecker haben sich nie mit Grenzen des Wissens oder Grenzen des Wachstums zufrieden gegeben. Stets waren sie bemüht, ein Wachstum dieser Grenzen so zu ermöglichen, dass dadurch die Stabilität der ineinandergreifenden Systemabläufe dieser Welt nicht infrage gestellt wurden. Immer mehr Geheimnisse wurden entschlüsselt. Immer tiefere Einblicke wurden gewonnen. Damit verbunden waren atemberaubende Fortschritte – Grundlagen für eine zivilisatorische Entwicklung der Menschheit.

Wissenschaft und Technik haben es geschafft, noch bis vor kurzem als gänzlich undenkbar gedachte Lösungen für das friedliche Zusammenleben der Menschen, für die Sicherstellung ihrer Ernährung, für genügend Wasser, für eine ungeahnte Mobilität, für die Gestaltung gänzlich neuer Siedlungsstrukturen und vieles mehr zu finden.

Ein entscheidender Schritt bei diesen Bemühungen der Menschen, den Heimatplaneten Erde zu erforschen, wurde mit der fantastischen Möglichkeit geschaffen, die Welt aus dem Weltall, mit Satelliten zu beobachten. Fantastische Bilder erreichten und erreichen uns – der blaue Planet Erde zeigte sich als Ikone einer zukunftsorientierten, verantwortlichen Menschheit. Die Bilder aus dem Weltall verpflichteten mehr als vieles andere zuvor, die Zerbrechlichkeit der Ökosysteme sehr genau zu respektieren und darauf hinzuarbeiten, dass nicht aus kurzfristigem Fortschrittsdenken heraus mittel- und langfristig irreversible Schäden in Kauf genommen werden. Über die Zeit hinweg kann mit der immer genaueren Satellitenbeobachtung verfolgt werden, wie sich die Erde durch die Eingriffe des Menschen verändert, wie etwa das Mündungsgebiet eines Flusses dadurch massiv umgestaltet wird, dass durch die Abholzung am Oberlaufe dieses Flusses außerordentlich viel Sediment abgespült wird, welches über den Fluss an die Mündung transportiert und dort abgelagert wird und dieses Mündungsgebiet permanent umgestaltet.

Durch Satellitenbeobachtung kann verfolgt werden, wie sich Städte

satisfied with reaching the limits of their knowledge or the frontiers of human development; they have always sought to exceed these confines, though without compromising the stability of the Earth's interdependent systemic processes. More and more enigmas have been unraveled and deeper insights have been achieved, and this has entailed breathtaking advances, forming the basis for the development of human civilization.

Science and technology have managed to find solutions that even only a little while ago would have been completely unthinkable. In doing so they have furthered mankind's peaceful cohabitation, the conservation of food supplies, the provision of sufficient water, undreamed-of mobility, the creation of entirely new settlement patterns, and much more.

Developing the potential to use satellites for the observation of Earth from space was a decisive step forward in mankind's attempts to explore his home planet. Fantastic images have been flooding down to us ever since and planet Earth has our consciousness as an icon for forward-looking, responsible humanity. The wealth of images that we now receive quite routinely from our constantly orbiting satellites represents a more insistent reminder than ever before that we should respect the fragile nature of our ecosystems and endeavour to prevent short-term thinking from exposing us to mid- and long-term harm, which may eventually prove irreversible. The refinements that have been made in satellite observation over time have revealed how the Earth has been altered by mankind's intervention—the extensive changes that occur in a river delta, for example, caused by the deforestation of the upper course of the waterway, so that substantial amounts of sediment are released, which are subsequently transported downriver and deposited at the estuary, thus changing it for ever.

Satellite observation also allows us to track the manner in which

savoir ni des limites de la croissance. Ils se sont toujours efforcés de repousser autant que possible ces limites, sans remettre en question la stabilité des processus interdépendants du système Terre. Ils ont percé des mystères toujours plus nombreux, et acquis des connaissances toujours plus profondes sur la Terre, suscitant ainsi des progrès extraordinaires, qui ont jeté les bases du développement de la civilisation humaine.

La science et la technique ont permis à l'humanité d'inventer des solutions encore récemment inimaginables pour coexister dans la paix, pour assurer son alimentation et son approvisionnement en eau, pour développer une mobilité insoupçonnée, et pour créer des structures d'habitation entièrement nouvelles.

Une étape s'est avérée décisive dans le déploiement de ces efforts humains pour explorer notre planète natale : la fantastique capacité d'observer le monde depuis l'espace, à l'aide de satellites. Des images sensationnelles nous ont été transmises, et continuent à nous parvenir. La planète bleue est alors apparue comme l'icône d'une humanité responsable tournée vers l'avenir. Les images spatiales ont, bien davantage que de nombreux efforts antérieurs, amené les habitants de la planète à respecter profondément la fragilité des écosystèmes et à refuser de causer des dommages irréversibles à moyen et long terme au profit d'une croissance à court terme. Au fil du temps, l'observation par satellite a gagné en précision, permettant de suivre les transformations de la Terre liées à l'activité humaine, telles que la modification radicale de l'estuaire d'un fleuve. En raison du déboisement des rives de son cours supérieur, ce fleuve emporte de grandes quantités de sédiments qui se déversent et se déposent dans son embouchure, entraînant une transformation incessante de son estuaire. Les satellites permettent d'observer la manière dont les villes, telles des cellules cancéreuses, ont envahi la campagne et continuent de se propager. Ils ont également montré comment les sols disparaissent irrémédiablement sous

Por eso, siempre se han esforzado por aumentar estos límites de tal forma que no pueda ponerse en cuestión la estabilidad del desarrollo interdependiente de los sistemas de este mundo. Se fueron descifrando cada vez más misterios. Se obtuvieron conocimientos cada vez más profundos. Al mismo tiempo, se dieron progresos impresionantes que constituyeron la base para un desarrollo civilizado de la humanidad.

La ciencia y la tecnología han conseguido encontrar soluciones teóricas para la coexistencia pacífica de la humanidad, para asegurar su alimentación, para disponer de suficiente agua, para una movilidad sin precedentes, para el diseño de patrones de asentamiento completamente nuevos y para muchas más que hasta hace poco tiempo eran completamente inconcebibles.

La fantástica oportunidad de observar el mundo desde el espacio con satélites, supuso un paso decisivo en este esfuerzo de la humanidad por explorar el planeta Tierra. Consiguieron, y nos consiguen, imágenes fantásticas en las que el planeta azul, la Tierra, se muestra como icono de una humanidad responsable orientada hacia el futuro. Las imágenes desde el espacio obligan, mucho más que antes, a respetar escrupulosamente la fragilidad de los ecosistemas y a no poner en práctica ideas de progreso a corto plazo que provoquen daños irreversibles a medio y largo plazo. Con las cada vez mejores observaciones por satélite se puede hacer un seguimiento a lo largo del tiempo de cómo va cambiando la Tierra debido a la intervención humana; por ejemplo, cómo se transforma radicalmente la desembocadura de un río debido a la deforestación de su curso superior y, con ello, a la liberación de abundantes sedimentos que son transportados hasta la desembocadura, donde se depositan y cambian la morfología el estuario. Mediante la observación desde los satélites también se puede seguir el desarrollo de las ciudades hacia el campo y cómo lo van devorando como si de un cáncer

kanzerogen in die Landschaft hineinfraßen und weiter hineinfressen. Es konnte verdeutlicht werden, wie Böden verlorengehen, unwiederbringlich durch die verschiedensten Formen der Erosion: nur einige Hinweise auf die fantastischen Möglichkeiten einer immer größeren Auflösungsfähigkeit der Bilder, die uns aus dem Weltall von Satelliten erreichen.

Vor allem aber: Diese fantastischen Bilder gaben eine neue Dimension der Verantwortung preis, führten dazu, dass viele Menschen einen neuen Zugang zu der Globalität fanden, in der ihre jeweilige Heimat, ihre unmittelbare Umgebung eingebettet ist und von deren Stabilität sie abhängig ist. Diese Bilder vermitteln auch Verständnis, tiefe Einblicke in die Schönheit, in die Ästhetik dieses Planeten Erde – nicht dadurch, dass von den Details abgelenkt wird, sondern dadurch, dass sich das Detail als wohlgeordneter Teil eines Ganzen präsentiert.

Das vorliegende Buch ist somit kein Bilderbuch – so schön, so ergreifend diese Bilder, diese Aufnahmen aus dem All auch sind. Es ist vielmehr ein Aufruf zur Entschlossenheit, diese Natur nicht durch menschliches Handeln aus dem Gleichgewicht zu bringen. Dieses Buch kann und wird hoffentlich viele Menschen aufrütteln, Partei zu ergreifen für die Schöpfung, für die Umwelt, für die Verantwortung mit Blick auf kommende Generationen.

Satellitenaufnahmen sind eben nichts Statisches, sie sind eingebunden in den Weg der Zeit, sie verfolgen Entwicklungen und Ereignisse – und sie sagen mehr, als tausend Worte je sagen können. Genau dies ist die Erwartung, die ich mit diesem Buch verbinde. Es ist mir eine Freude und eine Ehre, dass ich dazu ein Vorwort schreiben durfte. Dem Buch selbst wünsche ich außerordentlich viel Erfolg, viele Leserinnen und Leser, die dieses Buch nicht nur einmal in die Hand nehmen, sondern immer und immer wieder darin blättern und versucht sind, die weitere Entwicklung ebenfalls mitzuverfolgen.

cities continue to encroach upon the landscape like cancers and reveals to us just how land is irretrievably lost to every kind of erosion; these are but a few of the fantastic possibilities offered by the images that are transmitted to us from space by satellite, in increasingly sharp resolution.

Most importantly, however, these wonderful photographs have given rise to a new dimension to our responsibilities, evoking in many people an unprecedented sense of the global unity encompassing their individual homes and immediate surroundings, upon whose stability they are so dependent. These images also provide us with an awareness and understanding of the beauty and the aesthetic qualities of planet Earth—not by distracting us from the details, but rather by presenting them as part of a balanced whole.

This book is therefore not a picture book, however striking the images and photographs from space may be—it is instead a call for resolution in preventing human actions from upsetting the balance of nature. It is to be hoped that this book can and will inspire many people to take up the fight for the sake of all Creation, as well as for the sake of the environment and our responsibilities towards future generations. Satellite images are in no way static elements—they are part of the course of time, following developments and events and saying far more than a thousand words ever could. This is the hope that inspires me with this book, and it has been a pleasure and an honor to write this foreword. I wish the book itself every success in finding a wide readership who will return to this volume again and again as they too strive to keep up with the evolution of our planet.

l'action de diverses formes d'érosion. Il ne s'agit là que de quelques exemples des formidables possibilités offertes par les images que nous transmettent les satellites, avec une résolution de plus en plus fine.

Mais ces images fantastiques ont surtout révélé une nouvelle dimension de la responsabilité humaine. Elles ont fourni au plus grand nombre un nouvel accès à ce monde qui abrite leur terre natale, leur environnement immédiat, et dont la stabilité conditionne l'existence. Ces images transmettent également une meilleure connaissance et compréhension de la beauté, de l'esthétique de la planète Terre, non parce qu'elles nous détournent des détails, mais au contraire parce qu'elles présentent chaque détail comme la partie d'un tout bien ordonné.

Ce livre n'est pas pour autant un livre d'images, même si ces images, ces photographies spatiales sont d'une beauté émouvante. Il s'agit bien plus d'un appel à l'engagement pour respecter l'équilibre de la nature, sans intervention humaine. Ce livre interpellera, je l'espère, de nombreux lecteurs, les encouragera à prendre parti en faveur de notre planète et de l'environnement, et leur fera prendre conscience de la responsabilité des générations présentes à l'égard des générations futures.

Les images satellites ne sont pas statiques, elles s'inscrivent dans le cours du temps, enregistrant les transformations et les événements. Elles en disent ainsi beaucoup plus long que n'importe quel discours. C'est exactement l'espoir que ce livre fait naître en moi. Être invité à en écrire la préface est pour moi une joie et un honneur. Je souhaite à ce livre de rencontrer un franc succès, et de trouver de nombreux lecteurs qui ne se contenteront pas de le parcourir une seule fois, mais le feuilletteront encore et encore, et seront tentés de continuer à suivre l'évolution de notre planète.

se tratase, o explicar cómo se pierden los suelos irremediablemente debido a las diferentes formas de erosión. Estas son sólo algunas referencias sobre las fantásticas posibilidades de las imágenes, cada vez de una mayor resolución, que obtenemos con los satélites desde el espacio.

Pero, sobre todo, estas fantásticas imágenes revelaron una nueva dimensión de la responsabilidad y, como consecuencia, muchas personas encontraron un nuevo sentido a la globalización en la que están implicados sus respectivos países y su entorno inmediato, y de cuya estabilidad dependen. Estas imágenes también transmiten comprensión, visión profunda de la belleza, de la estética de este planeta Tierra, no porque los detalles supongan una distracción, sino por el hecho de que cada detalle se presenta como parte de un todo bien ordenado.

Este libro no es, pues, un libro de fotografías, por muy hermosas e impactantes que esas imágenes desde el espacio nos puedan parecer. Es más bien un llamamiento a la determinación de respetar el equilibrio de la naturaleza, sin intervención humana. Este libro puede remover, y espero que lo haga, las conciencias de muchas personas para que tomen partido por la creación, por el medio ambiente, por la responsabilidad con respecto a las generaciones futuras.

Las imágenes de los satélites no son, en absoluto, estáticas. Están integradas en el transcurso del tiempo, siguen el desarrollo y los resultados y dicen mucho más que mil palabras que se puedan pronunciar. Estas son precisamente las expectativas de este libro. Para mí es un placer y un honor haber podido prologarlo. Al libro le deseo un éxito extraordinario, que muchos lectores no se limiten a sostenerlo entre sus manos una sola vez, sino que vuelvan continuamente sobre él para hojearlo e intenten participar también en el desarrollo futuro.

Klaus Töpfer, founding director and Executive Director of the Institute for Advanced Sustainability Studies, IASS, Potsdam (Germany)

Die Erde im Blick
A View of the Earth
La Terre en vue
La Tierra a la vista

Volker Liebig

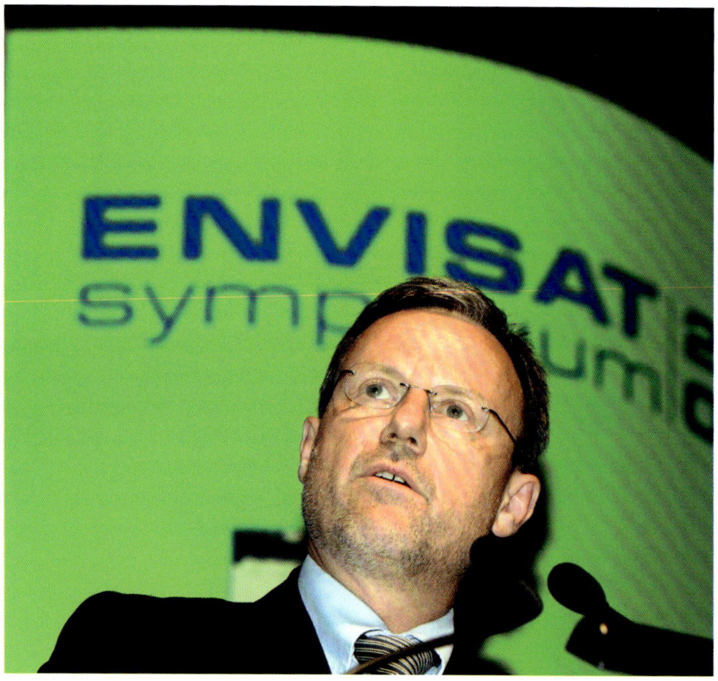

Fast alle Astronauten, die nach ihrer Rückkehr zur Erde befragt werden, was das beeindruckendste Erlebnis im Weltraum war, nennen den Blick zurück auf die Erde. Schon die ersten Bilder aus dem All in den sechziger Jahren änderten das Bewusstsein der Menschen. Unser Heimatplanet war plötzlich nicht mehr ein praktisch unendliches und unerschöpfliches Reservoir von Ressourcen, sondern wurde als das wahrgenommen, was er ist: ein kleiner blauer Planet mit einer dünnen Schicht von Atmosphäre, die uns vor Kälte, Strahlung und zumindest den meisten Meteoriten schützt. Das Foto der aufgehenden Erde über dem Horizont der Wüstenlandschaft des Mondes, aufgenommen 1968 von der Apollo-8-Crew, wurde oft als das einflussreichste Umweltfoto aller Zeiten bezeichnet. Es ist kein Zufall, dass die Vorhersagen des Club of Rome zu den Grenzen des Wachstums aus dieser Zeit stammen.

Seit den sechziger Jahren hat sich die Erdbeobachtung vom Satelliten enorm weiterentwickelt. Heute sind es nicht nur die faszinierenden Bilder von hochauflösenden digitalen Kameras, die von Satelliten zum Boden gefunkt werden, sondern Daten von hunderten physikalischer Parameter des Systems Erde, die von Wissenschaftlern benutzt werden, um das

When asked what experience made the greatest impression upon them in space, almost every astronaut gives the same reply—"Looking down and seeing the Earth." Taken in the 1960s, the first pictures from space changed mankind's awareness of our home planet for ever. No longer a practically infinite and inexhaustible reservoir of resources, it was seen for what it really is: a small, blue planet with a thin layer of atmosphere protecting us from cosmic radiation, the chill of space, and all but the largest meteorites. "Earthrise," the photograph taken in 1968 by the crew of Apollo 8 showing the Earth as it ascended above the desert horizon of the Moon, is often quoted as the most influential environmental image of all time, and it is no coincidence that the predictions contained in the Club of Rome's *Limits to Growth* report stem from the same period.

The use of satellites to observe the Earth has made huge progress since the 1960s—today, fascinating images taken by highresolution digital cameras are transmitted back to Earth accompanied by data relevant to the physical parameters of hundreds of terrestrial systems. Scientists are able to use these to research the Earth and document

Lorsque vous demandez aux astronautes de retour sur Terre ce qui les a le plus impressionné dans l'espace, la plupart répondent la vue de la Terre. Les premières images prises depuis l'espace ont, dès les années 1960, modifié notre façon d'appréhender la Terre. Notre planète est soudain apparue non plus comme un inépuisable réservoir de ressources, mais pour ce qu'elle est réellement : une petite planète bleue dotée d'une mince couche d'atmosphère qui nous protège du froid, du rayonnement spatial et des météorites. On s'accorde à penser que la photo du lever de Terre au-dessus de l'horizon lunaire, prise en 1968 au cours de la mission Apollo 8, est la photo environnementale qui a connu le plus grand retentissement. Ce n'est pas un hasard si la publication du rapport du Club de Rome sur les limites de la croissance date de la même époque.

L'observation de la Terre par satellite a considérablement évolué depuis les années 1960. De nos jours, les scientifiques ne se contentent plus des images fascinantes de caméras satellitaires à haute résolution pour étudier la Terre et suivre ses changements, ils étudient désormais des centaines de paramètres physiques distincts du système Terre. De nombreuses applications opérationnelles

Observar la Tierra desde el espacio es para la mayoría de astronautas la experiencia más impresionante. Ya desde los años sesenta, las primeras imágenes tomadas desde el espacio cambiaron la conciencia de la humanidad. Nuestro planeta dejó de ser visto prácticamente como un almacén infinito e inagotable de recursos para ser percibido como un pequeño planeta azul con una fina capa de atmósfera que nos protege del frío, de la radiación espacial y de la mayoría de meteoritos. La foto tomada por la tripulación del Apolo 8 en 1968 que muestra nuestra Tierra azul sobre el horizonte del desértico paisaje lunar, ha sido a menudo citada como la fotografía medioambiental más influyente de la historia. Fue durante este período cuando el Club de Roma redactó las previsiones de los Límites de Crecimiento.

Desde los años sesenta, la observación de la Tierra desde los satélitesha evolucionado enormemente. Hoy en día, para investigar el sistema Tierra y documentar sus cambios, los científicos no solo cuentan con fascinantes imágenes tomadas por cámaras digitales de alta resolución, sino también con cientos de parámetros físicos del sistema Tierra. Durante los últimos 50 años de teledetección por satélite se han desarrollado mu-

System Erde zu erforschen und seine Änderungen zu dokumentieren. Viele operationelle Anwendungen haben sich in den fast fünfzig Jahren der Satellitenfernerkundung entwickelt. Am Anfang stand die Meteorologie, und fast jeder sieht die Daten täglich im Wetterbericht. Durch Wettersatelliten und moderne Supercomputer erreicht die Wettervorhersage heute eine Genauigkeit in der Sechs-Tage-Vorhersage, die besser ist als vor einigen Jahrzehnten für drei Tage.

Neben der Entwicklung der Satelliteninstrumente hat die Satellitenfernerkundung aber auch von der sprunghaften Entwicklung der Computer und Datennetze profitiert. Ohne sie wären die vielen Terabyte von Daten, die laufend zur Erde gesendet werden, nicht zu verarbeiten. Heute erschließen Breitbandnetze den freien und schnellen Zugang zu Online-Datenarchiven der Raumfahrtagenturen. Unsere Partner aus den Entwicklungsländern sprechen deshalb auch von »Datendemokratie« und meinen damit den offenen und kostenfreien Zugang zu Umweltdaten.

Auch der private Zugang zu Satellitendaten hat sprunghaft zugenommen. Heute ist es möglich, sich per Internet bei der ESA zu informieren über die Wassertemperatur am Urlaubsort, die Schneelage in den Bergen oder wie viele Stunden man sich bei der augenblicklichen Ozonlage am Strand oder auf der Skipiste der Sonne aussetzen sollte. Google Earth ist ebenfalls ein gutes Beispiel, dass Satellitendaten den Haushalt der Menschen erreicht haben und täglich für viele Anwendungen genutzt werden.

Natürlich ist die Nutzung der Daten durch Wissenschaftler und öffentliche Dienste immer noch die wichtigste Anwendung der Erdbeobachtung. Das Spektrum ist dabei sehr weit, und dieses Buch soll einen Einblick geben.

Politisch von größter Bedeutung ist dabei die Beobachtung des Klimawandels und seine Auswirkung auf unseren Planeten. Zu diesem Zweck bereitet Europa weltweit das erste operationelle Umweltbeobachtungssystem mit dem Namen Copernicus vor. Wie vor mehr als dreißig Jahren

its changes. Many operational applications have been developed in the last 50 years of satellite remote sensing but meteorology has been there from the start, and this data is now used in daily weather reports watched by almost everyone. Weather satellites and today's supercomputers have brought to modern six-day weather forecasts an accuracy undreamed of in the three-day predictions of a few decades ago.

Besides improvements in satellite instruments, remote sensing has profited greatly from the rapid development of computers and data networks. Without these, it would be impossible to process and distribute the many terabytes of data constantly being downlinked to the Earth. Broadband networks have made available to all free and speedy access to the space agencies' online archives, providing our partners in emerging countries with "data democracy," their term for open and cost-free access to environmental data.

Private use of satellite data has also grown massively and it is now possible to surf the ESA's website to find out the water temperature at our holiday resorts, the snow cover in the mountains, or how long we should stay out in the sun, given the current ozone levels at the beach or on the ski slopes.

Google Earth is another good example of how satellite data has entered the home to be used in a variety of applications.

The most important function of satellite data is, of course, still to be found in its use by scientists and public bodies, but the spectrum of applications is extremely broad, and this book will allow an insight into many of these.

Politically, the most important of these is the observation of climate change and its impact on our planet. To this end, Europe is developing the world's first operational environmental monitoring system, to be called Copernicus. Much as in the field of meteorology for more than 30 years, by around 2021 a planned total of seven satellites will be observing

ont vu le jour au cours de ces quelque 50 années de télédétection par satellite. La météorologie a ouvert la voie, et les images satellite ne font aujourd'hui défaut à aucun bulletin météo quotidien. Grâce aux satellites météorologiques et aux superordinateurs modernes, la précision des prévisions météo à six jours est aujourd'hui meilleure que les prévisions à trois jours d'il y a seulement quelques décennies.

En plus du développement d'instruments sophistiqués de télédétection, l'observation de la Terre a largement profité du développement rapide de l'informatique et des réseaux de données, sans quoi il serait impossible de traiter les nombreux téraoctets de données brutes transmises par les satellites. De nos jours, les réseaux Internet à large bande assurent un accès libre et facile aux données des agences spatiales. C'est pourquoi nos partenaires des pays en voie de développement parlent de « démocratie des données », évoquant ainsi l'accès libre et gratuit aux données environnementales.

L'accès direct des citoyens aux données satellitaires a, lui aussi, rapidement progressé. Il est désormais possible de consulter les sites Internet de l'ESA pour connaître la température de l'eau de mer dans une station balnéaire, les conditions d'enneigement d'une station de ski, ou s'informer sur la durée recommandée d'exposition au soleil en fonction de l'indice UV. Google Earth est un autre exemple parmi les nombreuses applications basées sur les données satellitaires désormais largement entrées dans les foyers.

Scientifiques et services publics restent évidemment les premiers utilisateurs des données d'observation de la Terre. Le spectre des applications, dont ce livre donne un aperçu, est très large. Mais l'application primordiale, d'un point de vue politique, est l'observation du changement climatique et son impact sur notre planète. À cet effet, l'Europe prépare le premier système opérationnel au monde de surveillance de l'environnement, appelé Copernicus. Comme pour la météorologie depuis

chas aplicaciones operacionales. Todo comenzó con la meteorología y hoy casi todo el mundo obtiene esta información a diario en las noticias. Gracias a los satélites meteorológicos y a los modernos superordenadores, la previsión meteorológica actual a seis días vista es más precisa que la de a tres días vista de hace unas décadas.

La teledetección se ha beneficiado tanto del desarrollo de los instrumentos que llevan los satélites como de los ordenadores y de las redes de datos. Sin este desarrollo no se podrían procesar los muchos terabytes de datos que se transmiten continuamente a la Tierra.

Actualmente, las redes de banda ancha permiten un acceso libre y abierto a los archivos en línea de las agencias espaciales. Nuestros socios de los países en vías de desarrollo lo denominan «data democracy», es decir, acceso abierto y gratuito a datos medioambientales.

También el acceso privado de los ciudadanos a los datos de los satélites ha aumentado rápidamente. Hoy es posible acceder a la página web de la ESA y obtener información acerca de la temperatura del mar en la zona donde está planeando pasar las vacaciones, el estado de la nieve en su estación de esquí preferida o el tiempo máximo recomendable de exposición al sol dependiendo de dónde se encuentre y la situación de la capa de ozono. Google Earth es otro buen ejemplo de que los datos obtenidos por los satélites y sus múltiples aplicaciones ya forman parte de nuestra vida cotidiana.

Por supuesto que el uso de los datos por parte de los científicos y de las administraciones públicas es todavía una de las aplicaciones más importante de la observación de la Tierra. El espectro de aplicaciones es muy amplio y este libro le permitirá profundizar con gran detalle en ese campo.

Muy relevante para el ámbito político es la observación del cambio climático y su efecto en nuestro planeta. Con este propósito, Europa está preparando el primer sistema operacional de vigilancia medioambiental llamado Copérnico. Como hace

in der Meteorologie werden ab etwa 2021 täglich Umweltdaten von zur Zeit sieben geplanten Satelliten zur Erde gefunkt, die die Erde mit Kamera, Radar, Spektrometer und Radiometer beobachten – der erste Radarsatellit dieses Programms arbeitet bereits im All. Services zur Land-, Ozean- und Atmosphärenbeobachtung werden von der Europäischen Kommission vorbereitet, dazu noch die schnelle Erstellung von Karten für Rettungskräfte nach Katastrophen und ein Dienst für zivile Sicherheit, etwa um Schiffe aufzuspüren, die illegal Öl in europäische Gewässer ablassen.

ESA hat darüber hinaus vor sieben Jahren eine Klimainitiative gestartet, in der aus Archivdaten und den ständig neuen Beobachtungen globale Datensätze der »Essential Climate Variables« erstellt werden. Experten haben dazu einen Katalog von 45 wichtigen Klimavariablen definiert, von denen ESA 21 erstellen wird. Dazu gehören Meeresspiegelhöhe, Salzgehalt und Temperatur, Meereis- und Gletscherausdehnung, Änderung der Landnutzung, Biomasse etc. Die Datensätze werden allen Forschern zur Verfügung gestellt und sollen die Klimamodelle, aber auch die Beschreibung des Ist-Zustandes verbessern: Heute erkennt man aus dem Weltraum, dass die Vorhersagen des letzten Klimaberichtes des International Panel for Climate Change (IPCC) zum Teil schon überholt sind. So übertrifft der von den Satelliten beobachtete Rückgang des Meereises in der Arktis die Modellvorhersagen deutlich.

Europa hat mit Copernicus und seinen Wissenschaftssatelliten weltweit die Führung in der Erdbeobachtung übernommen. Viele der besten Ingenieure und Wissenschaftler arbeiten mit großer Begeisterung an den Projekten, die dem Schutz unseres Planeten dienen sollen. Zum Zeitpunkt des Erscheinens dieses Buches arbeitet ESA an rund zwei Dutzend Erdbeobachtungssatelliten für die Meteorologie, für Copernicus und für die wissenschaftliche Erkundung unseres Planeten.

Dieses Buch lässt einige dieser Leute zu Wort kommen, ohne deren Leistung und persönlichen Einsatz

the Earth with cameras, radar sensors, radiometers and spectrometers – the first radar satellite from this program is already operating in space.

The European Commission is developing land, ocean, and atmospheric observation services, allowing the rapid coordination of humanitarian relief efforts during natural disasters and providing a service for home security, including the ability to detect ships that illegally discharge oil in European waters.

In addition, the ESA started a climate change initiative seven years ago whereby global data sets of "Essential Climate Variables" as defined by the UN Climate Convention are produced from new and archived observations. Of a catalog of 45 important climate variables identified by these experts, some 21 will be produced by the ESA, including sea level rise, ocean salinity and temperature, sea ice and glacier cover, land use change and biomass, as well as numerous others. All these data sets will be made available to scientists throughout the world, both to improve climate models and to expand our understanding of the current state of the planet. Observation from space has already shown us that some of the predictions contained in the last climate report produced by the International Panel for Climate Change (IPCC) are already outdated—the observed retreat of Arctic sea ice far exceeds model predictions, for example.

Europe has taken the global lead with Copernicus and its other scientific satellites, and many of the best engineers and scientists are now working with great diligence on projects that will serve to protect our planet. At the time of this book's publication, the ESA is working on two dozen new Earth observation satellites for Copernicus, meteorology, and the scientific exploration of our planet.

Some of the people whose achievements and personal effort have

plus d'une trentaine d'années, sept satellites fourniront quotidiennement, à partir de 2021, des données environnementales. Ils observeront la Terre à l'aide de radars, caméras optiques, radiomètres et spectromètres – le premier satellite radar de ce programme est déjà opérationnel dans l'espace. La Commission européenne élabore actuellement de futurs services d'observation terrestre, maritime et atmosphérique, ainsi qu'un service de cartographie rapide destiné aux opérations de secours d'urgence suite aux catastrophes, et un service permettant aux organes de sécurité publique de repérer les dégazages illégaux dans les mers d'Europe.

L'ESA a par ailleurs lancé il y a sept ans une initiative sur le changement climatique visant à obtenir un relevé cohérent des « variables climatiques essentielles », à partir d'observations satellitaires passées et futures. Un panel d'experts a défini un catalogue de 45 variables climatiques, dont 21 seront générées par l'ESA. Parmi elles figurent l'élévation du niveau de la mer, la température et la salinité des océans, l'étendue des banquises et des glaciers, l'utilisation des sols ou la biomasse. Mises à la disposition de la communauté scientifique, elles permettront d'améliorer les modèles climatiques, et d'établir un diagnostic précis de l'état de la Terre. Les données satellitaires les plus récentes montrent en effet que les prévisions du dernier rapport sur le climat du Groupe d'experts intergouvernemental sur l'évolution du climat (GIEC) sont en partie dépassées. C'est le cas notamment pour le recul de la banquise arctique, nettement supérieur aux prédictions du modèle.

L'Europe s'est, grâce à Copernicus et à ses nombreux satellites scientifiques, hissée au premier rang mondial de l'observation de la Terre. Ses meilleurs ingénieurs et scientifiques contribuent avec passion aux projets consacrés à la protection de notre planète. Au moment de la publication de ce livre, l'ESA travaille sur environ deux douzaines de nouveaux satellites météorologiques, environnementaux (Copernicus) et d'exploration

más de 30 años con la meteorología, a partir de 2021 se transmitirán a la Tierra datos medioambientales obtenidos por los sietes satélites actualmente en proyecto, que observarán la Tierra con radares, cámaras ópticas, radiómetros y espectrómetros – el primer satélite radar de este programa ya está operando en el espacio.
La Comisión Europea está preparando diferentes servicios de observación del terreno, los océanos y la atmósfera para la elaboración rápida de mapas para los equipos de salvamento en desastres naturales y un servicio de seguridad civil para, por ejemplo, detectar barcos que viertan ilegalmente petróleo en aguas europeas.

Además, la ESA tomó hace siete años una iniciativa sobre el cambio climático por la que se elaboran registros globales de las «variables climáticas esenciales» a partir de datos de archivo y de las continuas nuevas observaciones. Un equipo de expertos ha definido para ello un catálogo de 45 variables climáticas importantes, de las cuales la ESA suministrará 21. Entre ellas se incluyen el nivel del mar, la salinidad y la temperatura de los océanos, la extensión del hielo marino y de los glaciares, los cambios en el aprovechamiento del suelo, la biomasa y muchas otras. Todos estos registros se pondrán a disposición de la comunidad científica con el fin de mejorar los modelos climáticos y la descripción del estado real de nuestro planeta. En la actualidad se puede detectar desde el espacio que algunas de las previsiones del último informe del Panel Internacional sobre el Cambio Climático (IPCC) ya están anticuadas. Así, el retroceso del hielo marino observado por los satélites en el Ártico supera claramente las previsiones del modelo.

El Copérnico y los satélites científicos han llevado a Europa al liderazgo mundial en la observación de la Tierra. Gran parte de los mejores ingenieros y científicos trabaja con entusiasmo en proyectos de protección de nuestro planeta. En el momento de la publicación de este libro, la ESA trabaja en unas dos docenas de nuevos satélites para la meteorología,

Apollo 8: Erstes Bild der Erde über dem Mondhorizont (1968)

Apollo 8: The first image of Earthrise over the Moon's horizon (1968)

Apollo 8 : première image de la Terre se levant sur l'horizon lunaire (1968)

Apolo 8: primera imagen de la Tierra sobre el horizonte lunar (1968)

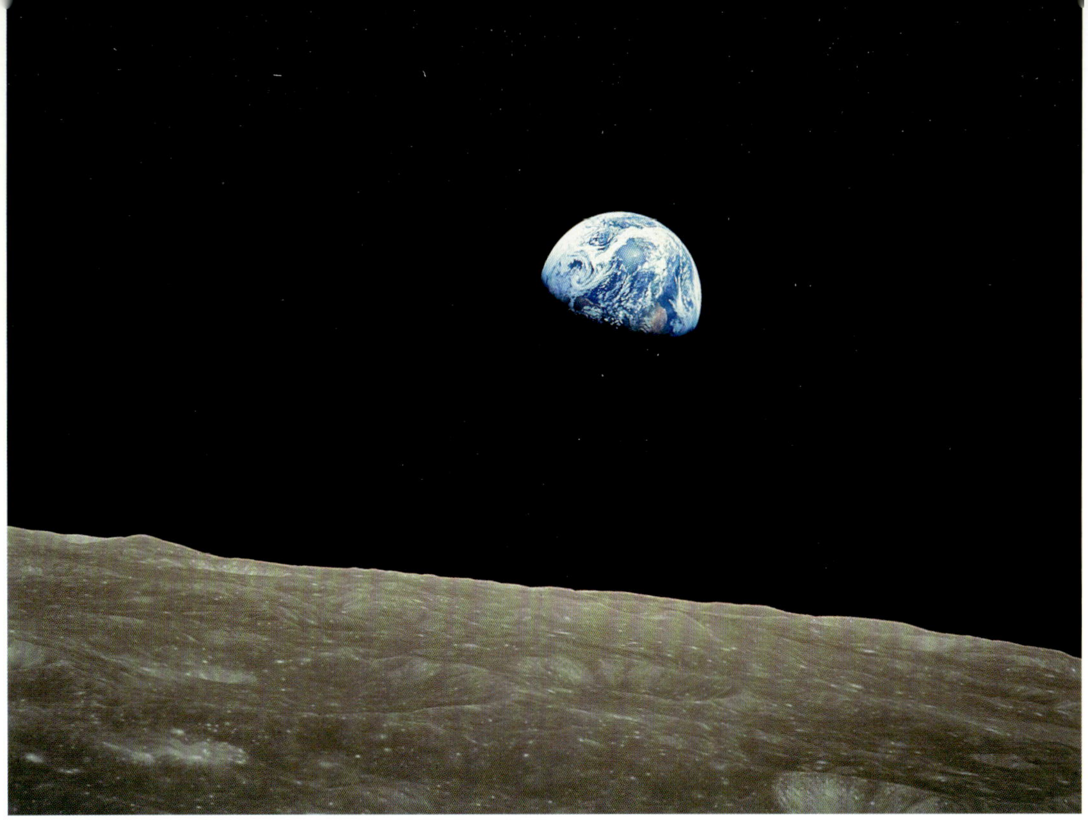

solche Missionen nicht möglich wären. Sie sind die Entdecker des 21. Jahrhunderts, und das Ziel heißt, unseren Heimatplaneten zu verstehen und dazu beizutragen, dass wir auch in hundert und zweihundert Jahren noch seine Schönheit und seinen Schutz genießen dürfen.

Auch ich persönlich habe mich vor Jahren dafür entschieden, in diesem faszinierenden Bereich zu arbeiten. Heute, 30 Jahre später, sehe ich den Aufschwung, den die Erdbeobachtung in diesem Zeitraum genommen hat, mit großer Freude. Ebenso wie unsere Astronauten hat auch mich die Begeisterung, unseren Planeten aus dem All zu beobachten, nie verlassen, und ich hoffe, dass sich diese Begeisterung auch auf die Leser dieses Buches beim Betrachten der außergewöhnlichen Aufnahmen übertragen wird.

Volker Liebig, Direktor für Erdbeobachtung der Europäischen Weltraumorganisation ESA

been invaluable in preparing those missions get a chance to speak in this book. They are the great explorers of the 21st century, and their objective is to understand our home planet, helping to ensure that we can still enjoy its beauty and protection hundreds of years from now.

Long ago, I decided that I too would work in this fascinating field. Today, 30 years later, it is a joy to behold the advances that have been made in Earth observation during this period. Much like our astronauts, I have never lost my love of observing our planet from space, and I hope that some of this enthusiasm will be conveyed to the readers of this book as they marvel at the extraordinary images it contains.

Volker Liebig, Director of Earth Observation Programmes of the European Space Agency ESA

scientifique de la Terre. Plusieurs personnes qui apportent une contribution majeure au succès de ces mis sions s'expriment dans ce livre. Ce sont les explorateurs du XXIe siècle. Leur objectif est de comprendre notre planète, pour contribuer à préserver sa beauté et la protection qu'elle nous offre, durant les prochains siècles.

J'ai moi aussi, il y a longtemps, pris la décision de me consacrer à ce domaine fascinant. Et 30 ans plus tard, je constate avec satisfaction la montée en puissance de l'observation de la Terre. Tout comme celui de nos astronautes, mon enthousiasme pour l'observation spatiale de notre planète n'a jamais faibli. Et je souhaite que cet enthousiasme gagne les lecteurs de ce livre, à la vue des images extraordinaires qu'il présente.

Volker Liebig, Directeur des programmes d'observation de la Terre à l'Agence spatiale européenne ESA

Copérnico y la exploración científica de nuestro planeta.

Este libro ofrece la visión personal de algunas personas sin cuyo trabajo y esfuerzo personal esas misiones hubieran sido imposibles. Ellos son los exploradores del siglo XXI y su objetivo es entender nuestro planeta y contribuir a que podamos disfrutar de su belleza y protegerlo durante los próximos siglos.

Hace años tomé la decisión de trabajar en esta área fascinante. Hoy, 30 años más tarde, me complace enormemente ver el impulso que ha tomado la observación de la Tierra. Como nuestros astronautas, no he perdido nunca el entusiasmo en la observación de nuestro planeta desde el espacio, y espero que las imágenes que se muestran en este libro transmitan esta satisfacción al lector.

Volker Liebig, Director de Programas de Observación de la Tierra de la Agencia Espacial Europea ESA

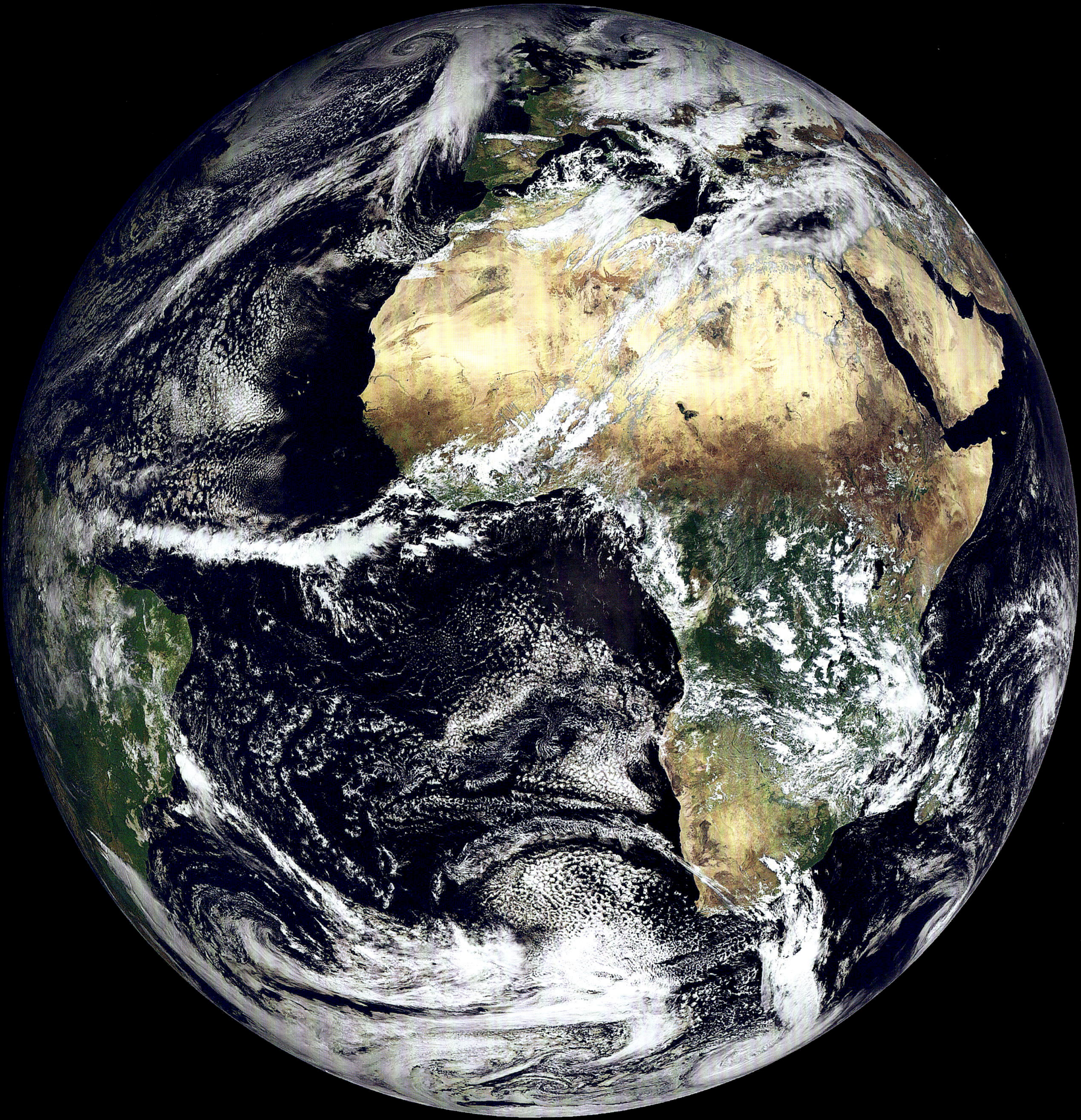

Oase im Weltall: Die Entdeckung der Erde
An oasis in space: discovering Earth
Une oasis dans l'espace : la découverte de la Terre
Un oasis en el universo: el descubrimiento de la Tierra

Woran denken Sie beim Stichwort Raumfahrt? An Flüge zum Mond? An Astronauten auf der Internationalen Raumstation? Oder an Planetensonden, die durch das Sonnensystem fliegen und spektakuläre Bilder vom Mars oder von den Saturnringen zur Erde funken? Solche und ähnliche Projekte prägen die Wahrnehmung der Raumfahrt in der Öffentlichkeit. Wenig bekannt ist, dass der schönste und für uns wichtigste aller Körper im Weltall am intensivsten beobachtet wird: die Erde.

Das hat vor vielen Jahrzehnten mit einfachen Wettersatelliten begonnen. Recht schnell hat das Militär die Erdbeobachtung aus sehr einseitigen Motiven fast zur Perfektion gebracht – jedenfalls was die Informationen betrifft, die man damals für strategisch wichtig erachtete. Doch inzwischen ist die Erde als Ganzes das wichtigste Thema der Raumfahrt. Um sie herum schwirren so viele Satelliten wie um keinen zweiten Himmelskörper. Kein anderer Ort im Sonnensystem wird so umfassend überwacht.

Das mag paradox klingen, ist es aber nicht: Die Schönheit eines Ortes offenbart sich oft erst aus der Entfernung. Auch die Raumfahrer haben dies schnell bemerkt. Ging es anfangs darum, möglichst schnell möglichst weit hinaus ins All vorzustoßen,

What does the word "space" make you think of? Trips to the Moon? Astronauts on the International Space Station? Or perhaps planetary probes, speeding through the solar system and beaming spectacular images of Mars or the rings of Saturn back to Earth? These are the kinds of projects that have shaped the public perception of space travel. However, what is less well known is that the most beautiful body in space and the most important planet for the human race is also the one that is the most closely observed: the Earth.

This scrutiny began decades ago with the launch of basic weather satellites, and the military soon more or less perfected the art of Earth observation for its own ends, at least as far as gathering strategically important intelligence was concerned. The Earth as a whole has since become the most important object in space travel—no other heavenly body has so many satellites whirring around it, and no other point in the solar system is subject to such constant surveillance.

That the beauty of a place only becomes apparent at a distance may sound paradoxical, but it is true, as the first space travelers soon found out. If the initial point of space

Qu'évoque pour vous la recherche spatiale ? Un voyage sur la Lune ? Un astronaute dans la Station spatiale internationale ? Ou bien les sondes planétaires qui parcourent le système solaire en quête d'images spectaculaires de Mars ou des anneaux de Saturne ? L'image de la recherche spatiale dans l'opinion publique se nourrit de projets de ce type. Ce que l'on sait moins, c'est que le plus beau des corps célestes, et pour nous le plus important, est celui que nous observons avec le plus d'attention : la Terre.

Tout a commencé il y a plusieurs décennies avec de simples satellites météorologiques. L'armée a pour des raisons tactiques rapidement atteint la quasi perfection dans le domaine de l'observation de la Terre, et plus particulièrement dans la collecte des informations alors considérées comme stratégiques. Depuis, la Terre est devenue la principale préoccupation des activités spatiales. Aucun autre corps céleste n'est entouré d'autant de satellites. Aucun autre endroit du système solaire n'est surveillé avec autant de précision.

Cela peut sembler paradoxal, et pourtant : c'est souvent le recul qui révèle pleinement la beauté d'un lieu. Les astronautes en ont rapidement pris conscience. S'il s'agissait

¿En qué piensa cuando escucha la palabra espacial? ¿En vuelos a la Luna? ¿En los astronautas de la estación espacial internacional? ¿O en sondas interplanetarias que navegan por el Sistema Solar y envían a la Tierra espectaculares imágenes de Marte o de los anillos de Saturno? Estos y otros proyectos similares marcan la percepción más habitual de lo espacial. Menos frecuente es pensar que lo observado con mayor intensidad sea el más hermoso y, para nosotros, más importante de todos los cuerpos del universo, la Tierra.

Todo comenzó hace décadas, con los sencillos satélites meteorológicos. Rápidamente, los militares llevaron casi a la perfección las técnicas de observación terrestre con un objetivo bastante parcial: la información considerada estratégicamente relevante. Pero desde esa época han cambiado las percepciones y ahora, lo más importante es la Tierra considerada como un todo. No hay lugar del Sistema Solar tan controlado como la Tierra.

Con frecuencia, y aunque pueda resultar paradójico, la belleza de un planeta sólo se revela a cierta distancia. También los astronautas lo advirtieron enseguida. Las

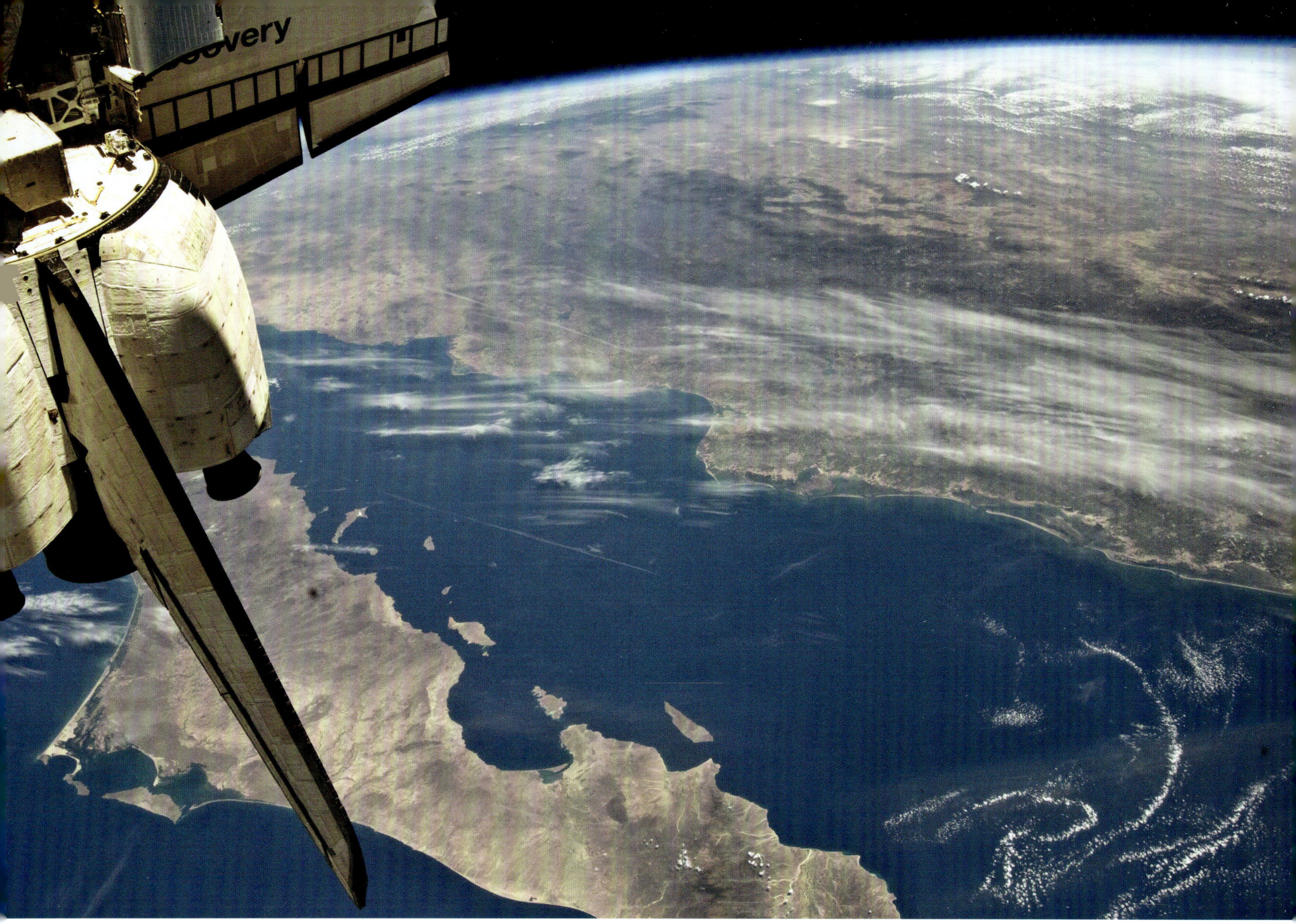

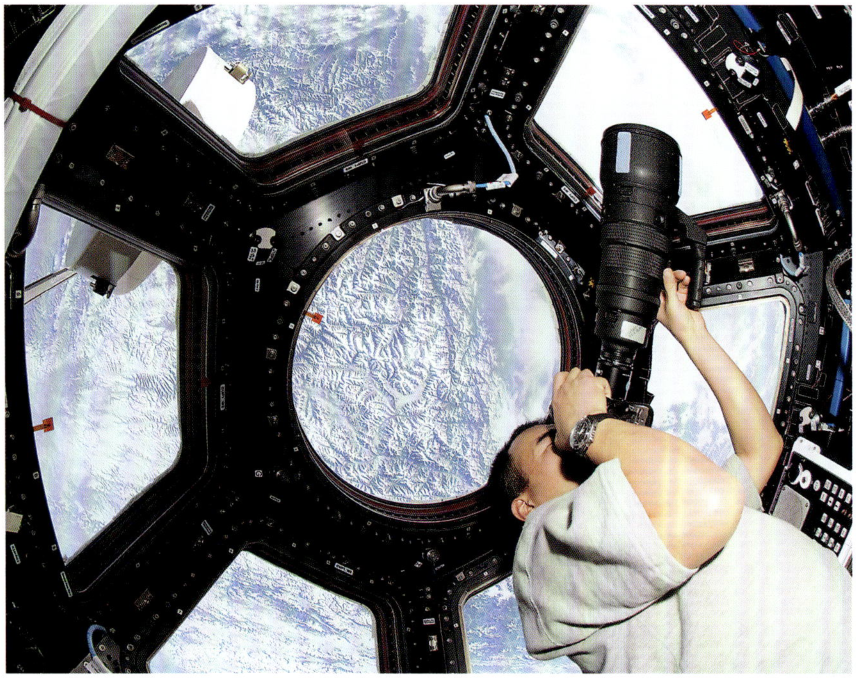

Oben: Der hauchdünne blaue Saum der Atmosphäre, gesehen von der Internationalen Raumstation (ISS)
Links: Das Fenster zur Welt in der Raumstation – die Cupola

Above: The wafer-thin blue blanket of the atmosphere, seen from the International Space Station
Left: The cupola, the Space Station's window on the world

Ci-dessus : vue du délicat voile bleu de l'atmosphère depuis la Station spatiale internationale (ISS).
À gauche : la Cupola, fenêtre sur le monde de la station spatiale

Arriba: La finísima orla azul de la atmósfera vista desde la estación espacial internacional (ISS).
Izda.: Ventana al mundo en la estación espacial: la Cupola

Seite 22: Unsere Erde mit Wolken, tropischem Regenwald, Wüsten und Meeresflächen. Afrika ist am auffälligsten (Meteosat)

Page 22: Our Earth with its clouds, tropical rainforest, deserts, and oceans. Africa is clearly visible (Meteosat)

Page 22 : la Terre, ses nuages, ses forêts tropicales, ses déserts et ses océans. L'Afrique est particulièrement bien visible (Météosat)

Página 22: La Tierra con nubes, bosque húmedo tropical, desiertos y la superficie de los mares. África es lo más llamativo (Meteosat)

zeigt sich nun, dass man vor allem deswegen ins All fliegt, um die himmlische Heimat im Auge zu behalten. Denn die großen Zusammenhänge auf unserer Erde lassen sich mit ein wenig Abstand viel besser erkennen als vom Boden aus, wo der Blick bestenfalls bis zur nächsten Bergkette reicht.

Jeder kennt den Wolkenfilm im täglichen Wetterbericht des Fernsehens. Aus 36.000 Kilometern Höhe hat der Satellit Meteosat-8 Europa, Afrika, den Atlantik, weite Teile von Asien und Amerika im Blick. Doch Erdbeobachtung ist mittlerweile viel mehr als das bloße Fotografieren von Wolken. Eine ganze Armada an spezialisierten Satelliten kreist um die Erde: Sie nehmen die Oberfläche auf und überwachen die Eisbedeckung an den Polen. Sie vermessen, wie viel es von welchen Gasen in der Erdatmosphäre gibt (das Ozonloch lässt grüßen) und wie sehr sich regional die Anziehungskraft unseres Planeten minimal ändert. Die Satelliten verfolgen, wie die großen Meeresströmungen durch die Ozeane pflügen und wie feucht der Boden ist. Radaraufnahmen zeigen auf den Meter genau, wo sich Gebäude oder Wald und Flüsse befinden. Bei Naturkatastrophen wie Erdbeben, Flutwellen oder Überschwemmungen liefern die Satelliten oft binnen Stun-

travel was to journey as far out into the cosmos as quickly as we could, it now seems that the principal reason for flights into space is to get a proper view of our home planet. A bigger picture of our Earth is gleaned more easily at a distance than from the surface, where you can see to the next mountain range at best.

Everyone has seen satellite images of cloud formations in the daily weather forecasts on television. The Meteosat-8 satellite covers Europe, Africa, the Atlantic, and large parts of Asia and America, but observing the Earth now involves much more than merely taking pictures of clouds. A whole armada of specialized satellites orbits the Earth, scrutinizing the surface and keeping an eye on the polar ice caps. They measure tiny variations in gravity across our planet and the concen-tration of various gases in the atmosphere (never forgetting the hole in the ozone layer). The satellites observe the path of great ocean currents across the high seas and record the moisture content of the Earth's surface. Radar images fix the locations of buildings, forests, and rivers to the inch, and satellites can provide a clear picture, often within a few hours, of those streets that are still passable and the areas that

au départ de pénétrer toujours plus loin et plus rapidement dans l'univers, les vols spatiaux visent désormais avant tout à veiller sur notre berceau céleste. Il est en effet plus facile d'étudier les grands phénomènes terrestres vus du ciel que sur le sol, où notre regard ne porte au mieux que jusqu'à la chaîne de montagne la plus proche.

Qui n'a jamais vu l'animation satellite des prévisions météo sur les chaînes télévisées ? À 36 000 kilomètres d'altitude, le satellite Météosat-8 a l'Europe, l'Afrique, l'océan Atlantique et de vastes régions asiatiques et américaines dans le collimateur. L'observation de la Terre est toutefois loin de se contenter de photographier les nuages. Une armada de satellites spécialisés est en orbite autour de la Terre. Ils photographient sa surface et surveillent les calottes glacières des pôles. Ils mesurent la quantité des gaz contenus dans l'atmosphère terrestre et les variations régionales minimales de l'attraction terrestre. Les satellites suivent avec attention la manière dont les courants marins sillonnent les océans et calculent l'humidité des sols. Les images radar indiquent au mètre près l'emplacement des bâtiments, fleuves et forêts. Lors

grandes interrelaciones de nuestra Tierra se aprecian mejor tomando un poco de distancia que desde el suelo.

Todos conocemos la previsión meteorológica diaria de la televisión. Desde 36.000 km de altura, el satélite Meteosat 8 ofrece una panorámica de Europa, África, el Atlántico y amplias regiones de Asia y América. Pero, en la actualidad, la observación terrestre es mucho más que la mera toma de fotografías de las nubes. En torno a la Tierra gira una completa armada de satélites especializados que fotografían su superficie y vigilan la cubierta de hielo de los polos, analizan los gases de la atmósfera terrestre, en qué cantidad se encuentran y cómo varía regionalmente la fuerza de gravedad de nuestro planeta. Los satélites siguen el rastro de las corrientes marinas a través de los océanos y miden la humedad del suelo. Los levantamientos de radar muestran con todo con gran precisión. En caso de catástrofes naturales, como terremotos, tsunamis o inundaciones, los satélites ofrecen en cuestión de horas una imagen clara de las vías que aún son transitables y de cuáles son las regiones completamente devastadas.

Europas SMOS-Satellit kreist seit November 2009 um die Erde und misst Bodenfeuchte und Salzgehalt der Meere

Europe's SMOS satellite has been measuring soil moisture and ocean salinity from orbit since November 2009

Le satellite européen SMOS gravite depuis novembre 2009 autour de la Terre, mesurant l'humidité des sols et la salinité des océans

El satélite europeo SMOS gira en torno a la Tierra desde noviembre de 2009 y mide la humedad del suelo y la salinidad de los mares

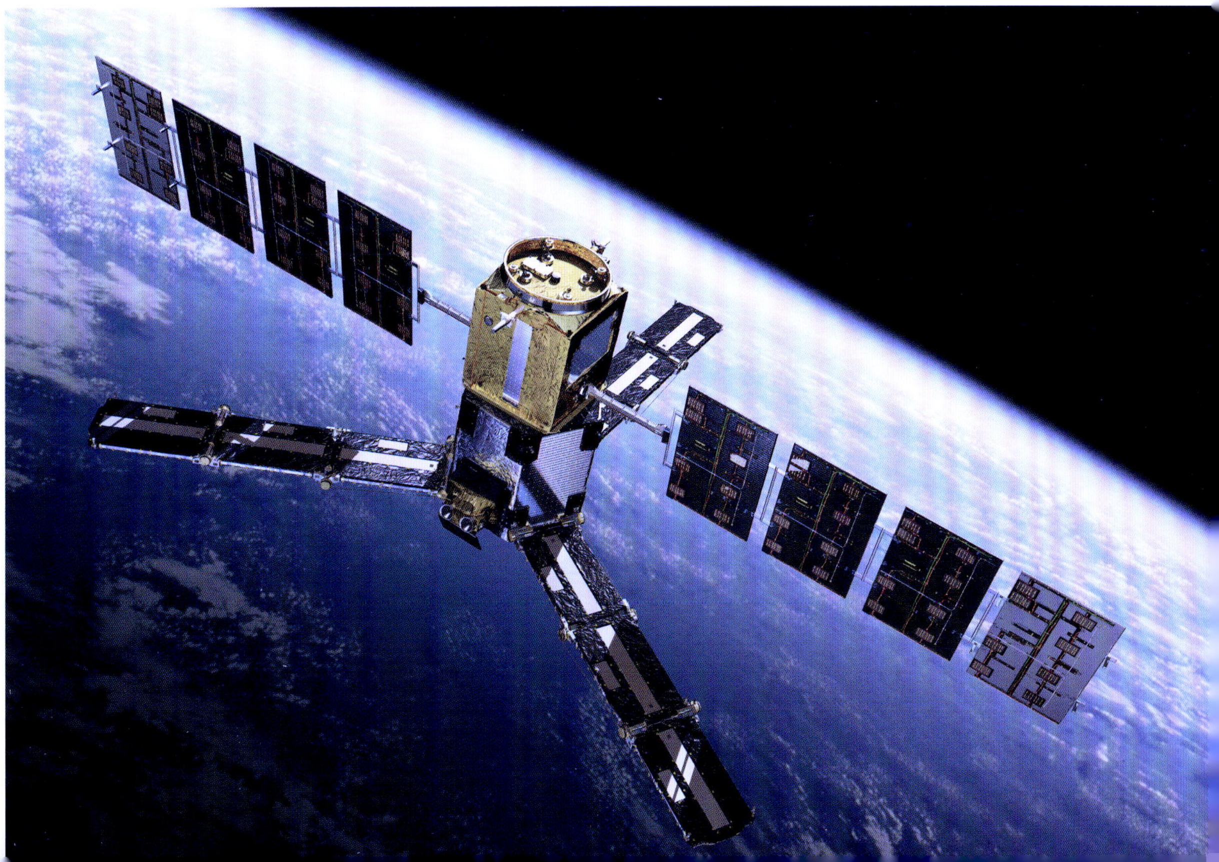

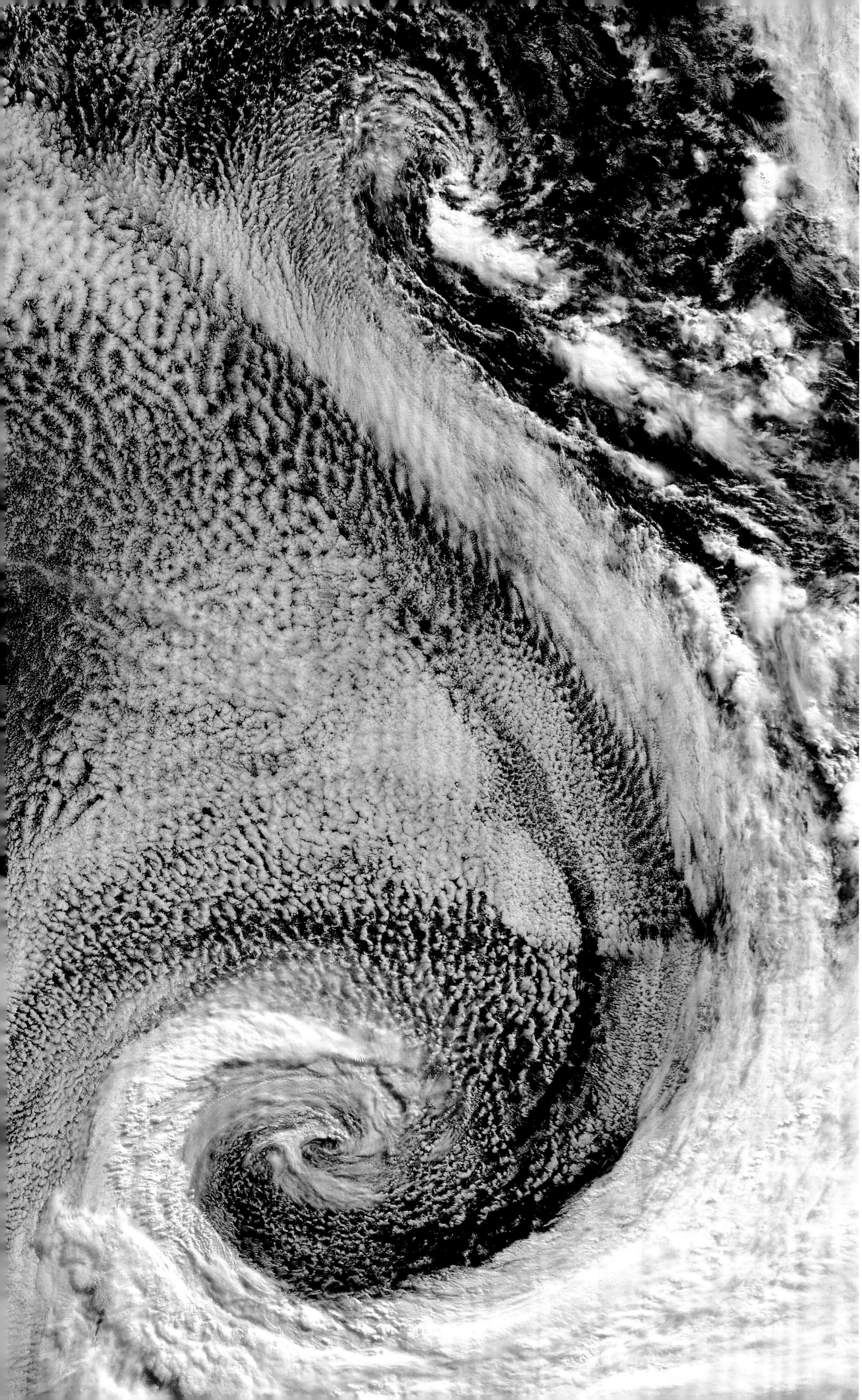

Dieser Wolkenwirbel über dem Atlantischen Ozean erstreckt sich über 4000 km. Die Wolken vor (unten) und hinter (Mitte) der Front unterscheiden sich deutlich (Terra)

This swirl of clouds above the Atlantic Ocean is 2,500 miles (4,000 km) long. The difference between the cloud formations ahead of and behind the front (i.e. the bottom and top of the image) is readily apparent (Terra)

Ce tourbillon nuageux s'étend sur 4 000 km au-dessus de l'océan Atlantique. On distingue clairement les nuages situés à l'avant (en bas) et à l'arrière (au centre) du front (Terra)

Este remolino de nubes sobre el océano Atlántico se extiende a lo largo de 4.000 km. Se distinguen claramente las nubes de delante del frente (abajo) y las de detrás (centro, Terra)

Rechte Seite: Flugzeuge und Schiffe hinterlassen Spuren. Kondensstreifen über der Adria zeigen den Flugverkehr (links, Envisat), vor Alaska sind die Rauchfahnen auf den Schiffsrouten zu erkennen (rechts, Terra)

Right: boats and planes leave smoke and vapor trails. Contrails over the Adriatic are indicators of air traffic (near right, Envisat). The shipping lanes off Alaska are easy to pick out (far right, Terra)

Page de droite : trainées laissées par les avions et navires. Les trainées de condensation révèlent le trafic aérien au-dessus de l'Adriatique (à gauche, Envisat). On distingue les navires au large de l'Alaska (à droite, Terra)

Página dcha.: Los aviones y los barcos dejan rastros. Las estelas sobre el Adriático muestran el tráfico aéreo (izda., Envisat). Delante de Alaska se pueden ver las rutas marítimas (dcha., Terra)

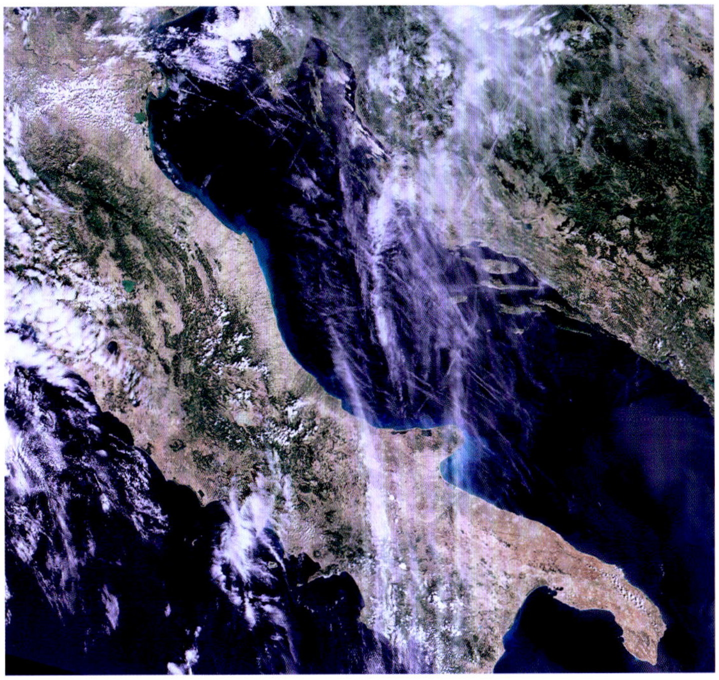

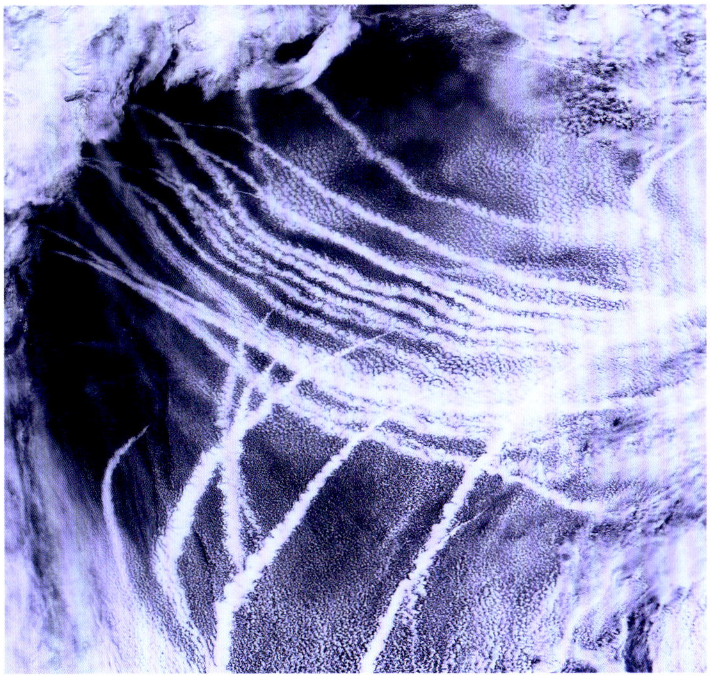

den ein klares Bild, welche Straßen noch passierbar und welche Regionen völlig verwüstet sind.

Augen für fast alles

Aus dem All lassen sich Schiffe erfassen, die auf hoher See ihre Tanks reinigen und mit Ölteppichen das Meer verschmutzen – oft sogar bei Nacht und Nebel, weil viele Instrumente an Bord der Satelliten pfiffig konstruiert sind und viel mehr registrieren, als das menschliche Auge wahrnehmen kann. Im letzten Winkel des tropischen Regenwalds im Amazonasbecken mag man sich bei illegalen Rodungen unbeobachtet fühlen. Es ist sicher so, dass man sich den Blicken der lokalen Behörden entziehen kann – aber die automatischen Zuschauer in 800 Kilometern Höhe ziehen alle hundert Minuten um die Erde. Ihren Augen entgeht fast nichts.

Was könnte ein besserer Sinn der Raumfahrt sein als die Beobachtung der Erde? Beobachtung im wissenschaftlichen Sinne hat nichts mit Überwachung im polizeilichen oder militärischen Sinne zu tun. Nicht »Big Brother« hat uns im Blick, kleine Satelliten behalten die große Erde im Auge. Denn unser Planet ist viel verletzbarer, sensibler und komplexer,

have been completely devastated by natural catastrophes such as earthquakes, tsunamis, and floods.

Eyes in the Sky

Ships which flush their tanks at sea, polluting the ocean with oil slicks, can be spotted from space, even at night or in bad weather—many of the instruments on board the satellites are of such sophisticated construction that they capture far more than the human eye can perceive. Illegal loggers in the furthest corners of the Amazon rainforest might think themselves unobserved, but although they may certainly be able to evade the eyes of the local authorities, hardly anything escapes the attention of the automated observers orbiting the Earth every 100 minutes at an altitude of 500 miles (800 kilometers).

What better use for space travel than the observation of the Earth? Observation in the scientific sense has nothing to do with the surveillance undertaken by the police or the military. "Big Brother" isn't watching us, but tiny satellites do keep an eye on the vast expanse of the Earth. It turns out that our planet is far more complex, exposed, and

des catastrophes naturelles, séismes, raz-de-marée ou inondations, les satellites fournissent souvent en l'espace de quelques heures une image détaillée des axes routiers dégagés et des régions dévastées.

Des yeux presque partout

Depuis l'espace, il est possible de repérer les navires qui dégazent au large et les nappes pétrolières qui souillent la mer, même la nuit et par temps de brouillard, grâce aux instruments sophistiqués embarqués à bord des satellites et capables d'enregistrer bien plus d'informations que n'en perçoit l'œil humain. Dans les profondeurs de la forêt amazonienne, les défricheurs illégaux peuvent se croire à l'abri des regards. Et il leur est certes possible de se soustraire à la surveillance des autorités locales, mais peu de choses échappent aux spectateurs automatiques qui, perchés à 800 kilomètres d'altitude, font le tour de la Terre toutes les cent minutes.

Peut-on imaginer pour les activités spatiales mission plus noble que l'observation de la Terre ? L'observation scientifique n'a rien à voir avec la surveillance policière ou militaire. Nous ne sommes pas épiés par « Big

Ojos para casi todo

Desde el espacio se pueden detectar barcos que limpian sus depósitos en alta mar y lo contaminan con vertidos de petróleo, incluso de noche y con niebla. También los desmontes ilegales que se realizan en el último rincón de la selva lluviosa tropical de la cuenca del Amazonas. Es evidente que se pueden ocultar a la mirada de las autoridades locales, pero los espectadores automáticos que hay a 800 km de altura giran cada 100 minutos en torno a la Tierra y casi nada escapa a sus ojos.

¿Qué mejor sentido podría tener lo espacial que el de la observación de la Tierra? Observación, en el sentido científico de la palabra, no es asimilable a una vigilancia policial o militar. No es un «Gran Hermano» vigilante, sino pequeños satélites que no pierden de vista la gran Tierra. Porque nuestro planeta es mucho más vulnerable, sensible y complejo de lo que desde su superficie queremos admitir. La atmósfera se asemeja a un gigantesco mar de aire que flota sobre nuestras cabezas. Al observar la Tierra desde la órbita, se puede distinguir la finísima orla azul que rodea al planeta. A solo 10 km de altura ya no podríamos respirar y eso en un planeta que tiene un diámetro

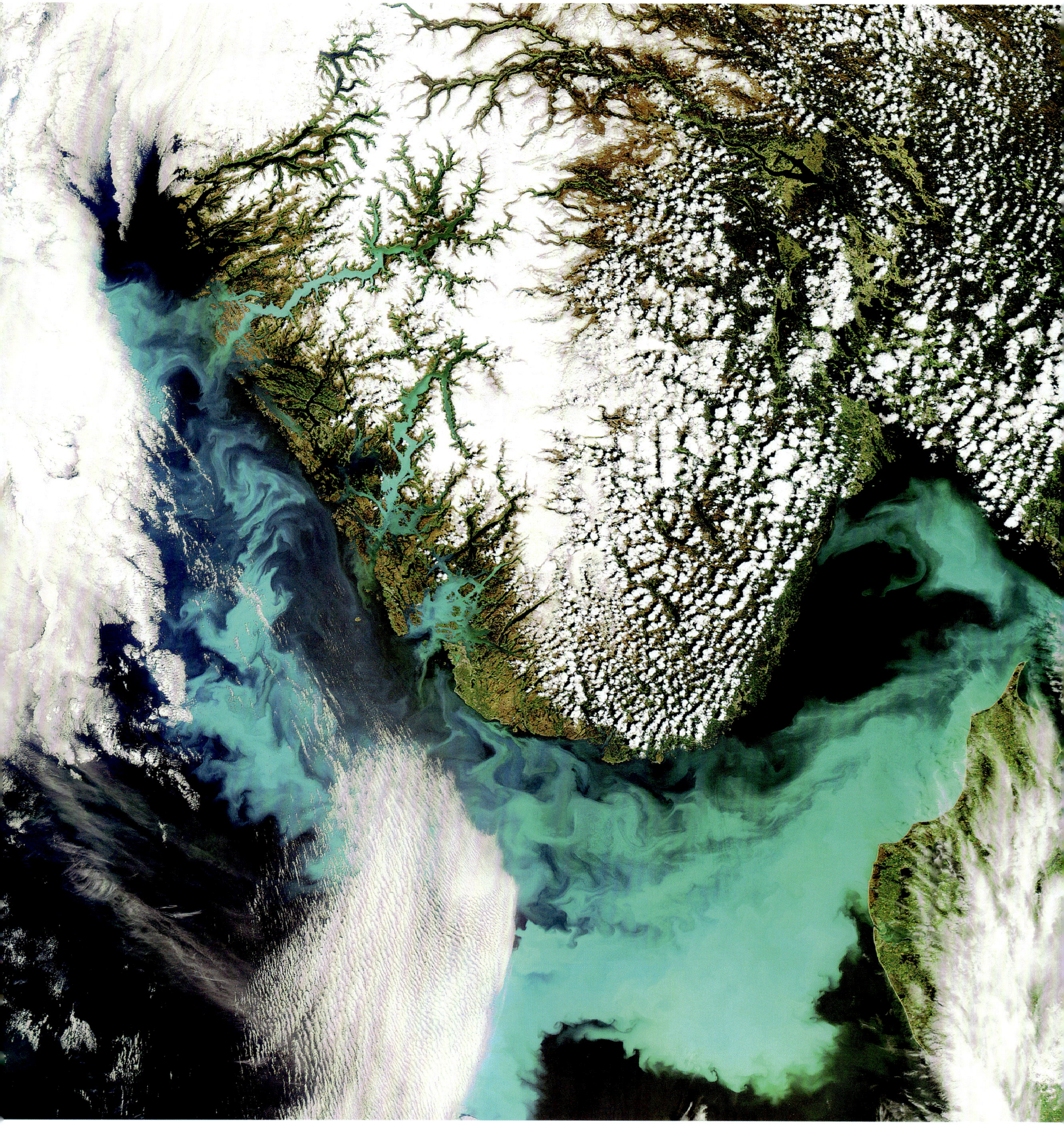

als wir auf dem Erdboden wahrhaben wollen. Die Atmosphäre scheint ein gigantisches Luftmeer über uns zu sein. Ein fataler Eindruck, wenn man auf dem Boden der Atmosphäre lebt. Wer aus der Umlaufbahn auf die Erde blickt, erkennt den hauchdünnen blauen Saum, der unseren Planeten umgibt. Schon in zehn Kilometern Höhe könnten wir nicht mehr atmen – und das auf einem Planeten, der einen Durchmesser von 12.756 Kilometern hat. Eine Apfelschale ist verglichen zur Größe des Apfels viel dicker als die Lufthülle, die die Erde umgibt. Ulf Merbold, nach dem Kosmonauten Sigmund Jähn der zweite Deutsche im All, hat nach der Rückkehr von seinem Flug in einer US-Raumfähre 1983 oft betont, dass ihn auf seiner Weltraumreise am meisten beeindruckt habe, mit eigenen Augen zu sehen, wie dünn die Erdatmosphäre ist. Die Ausmaße von Erde und Lufthülle waren dem Physiker sicher vorher klar. Aber erst das Betrachten mit eigenen Augen hat ihm diesen Umstand bewusst gemacht: Erdbeobachtung, wie sie schöner, unmittelbarer und nachhaltiger kaum sein kann.

Das Privileg, mit eigenen Augen die Erde aus großer Höhe zu betrachten, wurde bisher nur etwa 500 Menschen zuteil. Mehr Personen waren noch nicht in der Erdumlaufbahn. Selbst

vulnerable than we who live on its surface would care to admit. From the ground, it is easy to presume that our atmosphere immerses us in a gigantic ocean of air, but look at the Earth from orbit and you will see how wafer-thin the blue film surrounding our planet really is. We are no longer able to breathe at an altitude of only 6 miles (10 kilometers) above a planet with a diameter of 7,926 miles (12,756 kilometers). The skin of an apple is, comparatively speaking, far thicker than the blanket of air surrounding the Earth. After his return from a US Space Shuttle mission in 1983, Ulf Merbold, the second German in space after the cosmonaut Sigmund Jähn, would often recount how the most impressive part of his journey across the heavens had been seeing with his own eyes just how thin the Earth's atmosphere is. As a physicist, he had been very well aware of the dimensions of the Earth and the atmosphere that encircles it, but its true size was only brought home to him when he saw it with his own eyes—observation of the Earth could hardly be more immediate, more potent, or more beautiful than this.

Only about 500 people have had the privilege of observing the Earth

Brother ». Ces petits satellites veillent au contraire sur la Terre, qui est beaucoup plus fragile, sensible et complexe que nous ne voulons bien le croire. L'atmosphère nous apparaît comme un vaste océan d'air au-dessus de nos têtes. Mais il s'agit d'une impression trompeuse. Si on regarde la Terre depuis l'espace, on aperçoit le délicat voile bleu qui la recouvre. À dix kilomètres d'altitude, nous serions déjà incapables de respirer, alors que notre planète a un diamètre de 12 756 kilomètres. La peau d'une pomme est, comparée à la taille du fruit, bien plus épaisse que la pellicule d'air qui entoure la Terre. Ulf Merbold, le deuxième allemand dans l'espace après le cosmonaute Sigmund Jähn, a souvent mentionné après son vol à bord d'une navette spatiale américaine en 1983, à quel point il avait été impressionné par la minceur de l'atmosphère terrestre. Ce physicien connaissait parfaitement les dimensions de la Terre et de l'atmosphère, mais c'est en les observant de ses propres yeux qu'il a réellement pris conscience du phénomène.

Seulement 500 personnes environ ont eu le privilège d'observer la Terre de leurs propres yeux depuis l'espace. Même à 300 ou 400 kilomètres d'altitude, il est impossible d'avoir

de 12.756 km. La piel de una manzana es, en comparación con el tamaño de la misma, mucho más gruesa que la cubierta gaseosa que envuelve la Tierra. Tras regresar, en 1983, de su vuelo en una nave espacial estadounidense, Ulf Merbold, el segundo alemán que viajó al espacio tras el cosmonauta Sigmund Jähn, ha destacado que lo más impresionante de su viaje por el espacio sideral fue ver con sus propios ojos lo delgada que es la atmósfera terrestre. Con toda seguridad, el físico ya era consciente con anterioridad de las dimensiones de la Tierra y de la capa gaseosa. Pero contemplarla con sus propios ojos le permitió darse cuenta de que la observación de la Tierra no podía ser más bella, inmediata y eficaz.

Hasta ahora, sólo 500 personas han podido disfrutar del privilegio de observar la Tierra desde gran altura con sus propios ojos. Aunque a 300 o 400 km de altura aún no puede abarcarse completamente la Tierra con la vista, sin embargo sí pueden distinguirse los países. Además, se aprecia la curvatura de la Tierra y, al otro lado del arqueado horizonte, la fascinante y al mismo tiempo amenazadora negrura del espacio sideral. A partir de la órbita geoestacionaria, aproximadamente a 36.000 km de altura, la Tierra se

Blaue Wirbel vor Norwegen: Die Planktonblüte sorgt in der Nordsee für eine charakteristische Färbung (Envisat)

Blue algae thrive off Norway; plankton growth lends the North Sea its characteristic coloration (Envisat)

Volutes bleues sur la côte norvégienne : la floraison planctonique est à l'origine d'une couleur singulière dans la mer du Nord (Envisat)

Torbellino azul ante Noruega: el florecimiento del plancton produce una característica coloración en el Mar del Norte (Envisat)

Weiße Wolken, türkise Planktonblüte, grüne Insel: Schnappschuss von Irland durch Europas Satelliten Envisat

White clouds, turquoise plankton growth, and a green island; a snapshot of Ireland from the European Envisat satellite

Nuages blancs, ile verte, floraison planctonique turquoise : cliché de l'Irlande pris par le satellite européen Envisat

Nubes blancas, floración turquesa de plancton, isla verde: instantánea de Irlanda tomada por el satélite europeo Envisat

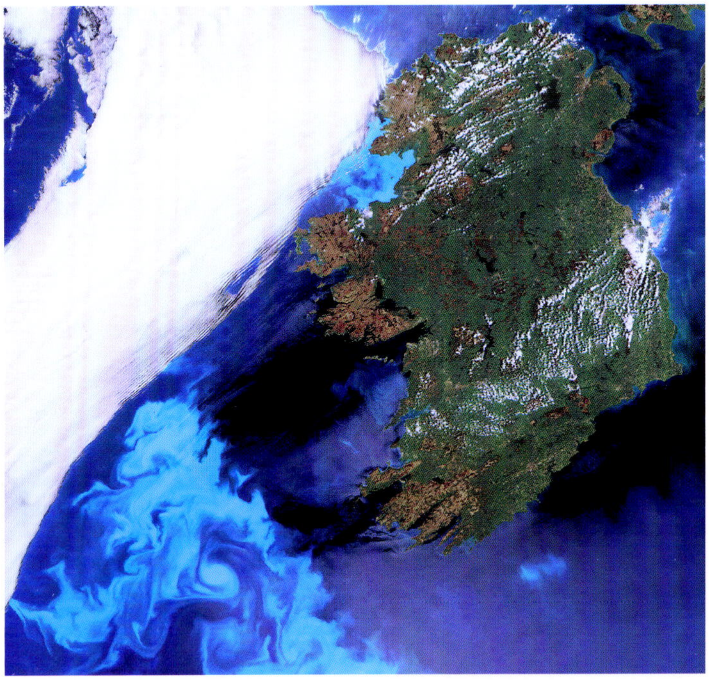

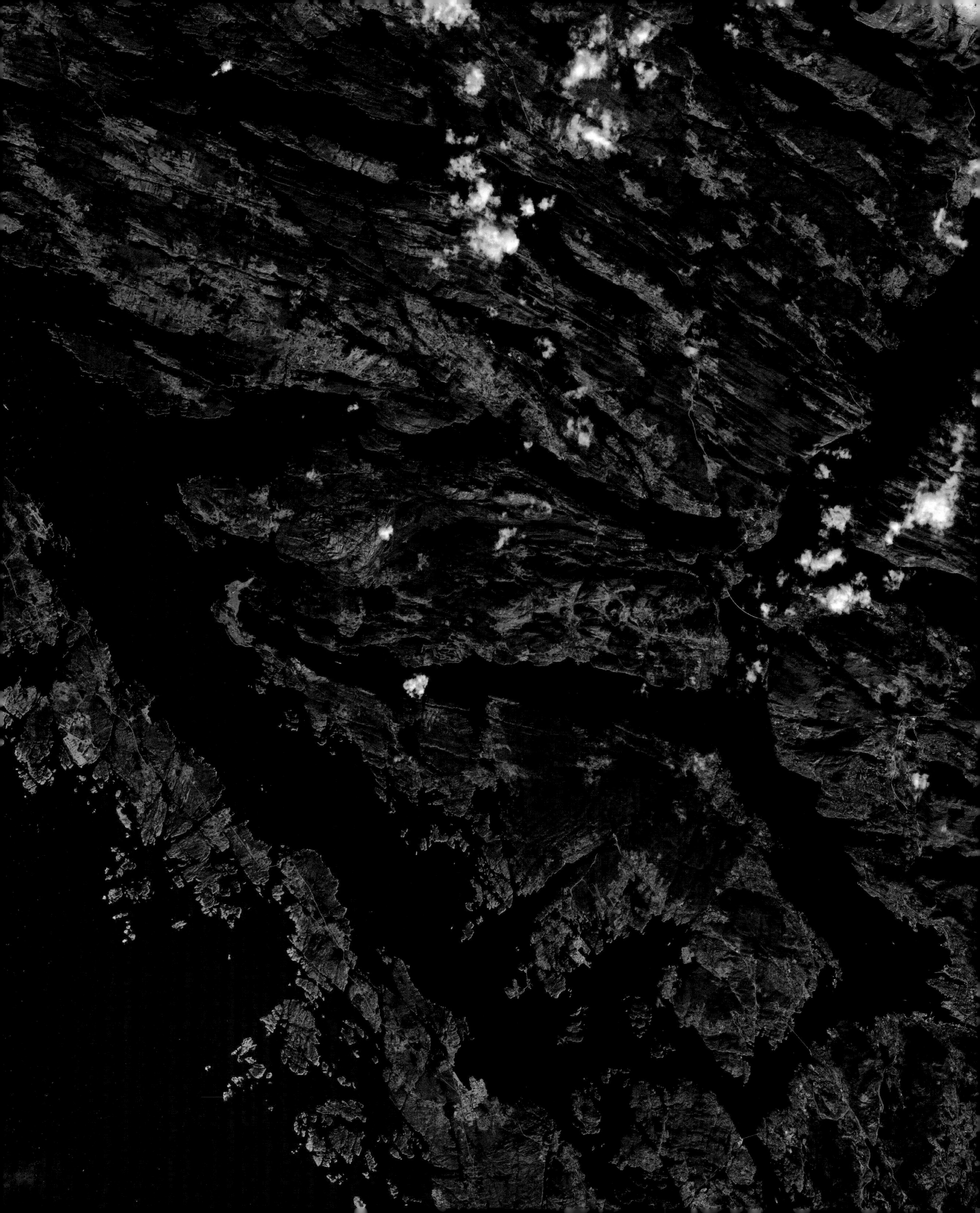

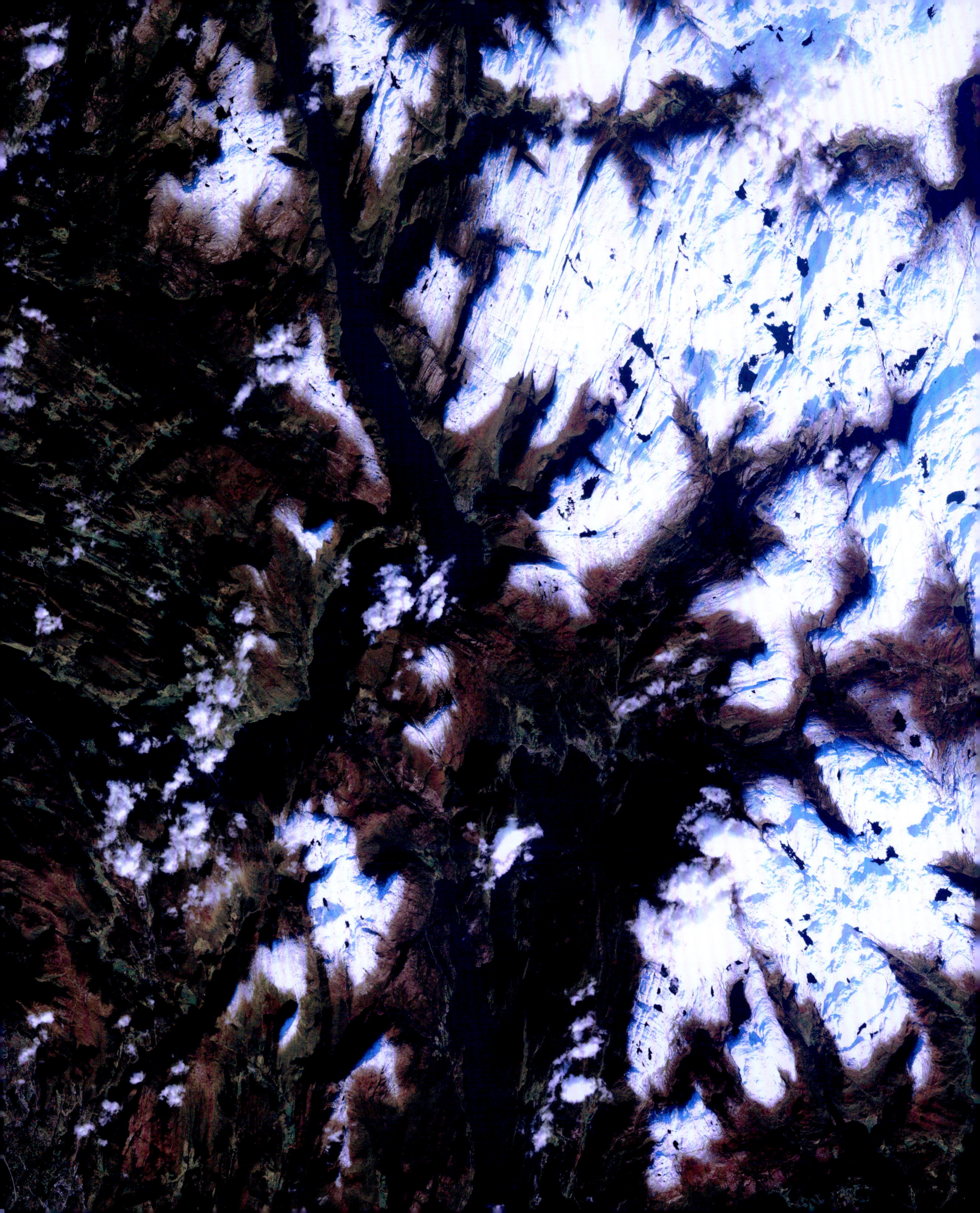

aus 300 oder 400 Kilometern Höhe hat man die Erde nicht komplett im Blick, immerhin aber ganze Länder. Zudem sieht man die Krümmung der Erde und jenseits des geschwungenen Horizonts die ebenso faszinierende wie bedrohliche Schwärze des Weltraums. Als blaue Kugel im All zeigt sich die Erde erst ab etwa der geostationären Umlaufbahn in 36.000 Kilometer Höhe. Noch besser ist es, man reist zum Mond. Von dort lässt sich wirklich die ganze Erde in Augenschein nehmen bzw. die ganze dem Betrachter zugewandte Seite. Diesen grandiosen Anblick haben nur die vierundzwanzig Apollo-Astronauten genossen, die zum Mond geflogen sind. Viele kamen verändert wieder – aber weniger wegen der Erlebnisse auf dem Mond, sondern weil sie bei und nach dieser Reise die Erde ganz anders wahrgenommen haben.

Mondflug zur Erde

Zu Weihnachten 1968 umrundete das Raumschiff Apollo 8 den Mond. Erstmals hatten Menschen den unmittelbaren Anziehungsbereich der Erde verlassen. Nie zuvor hatten Bewohner des Planeten Erde ihre kosmische Heimat aus so großer Entfernung gesehen. Die Astronauten lasen in der Mondumlaufbahn aus der Genesis, der biblischen Schöpfungsgeschichte. Man braucht nicht religiös veranlagt zu sein, um selbst heute noch einen Schauer auf dem Rücken zu spüren, wenn man sich die Tondokumente von damals anhört. Man fühlt unmittelbar, wie sehr dieser Moment die Astronauten, von der Ausbildung her abgebrühte Militärpiloten, bewegt hat. Die ersten Bilder der gesamten Erde vermittelten zugleich Erhabenheit und Verwundbarkeit jenes Körpers in den Weiten des Universums, der unser Zuhause ist und dem wir nicht entkommen können. Plötzlich wurde vielen Menschen bewusst, dass die Erde unsere Oase ist in den Lichtjahre weiten lebensfeindlichen Weiten des Kosmos. Nicht wenige meinen, mit den Apollo-Bildern habe die Umweltbewegung begonnen – mit den Bildern der Erde, wohlgemerkt!

from such a great height, and no more than this have been in orbit. Even at an altitude of 190–250 miles (300–400 kilometers), it is impossible to see the entire Earth, although you can see complete countries. From here it is also possible to see the curvature of the Earth and, beyond this arcing horizon, the blackness of space itself, fascinating and threatening in equal measure. However, the Earth as a whole only begins to reveal itself as a blue sphere floating in space from a geostationary orbit at a distance of 22,400 miles (36,000 kilometers). Better still, take a trip to the Moon (almost 250,000 miles/400,000 kilometers away)—from there it is possible to see the entire Earth all at once, or at least the side of Earth that happens to be turned toward the observer. The only humans privileged to enjoy this magnificent view have been the 24 Apollo astronauts who visited the Moon, and many of these came back changed people—not so much as a result of their experiences on the Moon, but rather because they saw the Earth differently after their voyage.

A Moon mission to Earth

The first human beings to escape the Earth's immediate gravitational field were aboard the spaceship Apollo 8, which orbited the Moon at Christmas 1968. Never before had inhabitants of our planet viewed their terrestrial home from such a great distance. The astronauts read aloud the Biblical story of the Creation from the book of Genesis as they orbited the Moon, and you don't have to be religious to feel a shiver run down your spine as you listen to the recording of this—it is easy to hear how the astronauts, who were all hard-bitten military pilots, were moved by the experience. The first images of the whole Earth gave a sense of both the sublimity of this home to which we are bound and the vulnerability of this single body in the immensity of the universe. Many soon came to realize that the

un panorama intégral de la Terre, cependant on peut apercevoir des pays tout entiers. On distingue également la courbure de la Terre et, au-delà de la ligne d'horizon, l'obscurité aussi fascinante que menaçante de l'univers. La Terre ne se montre sous les traits d'une sphère bleue qu'à partir d'une orbite de 36 000 kilomètres d'altitude. La vue est encore meilleure depuis la Lune, d'où l'on aperçoit vraiment la Terre dans toute sa splendeur, ou du moins la face qu'elle présente à l'observateur. Seuls les vingt-quatre astronautes des missions lunaires Apollo ont pu jouir de ce spectacle époustouflant. Ils sont d'ailleurs revenus transformés, non pas tant par leur expérience lunaire, que par leur nouvelle manière d'appréhender la Terre après ce voyage.

De la Lune à la Terre

À la veille de Noël 1968, l'astronef Apollo 8 a opéré plusieurs révolutions autour de la Lune. C'était la première fois que des hommes sortaient du champ de gravitation de la Terre. Jamais auparavant des terriens n'avaient eu le loisir d'observer leur berceau cosmique d'aussi loin. En orbite autour de la Lune, les astronautes ont relu la Genèse, le récit biblique de la création du monde. Il n'est pas nécessaire d'être croyant pour être aujourd'hui encore parcouru d'un frisson à l'écoute des documents sonores de l'époque, et ressentir immédiatement combien cet instant a pu émouvoir les astronautes pourtant aguerris par leur formation de pilotes militaires. Les premières vues globales de la Terre révélèrent le caractère à la fois sublime et fragile de ce corps céleste qui nous abrite dans l'immensité de l'univers, et auquel nous ne pouvons nous soustraire. La Terre est alors apparue à de nombreux terriens comme une oasis perdue dans les années lumières d'une immensité cosmique hostile. Et il est largement admis que ce sont les images d'Apollo qui sont à l'origine du mouvement écologique.

muestra como una esfera azul en el espacio. Todavía es mejor si se viaja a la Luna. Desde allí se puede examinar toda la Tierra o, al menos, la totalidad del lado vuelto hacia el espectador. Esta grandiosa vista únicamente la han disfrutado los 24 astronautas del proyecto Apolo que volaron hasta el satélite lunar. Muchos de ellos volvieron completamente transformados, pero no precisamente por sus experiencias en la Luna, sino porque en este viaje, y tras el mismo, empezaron a percibir la Tierra de un modo completamente distinto.

Vuelo de la Luna a la Tierra

En las navidades de 1968, la nave espacial Apolo 8 orbitó alrededor de la Luna. Por primera vez, los seres humanos abandonaban el campo gravitatorio de la Tierra. Nunca antes los habitantes del planeta habían visto su patria cósmica desde una distancia tan grande. Ya en la órbita lunar, los astronautas leyeron el comienzo del Génesis, el relato bíblico de la creación. No es necesario tener convicciones religiosas para sentir un escalofrío. Incluso en la actualidad, cuando se escuchan los documentos sonoros de entonces, se perciben los sentimientos que conmovieron a los astronautas en esos momentos, a pesar de su formación como pilotos experimentados. Las primeras imágenes de la Tierra transmiten lo sublime y lo vulnerable de aquel cuerpo en la inmensidad del universo, un cuerpo que es nuestro hogar y del que no podemos escapar. Repentinamente, muchas personas se dieron cuenta de que la Tierra es nuestro oasis en medio de los años luz del amplio espacio hostil para la vida que representa el cosmos. No pocos piensan que el movimiento ecologista comenzó con las imágenes de la Tierra que se tomaron desde el Apolo.

Nuestra meta es poner un hombre en la Luna antes de que termine la década y traerlo de nuevo a salvo a la Tierra, para que sus habitantes tomen conciencia de su propio planeta.

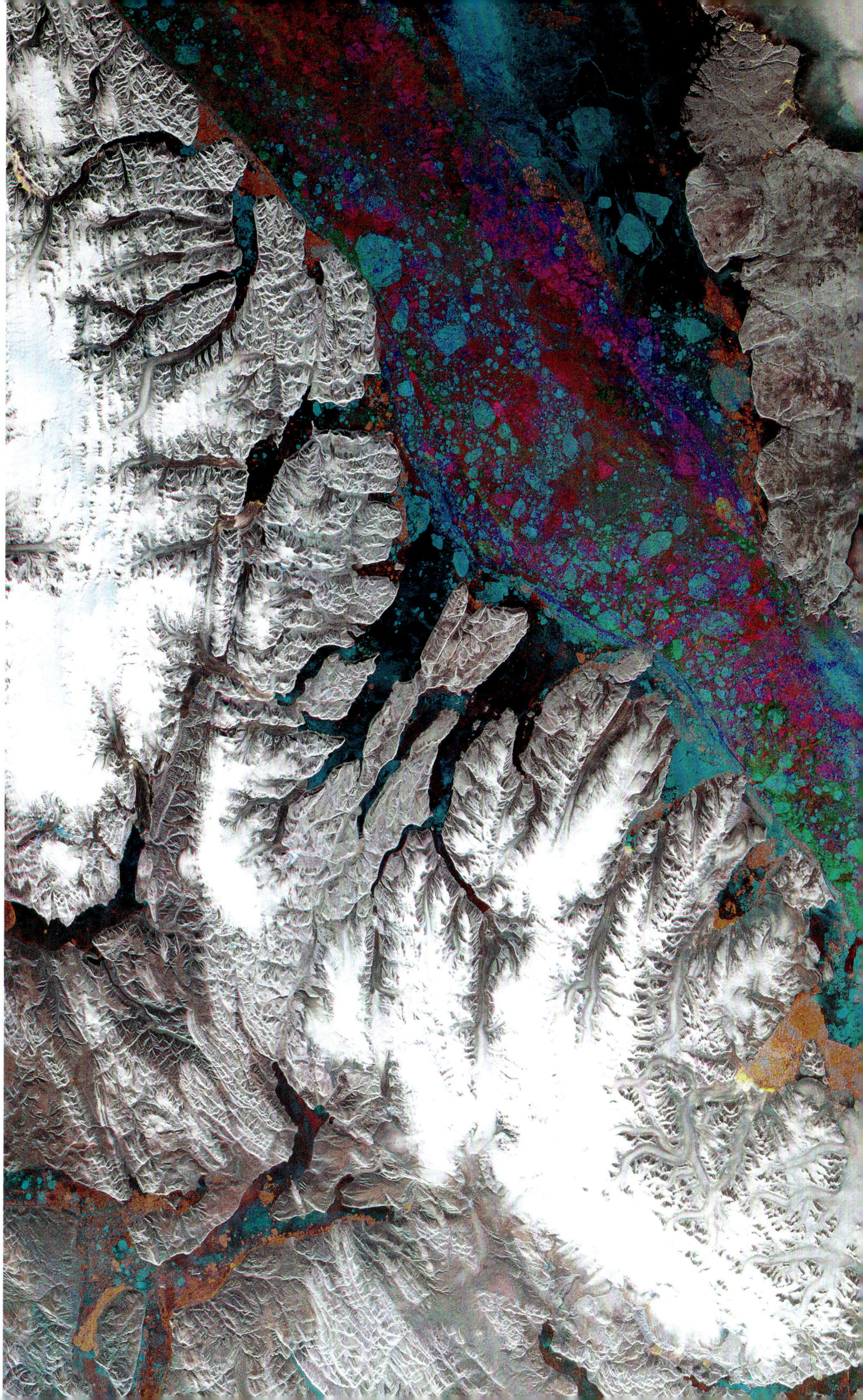

Eisschollen auf der Wasserstraße im Norden Kanadas. Unterschiedliche Oberflächen reflektieren Radarsignale auf jeweils andere Weise (TerraSAR-X)

Ice floes in the straits of northern Canada. Every surface reflects radar signals in a different way (TerraSAR-X)

Blocs de glace sur la voie navigable au nord du Canada. Les différentes surfaces reflètent chacune les ondes radar de manière spécifique (TerraSAR-X)

Témpanos de hielo en la vía de navegación del norte de Canadá. Las diferentes superficies reflejan de maneras distintas las señales del radar (TerraSAR-X)

Seite 30/31: Berge, Schnee und Fjorde: Die Stadt nahe dem unteren Bildrand (etwas links der Mitte) ist Bergen in Norwegen (Envisat)

Pages 30–1: Mountains, snow and fjords: the city near the bottom edge of the image (just left of center) is Bergen in Norway (Envisat)

Pages 30 et 31 : montagnes, neiges et fjords. Au bas de la photo (légèrement à gauche du milieu de page), la ville norvégienne de Bergen (Envisat)

Páginas 30/31: Montañas, nieve y fiordos: la ciudad cercana al borde inferior de la imagen (ligeramente a la izquierda) es Bergen, en Noruega. (Envisat)

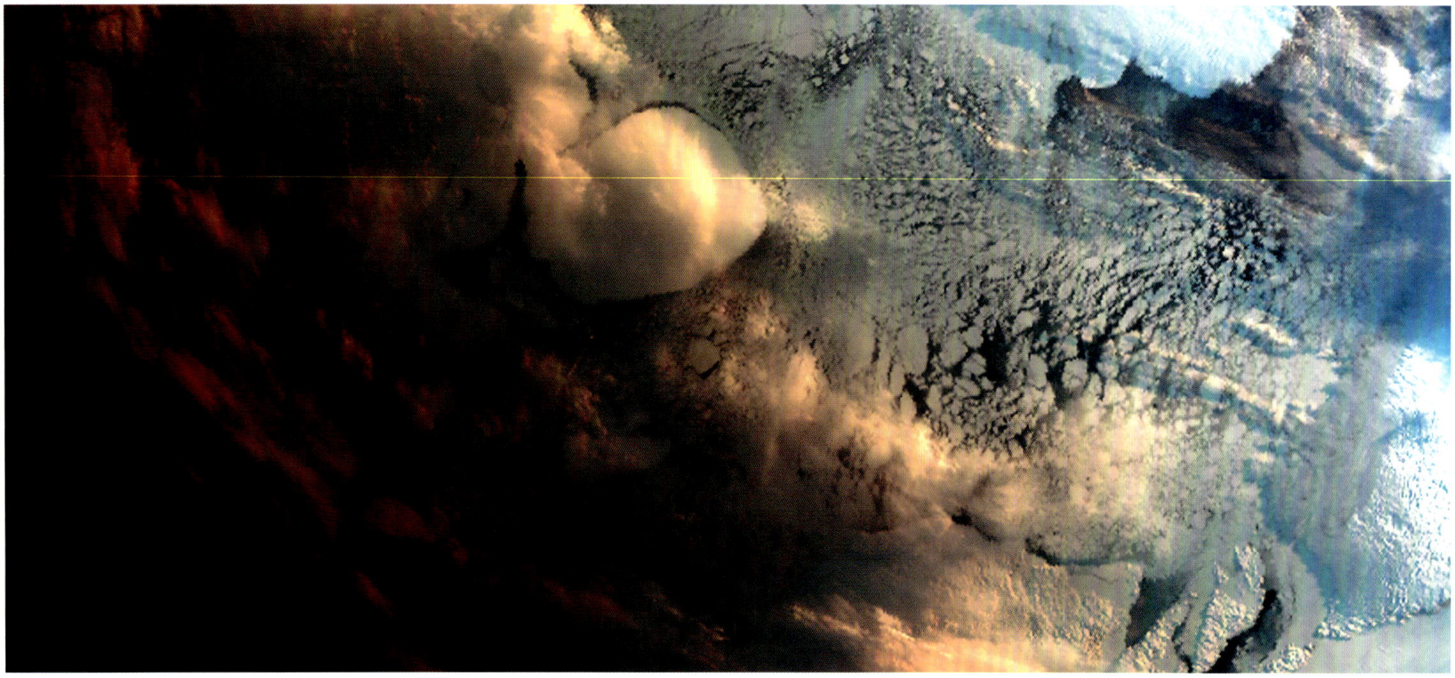

Unser Ziel ist es, bis zum Ende des Jahrzehnts einen Menschen zum Mond und wieder sicher zurück zur Erde zu bringen, um so unseren Planeten in das Bewusstsein seiner Bewohner zu rücken. Nein, den zweiten Halbsatz hat US-Präsident John F. Kennedy in seiner legendären Mondrede nicht gesagt. Er hat wohl auch kaum ahnen können, dass der Blick auf die Erde fast zum wichtigsten Aspekt des gigantischen Programms werden würde, auf jeden Fall zu einem der Aspekte, die am längsten nachwirken. Man muss den Flug zum Mond nicht gut finden. Aber man muss anerkennen, dass es letztlich eine Mission zur Erde war.

Ging es anfangs einfach darum, überhaupt zu verstehen, dass die Erde ein nach menschlichen Maßstäben zwar sehr großer, aber eben doch begrenzter und keineswegs unverletzlicher Körper ist, so hat sich die Erdbeobachtung mittlerweile zu einer facettenreichen Disziplin entwickelt. Die Apollo-Astronauten oder auch Ulf Merbold haben die Erde im Wesentlichen mit bloßen Augen betrachtet und mit der damals üblichen Ausrüstung fotografiert. Heute geschieht das mit einer Vielzahl unterschiedlicher Instrumente und keineswegs mehr nur im sichtbaren Licht, das das menschliche Auge wahrnimmt – und es geschieht mit Maschinen und

Earth is no more than a tiny oasis in the hostile expanses of a cosmos that is light years across, and some believe that the environmental movement started with the images taken from Apollo 8—images of the Earth itself.

First, I believe that this nation should commit itself to achieving the goal, before this decade is out, of landing a man on the Moon and returning him safely to the Earth, in order to make our planet's inhabitants more aware of their home. Of course, US President John F. Kennedy didn't really include the last part of that sentence in his legendary Moon speech—he could hardly have imagined how observation of the Earth would become perhaps the most important aspect of the massive program, and certainly the one that would have the greatest repercussions. Whether they approved of the concept of flying to the Moon or not, most people would have to admit that it turned out to be a mission to the Earth.

If the initial outcome was simply the understanding that, although the planet was very big by human standards, it was also a limited and even vulnerable body, observing the Earth has now become a multidisciplinary undertaking. The Apollo

Notre nation doit s'engager à faire atterrir un homme sur la Lune et à le ramener sur la Terre sain et sauf avant la fin de la décennie, afin que les Hommes prennent conscience de la fragilité de notre planète. Bien sûr, John F. Kennedy, alors président des États-Unis, n'a pas prononcé la deuxième partie de cette phrase dans son discours légendaire sur la conquête spatiale. Il faut dire qu'il était loin d'imaginer que le regard porté sur la Terre compterait parmi les aspects les plus importants de ce gigantesque programme, et dont les effets seraient les plus durables. On peut ne pas approuver le voyage sur la Lune. Il faut toutefois reconnaître que cette aventure s'est finalement avérée être une mission vers la Terre.

S'il s'agissait simplement au départ de comprendre que la Terre est un corps céleste qui à l'échelle humaine paraît certes très grand, mais est en réalité limité et vulnérable, l'observation de la Terre est devenue avec le temps une discipline aux multiples facettes. Les astronautes d'Apollo, tout comme Ulf Merbold, ont pour l'essentiel observé la Terre à l'œil nu, et l'ont photographiée avec le matériel alors disponible. De nos jours, les prises de vue sont réalisées à l'aide d'une multitude d'instruments

No, la segunda mitad de la frase no la pronunció el presidente de Estados Unidos John F. Kennedy en su legendaria alocución sobre la Luna. Tampoco pudo siquiera sospechar que la mirada puesta sobre la Tierra se iba a convertir casi en el aspecto más importante del gigantesco programa espacial o, en todo caso, en uno de los aspectos de más amplia repercusión. En último término, se puede pensar que el viaje a la Luna no obtuvo éxito en su objetivo, ya que en realidad fue una misión a la Tierra.

Al principio, sobre todo, se trataba de entender que, si bien la Tierra era muy grande según la escala humana, también era un cuerpo limitado y vulnerable. En la actualidad, la observación de la Tierra se ha convertido en una disciplina polifacética. Los astronautas del programa Apolo vieron la Tierra a simple vista y la fotografiaron con el equipo del que entonces disponían. Ahora esto se hace con muchos instrumentos diferentes y, por supuesto, no sólo con la luz visible. Se hace con máquinas y no con personas. Sin embargo, la tripulación de la estación espacial internacional ISS lleva a cabo observaciones diarias de la Tierra. Y lo hace, sobre todo, para su propia satisfacción. Los astronautas insistieron

Links: Winterliche Dämmerung: Rechts scheint noch die Sonne, links herrscht Polarnacht. Bildausschnitt 300 x 700 km (Envisat)

Left: Dusk in winter: the sun is still shining to the right but the left side is already plunged into polar darkness. The region shown is 185 x 435 miles (300 x 700 km) in area (Envisat)

À gauche : crépuscule hivernal. À droite, le soleil brille, à gauche règne la nuit polaire. Image couvrant une surface de 300 x 700 km (Envisat)

Izda.: Crepúsculo invernal: a la derecha aún brilla el Sol, a la izquierda reina la noche polar. Encuadre de 300 x 700 km (Envisat)

Unten: Im Petermann-Gletscher in Grönland ist ein 25 km großer Eisberg abgebrochen und driftet ab (Terra)

Below: A 15-mile-wide (25-km) iceberg breaks away from the Petermann Glacier in Greenland and drifts with the current (Terra)

Ci-dessous : un iceberg de 25 km détaché du glacier Petermann au nord du Groenland dérive dans l'océan (Terra)

Abajo: En el glaciar Petermann, de Groenlandia, se desprendió y quedó a la deriva un gran iceberg de 25 km (Terra)

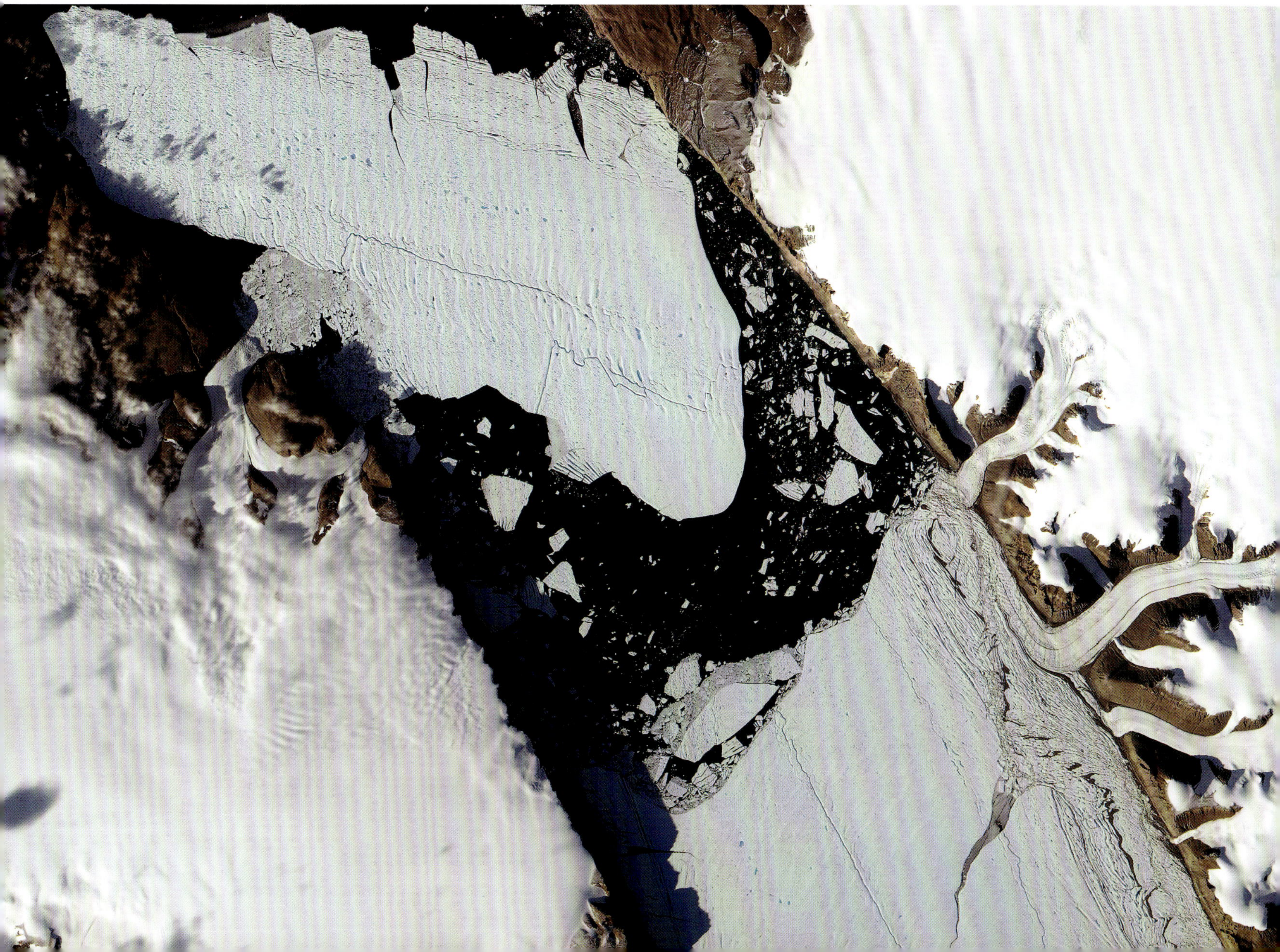

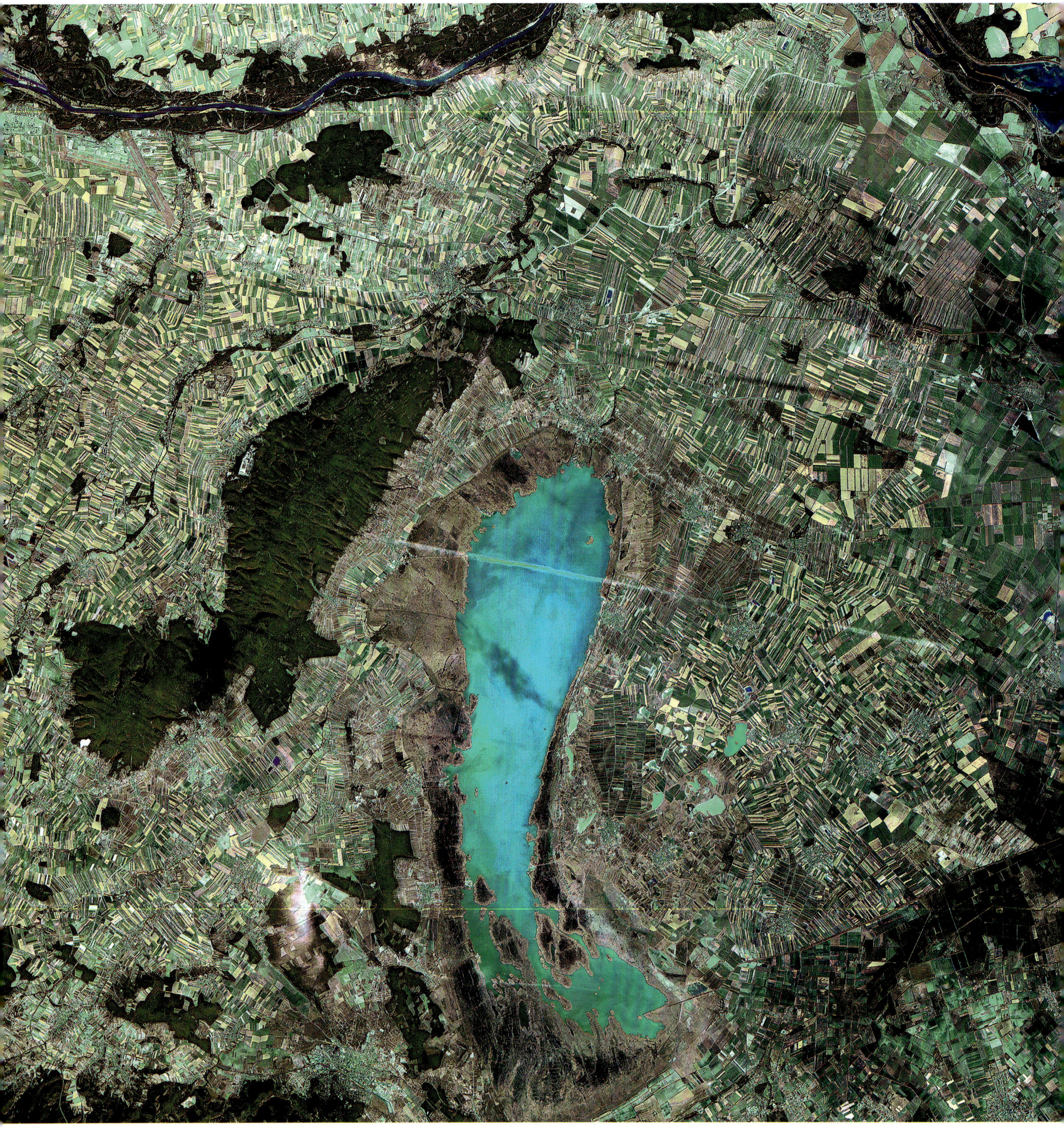

nicht mehr mit Menschen. Die Besatzung der Internationalen Raumstation ISS führt zwar täglich Erdbeobachtung durch. Das geschieht vor allem zur eigenen Freude. Die Astronauten hatten massiv auf den Einbau der Cupola gedrängt, einer Aussichtskuppel, die einen Panoramablick über die fußballfeldgroße Raumstation ermöglicht und dabei immer auch einen großen Teil der Erde zeigt. Im Himmel-und-Erde-Kino auf der ISS läuft ein besonderer Film in der Endlosschleife. Blickrichtung und Beleuchtung bleiben nie gleich: Denn die Raumstation rast in gut neunzig Minuten um unseren Planeten herum.

Die Entdecker von heute

Für uns Normalbürger, die wir am Boden bleiben müssen, übernehmen mittlerweile Dutzende von Satelliten das Beobachten im All. Die Erde im Blick zu behalten hat ganz praktischen Nutzen, etwa um einen Wetterbericht zu erstellen, Klimamodelle auszuarbeiten, Anbauflächen zu überwachen oder Fangverbote zu kontrollieren. Noch häufiger geht es um das grundlegende Erforschen unseres Heimatplaneten. Um den Zustand der Atmosphäre zu untersuchen, nützt das beste Messsystem in Europa, in den

astronauts looked down on the Earth with their own eyes and photographed it with the equipment of their day, but such images are now captured with a host of different instruments extending far beyond what is visible to the naked human eye. The crew of the International Space Station (ISS) carry out daily observations of the Earth, but this is done largely for their own entertainment—the astronauts lobbied hard for the construction of the "cupola," an observation dome with a panoramic view of the football-field-sized space station and thus a large part of the Earth. This "3-D cinema" shows a very special film on an endless loop, with the point of view and lighting always changing, as the station completes its orbit around the Earth in about 90 minutes.

Great modern explorers

Satellites in space have also taken over the role of observers for us, the normal citizens who have to stay Earthbound. Keeping an eye on the Earth has entirely practical benefits—when preparing weather forecasts, computing climate models, monitoring agriculture, or checking fishing

qui dépassent largement les capacités de perception de l'œil humain. Elles sont réalisées par des machines, et non plus par des hommes. L'équipage de la Station spatiale internationale ISS continue certes à observer la Terre quotidiennement, mais c'est par pur plaisir. Les astronautes avaient en effet demandé avec insistance la construction de la Cupola, une coupole d'observation offrant une vue panoramique de la station spatiale, dont la taille est comparable à celle d'un terrain de football, et d'une grande partie de la Terre. Le cinéma entre ciel et Terre de l'ISS passe en boucle un film très particulier. Angle de vue et éclairage varient sans cesse car la station spatiale fait le tour de la Terre en seulement quatre-vingt-dix minutes.

Les explorateurs d'aujourd'hui

Pour le commun des mortels contraint de rester au sol, des dizaines de petits satellites se chargent d'observer la planète depuis l'espace. La surveillance de la Terre a une utilité pratique puisqu'elle permet de préparer les bulletins météo, d'élaborer des modèles climatiques, de surveiller les surfaces de culture, ou de contrôler les captures de pêche.

mucho en la instalación de una estructura bautizada como Cupola, en realidad una cúpula de observación, que permite una vista panorámica sobre la estación espacial, tan amplia como un campo de fútbol, al mismo tiempo que sobre una gran parte de la Tierra. En el cine «cielo y tierra» de la ISS se proyecta una película extraordinaria en sesión continua. La perspectiva y la iluminación no son siempre las mismas porque la estación espacial rodea nuestro planeta en apenas 90 minutos.

Los descubridores de hoy en día

Mientras tanto, docenas de satélites se encargan de observar el universo por nosotros. Mantener una continua observación sobre la Tierra tiene una utilidad enormemente práctica, ya sea para elaborar predicciones meteorológicas, ya para calcular modelos climáticos, vigilar superficies cultivadas o controlar prohibiciones de pesca. Mucho más frecuente que todas las demás es la investigación básica sobre nuestro planeta. Para examinar el estado de la atmósfera, el mejor sistema de medición utilizado en Europa, Estados Unidos o Asia oriental sirve de poco si no se rellenan las lagunas que aún quedan.

Links: Neusiedler See, viele Felder und oben die Donau (Envisat)
Rechts: Mondspiegelung im Mittelmeer; vorn Lyon, links Turin (ISS)

Left: Lake Neusiedl, surrounded by fields, with the Danube at the top (Envisat)
Right: The Moon reflected in the Mediterranean Sea; Lyon lies at the bottom of the picture, Turin to the left (ISS)

À gauche : le lac de Neusiedl, à la frontière austro-hongroise, et le Danube au milieu des champs (Envisat)
À droite : reflet de la Lune dans la Méditerranée ; en bas Lyon, à gauche Turin (ISS)

Izda.: Lago Neusiedl, campos de labor con el Danubio en la parte superior (Envisat)
Dcha.: Reflejo de la luna en el Mediterráneo; en primer plano Lyon, a la izquierda Turín (ISS)

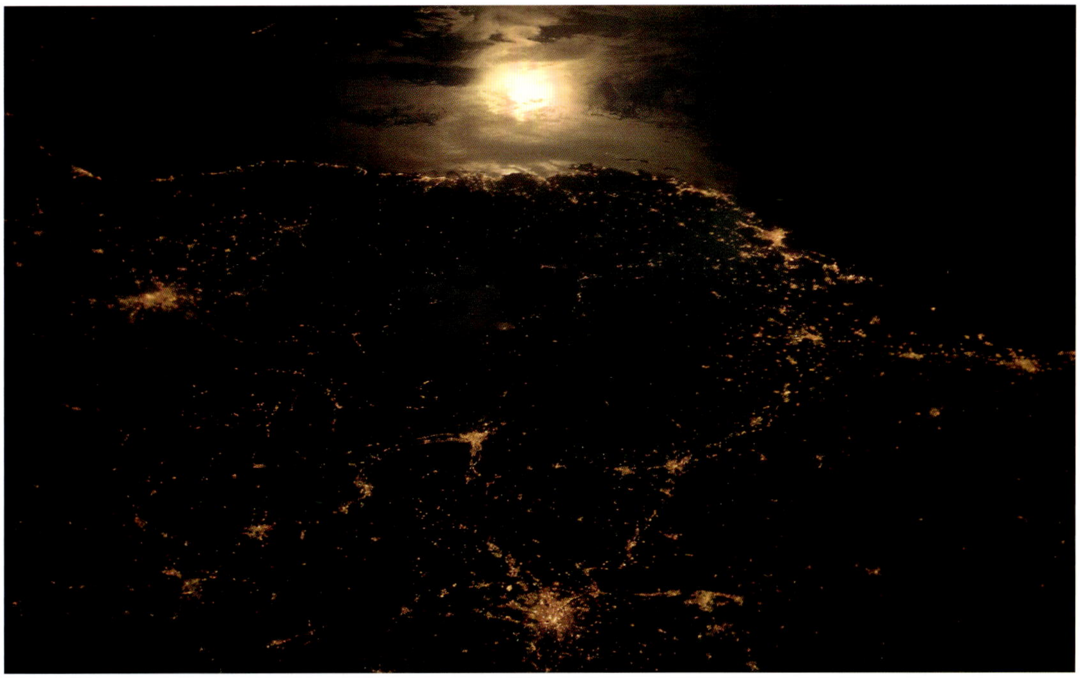

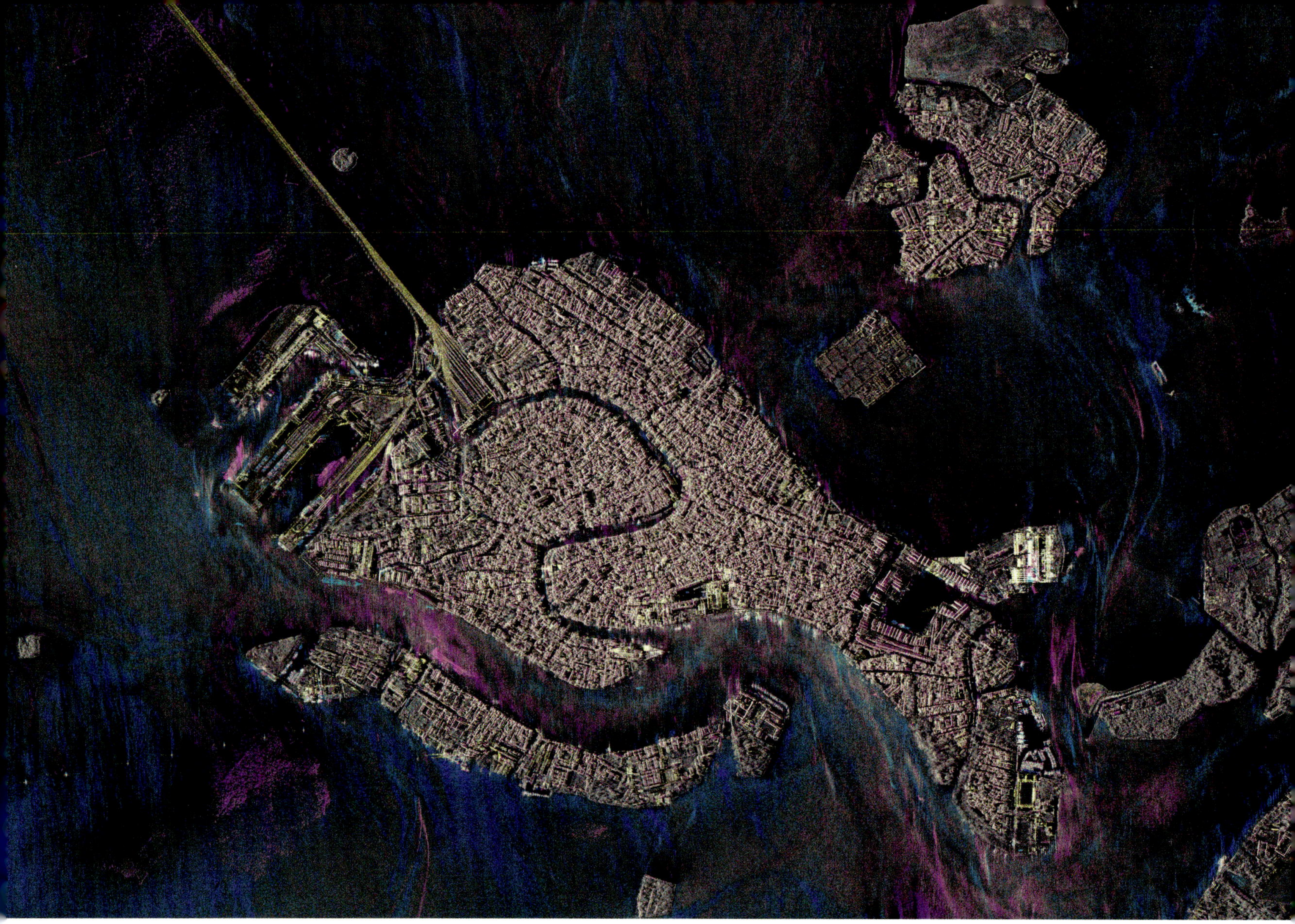

Radarbild von Venedig: Solche Aufnahmen zeigen Experten Wellenfelder, Erosion und mögliche Gebäudesenkungen (TerraSAR-X)

A radar image of Venice: such pictures reveal wave fields, erosion, and potential subsidence to scientists (TerraSAR-X)

Image radar de Venise : sur ces photos, les experts repèrent les champs de vagues, l'érosion et les possibles affaissements de bâtiments (TerraSAR-X)

Imagen de radar de Venecia: estas tomas muestran a los expertos los bancos de olas, la erosión y los posibles hundimientos de edificios (TerraSAR-X)

Rechte Seite: Rushhour zwischen den Inseln vor Venedig. Gut zu erkennen: die Boote und ihre Bugwellen (Ikonos)

Right: Rush hour between Venice's islands; the boats and their bow waves are clearly visible (Ikonos)

Page de droite : heure de pointe entre les îles de Venise. On aperçoit les vaporetti et leurs vagues d'étrave (Ikonos)

Página dcha.: Hora punta entre las islas de Venecia. Fácil de reconocer: las lanchas y sus olas de proa (Ikonos)

USA und Ostasien wenig, wenn man nicht die weißen Flecken dazwischen füllt. Es ist wenig praktikabel und vor allem unbezahlbar, alle Kontinente mit einem dichten Netz von Messgeräten und Sensoren zu überziehen – von den Weltmeeren, die siebzig Prozent der Erdoberfläche ausmachen, ganz zu schweigen. Eine ideale Lösung sind Satelliten: Je nach Umlaufbahn fliegen sie im Laufe der Zeit über jeden Punkt der Erde. Satelliten liefern so innerhalb kurzer Zeit riesige Datenmengen. Da alle Messungen, ob über Chile, Dänemark oder Palau, mit denselben Instrumenten gemacht werden, lassen sich die Daten auch bestens vergleichen. Für Wissenschaftler werden so Träume wahr.

Mussten Alexander von Humboldt und sein Kollege Aimé Bonpland damals ihre Messgeräte mühsam den Hang des Chimborazo in Ecuador hinaufschleppen, um Temperatur und Luftdruck zu bestimmen, so lassen heute die Forscher die meisten Geräte schleppen: Satelliten haben zahlreiche Messinstrumente in die Umlaufbahn gehievt und führen die Messungen buchstäblich im Fluge durch.

Nirgendwo sonst kann man so viel über die Erde lernen wie im Weltall. Für die Forscher ist die Erde ein riesiger Organismus, den sie aus der Umlaufbahn überwachen lassen. Die

quotas, for example—but its most common use is in fundamental research exploring our home planet. There is little point in using the best measuring devices in Europe, the USA, and East Asia to monitor the condition of the atmosphere if all the other areas of Earth are left unexamined. It is hardly practical to cover every continent in a thick web of detectors and sensors—and that's without considering the oceans, which make up 70 percent of the Earth's surface—so satellites are an ideal solution. Depending on their orbit, they will eventually cover every point on Earth. Satellites can thus provide large amounts of data in a short time, and as all the measurements, whether over Chile, Denmark, or Palau, are made with the same instruments, it is very easy to compare them—a dream come true for scientists.

Although Alexander von Humboldt and his colleague Aimé Bonpland had to carry their own instruments up the arduous slopes of Chimborazo in Ecuador, modern researchers have their measuring devices transported for them: satellites have transported any number of sensors into orbit, and these are now taking measurements literally on the fly.

Nowhere else can teach us more

Mais il s'agit bien plus souvent de poursuivre l'exploration de notre planète. Pour mesurer l'état de l'atmosphère, il serait difficile, et surtout hors de prix, de tapisser l'ensemble des continents d'un réseau dense de capteurs et d'appareils de mesure, sans parler des océans qui recouvrent soixante-dix pour cent de la surface terrestre. Ainsi les satellites fournissent la solution idéale : suivant leur orbite, ils survolent chacune des parcelles de la Terre. Ils livrent donc en très peu de temps une énorme quantité d'informations. Les mesures prises au Chili, au Danemark ou aux Palaos étant réalisées avec les mêmes instruments, les données sont parfaitement comparables. C'est ainsi que se réalisent les rêves les plus fous des scientifiques.

Alors qu'Alexander von Humboldt et son collègue Aimé Bonpland ont à l'époque péniblement traîné leurs instruments de mesure de la température et de la pression sur la pente du Chimborazo en Équateur, les chercheurs d'aujourd'hui confient leurs appareils à des tiers : les satellites ont hissé d'innombrables instruments scientifiques en orbite et opèrent littéralement les mesures en vol.

On ne peut acquérir autant de connaissances sur la Terre que dans l'espace. Pour les chercheurs notre

Es mucho menos factible, y sobre todo carísimo, cubrir todos los continentes con una densa red de equipos de medición y sensores, sin mencionar que los océanos representan el setenta por ciento de la superficie terrestre. La solución ideal son los satélites que orbitan sobre cada punto de la Tierra. De este modo, los satélites proporcionan grandes cantidades de datos en muy poco tiempo. Como todas las mediciones, ya sea sobre Chile, Dinamarca o Palaos, se realizan con los mismos instrumentos, este sistema también resulta muy eficaz para la comparación de los datos. Un sueño hecho realidad para los científicos.

Si Alexander von Humboldt y su colega Aimé Bonpland tenían que cargar penosamente con sus equipos de medición hasta la cumbre del Chimborazo, en Ecuador, para determinar la temperatura y la presión del aire, en la actualidad los investigadores utilizan la mayoría de los equipos de otro modo: los satélites mantienen en órbita numerosos equipos de medición y realizan las mediciones, literalmente, en vuelo.

Por regla general, en ninguna otra parte se puede aprender tanto sobre la Tierra como desde el universo. Para los investigadores, la Tierra es un organismo colosal que se puede

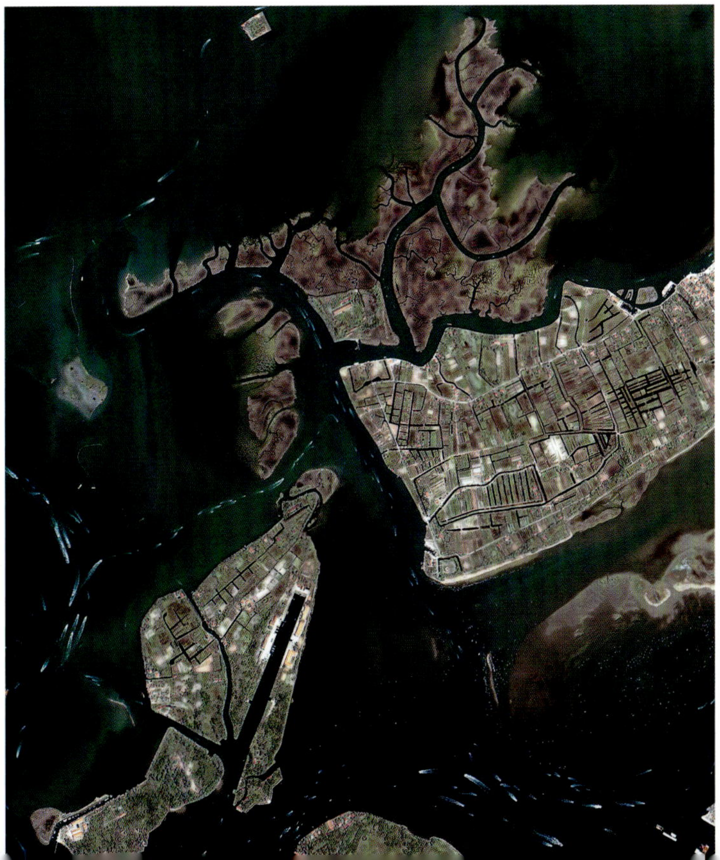

himmlischen Späher messen, bildlich gesprochen, den Blutdruck der Erde, bestimmen ihre Leberwerte, ermitteln den Cholesterinspiegel, verfolgen die Funktion der Herzkammern und messen die Hirnströme. Dies alles, um den Zustand unseres Heimatplaneten so genau wie möglich zu bestimmen – und um mögliche Therapien ermitteln und ihren Erfolg abschätzen zu können. Überwachung nicht als Kontrolle von oben, sondern aus Sorge um die blaue Kugel im All und die Menschen, die oft wenig pfleglich mit ihr umgehen, aber doch noch lange Zeit auf ihr leben wollen.

Am Anfang war das Wetter

Am 9. Dezember 1977 hat Meteosat-1 sein erstes Bild zur Erde geschickt. Die Schwarzweißaufnahme markiert den Beginn der systematischen europäischen Erdbeobachtung. Die US-Amerikaner hatten den Wettersatelliten mit einer ihrer Delta-Raketen ins All geschossen – die europäische Ariane-Rakete gab es damals noch nicht. Für Europa begann eine neue Ära. Das Wetter ist durch die Meteosats vermutlich nicht besser geworden, sicher aber die Wetterprognosen. Eine katastrophale Sturmflut wie 1962 in Hamburg, die mehr als

about the Earth than space. For researchers, the Earth is like a giant organism which can be observed from space. These watchers in the sky can—speaking figuratively—monitor the Earth's blood pressure, check its vital signs, record its cholesterol level, scan its heart valve function, and track its brain activity. All this is done to determine as accurately as possible the state of our home planet's health, prescribing possible treatments and recording the success of the outcomes. Such observation isn't surveillance from above—it is a matter of taking care of this blue dot in space and the people living on it, people who often treat it rather carelessly but nonetheless wish to continue living on it for many years to come.

In the beginning was the weather

Meteosat-1 transmitted its first image to the Earth on December 9, 1977, and this black-and-white picture marked the dawn of systematic observation of the Earth. The USA sent this weather satellite into space on one of its Delta rockets, as the European Ariane rocket series was yet to be developed. A new era

planète est un organisme géant, veillé depuis l'espace. Les gardes célestes, pour utiliser une métaphore, mesurent la tension artérielle de la Terre, réalisent son bilan hépatique, déterminent son taux de cholestérol, surveillent les battements de son cœur et son activité cérébrale. Toutes ces mesures visent à établir aussi précisément que possible le bilan de santé de la Terre, afin de proposer des thérapies et d'évaluer leur chance de réussite. Cette surveillance n'a pas pour but de contrôler le monde, mais de veiller depuis l'espace sur la planète bleue et sur ses habitants qui ont souvent tendance à la négliger, même s'ils souhaitent y vivre encore longtemps.

Au début était la météo

Le 9 décembre 1977, Météosat-1 transmet sa première image vers la Terre. Cette photographie noir et blanc marque le début de l'observation systématique de la Terre par les Européens. Le satellite météorologique avait été lancé dans l'espace par une fusée Delta américaine, la fusée européenne Ariane n'ayant alors pas encore vu le jour. Pour l'Europe, c'était le début d'une ère

controlar desde la órbita. Los exploradores celestes miden, en sentido figurado, la presión arterial de la Tierra, determinan los valores hepáticos, averiguan el nivel de colesterol, observan el funcionamiento de los ventrículos y miden la actividad eléctrica cerebral. Y todo esto para determinar, lo más exactamente posible, el estado de nuestro planeta, establecer las posibles terapias y cuantificar su expectativa de éxito.

En el principio era el tiempo

El 9 de diciembre de 1977, el Meteosat 1 envió su primera imagen a la Tierra. Aquella fotografía en blanco y negro marcó el comienzo de la observación sistemática del planeta por parte de Europa. Los estadounidenses lanzaron al espacio el satélite meteorológico con uno de sus cohetes Delta, ya que en aquel momento aún no existían los cohetes Ariane europeos. Para Europa significó el comienzo de una nueva era.
Es de suponer que el tiempo no ha mejorado gracias al Meteosat, pero sí que lo han hecho los pronósticos meteorológicos. Aun hoy es posible una catastrófica marea viva como la de 1962 en Hamburgo, que costó la vida a más de 300 personas.

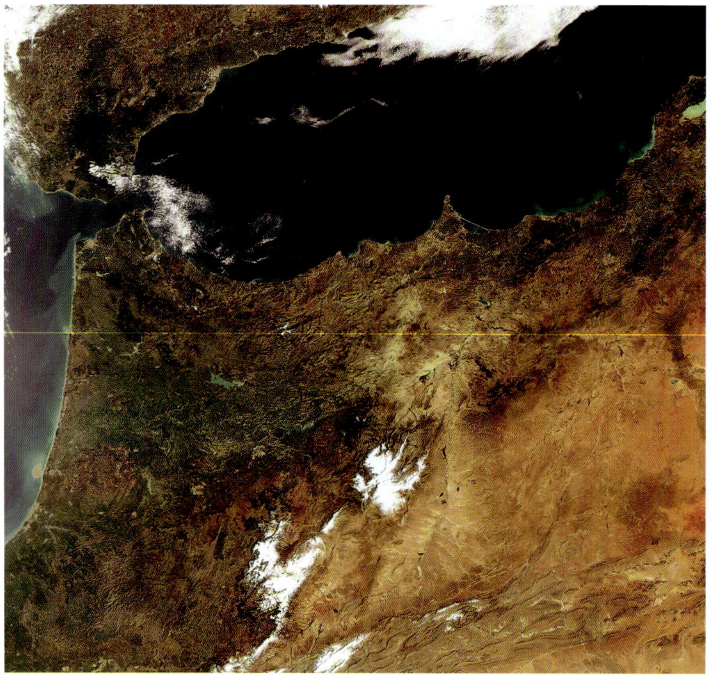

Europa auf den Schultern: das Atlas-Gebirge im Norden Marokkos. Links oben die Straße von Gibraltar (Envisat)

Europe on their shoulders: the Atlas Mountains in northern Morocco. The Strait of Gibraltar is top left (Envisat)

L'Europe sur les épaules : la chaîne de l'Atlas au nord du Maroc. En haut à gauche, le détroit de Gibraltar (Envisat)

Europa a la espalda: cordillera del Atlas en el norte de Marruecos. Arriba a la izquierda, el estrecho de Gibraltar (Envisat)

Rechte Seite: Die von Norden zwischen den Kanarischen Inseln hindurch strömende Luft bildet auffällige Wolkenwirbel (Envisat)

Right: Air flowing south between the Canary Islands gives rise to striking cloud formations (Envisat)

Page de droite : les masses d'air venues du nord qui circulent entre les îles Canaries forment des volutes de nuages fort singulières (Envisat)

Página dcha.: El aire que desde el norte circula entre las islas Canarias forma llamativos remolinos de nubes (Envisat)

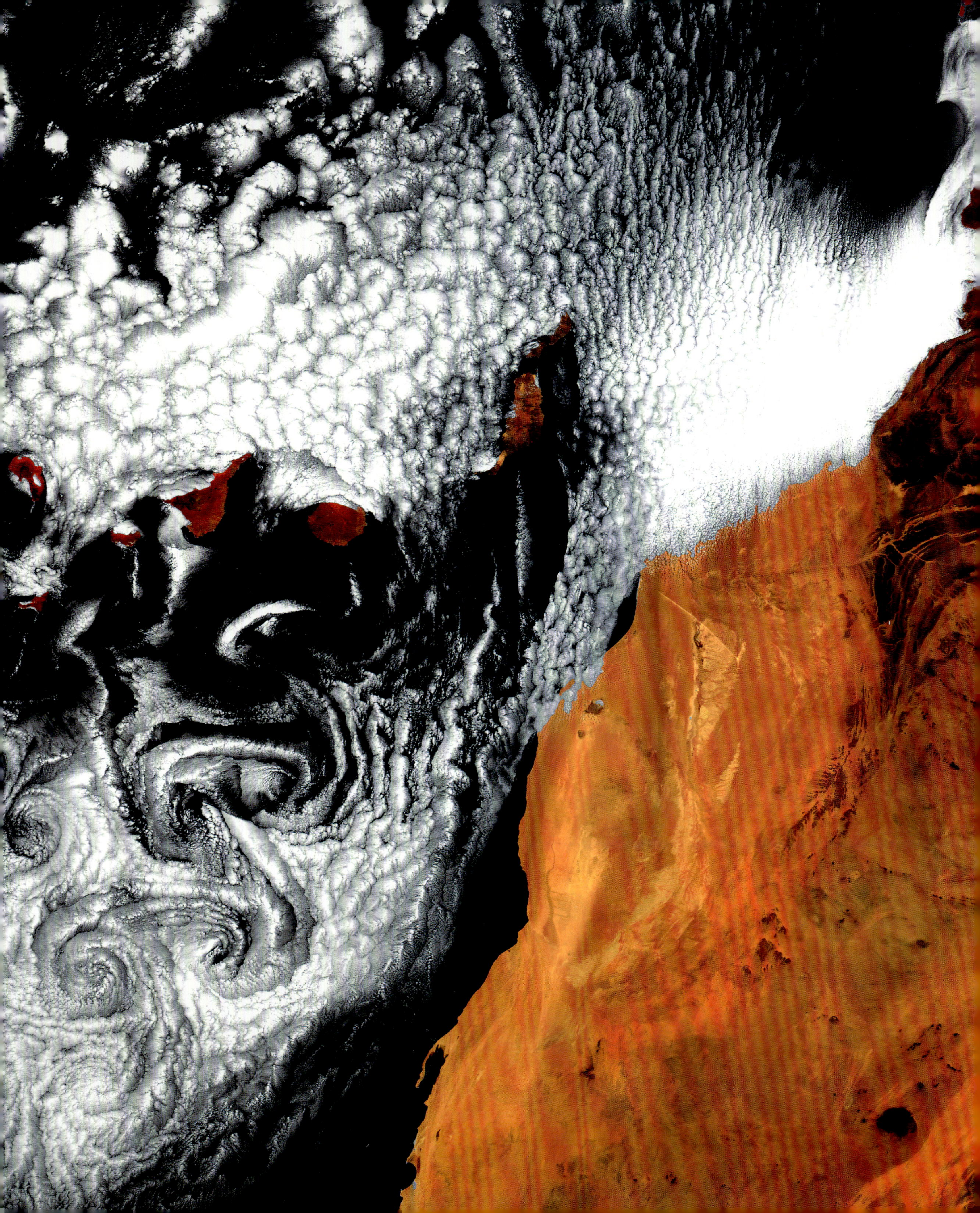

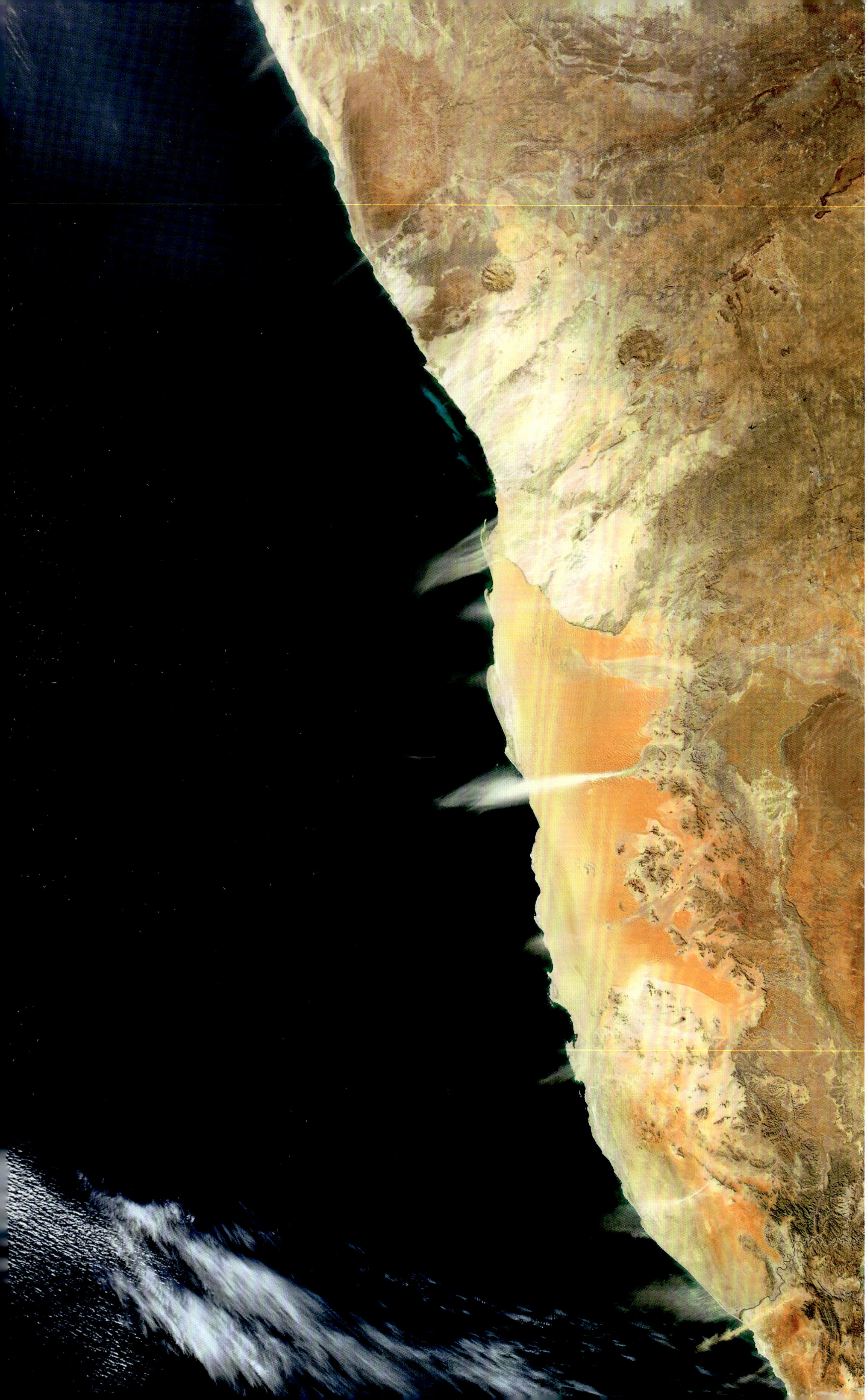

Links und Seite 44/45: Aus der Wüste Namibias wehen starke Winde Staubwolken über den Atlantischen Ozean (Envisat). Auf der folgenden Doppelseite ist der Ursprungsbereich der stärksten Staubfahne etwas unterhalb der Bildmitte zu sehen. Dort gibt es bis zu 300 m hohe Dünen (Landsat)

Left and pages 44–5: Strong winds blow dust clouds from Namibia's deserts far out over the Atlantic Ocean (Envisat). The double page spread (44–5) shows where the largest trail of dust, visible just below the middle of the picture here, originated. The local dunes can reach heights of up to 1,000 ft (300 m) (Landsat)

À gauche ainsi que pages 44 et 45 : les forts vents du désert namibien poussent des nuages de poussière au-dessus de l'océan Atlantique (Envisat). On distingue sur la double page suivante, dans la partie inférieure de l'image, l'origine du plus puissant ruban de poussière. À cet endroit, les dunes peuvent atteindre 300 m (Landsat)

Izda. y páginas 44/45: Desde el desierto de Namibia, los fuertes vientos proyectan nubes de polvo sobre el océano Atlántico (Envisat). En la siguiente doble página se puede ver la zona de origen del remolino de arena más intenso, un poco por debajo del centro de la imagen. Allí hay dunas de hasta 300 m de altura (Landsat)

300 Menschen das Leben kostete, ist auch heute noch möglich. Aber heute würde niemand mehr von dem unerwartet auftauchenden Orkantief überrascht, das mitten in der Nacht die Deiche brechen ließ. Denn die Raumfahrtingenieure haben Satelliten und Instrumente im Einsatz, die rund um die Uhr in die Töpfe der Wetterküchen blicken. Satelliten haben komplette Sturmsysteme mit allen Details im Blick, auch wenn diese mehr als fünfhundert Kilometer Durchmesser haben.

Die Meteosats sind in der geostationären Umlaufbahn postiert. Dort dauert ein Umlauf um die Erde genau solange, wie die Erde braucht, sich einmal um sich selbst zu drehen – also 23 Stunden und 56 Minuten. Die Wettersatelliten scheinen über einem Ort der Erde stillzustehen. Die geostationäre Umlaufbahn hat große Vorzüge, weil die Satelliten damit stets denselben Bereich der Erde im Blick haben. Zudem braucht man die Empfangsantennen am Erdboden nur einmal auf die Satelliten auszurichten. Der Kontakt bleibt immer bestehen. Doch auch aus 36.000 Kilometern Höhe ist nicht eine komplette Hälfte der Erde zu beobachten. Von dort aus sind nur die Gebiete bis etwa 81 Grad nördlicher und südlicher Breite zu sehen. Die Polargebiete im Norden und Süden entziehen sich den Blicken aus der berühmten Umlaufbahn genau über dem Äquator.

Natürlich kann niemand auf die Erforschung und Überwachung der Polargebiete verzichten. Sie sind im wahrsten Sinne des Wortes die Klimakammern unserer Erde. Die dortige Eisbedeckung, Temperatur, Niederschlag, Windbewegung etc. haben Einfluss auf die gesamte Erde. Daten aus den Polargebieten sind von größter Bedeutung. Gerade dort sind die Wissenschaftler ganz besonders auf den Einsatz von Satelliten angewiesen. Denn auf den Eisschollen im Nordpolarmeer oder in den weißen Wüsten der Antarktis ist nur selten ein Mensch, ein Schiff oder ein Flugzeug unterwegs, um die Bedingungen vor Ort zu erkunden. Für die Forscher muss dort also der Kollege Satellit die Arbeit übernehmen.

began for Europe—the weather hasn't been noticeably improved by the Meteosats, but the forecasts certainly have. A catastrophic flood, such as that in Hamburg in 1962, which cost 300 lives, could still happen today, but now no one would be caught off-guard, or not to quite the same degree, by the unexpected low-pressure system and hurricane-force winds which breached the dykes in the middle of the night. Aerospace engineers have installed instruments in satellites which keep a constant watch on what the weather is cooking up, and even storm systems with diameters of more than 300 miles (500 kilometers) can be watched in every detail.

The Meteosats are located in geostationary orbits, which means that one circuit of the Earth takes as long as the planet itself takes to complete one rotation—about 23 hours and 56 minutes. The weather satellites thus seem to stand still above one point on Earth. Geostationary orbits have great benefits: the satellite always has the same area of the Earth under surveillance and the antennae of the receivers on the surface need to be adjusted to aim at the satellite only once for constant contact to be maintained. Nonetheless, less than half of the Earth can be seen from an altitude of 22,400 miles (36,000 kilometers). Only the areas between latitudes of about 81° north and south are visible; the northern and southern polar regions cannot be seen from this well-known orbit directly above the equator.

Full Earth observation, of course, is impossible to do without researching and monitoring the poles—in the truest sense, they are our Earth's climatic chambers and the ice-cover, temperature, precipitation, and wind activity there influences the whole planet. Data from the polar regions is of crucial importance, and this is where scientists are particularly dependent on the use of satellites; people, planes, and boats are generally unlikely to be around among the ice floes of the Arctic Ocean or in the white wilderness of Antarctica

nouvelle. Météosat n'a sans doute pas amélioré la météo, mais il a certainement perfectionné les prévisions météorologiques. Nous ne sommes pas à l'abri d'un raz-de-marée semblable à celui qui a coûté la vie à plus de 300 personnes à Hambourg en 1962, mais aujourd'hui, personne ne se laisserait plus surprendre par l'apparition subite du cyclone qui a brisé les digues de la ville au milieu de la nuit. Les ingénieurs spatiaux utilisent des satellites et instruments qui surveillent sans répit la danse des éléments climatiques. Les satellites observent avec précision les systèmes cycloniques dont le diamètre dépasse parfois cinq cents kilomètres.

Météosat évolue sur l'orbite géostationnaire terrestre. Sur cette orbite, la durée d'une révolution autour de la Terre correspond à une rotation de la Terre sur elle-même, c'est-à-dire 23 heures et 56 minutes. L'orbite géostationnaire présente un avantage de taille, parce qu'elle permet aux satellites d'observer continuellement les mêmes territoires. En outre, il suffit de braquer une seule fois les antennes réceptrices sur les satellites, pour que le contact soit à jamais établi. Cependant, même à 36 000 kilomètres d'altitude, il est impossible d'observer entièrement un hémisphère. À cette altitude, seules sont visibles les régions situées à une latitude comprise entre le 81e degré nord et le 81e degré sud. Les régions polaires échappent au regard de la fameuse orbite située dans le plan de l'équateur.

Nous ne pouvons bien sûr nous permettre de renoncer à étudier les pôles. Ils sont au sens propre du terme les chambres climatiques de la Terre. Leur couverture de glace, température, précipitations, et courants aériens ont un impact sur l'ensemble de la planète. Les données provenant les régions polaires sont essentielles, et dans ces régions, le travail des scientifiques dépend fortement des satellites. Rares sont en effet les hommes, les navires ou les avions qui s'aventurent sur les plaques de banquise de l'océan Arctique, ou dans le désert blanc de

Pero en la actualidad ya nadie se vería sorprendido por una inesperada borrasca ciclónica capaz de romper los diques en plena noche. Porque los ingenieros astronáuticos tienen satélites e instrumentos que husmean las 24 horas del día en los pucheros de las cocinas meteorológicas. Los satélites ofrecen una completa y detallada panorámica de los sistemas de tormentas, aunque estos tengan más de 500 km de diámetro.

Los Meteosat están apostados en una órbita geoestacionaria. Allí, una rotación en torno a la Tierra dura tanto tiempo como el que la Tierra necesita para girar una vez sobre sí misma, es decir, 23 horas y 56 minutos. Los satélites meteorológicos parecen estar inmóviles sobre un lugar de la Tierra. La órbita geoestacionaria tiene grandes ventajas, porque los satélites dominan la zona de la Tierra que se muestra ante ellos. Además, las antenas terrestres de recepción sólo se tienen que orientar una vez. El contacto se mantiene siempre. Pero también hay que tener en cuenta que desde 36.000 km de altura no se puede observar una mitad completa de la Tierra. Desde allí sólo se pueden ver las regiones que llegan hasta, aproximadamente, los 81° de latitud norte y sur. Las regiones polares de los dos hemisferios se ocultan de las miradas desde la célebre órbita situada exactamente encima del ecuador.

Naturalmente, nadie puede renunciar a la investigación y la observación de las regiones polares. Son, en el sentido más estricto de la expresión, las cámaras climáticas de nuestra Tierra. Sus cubiertas de hielo, temperatura, precipitaciones, vientos y otros factores, influyen directamente en toda la Tierra. Los datos de las regiones polares revisten gran trascendencia. En este aspecto los científicos dependen especialmente del uso de satélites. Porque a los témpanos de hielo del océano Ártico o a los blancos desiertos de la Antártida sólo en raras ocasiones se envía una persona, un barco o un avión para explorar in situ sus condiciones. El colega satélite debe asumir entonces el trabajo de los investigadores.

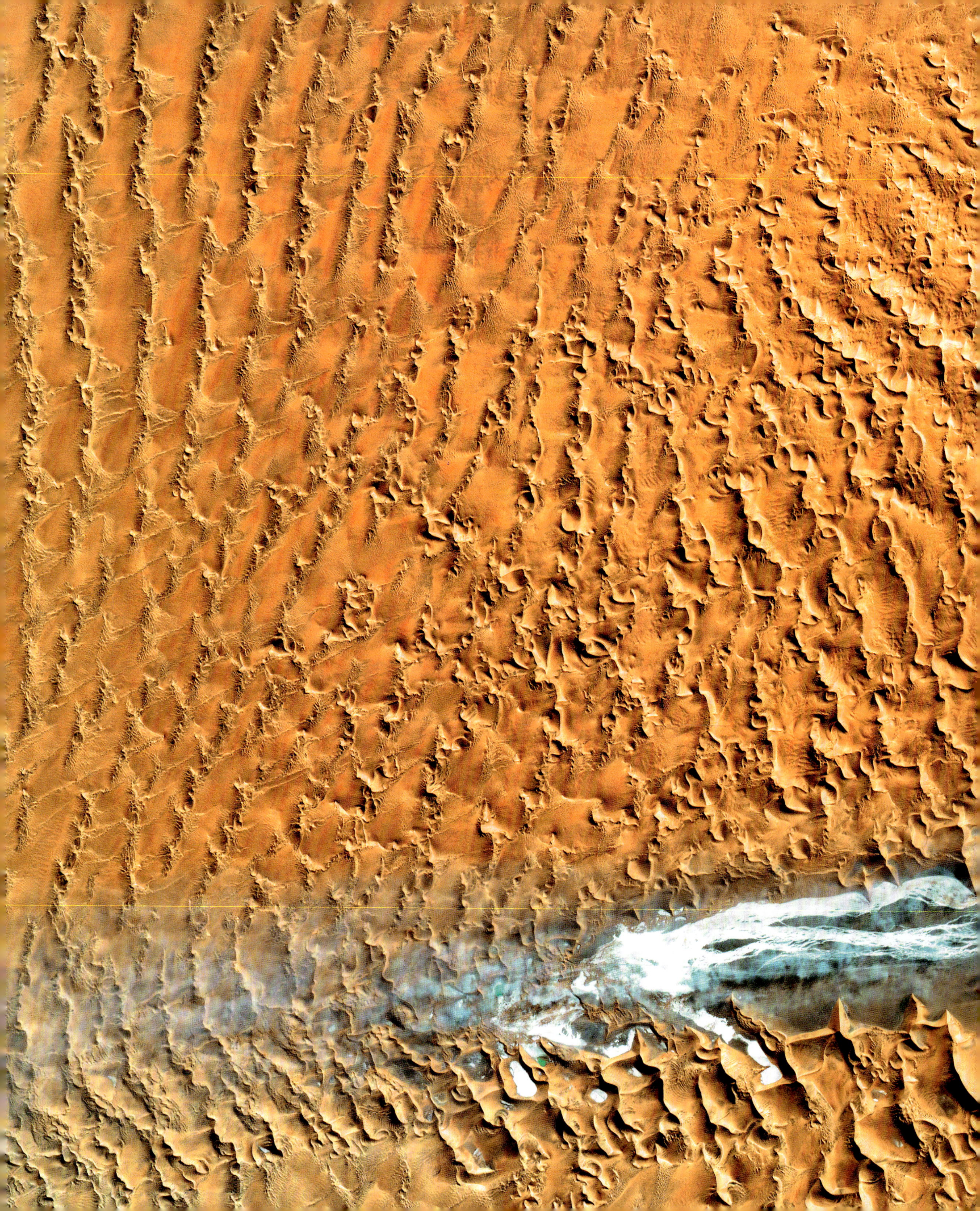

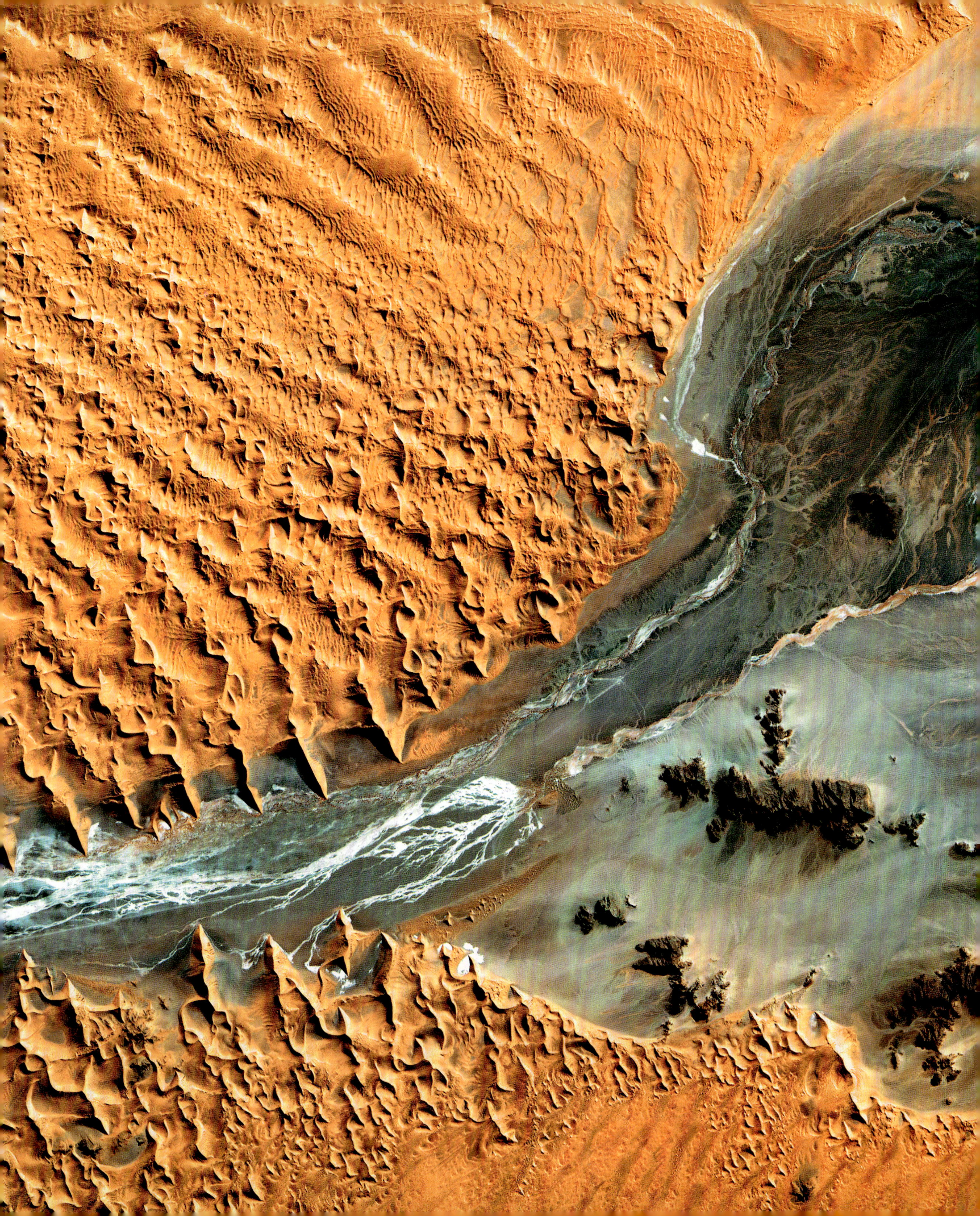

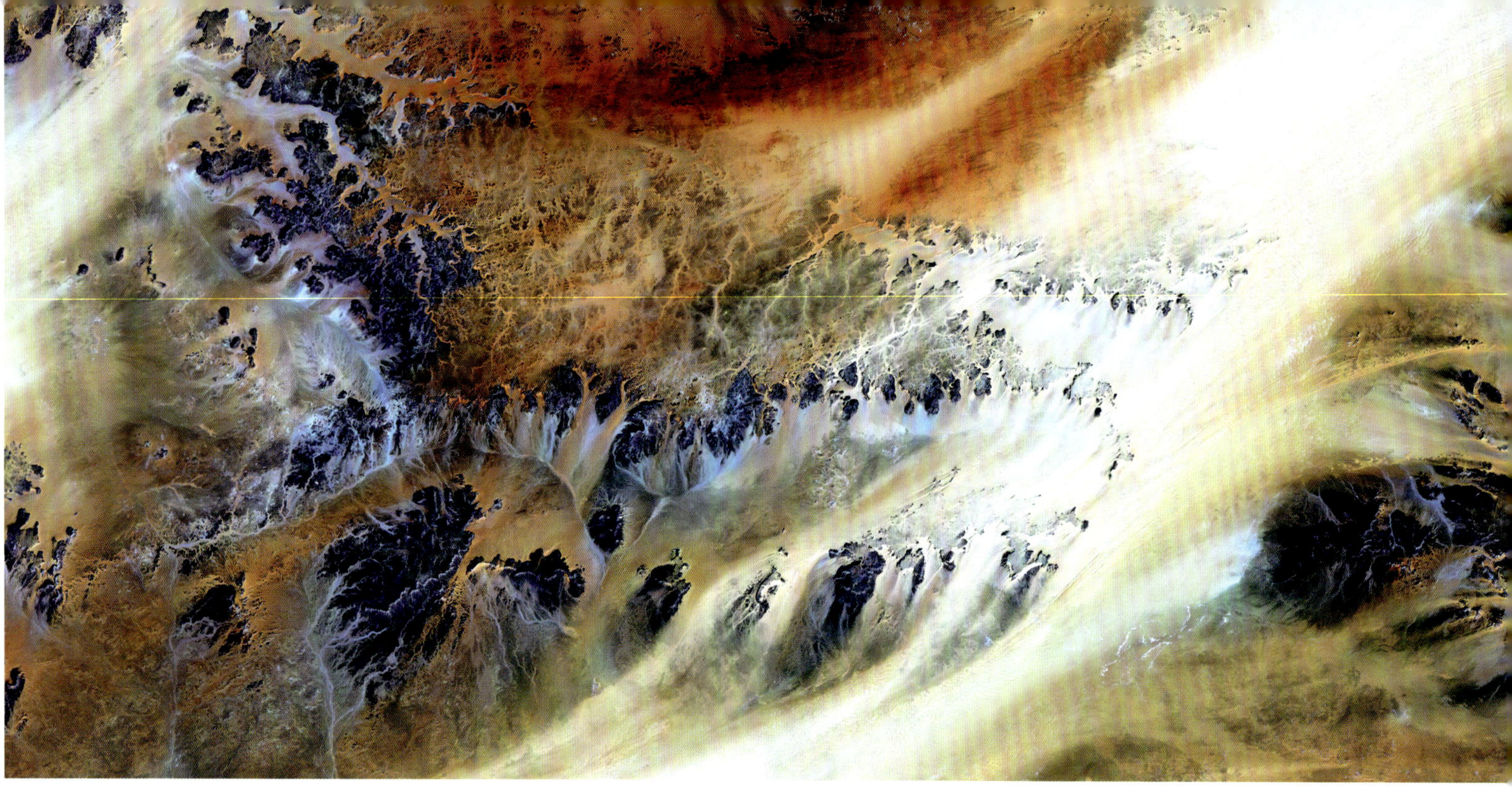

Ein Satellit für alle Fälle

Am 17. Juli 1991 stieg eine Ariane-4-Rakete in den Himmel über Französisch-Guayana. An Bord wertvolle Fracht: Von Europas Weltraumbahnhof in Kourou aus startete ein Satellit, der auf einer Bahn über die Polargebiete hinweg die Erde erforschte. ERS-1 (European Remote Sensing) war der erste Fernerkundungssatellit der Europäischen Weltraumorganisation ESA. Er verfügte über vier Instrumente: Ein Radargerät bestimmte die Höhe von Meereswellen und Eismassen sowie die Form des überflogenen Geländes. Ein Infrarotmessgerät erfasste die Temperatur von Wolken, Erdoberfläche und Ozeanen. Ein Mikrowellengerät untersuchte den Wasserdampfgehalt der Atmosphäre, und schließlich konnte ERS-1 sogar Windrichtung und -stärke über dem Meer messen. Der Satellit wurde auf eine 800 Kilometer hohe Umlaufbahn gesetzt, die über die Polargebiete führt. Auf so einer Bahn braucht ein Satellit hundert Minuten, um die Erde zu umrunden.

Gemessen an heutiger Technik verfügte ERS-1 über eher geringe Auflösung, und der Blick nach unten war so schmal, dass es meist 35 Tage dauerte, bis der Satellit beim Überflug wieder dieselbe Gegend zu sehen be-

to test local conditions for themselves. Researchers have to let the satellites take over in order to get the work done.

A satellite for all seasons

An Ariane 4 rocket launched from the European spaceport near Kourou took to the skies over French Guiana on July 17, 1991 with a valuable payload on board: a satellite intended to explore the Earth from an orbit over the poles. ERS-1 (European Remote-Sensing Satellite), the European Space Agency's (ESA) first Earth observation satellite, carried four main instruments: a radar detector to measure the heights of waves and ice formations as well as the elevation of the land below, an infrared device to record the temperature of the Earth's surface and oceans, a microwave radiometer to investigate the humidity of the atmosphere, and apparatus to determine the direction and strength of sea winds. The satellite was placed in a polar orbit at an altitude of 500 miles (800 kilometers) and required 100 minutes to circuit the Earth.

Compared with modern technology, ERS-1 had such low resolution and

l'Antarctique. C'est donc le collègue satellite qui se charge du travail pour le compte des chercheurs.

Un satellite omnipotent

Le 17 juillet 1991, une fusée Ariane 4 s'élançait dans le ciel au-dessus de la Guyane emportant une charge précieuse : un satellite quittait en effet le centre spatial européen de Kourou pour explorer la Terre à partir d'une orbite polaire. ERS-1 (European Remote Sensing) fut le premier satellite européen d'observation de la Terre de l'Agence spatiale européenne (ESA). Il avait à son bord quatre instruments. Un radar mesurait la hauteur des vagues et des masses de glace ainsi que la forme du terrain survolé. Un radiomètre infra-rouge enregistrait la température des nuages, de la surface terrestre et des océans. Un radiomètre à micro-ondes analysait la concentration de vapeur d'eau dans l'atmosphère. Enfin, ERS-1 était capable de mesurer la force et la direction des vents sur la mer. Le satellite avait été placé sur une orbite située à 800 kilomètres d'altitude autour des pôles, et effectuait un tour de la Terre en cent minutes. Comparé aux techniques actuelles, ERS-1 fournissait

Un satélite para todo

El 17 de julio de 1991, un cohete Ariane 4 asciende al cielo sobre la Guayana francesa. A bordo va una valiosísima carga: un satélite para investigar la Tierra en una órbita sobre las regiones polares. ERS-1 (European Remote Sensing) fue el primer satélite de teledetección de la Agencia Espacial Europea ESA. Disponía de cuatro instrumentos: un equipo de radar determinaba la altura de las olas del mar y de las masas de hielo, así como la forma del terreno que sobrevolaba; otro equipo de medición infrarroja registraba la temperatura de las nubes, la superficie terrestre y los océanos; un equipo microondas analizaba el contenido en vapor de agua de la atmósfera y, finalmente, había otro equipo que podía medir la dirección y la fuerza de los vientos sobre el mar. El satélite se situó en una órbita a 800 km de altura, que pasaba por encima de las regiones polares. En una trayectoria de este tipo, un satélite necesita cien minutos para completar la órbita alrededor de la Tierra.

En comparación con la tecnología actual, el ERS-1 disponía de una resolución muy baja y la amplitud de su visión era reducida, por lo que el satélite solía tardar 35 días en poder

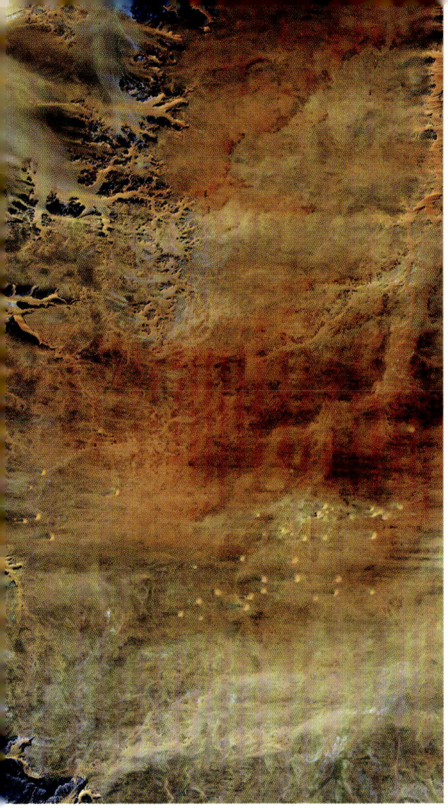

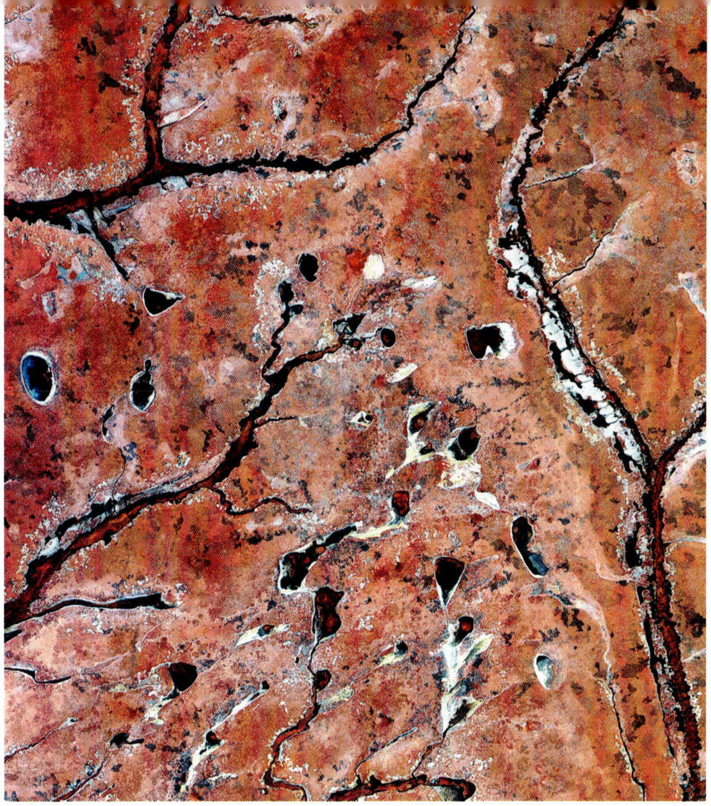

Farbenfrohes Afrika: Nahe der Oase Terkezi im Tschad (links, Landsat) und an den Zuflüssen des Sambesi in Sambia (rechts, Alos)

The colors of Africa: around the Terkezi Oasis in Chad (left, Landsat) and the tributaries of the Zambezi in Zambia (right, Alos)

Les couleurs de l'Afrique : aux alentours de l'oasis de Terkezi au Tchad (à gauche, Landsat) et des affluents du Zambèze en Zambie (à droite, Alos)

África multicolor: cerca del oasis de Terkezi, en el Chad (izda., Landsat), y en los afluentes del Zambeze, en Zambia (dcha., Alos)

kam. Dennoch hat ERS-1, gemeinsam mit seinem 1995 gestarteten Schwestersatelliten ERS-2, der zudem ein Instrument zur Messung des Gehalts von Ozon und anderen Spurengasen in der Atmosphäre an Bord hatte, die Erdbeobachtung revolutioniert. Die Federführung beider Missionen lag bei der damaligen Firma Dornier in Friedrichshafen, die heute zu Airbus gehört.

Im Jahr 2000 ist ERS-1 ausgefallen. Er hatte seine ursprünglich geplante Lebenszeit um das Doppelte überschritten. Gut 50.000-mal hat er die Erde umrundet, Daten gesammelt und später zur Bodenstation gefunkt. Für die Erde lief er meilenweit, viele Meilen weit. Mehr als zwei Milliarden Kilometer hat ERS-1 in seiner aktiven Zeit zurückgelegt. Wäre er nicht um die Erde, sondern hinaus in die Weiten des Alls geflogen, so hätte er den Planeten Saturn hinter sich gelassen. Doch jeder Umlauf war seinem Heimathafen gewidmet: der Erde. ERS-1 kreist noch immer um die Erde, allerdings ist er nicht mehr in Betrieb. Gleiches gilt für seinen Schwestersatelliten, der 2011 abgeschaltet worden ist.

An die Weltspitze der Erdbeobachtung hat sich Europa dann mit Envisat katapultiert. Es ist der größte je gebaute Satellit, der der Erforschung

the scope of its receivers was so limited that it generally took about 35 days until the satellite was in a position to observe the same region again. Nonetheless, in tandem with its sister satellite ERS-2, which contained an instrument to measure ozone levels and other trace gases in the atmosphere and was launched in 1995, ERS-1 revolutionized the observation of the Earth. Overall control of the project was the responsibility of the Dornier company of Friedrichshafen, Germany, now part of Airbus.

ERS-1 finally failed in the year 2000, having exceeded its original planned lifespan by more than 100 percent. It had circled the Earth over 50,000 times, collecting data and transmitting it to the surface station, and covering unimaginable distances for the sake of the Earth—more than 1.25 billion miles (2 billion kilometers), in fact, during its active service. Had it been sent out into space rather than into orbit, it would have passed the planet Saturn, but every circumnavigation was dedicated to its home port, the Earth.

ERS-1 is still circling the planet, but it is no longer in use; its sister satellite ran into the same bad luck when it was deactivated in 2011.

des données à faible résolution et son champ visuel était si étroit qu'il lui fallait 35 jours avant de pouvoir observer une nouvelle fois la même région. Pourtant, ERS-1 et son satellite jumeau ERS-2 lancé en 1995 avec à son bord un instrument destiné à mesurer la répartition de l'ozone et autres gaz traces dans l'atmosphère, ont révolutionné l'observation de la Terre. La maîtrise d'œuvre de ces deux missions était assurée par la société Dornier de Friedrichshafen, en Allemagne, qui appartient aujourd'hui à Airbus.

ERS-1 s'est éteint en l'an 2000, après avoir largement doublé sa durée de vie initialement prévue. Il a fait le tour de la Terre près de 50 000 fois, collectant des données qu'il transmettait au sol. Au cours de sa durée de vie opérationnelle, ERS-1 a parcouru pour la Terre une distance incroyable de plus de deux milliards de kilomètres. S'il avait poursuivi sa course dans l'espace au lieu de graviter autour de la Terre, il aurait dépassé Saturne. Cependant, chacune de ses rotations était dédiée à son port d'attache : la Terre. ERS-1 tourne toujours autour de la Terre, mais il n'est plus en service. Son satellite jumeau connaît le même sort depuis qu'il a été désactivé en 2011.

observar de nuevo la misma región. Pero el ERS-1 revolucionó la observación terrestre junto con su satélite gemelo, el ERS-2, lanzado en 1995, que incluía un instrumento para medir el contenido de ozono y de otros gases traza en la atmósfera. La responsabilidad de ambas misiones fue de la empresa Dornier de Friedrichshafen, que hoy día se integra en Airbus.

El ERS-1 suspendió sus operaciones en el año 2000, habiendo sobrepasado el doble de su vida útil inicialmente proyectada. Completó más de 50.000 órbitas alrededor de la Tierra, recogiendo datos y transmitiéndolos más tarde a la estación terrestre. Una vasta singladura al servicio de la Tierra. Durante su etapa de actividad, el ERS-1 recorrió más de dos mil millones de km. Si en vez de viajar alrededor de la Tierra lo hubiese hecho a las profundidades del universo, habría llegado más allá del planeta Saturno. El ERS-1 continúa girando alrededor de la Tierra, pero ya no está en servicio. Su hermano gemelo corrió la misma suerte cuando quedó desactivado en 2011.

Tras ese satélite, Europa se catapultó a la cabeza mundial de la observación terrestre con el Envisat, que es el mayor satélite de observación jamás construido. El 1 de marzo de 2002, el satélite de ocho tonela-

unserer Erde dient. Am 1. März 2002 wurde der acht Tonnen schwere Satellit von den Ausmaßen eines kleinen Reisebusses von einer Ariane-5-Rakete im All abgesetzt. Envisat läuft auf einer ganz ähnlichen Bahn wie die ERS-Satelliten um die Erde: 800 Kilometer hoch über die Polargebiete hinweg. Auch bei ihm dauert ein Umlauf hundert Minuten.

Der Name ist Programm: Envisat steht für Environmental Satellite, der Satellit hat bis zu seinem Ausfall 2012 Umweltbedingungen auf der Erde so genau wie möglich erfasst. Zehn Instrumente an Bord machten Envisat zur Eier legenden Wollmilchsau. Präzise hat er Form und Beschaffenheit der Erdoberfläche vermessen – egal ob Festland, Meer oder Eispanzer. Eine große Kamera hat die Ozeanflächen in verschiedenen Farben aufgenommen. Damit ließ sich bestimmen, wie viel Chlorophyll und Sedimentstoffe in Wasser vorkommen. Die Kamera erfasste einen bis zu mehr als tausend Kilometer breiten Streifen. Binnen drei Tagen ließ sich jedes Gebiet erneut inspizieren. Instrumente zur Messung von Spurengasen lieferten dreidimensionale Informationen darüber, wie sich Ozon, Kohlenwasserstoffe, Stickoxide und andere Stoffe in der Atmosphäre verteilen. Das Radargerät an Bord schickte bei jedem Wetter und bei Tag und Nacht genaue Daten über Oberflächendetails. Die Daten ließen sich vielseitig anwenden. Sie flossen ein in Klima- und Wettermodelle, mit ihnen lassen sich die Vegetations- und Erosionsprozesse in aller Welt erfassen, Umweltverschmutzungen an Land, in der Luft und auf See nachweisen, um bei Unglücken und Naturkatastrophen Hilfe schnell koordinieren zu können, sie ermöglichen, den Schiffsverkehr und die Eisbedeckung der Meere zu überwachen usw.

Europa führt

Es ist kaum bekannt, dass die Anfänge des Envisat-Projekts auf das Columbus-Programm der ESA zurückgehen. Von diesem ambitionierten Raumfahrtprogramm mit bemannten

Europe shot to the forefront of Earth observation with Envisat, the largest satellite ever built to explore the Earth. The device, which weighs 8 tons (7 tonnes) and was the size of a small bus, was launched aboard an Ariane 5 rocket on March 1, 2002. Orbiting 500 miles (800 kilometers) above the poles, Envisat followed a similar trajectory to the ERS satellites and also completed a circuit in 100 minutes.

Nomen est omen: Envisat stood for Environmental Satellite, and the device was intended to measure environmental conditions on Earth as accurately as possible, until in 2012 it went out of service because of a failure. The ten different kinds of instruments on board made Envisat a Jack-of-all-trades: it precisely measured the form and composition of the Earth's surface, whether land, sea, or ice, and a large camera photographs the oceans in various colors to determined how much chlorophyll and sediment was present in the water. The camera could capture an area more than 600 miles (1,000 kilometers) wide and every area could be inspected again within three days. Instruments to measure trace gases provided 3-D models of the distribution of ozone, hydrocarbons, nitrogen oxides, and other compounds in the atmosphere, and the various onboard radars provided accurate data about the surface below by day or night, whatever the weather. Such data had a variety of uses: for example, it could be incorporated into climate and weather models, used to examine vegetation and erosion processes around the world, helped to examine land, sea, and air pollution, or allowed the speedy coordination of aid during accidents or natural disasters. It made shipping traffic possible and was also used to monitor sea ice, among other uses.

Europe leads the field

It is not generally known that the roots of the Envisat project are to be found in the ESA's Columbus

L'Europe s'est ensuite propulsée à la pointe de l'observation de la Terre avec Envisat. Il s'agit du plus grand satellite jamais construit pour explorer la Terre. Ce satellite de huit tonnes aussi grand qu'un autocar a été envoyé dans l'espace le 1er mars 2002 par une fusée Ariane 5. Envisat suivait la même orbite que les satellites ERS, à 800 kilomètres d'altitude au-dessus des pôles, et tournait également autour de la Terre en cent minutes.

Son nom était déjà tout un programme : Envisat est l'abréviation d'Environmental Satellite. Il a en effet collecté des données aussi précises que possible sur la Terre et son environnement jusqu'à son arrêt en 2012 dû à une défaillance. Ses dix instruments embarqués faisaient d'Envisat une véritable panacée. Il a mesuré avec précision la structure et les propriétés de la surface du globe, terrestre, maritime ou glaciaire. Un énorme appareil photo reproduisait la surface des océans en plusieurs couleurs qui révélaient la concentration de l'eau en chlorophylle et en sédiments. Cet appareil couvrait une bande de plusieurs milliers de kilomètres, inspectant ainsi l'ensemble des territoires en trois jours. Les instruments de mesure des gaz traces fournissaient des informations tridimensionnelles sur la répartition de l'ozone, des hydrocarbures, oxydes d'azote et autres substances dans l'atmosphère. Son radar livrait de jour comme de nuit et par tous les temps des données précises sur certaines particularités du relief.
Ces données se sont prêtées à des utilisations multiples. Elles ont été intégrées aux modèles climatiques, ont permis d'étudier les processus végétaux et l'érosion, de repérer la pollution de la terre, de l'air et des mers, et de coordonner rapidement les secours en cas d'accidents ou de catastrophes naturelles. Elles assuraient enfin le contrôle du trafic maritime et de la banquise.

L'Europe en tête

Peu de gens savent que le projet Envisat a été lancé dans le cadre du

das de peso y las dimensiones de un pequeño autocar, fue lanzado al espacio con un cohete Ariane 5. Envisat describía una órbita alrededor de la Tierra muy similar a la de los satélites ERS: a 800 km de altura sobre las regiones polares. También en su caso la rotación duraba cien minutos.

Su nombre era el símbolo de todo un programa: Envisat significa Environmental Satellite, es decir, satélite medioambiental. Y es que este vehículo registraba, con la mayor fiabilidad posible, datos sobre las condiciones medioambientales y la Tierra, hasta que en 2012 quedó fuera de servicio por un fallo. Los diez instrumentos que el Envisat llevaba a bordo lo convertían en un multitalento imprescindible. Medía con precisión la forma y la estructura de la superficie terrestre, ya fuese tierra firme, mar o casquete polar. Una gran cámara captaba la superficie de los océanos en diferentes colores. Con ello se podía determinar cuánta clorofila y materia sedimentaria hay en el agua. Esta cámara captaba una banda de más de 1.000 km de ancho. Cada región se inspeccionaba de nuevo en un plazo de tres días. Los instrumentos para medición de gases traza proporcionaban información tridimensional sobre la distribución en la atmósfera de ozono, hidrocarburos, óxido nítrico y otras sustancias. El equipo de radar ofrecía datos exactos sobre los detalles de la superficie, sin importar las condiciones meteorológicas o que fuese de día o de noche. Todos estos datos han tenido múltiples aplicaciones: han incidido en los modelos climáticos y meteorológicos, han permitido comprender los procesos de vegetación y erosión de todo el mundo, han detectado la contaminación ambiental en tierra, mar y aire para poder coordinar rápidamente la ayuda en caso de desgracias y catástrofes naturales, han facilitado el tráfico marítimo y la vigilancia de la cubierta de hielo de los mares, etc.

Europa lidera

Apenas se conoce que los comienzos del proyecto Envisat se basan en el

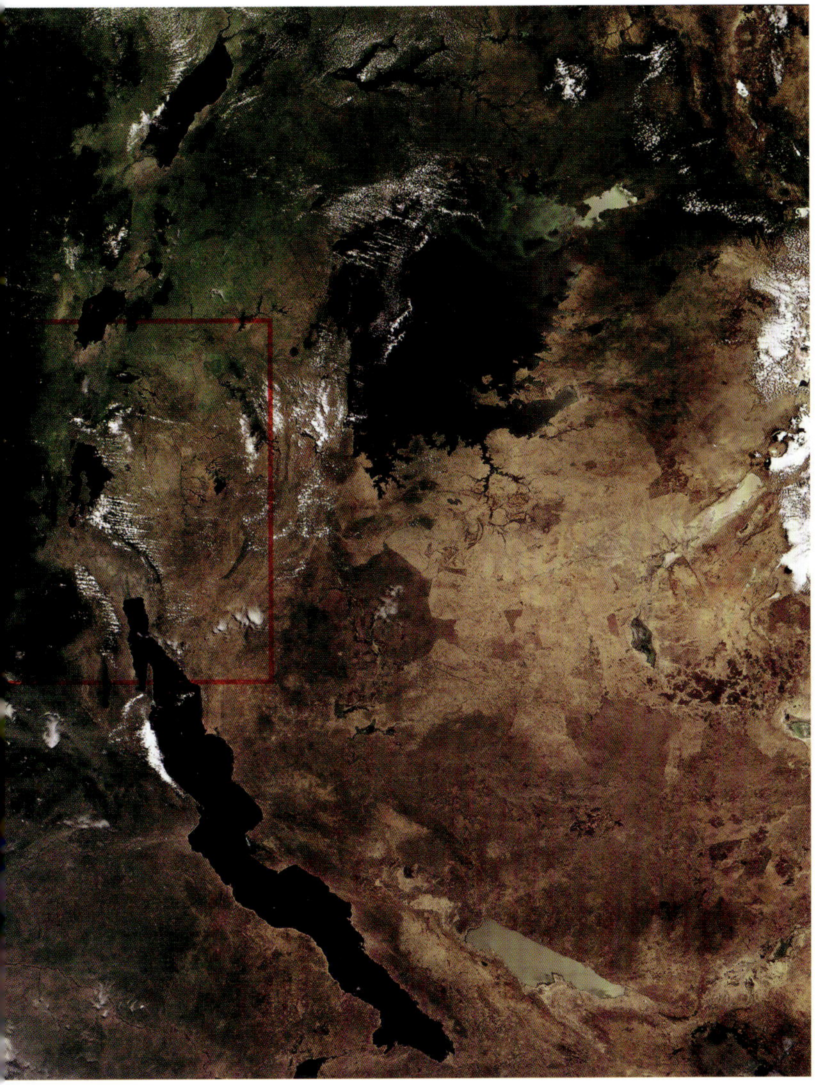

Der Victoriasee (Mitte) und
der Tanganjikasee (links unten)
im Südosten Afrikas (Envisat)

Lake Victoria (middle) and Lake
Tanganyika (bottom left) in south-
eastern Africa (Envisat)

Le lac Victoria (au centre) et le
lac Tanganyika (en bas à gauche)
en Afrique orientale (Envisat)

Lago Victoria (centro) y lago
Tanganica (izda. abajo),
en el sudeste de África (Envisat)

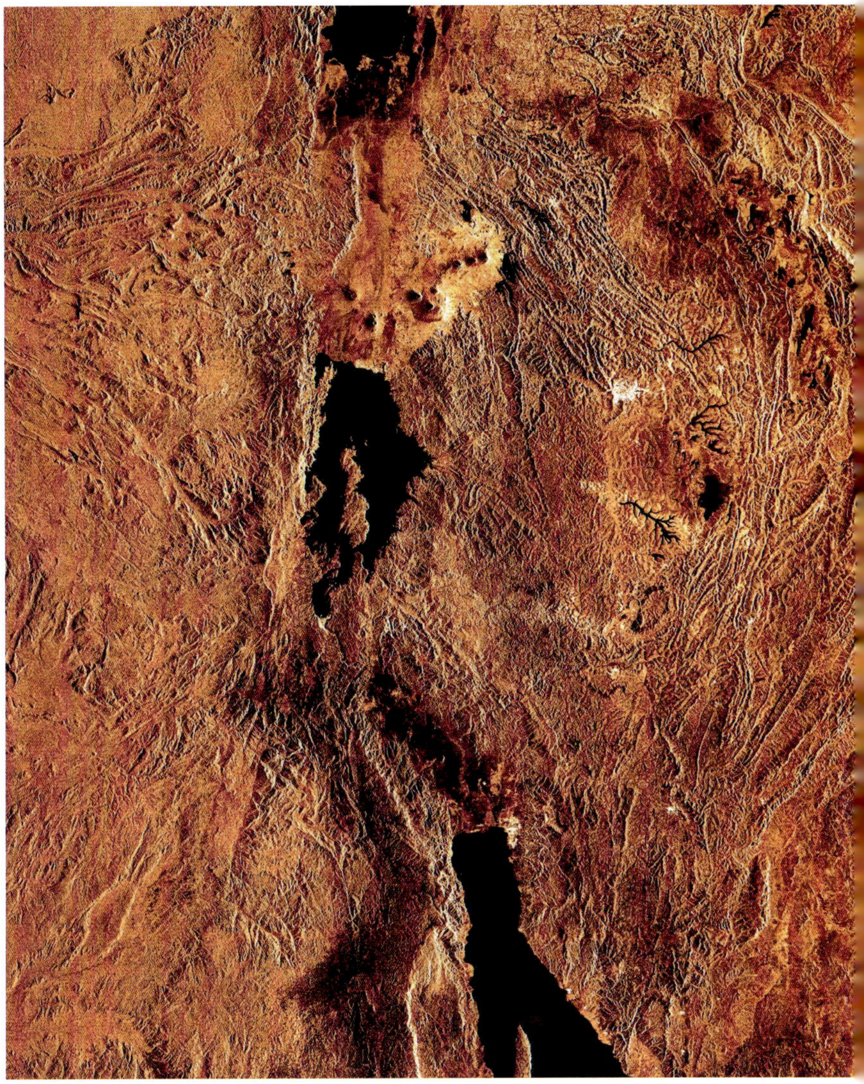

Die Radaraufnahme vom Ausschnitt
des linken Bildes zeigt Bergketten
und Vulkane (Envisat)

The radar image of a detail from
the photograph on the left reveals
mountain chains and volcanoes
(Envisat)

Image radar du détail de la photo
de gauche indiquant chaînes de
montagnes et volcans (Envisat)

La toma de radar del detalle de la
imagen izquierda muestra cadenas
montañosas y volcanes (Envisat)

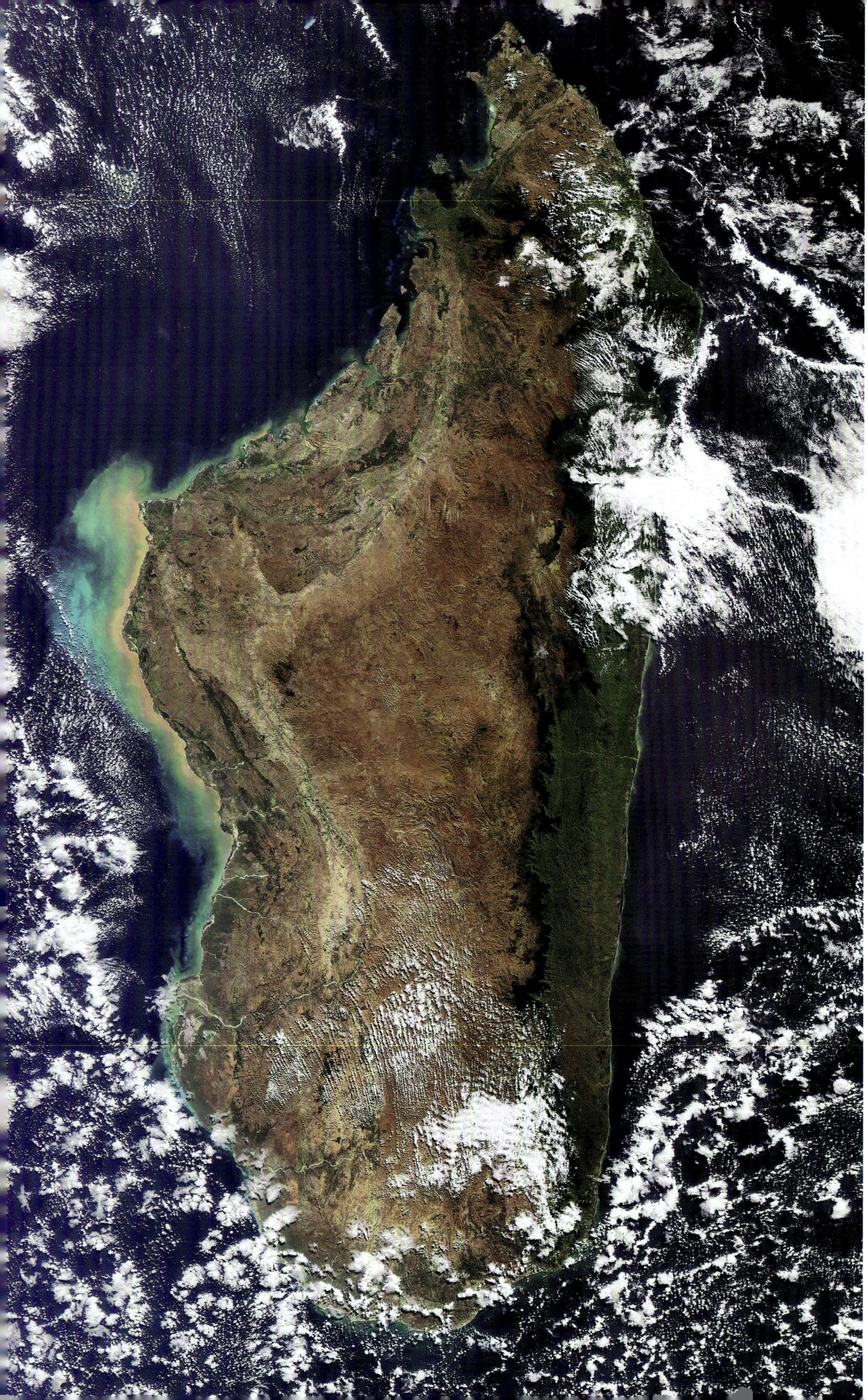

Die Ostküste der 1500 km langen Insel Madagaskar im Indischen Ozean ist üppig bewachsen (Envisat)

The 900-mile-long (1,500-km) east coast of the island of Madagascar in the Indian Ocean is covered in thick forest (Envisat)

La côte orientale luxuriante de Madagascar, dans l'océan Indien, s'étend sur 1 500 km (Envisat)

Los 1.500 km de la costa oriental de Madagascar, en el océano Índico, están cubiertos de exuberante vegetación (Envisat)

Rechte Seite: Künstliche Inseln im Persischen Golf vor der Küste Dubais. Rechts oben der Flughafen (Terra)

Right: Artificial islands in the Persian Gulf off the coast of Dubai. The airport can be seen top right (Terra)

Page de droite : les îles artificielles de Dubaï dans le golfe Arabo-Persique. En haut à droite, l'aéroport (Terra)

Página dcha.: Islas artificiales en la costa del golfo Pérsico, en Dubai. Arriba a la derecha, el aeropuerto (Terra)

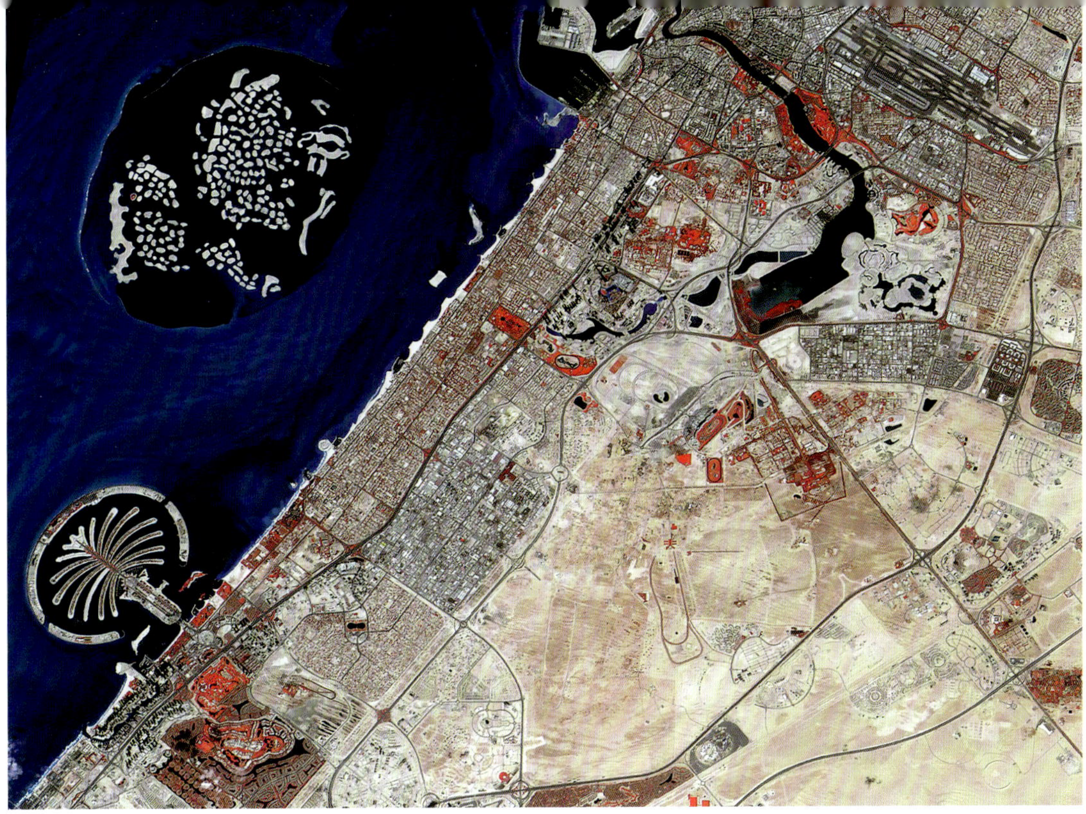

und unbemannten Missionen ist nur das Raumlabor Columbus übrig geblieben, das heute das wichtigste Forschungsmodul an Bord der ISS ist. Zum Columbus-Programm gehörte einst auch eine große polare Plattform: eine fliegende Forschungsstation, die vollautomatisch die Erde und ihre Umgebung erkunden sollte. Aus der polaren Plattform wurde Envisat. Er startete sogar sechs Jahre vor dem Columbus-Modul, dem Herzstück des ursprünglichen Programms, das Anfang 2008 endlich an die ISS angekoppelt wurde.

Im Oktober 2010 waren die Treibstoffvorräte an Bord von Envisat verbraucht. Seitdem ließ sich der große Satellit nicht mehr so präzise steuern und ausrichten wie in den ersten achteinhalb Jahren seiner Mission. Der Reservebetrieb dauerte nicht lange: Im April 2012 ging mitten im Messbetrieb urplötzlich der Kontakt zu Envisat verloren. Die Ingenieure rätseln, woran das liegt. Seitdem kreist Envisat nur noch als Weltraummüll um die Erde – ein bitteres Ende für den legendären Erdbeobachtungssatelliten.

Mit mehr als zwei Milliarden Euro Gesamtkosten für Bau, Start und Betrieb des Satelliten inklusive der Bereitstellung aller Daten ist Envisat die teuerste Satellitenmission, die

program; only the Columbus science laboratory, the most important research module on board the ISS, now remains of this ambitious space program, which included both manned and unmanned missions. The Columbus program was once meant to include a large polar platform—a fully automated aerial research station intended to explore the Earth and its near environs. This polar platform became Envisat and was launched six years before the Columbus module, the heart of the original program, was finally connected to the ISS at the beginning of 2008.

By October 2010, Envisat had exhausted its onboard fuel reserves and the great satellite could then no longer be steered with the same precision as during the first eight-and-a-half years of its mission. Its operation didn't last much more time: in April, 2012, contact was lost with Envisat, the reason why was never understood by engineers. Since then, the great satellite has been orbiting the Earth, though it isn't more than a remnant space; without a doubt, a sad end for the legendary observation satellite.

With final costs amounting to more than 2 million euros for construction, launch, and maintenance, including the processing of all the

programme Columbus de l'ESA. Le laboratoire spatial Columbus, le principal module de recherche actuellement à bord de l'ISS, est la seule composante de cet ambitieux programme spatial qui ait subsisté. Le programme Columbus prévoyait à l'origine la construction d'une plate-forme polaire devant assurer l'exploration automatique de la Terre et de son environnement. C'est de cette plate-forme polaire qu'est né Envisat. Le satellite a d'ailleurs précédé de six ans le module Columbus, pièce maîtresse du programme originel qui devait enfin être amarrée à l'ISS début 2008.

Depuis qu'il avait épuisé ses réserves de carburant en octobre 2010, le contrôle et l'orientation d'Envisat étaient moins précis qu'au cours des huit ans et demi précédents. Le fonctionnement sur réserve n'a pas duré longtemps : en avril 2012, le contact avec Envisat est soudainement perdu au milieu d'une mesure, sans que les ingénieurs arrivent à en trouver la cause. Depuis, Envisat reste en orbite autour de la Terre mais n'est plus qu'un déchet dans l'espace – une triste fin pour ce satellite d'observation légendaire.

Avec un budget global de deux milliards d'euros pour la construction, le lancement et l'exploitation

programa Columbus de la ESA. De este ambicioso programa espacial con misiones tripuladas y no tripuladas solo quedó el laboratorio espacial Columbus, que en la actualidad es el módulo de investigación más importante que hay a bordo de la ISS. También formaba parte del programa Columbus una gran plataforma polar: una estación volante de investigación que debía explorar, automáticamente, la Tierra y su entorno. De esta plataforma polar surgió Envisat. Fue lanzada seis años antes que el módulo Columbus, el núcleo del programa original que se acopló finalmente a la ISS a comienzos de 2008.

En octubre de 2010 se agotaron las reservas de combustible a bordo del Envisat. Desde entonces, el gran satélite ya no se pudo pilotar y orientar con la misma precisión de los primeros ocho años y medio de su misión. Su funcionamiento no duró mucho tiempo más: en abril de 2012 se perdió el contacto con el Envisat, sin que los ingenieros lograran encontrar la causa. Desde aquel momento, el gran satélite permanece en órbita alrededor de la Tierra, aunque ya no es más que un residuo en el espacio; sin duda, un triste final para este legendario satélite de observación.

Con más de dos mil millones de euros de coste total en construcción,

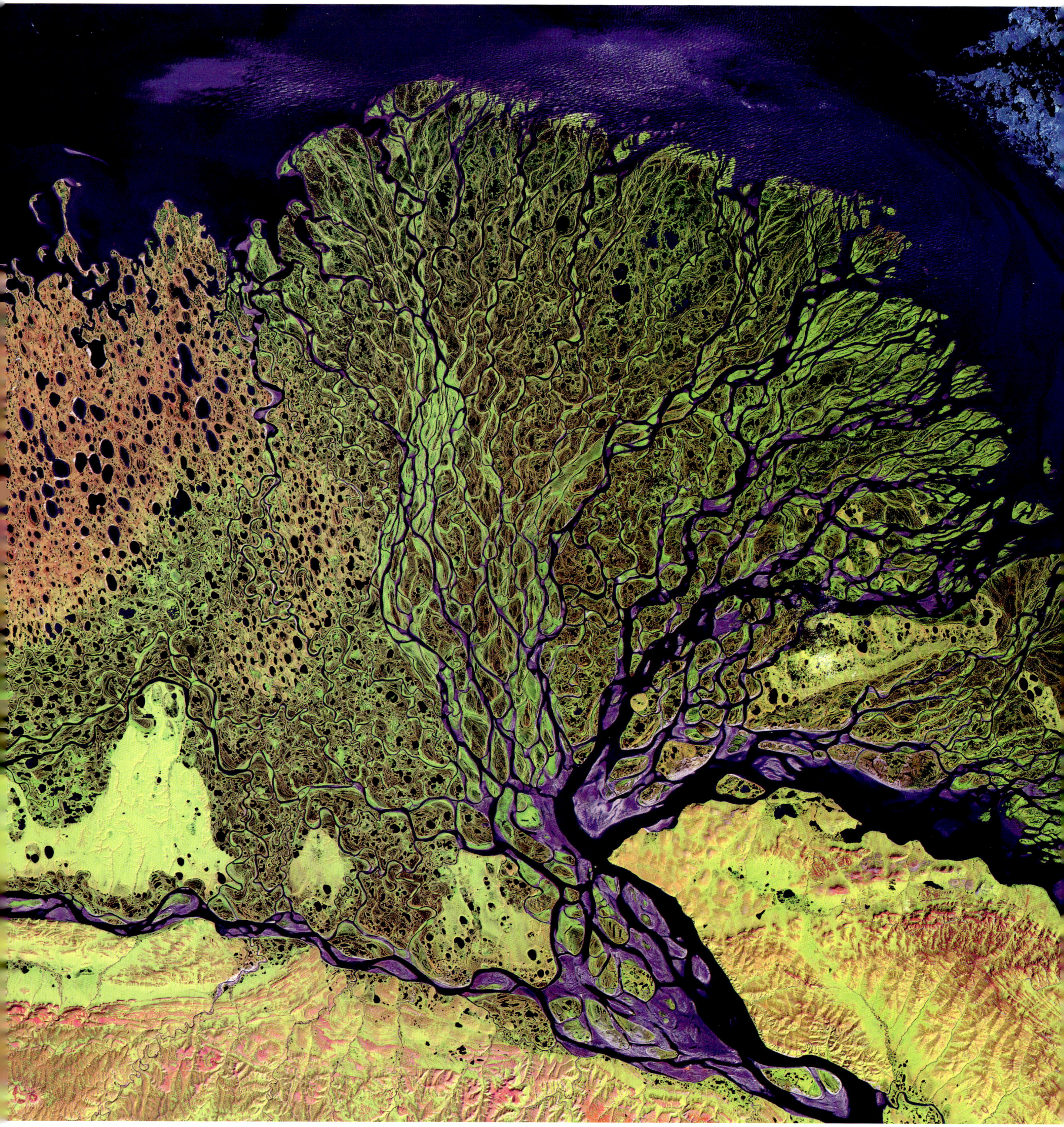

die ESA je durchgeführt hat. Zu viel für die Erde? Wohl kaum. Zudem hat Envisat das Geld nicht auf seiner Bahn im All verteilt. Es wurde auf dem Boden ausgegeben und vor allem in Europa. Denn der Satellit mit seinen zahlreichen Instrumenten ist von einem Konsortium aus gut hundert Firmen in vielen Ländern Europas gebaut worden. Mag in der Raumfahrt oft vom »faster, cheaper, better« die Rede sein, also dass Missionen schneller, billiger und besser werden sollen, so ist eine Mission wie Envisat der schlagende Beweis dafür, dass große ambitionierte Projekte ebenfalls zum Erfolg führen. Der große Satellit hatte seine geplante Lebensdauer um viele Jahren überschritten. Von der Komplexität der Aufgabenstellung, vom Budget und der Missionsdauer her reiht sich Envisat nahtlos ein in die Liste der großen Planetenmissionen. Was die Raumsonden wie Voyager, Galileo und Cassini für die Planeten Jupiter, Saturn, Uranus und Neptun sind, das ist Envisat für die Erde. Envisat ist die Planetenmission für die Oase im All.

Ebenso wichtig wie die wissenschaftlichen Botschaften, die der Satellit zu Boden funkt, ist die politische Aussage der Mission: Europa hat sich mit Envisat zum Weltmarktführer

data, Envisat is the most expensive satellite mission ever undertaken by the ESA. But this should not be considered too expensive for Earth. What's more, this money wasn't just shot into space with Envisat, it was spent on the ground, predominantly in Europe—the satellite and its myriad instruments were built by a consortium of over 100 companies in a number of European countries. Although the talk in space travel is often of "faster, cheaper, better," Envisat is compelling proof that large-scale, ambitious projects can also be successful. Envisat outlived its anticipated lifespan by years, and still it continues to fly. The organizational complexity, budget, and mission length of the Envisat project easily allow it to take its place among the other great planetary missions. What space probes such as Voyager, Galileo, and Cassini have done for planets such as Jupiter, Saturn, Uranus, and Neptune, Envisat has done for the Earth. Envisat was the planetary mission to our oasis in space.

The scientific information beamed to the surface by the satellite is rivaled in importance only by the mission's political statement: Envisat has made Europe the world market leader in Earth observation. In the

du satellite et la mise à disposition des données, Envisat est la mission satellitaire la plus onéreuse jamais menée par l'ESA. Trop cher pour la Terre ? Certainement pas. D'autant moins que l'argent d'Envisat n'a pas été disséminé dans l'espace, mais réparti entre la centaine de sociétés européennes membres du consortium chargé de la construction du satellite et de ses nombreux instruments. Même si la communauté spatiale cherche à réaliser des missions toujours plus rapides, moins onéreuses et plus performantes, une mission telle qu'Envisat est la preuve irréfutable que les grands projets ambitieux sont également porteurs de succès. Ce grand satellite a poursuivi son vol bien au-delà de sa durée de vie initialement prévue. Au vu de la complexité, du budget et de la durée de mission, Envisat figure incontestablement parmi les plus grandes missions planétaires. Il est à la Terre ce que les sondes Voyager, Galileo et Cassini sont à Jupiter, Saturne, Uranus et Neptune. Envisat est la mission planétaire d'exploration de l'oasis de l'espace.

Le message politique véhiculé par cette mission est tout aussi important que les données scientifiques transmises par le satellite : grâce à Envisat, l'Europe s'est hissée au rang

despegue y funcionamiento del satélite, incluyendo la preparación de todos los datos, Envisat es la misión de satélites más cara que ha llevado a cabo la ESA. ¿Demasiado para la Tierra? No tanto. Además, Envisat no ha esparcido el dinero por su órbita espacial. Se gastó todo en tierra firme, sobre todo en Europa. Porque el satélite, con sus numerosos instrumentos, fue construido en muchos países europeos por un consorcio de más de cien empresas. Aunque a menudo pueda parecer que el discurso en lo espacial sea «faster, cheaper, better» (rápido, barato, mejor), una misión como Envisat es una prueba irrefutable de que los proyectos grandes y ambiciosos también pueden conducir al éxito. Este gran satélite continuó volando más allá de su vida útil inicialmente prevista. Y sigue volando y volando. Por la complejidad de su cometido, el presupuesto y la duración de la misión, Envisat se incluye en la lista de grandes misiones planetarias. Envisat es a la Tierra lo que las sondas espaciales Voyager, Galileo y Cassini representan para los planetas Júpiter, Saturno, Urano y Neptuno.

Tan importante como la información científica que el satélite transmite, es el mensaje político de la misión: con Envisat, Europa se ha convertido

Die Lena mündet in Sibirien in einem 200 km breiten Delta ins Nordpolarmeer (Landsat)

The Lena River in Siberia flows into the Arctic Ocean through a delta 125 miles (200 km) wide (Landsat)

La Lena se jette dans l'océan Arctique en Sibérie en formant un delta de 200 km de large (Landsat)

El río Lena desemboca en Siberia formando un delta de 200 km de ancho sobre el océano Ártico (Landsat)

Das Mekong-Delta; oben Foto im sichtbaren Licht, unten eine Radaraufnahme (Envisat)

The Mekong Delta: the photograph above was taken using visible light, the image below used radar (Envisat)

Le delta du Mékong : en haut, photo en lumière visible, en bas image radar (Envisat)

Delta del Mekong: foto superior con luz visible, inferior de toma de radar (Envisat)

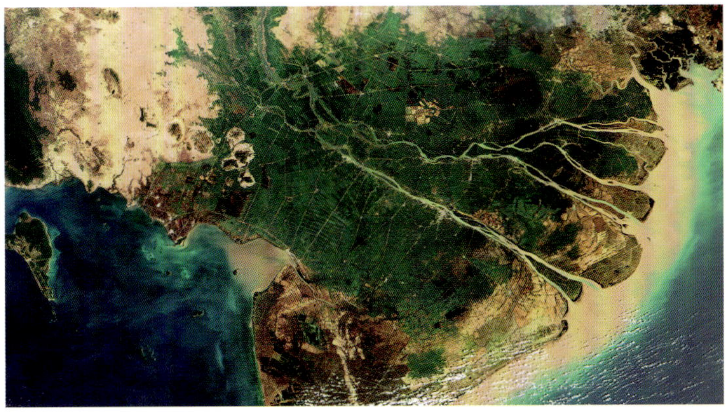

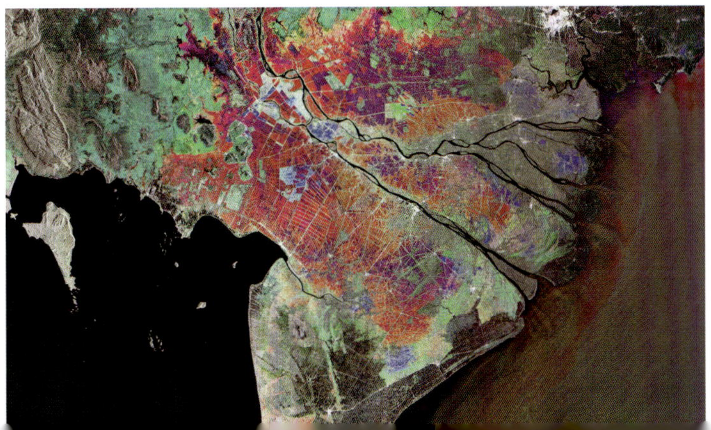

in Sachen Erdbeobachtung gemacht. Aus den USA, wo man auf politischer Seite in den vergangenen Jahren die Bedeutung dieser Disziplin verkannt und sich so weit ins Hintertreffen manövriert hat, blicken viele Forscher neidvoll an den Himmel, wenn Europas Satelliten 800 Kilometer hoch über den Mittleren Westen oder Neuengland ziehen.

»Living Planet«

Auch in Europa selbst haben die beiden ERS-Satelliten und Envisat viel bewegt: Sie brachten Ende der neunziger Jahre die ESA, die Europäische Union sowie Eumetsat (die Betreiberorganisation der europäischen Wettersatelliten) auf die Idee, ein großes Informationssystem für alle Aspekte der Umweltüberwachung und Sicherheit aufzubauen. Daraus wurde GMES (Global Monitoring for Environment and Security), die Globale Überwachung für Umwelt und Sicherheit, die inzwischen Copernicus heißt. Im Rahmen von Copernicus wird in den kommenden Jahren eine ganze Reihe von Satelliten starten. In Instituten und Datenzentren sollen die himmlischen Messungen möglichst in Echtzeit analysiert und allen Interessierten zur Verfügung gestellt werden. Dabei geht es vor allem darum, den Klimawandel zu überwachen und vor möglichen Gefahren gewarnt zu sein. Zu den weiteren Aufgaben von Copernicus gehört die Überwachung der Meere, z.B. Ölteppiche frühzeitig zu erken-nen und die Eisbedeckung auf Schifffahrtsrouten zu erfassen, die Landwirtschaft mit Hilfe von Daten über Klimageschehen, Erosion, Bodenfeuchte, Vegetationsentwicklung etc. zu optimieren, laufend den Zustand der Atmosphäre zu vermessen, Verschmutzungen zu erkennen, die Be-lastung durch Ultraviolettstrahlen zu bestimmen und Extremwetter besser vorherzusagen. Bei Katastrophen sollen Copernicus-Daten binnen kürzester Zeit Orientierung über Ausmaße und Folgen von Waldbränden, Erdbeben, Vulkanausbrüchen, Überflutungen etc. geben. Schließlich wird der Schiffsverkehr überwacht, und Co-

USA, where political developments in recent years have tended to ignore the importance of the discipline, allowing it to fall behind, many scientists now look with envy to the skies as Envisat and other satellites skim past the Midwest or New England at an altitude of 500 miles (800 kilometers).

"Living Planet"

The two ERS satellites and Envisat have achieved a great deal in Europe as well: at the end of the 1990s they gave the ESA, the European Union, and EUMETSAT (the umbrella organization for European weather satellites) the idea of constructing a large information system to cover every aspect of environmental monitoring and security, and this became GMES (Global Monitoring for Environment and security, later rebaptised as Copernicus). A number of satellites are due to be launched in the next few years as part of Copernicus and the results will be transmitted to institutions and data centers to be analyzed as much as possible in real time before being made generally available. The primary concern is to monitor climate change and note any possible dangers, but Copernicus' other tasks include keeping an eye on the oceans, giving timely warnings of oil slicks and ice floes on shipping routes, optimizing agriculture using climate, erosion, surface moisture, and vegetation growth data, continually scanning atmospheric conditions, spotting pollution, measuring ultraviolet ray levels, and improving the prediction of extreme weather events. When catastrophe strikes, Copernicus data should need little or no time to build up a proper picture of the extent and consequences of forest fires, earthquakes, volcanic eruptions, floods, and the like, and is also used to guide shipping and support military operations in areas of conflict.

This great program is no longer just a dream: several satellites were under construction for some time and the first mission launched in

de leader mondial de l'observation de la Terre. Aux États-Unis, alors que les responsables politiques ont, au cours des dernières années, sous-estimé et négligé cette discipline, nombreux sont les chercheurs qui regardent le ciel avec envie lorsque les satellites européens survolent la Nouvelle-Angleterre à 800 kilomètres d'altitude.

« Living Planet »

Mais Envisat et les satellites ERS ont également influencé l'Europe : ils ont dans les années 1990 encouragé l'ESA, l'Union européenne et Eumetsat (l'opérateur des satellites météorologiques européens) à lancer le programme européen de surveillance mondiale pour l'environnement et la sécurité GMES (Global Monitoring for Environment and Security, depuis rebaptisé Copernicus), un système d'information complet intégrant l'ensemble des données relatives à l'environnement et à la sécurité. Ce programme prévoit le lancement de toute une série de satellites dans les années à venir. Instituts et centres de données seront ensuite chargés d'analyser les mesures célestes en temps réel avant de les mettre à la disposition des utilisateurs. Ce système vise avant tout à observer les changements climatiques afin de prévenir les dangers potentiels. Parmi les missions de Copernicus figurent en outre la surveillance maritime, qui consiste par exemple à repérer les marées noires et les plaques de banquise entravant les voies maritimes, l'optimisation de l'agriculture par la mise à disposition de données relatives au climat, à l'érosion, à l'humidité des sols et à la végétation, l'analyse permanente de l'état de l'atmosphère, l'observation de la pollution et des nuisances causées par le rayonnement ultra-violet, et l'amélioration des prévisions météo. En cas de catastrophe, Copernicus est censé fournir rapidement les données permettant d'estimer l'ampleur des feux de forêt, tremblements de terre, éruptions volcaniques, inondations, etc. Enfin, le

en líder mundial en cuestiones de observación terrestre. Desde Estados Unidos, donde en los últimos años el área política menospreció el significado de esta disciplina y maniobró para relegarla, muchos investigadores elevan envidiosos sus ojos al cielo cuando los satélites europeos sobrevuelan Nueva Inglaterra a 800 km de altura.

«Living Planet»

Incluso en la misma Europa, los satélites ERS y Envisat han inspirado otros grandes proyectos: a finales de la década de 1990 dieron la idea a la ESA, a la Unión Europea y a la Eumetsat (Organización Europea de Explotación de Satélites Meteorológicos) de construir un gran sistema de información para la supervisión del medio ambiente y la seguridad. De ahí surgió el GMES (Global Monitoring for Environment and Security, más tarde rebautizado como Copérnico), el sistema de vigilancia global para el medio ambiente y la seguridad. En el marco del Copérnico, en los próximos años se lanzará una serie completa de satélites. Las mediciones celestes deben ser analizadas por institutos y centros de datos, a ser posible en tiempo real, y ser puestas a disposición de todos los interesados. Con esto se trata, sobre todo, de vigilar el cambio climático y de estar prevenidos frente a posibles peligros. Otras misiones del Copérnico son la vigilancia de los mares, la detección temprana de vertidos de petróleo y el registro de la formación de hielo en las rutas marítimas, la optimización de la agricultura con ayuda de datos sobre acontecimientos climáticos, erosión, humedad de los suelos, evolución de la vegetación y otros factores, la medición continua del estado de la atmósfera, la detección de focos de contaminación, la determinación de la incidencia de la radiación ultravioleta y la predicción anticipada de los fenómenos meteorológicos extremos. En caso de catástrofes, los datos del Copérnico deben ofrecer orientación en el lapso más breve de tiempo. Finalmente, los

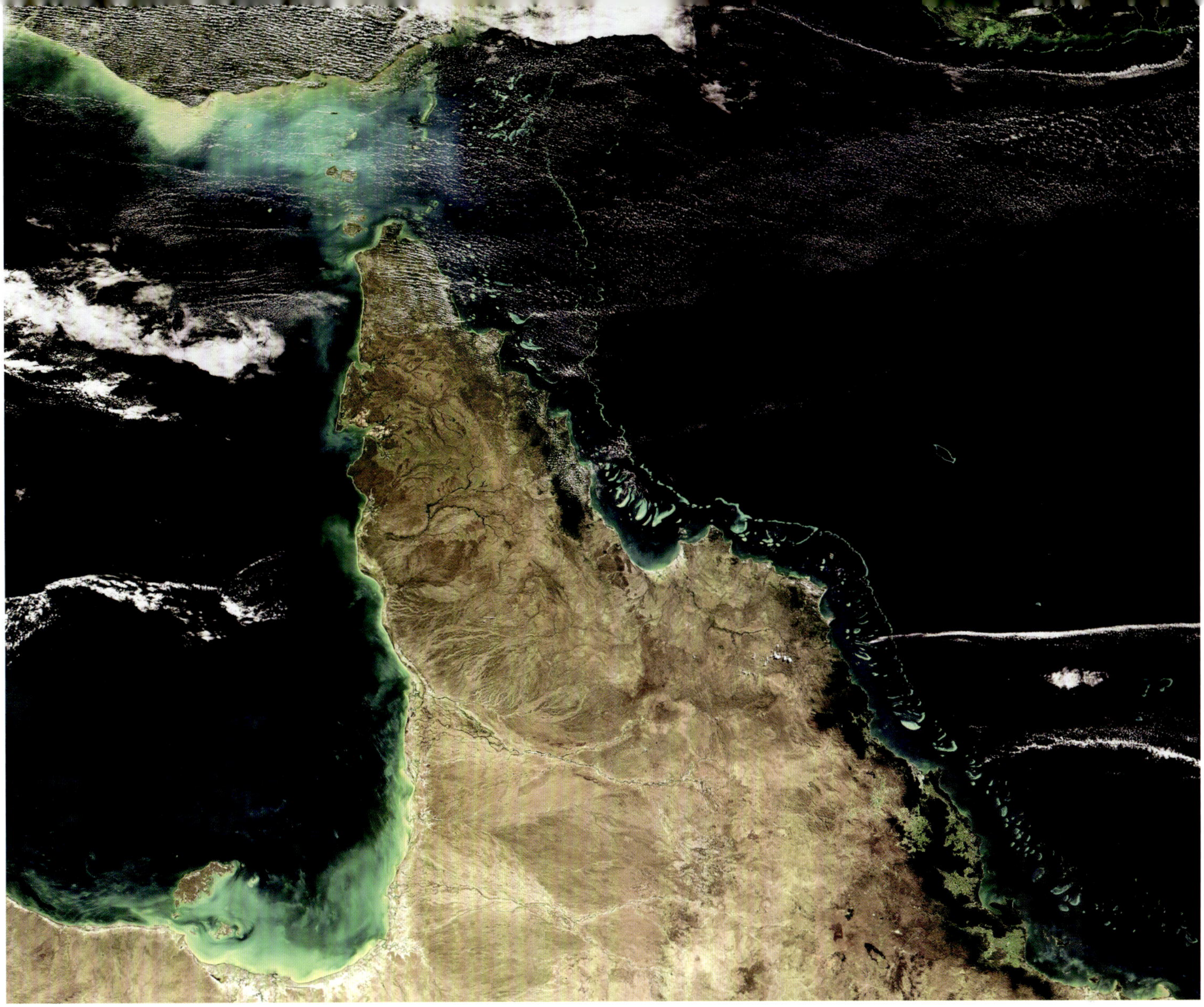

Oben: Die australische Küste mit Korallenriffen, die sich bis nach Neuguinea erstrecken (Envisat)

Above: The Australian coast with its coral reefs, which stretch all the way to New Guinea (Envisat)

Ci-dessus : la côte australienne et ses récifs coralliens qui s'étendent jusqu'à la Nouvelle-Guinée (Envisat)

Arriba: La costa australiana con arrecifes coralinos que se extienden hasta Nueva Guinea (Envisat)

Rechts: Spektakuläre Farben und lang gestreckte Wolken im Landesinneren Australiens (Envisat)

Right: Spectacular colors and long wisps of cloud in Australia's outback (Envisat)

À droite : couleurs spectulaires et longs nuages effilés au cœur de l'Australie continentale (Envisat)

Dcha.: Espectaculares colores y nubes alargadas en el interior de Australia (Envisat)

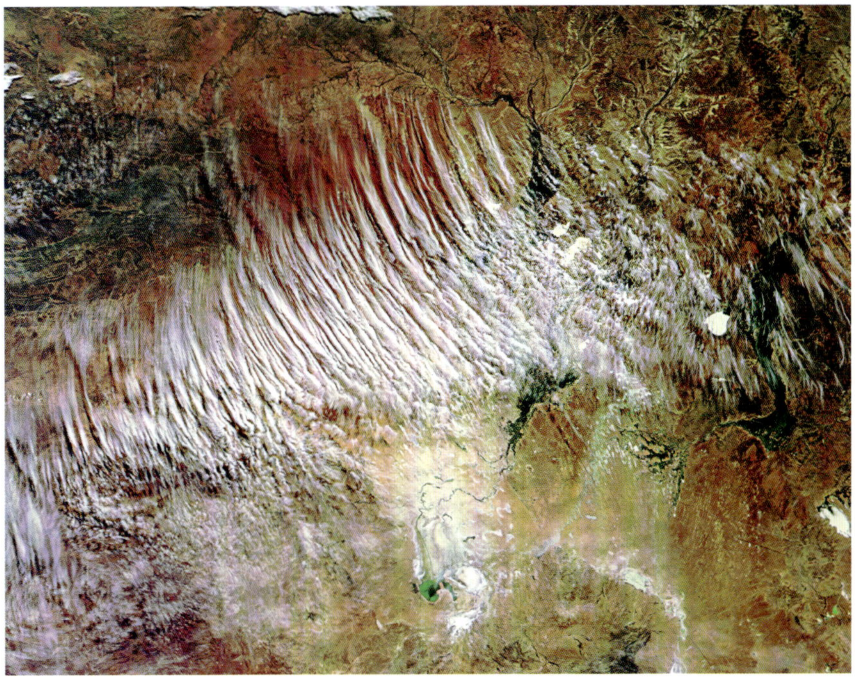

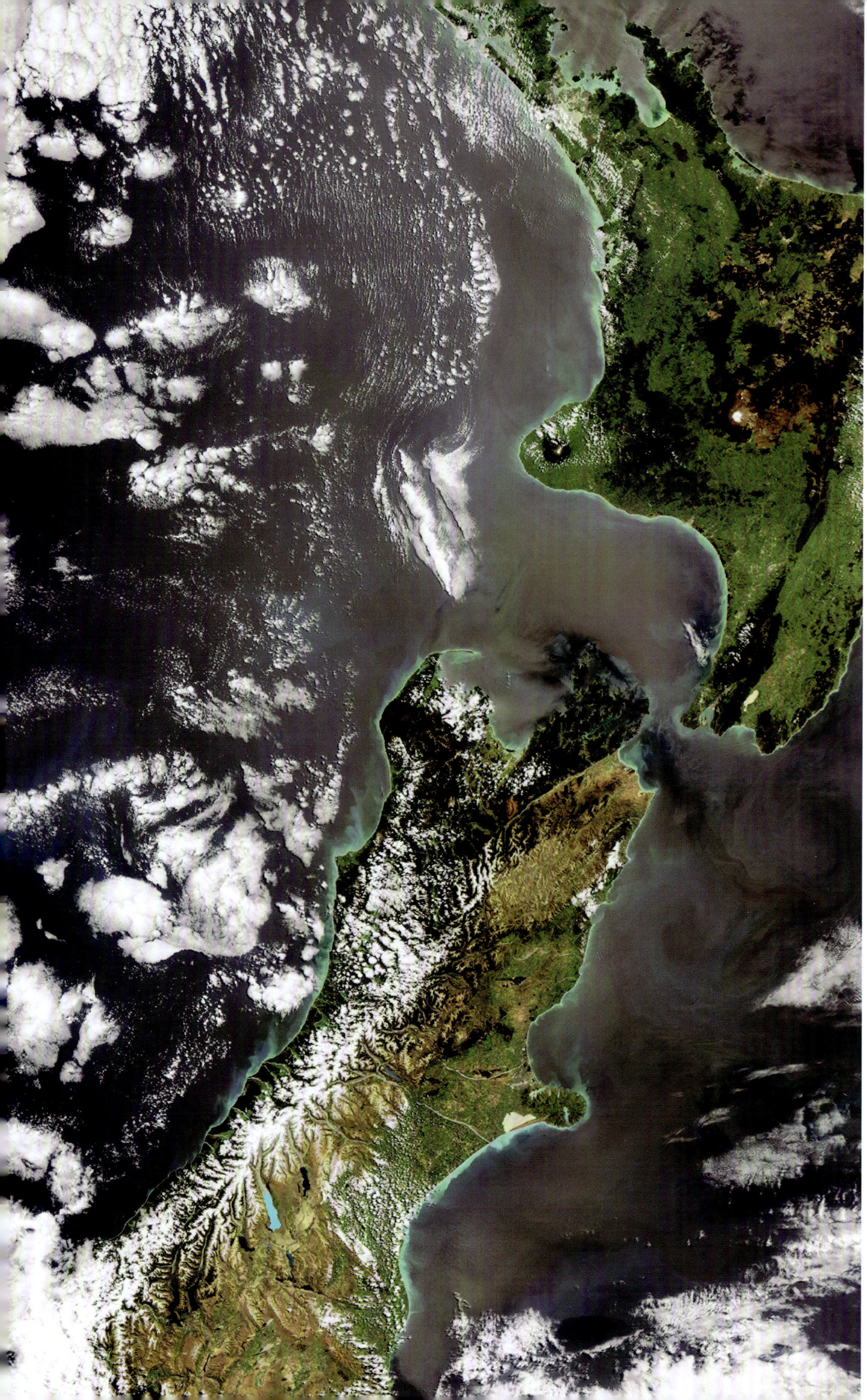

Wo Neuseelands Nord- und Südinsel fast zusammentreffen, verschneite Berge, viel Grün, Wolken und Wasserwirbel (Envisat)

New Zealand's North and South Islands don't quite meet: snow-capped mountains, lush vegetation, clouds, and whirlpool currents (Envisat)

À l'endroit où se frôlent les îles du nord et du sud de la Nouvelle-Zélande, montagnes enneigées, verdure, nuages et vortex (Envisat)

Montañas nevadas, mucho verdor, nubes y remolinos de agua donde casi se tocan las islas norte y sur de Nueva Zelanda (Envisat)

Rechte Seite: Patagonien mit einer an Norwegen erinnernden Fjordlandschaft. Links liegt Chile, rechts Argentinien (Envisat)

Right: Patagonia's fjord landscape is reminiscent of Norway's. Chile lies to the left, Argentina to the right (Envisat)

Page de droite : en Patagonie, un paysage de fjords qui rappelle la Norvège. À gauche le Chili, à droite l'Argentine (Envisat)

Página dcha.: Patagonia con un paisaje de fiordos que recuerda a Noruega. A la izquierda se encuentra Chile, a la derecha, Argentina (Envisat)

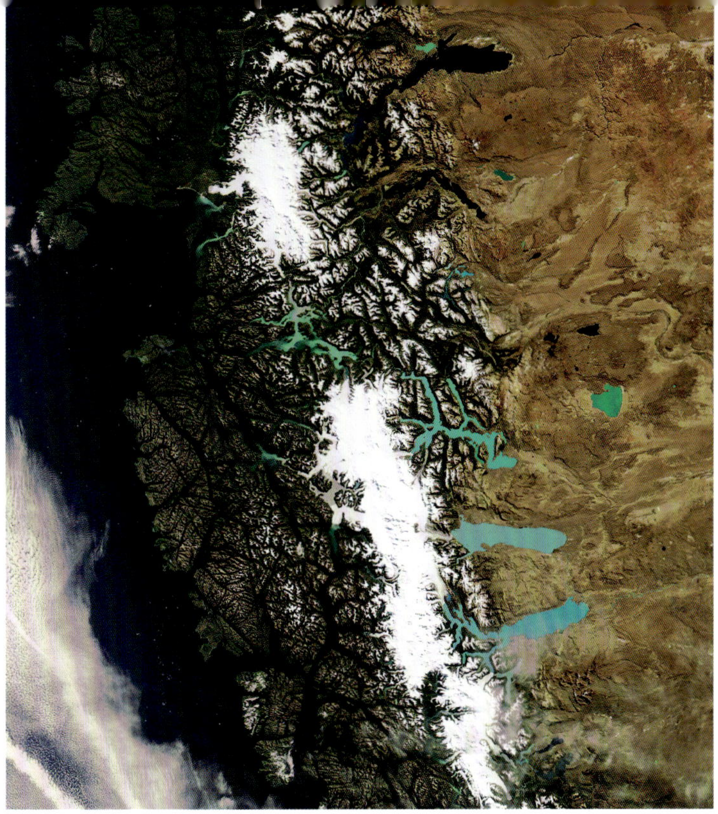

pernicus-Daten unterstützen militärische Einsätze in Krisenregionen.

Dieses große Programm ist keine reine Zukunftsmusik mehr. Einige Satelliten sind längst im Bau. Im April 2014 ist der erste ins All geflogen. Weitere Satelliten mit hoch spezialisierter Aufgabenstellung werden mindestens im Jahrestakt starten. Hinzu kommt die Flotte der Earth Explorer. Von diesen Erdbeobachtungssatelliten befindet sich bereits etwa die Hälfte im All. Sie vermessen zentimetergenau die Eismassen der Erde, bestimmen die Verteilung der Materie auf der Erdoberfläche und tief im Erdinneren, und sie erkunden die Bodenfeuchte und den Salzgehalt der Meere. Weitere Satelliten folgen, die das Magnetfeld der Erde untersuchen, die chemischen Prozesse in unserer Atmosphäre äußerst präzise verfolgen und die Arbeit von Envisat übernehmen sollen. Europa führt die Erdbeobachtung weltweit. Die Erdbeobachtung ist das größte Teilprogramm des europäischen Raumfahrtprogramms. Die Flotte aus den beiden Allroundsatelliten ERS-2 und Envisat und den aktuell sieben Spezialsatelliten der ESA sucht weltweit ihresgleichen.

Die ESA nennt ihr Erdbeobachtungsprogramm bewusst »Living Planet«. Denn die Forscher verfolgen das Ge-

April, 2014. At least one further satellite with highly specialized functions will launch every year after, and there is also the fleet of Earth Explorers: about half of these Earth observation satellites are already in space, measuring to the inch the Earth's ice shelves, examining the distribution of material on the planet's surface and deep beneath it, recording soil moisture content, and charting the salinity of the oceans. Other satellites will follow to take up Envisat's baton, investigating the Earth's magnetic field and keeping a precise record of the chemical processes taking place in our atmosphere. Europe leads the world in Earth observation. Observing the Earth is the largest single component of the European space program. The Earth Explorers fleet, which includes the two all-rounder satellites, ERS-2 and Envisat, and the ESA's current four specialized satellites, is the finest in the world.

The ESA has quite consciously called its Earth observation program "Living Planet" and scientists are following events as they happen on our planet, and more closely than ever before. An annual budget of 800 million euros has been made available, and no less than 12 further satellites are due to be launched by 2020, by

système contrôle le trafic maritime et facilite les interventions militaires dans les zones de conflit.

Ce grand programme n'est plus une utopie. Le premier satellite a été lancé en avril 2014. D'autres satellites spécialisés suivront dans les années suivantes. À cela s'ajoute la flotte des satellites d'observation de la Terre appelés Earth Explorer, dont la moitié est déjà sur orbite. Ils mesurent au centimètre près les couches de glace, étudient la distribution des masses à la surface comme au cœur de la Terre, et analysent l'humidité des sols et la salinité des océans. Suivront d'autres satellites chargés d'examiner le champ magnétique de la Terre et les processus chimiques régissant l'atmosphère, et de prendre le relais d'Envisat. L'Europe est le leader mondial de l'observation de la Terre, qui constitue par ailleurs le plus grand projet du programme spatial européen. Les satellites ERS-2, Envisat et les sept satellites spécialisés de l'ESA n'ont pas leur pareil dans le monde.

« Living Planet » est le nom donné par l'ESA à son programme d'observation de la Terre, dont le budget annuel s'élève à 800 millions d'euros, car les chercheurs scrutent les phénomènes qui régissent notre « planète vivante » avec une précision sans

datos del Copérnico también sirven para el control del tráfico marítimo y ofrecen respaldo a las operaciones militares en regiones en crisis.

Este gran programa ya no es una quimera. El primer satélite se lanzó en abril de 2014. Con frecuencia anual, se lanzarán otros satélites con cometidos altamente especializados A ellos hay que añadir la flota del Earth Explorer. Casi la mitad de estos satélites de observación terrestre se encuentra ya instalada en el espacio. Miden con una precisión de centímetros las masas de hielo de la Tierra, determinan la distribución de la materia en la superficie terrestre y la profundidad en su interior y examinan la humedad del suelo y el contenido de sal de los mares. Siguen otros satélites que analizan el campo magnético de la Tierra, observan con extremada precisión los procesos químicos que se producen en nuestra atmósfera y asumen el trabajo del Envisat. Europa lidera mundialmente la observación terrestre. La observación de la Tierra es el mayor subprograma del programa espacial europeo. La flota de ambos satélites multifuncionales ERS-2 y Envisat y los siete satélites especiales de la ESA, no tiene parangón en el mundo. La ESA denomina conscientemente «Living Planet» a su programa de observación terrestre.

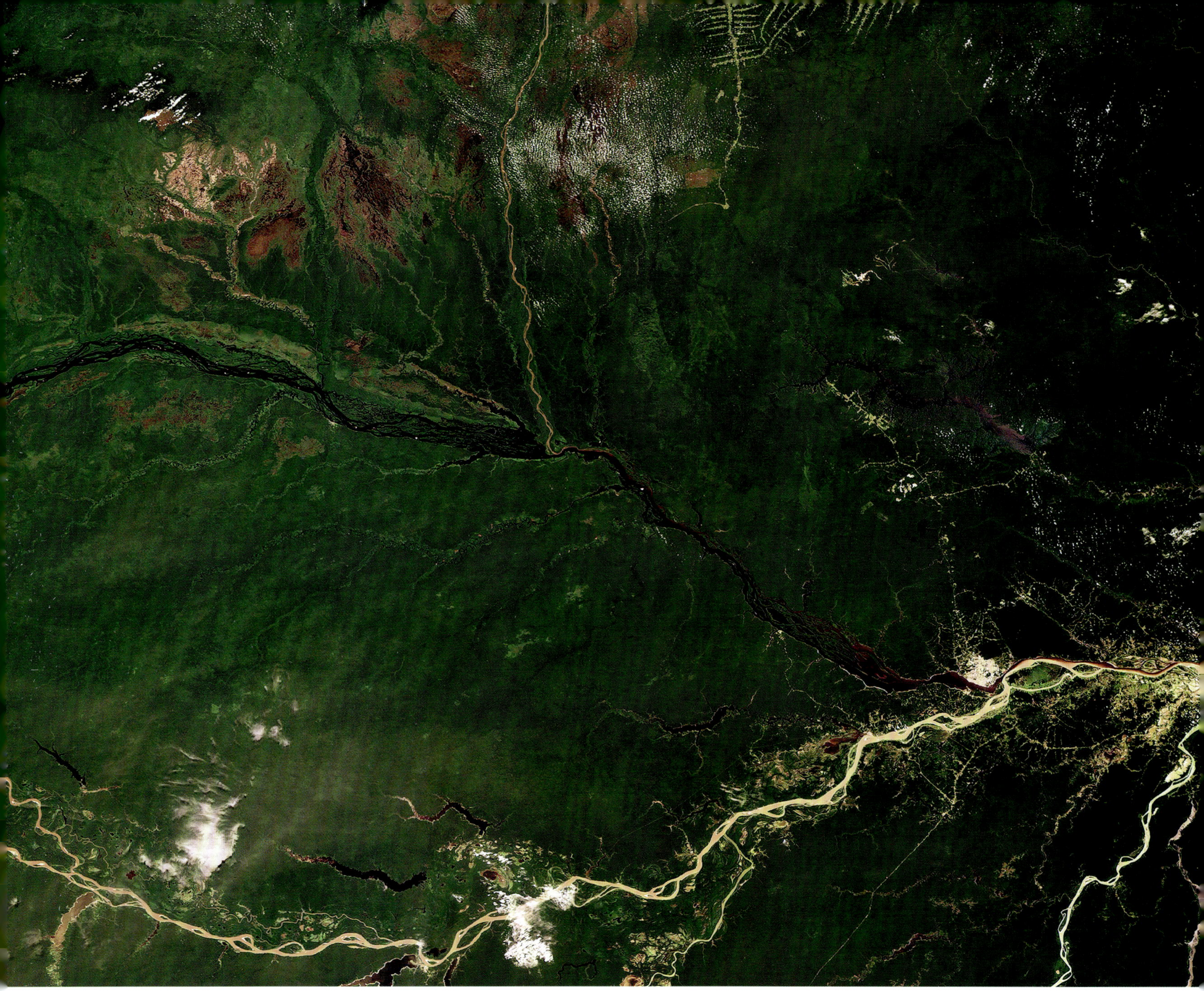

Schwarzes und braunes Wasser: Bei Manaus mündet der Rio Negro in den Amazonas (Envisat)

Black and brown water: the Rio Negro flows into the Amazon near Manaus (Envisat)

Eaux noires et eaux brunes : le Río Negro se jette dans l'Amazone près de Manaus (Envisat)

Agua negra y marrón: Manaos, desembocadura del río Negro en el Amazonas (Envisat)

Rechte Seite: Die Alexander-Selkirk-Insel im Südpazifik lässt die Luftmassen verwirbeln (Landsat)

Right: Alexander Selkirk Island stirring up cloud turbulence over the South Pacific (Landsat)

Page de droite : l'île Robinson Crusoé, dans le Pacifique sud, fait tourbillonner les masses d'air (Landsat)

Página dcha.: La isla Alejandro Selkirk, en el Pacífico Sur, provoca remolinos en las masas de aire (Landsat)

schehen auf und in unserem »lebendigen Planeten« so genau wie nie zuvor. Derzeit stehen gut 800 Millionen Euro im Jahr dafür zur Verfügung. Bis 2020 könnten mehr als ein Dutzend weitere Satelliten ins All starten. Dann würde der Rund-um-die-Uhr-Check der Erde Wirklichkeit. Noch vor wenigen Jahren erschien so etwas als völlige Fantasterei. Auch andere Weltraumnationen, etwa die USA, rücken die Erde mehr und mehr in den Brennpunkt ihres Raumfahrtinteresses. Die Daten der Missionen aus aller Welt werden sich bestens ergänzen.

Unsere Erde kann man kaum zu genau beobachten. Dafür ist unser wunderbarer blauer Planet viel zu wichtig – für uns Menschen. Flüge in die Weiten des Kosmos sind faszinierend. Sie befriedigen unsere Neugier und stacheln unseren Forscherdrang stets aufs Neue an. Die Raumsonden zeigen uns die Wunder im Planetensystem und lassen uns manchmal ob der arg unwirtlichen Bedingungen dort draußen schaudern. Satelliten, die die Erde beobachten, liefern mindestens ebenso faszinierende Daten. Wir alle sind Ziel dieser Missionen. Denn Envisat & Co. zeigen uns, wie robust unsere Erde, aber wie zerbrechlich die Basis unseres Lebens ist.

Steigen Sie ein, bzw. blättern Sie weiter! Fliegen Sie mit, und erkunden Sie Ihre kosmische Heimat auf eine Weise, wie Sie sie noch nie zuvor gesehen haben. Entdecken Sie mit uns den schönsten Ort im Universum!

which point 24-hour surveillance of the Earth will be a reality. Only a few years ago, such talk would have been pure fantasy, but now other space-going nations such as the USA are also paying more and more attention to the Earth in their space programs. Data gathered by missions from all over the world should tie in together nicely.

It is almost impossible to examine the Earth too closely—our wonderful blue planet is much too important to us humans. Missions into the depths of the cosmos are fascinating, continually satisfying our curiosity and stimulating our desire to know more, and space probes show us the wonders of the solar system, sending a thrill through us as they reveal the hostile conditions out there, but the satellites scanning the Earth provide equally fascinating data. We are all the object of these missions—Envisat and its cohorts are showing us how robust the Earth is, and yet also how tenuous the basis of our life here really is.

Climb aboard, turn the page—join the mission and investigate your home in space in ways you have never seen before. Come with us to explore the most beautiful place in the universe!

précédent. Plus d'une douzaine de nouveaux satellites s'élanceront dans l'espace d'ici 2020, pour garantir la surveillance continue de la Terre. Une vision qui semblait encore fantasque il y a quelques années. D'autres nations spatiales, dont les États-Unis, s'intéressent de plus en plus à la Terre, et les données collectées par toutes ces missions sont amenées à se compléter.

On ne peut être trop précis dans l'observation de la Terre. Notre merveilleuse planète bleue est bien trop importante, pour nous autres humains. L'exploration des profondeurs de l'univers fascine. Elle nourrit notre curiosité et ne cesse de stimuler notre soif de découverte. Les sondes spatiales nous révèlent les merveilles du système planétaire et nous font parfois frémir à la vue des conditions hostiles qui y règnent. Les données transmises par les satellites observant la Terre sont tout aussi passionnantes. Nous sommes l'objectif de ces missions, car Envisat et les autres satellites nous dévoilent la robustesse de la Terre, mais aussi la fragilité du berceau de la vie humaine.

Embarquez dans cette aventure qui vous fera découvrir votre planète natale comme vous ne l'avez encore jamais vue, et admirez le plus bel endroit de l'univers !

Porque los investigadores observan los acontecimientos que ocurren en y dentro de nuestro «planeta vivo» con una precisión. Actualmente disponen de más de 800 millones de euros al año. Hasta 2020 se podrían lanzar al espacio más de una docena de nuevos satélites. Entonces se haría realidad la supervisión terrestre durante las 24 horas del día. No hace demasiados años, algo así parecía demasiado fantasioso. También otras naciones con tradición espacial, como Estados Unidos, van convirtiendo a la Tierra en el centro de su interés en investigación espacial.

Nuestra Tierra apenas se empieza a observar con esa precisión. Por eso, para nosotros, los seres humanos, nuestro maravilloso planeta azul es demasiado importante. Los vuelos a las profundidades del cosmos resultan fascinantes. Satisfacen nuestra curiosidad y estimulan nuestra sed investigadora. Las sondas espaciales nos muestran las maravillas del sistema planetario y, a veces, nos hacen estre mecer ante las inhóspitas condiciones que se dan allá afuera. Los satélites que observan la Tierra nos envían datos, al menos, igualmente fascinantes. Porque Envisat y el resto de satélites nos muestran la fortaleza de la Tierra, pero también la fragilidad de la base de la vida humana.

Embarque con nosotros en esta aventura. Vuele y explore su patria cósmica de una manera que nunca hasta ahora había visto. Descubra con nosotros el lugar más hermoso del universo!

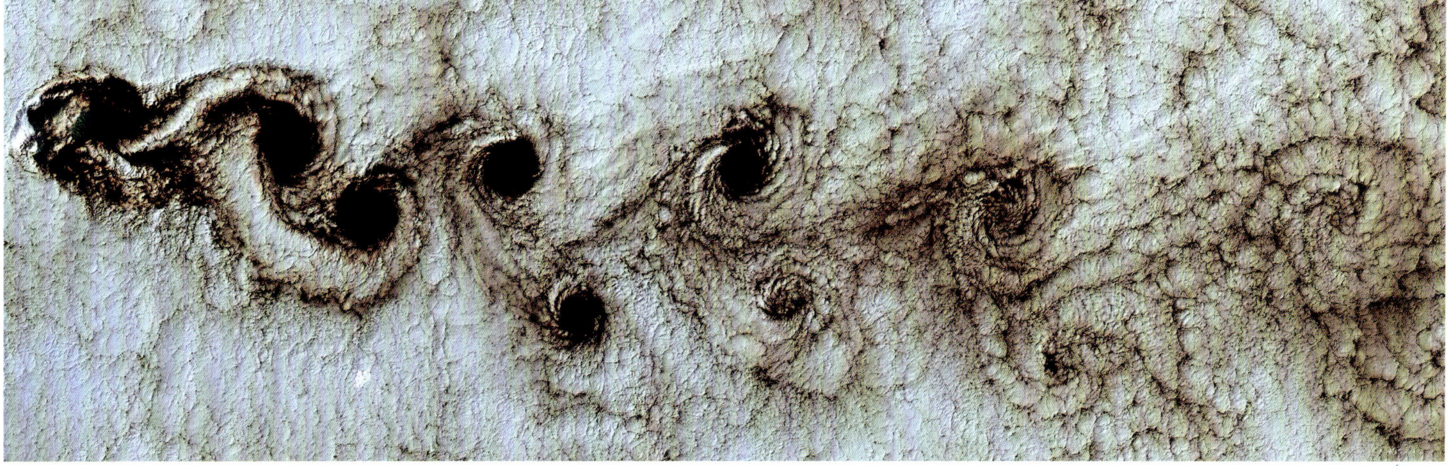

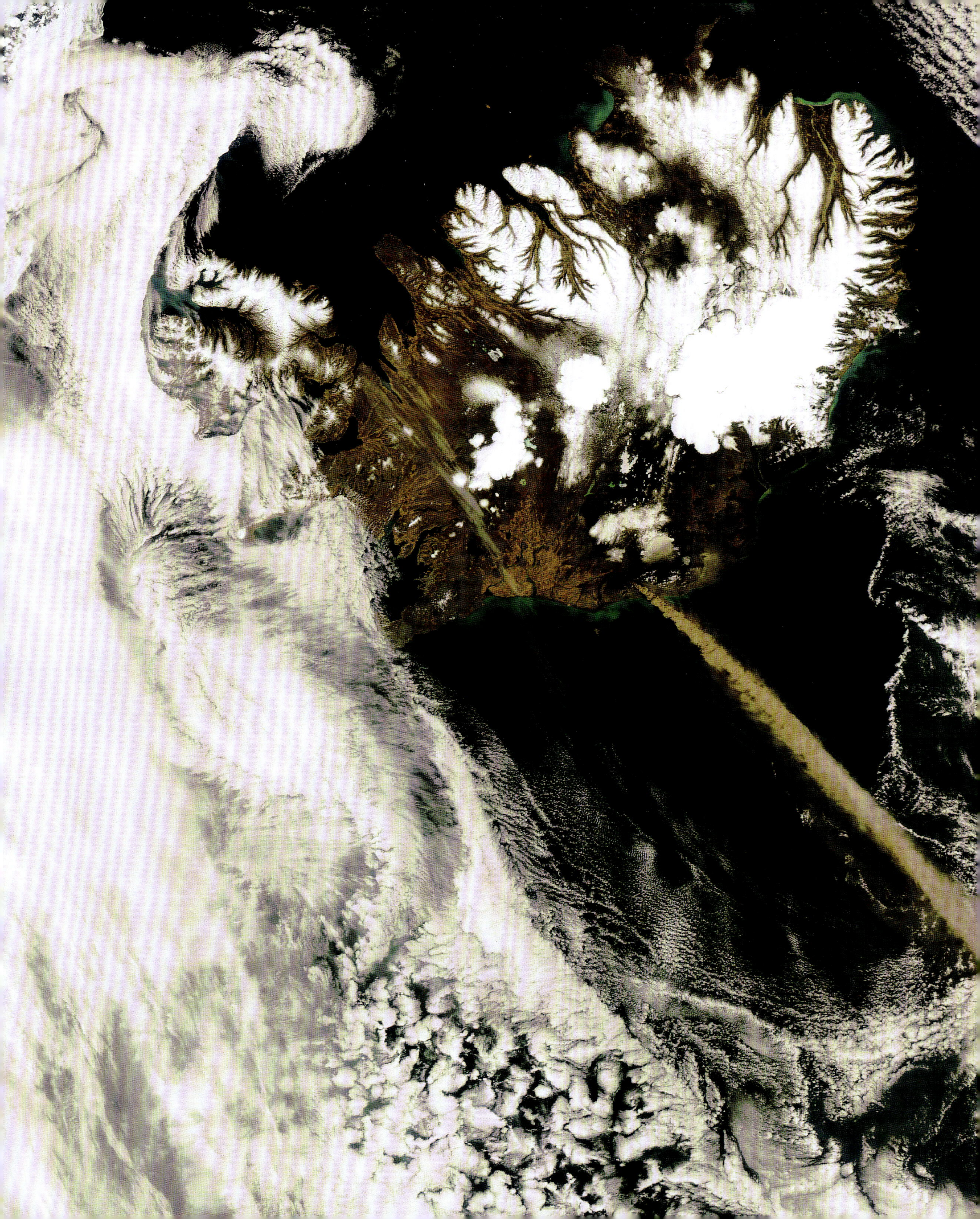

Eis, Wasser, Feuer: Was Satelliten sehen

Ice, water, fire: what do satellites see?

Le feu, l'eau et la glace : ce qu'observent les satellites

Hielo, agua, fuego: lo que los satélites ven

Die Bäume scheinen fast in den Himmel zu ragen, dichtes Unterholz aus Buschwerk, Farnen, Lianen und Moos macht den Urwald fast unpassierbar. Die heiße, feuchte, schwere Luft ist fast unerträglich. Grillen zirpen, in der Ferne kreischen Affen. Um die Baumwipfel wabern Wolkenfetzen. Plötzlich zerreißt das Kreischen von Motorsägen das Idyll im tropischen Regenwald irgendwo im Amazonasbecken. Minuten später stürzt krachend ein mächtiger Mahagonistamm zu Boden. In den folgenden Stunden kommen viele weitere hinzu. Die Mitarbeiter der Holzfirma fühlen sich unbeobachtet. Der nächste Ort ist Dutzende Kilometer entfernt. Niemand scheint von der Abholzung Notiz zu nehmen.

Doch unbemerkt von den Männern am Boden taucht am Südhorizont ein Satellit auf. Mit 27.000 Kilometern pro Stunde rast er um die Erde. Nach drei Minuten hat er einen Logenplatz, um das Treiben im Regenwald zu dokumentieren. Auch die Nebelschwaden, die im hügeligen Gelände hängen, nützen den Baumfrevlern nichts. Niemand spürt die lichtschnellen Radarwellen, die der Satellit zu Boden schickt und deren Echos unbestechliche Belege für das Abholzen liefern. Was vorgestern noch ein komplettes Waldstück war, ist jetzt

The trees seem to reach the sky and thick undergrowth of shrubs, ferns, lianas, and moss make this primeval forest practically impassable. The hot, damp, heavy air is almost unbearable. Crickets chirp and monkeys call in the distance; wisps of cloud cling to the treetops. Suddenly the roar of chainsaws interrupts this idyll in the tropical rainforest somewhere in the Amazon Basin. Minutes later, a mighty mahogany trunk crashes to the ground, to be followed by many others in the next few hours. The logging firm's employees think they are acting unobserved—the next town is hours away and no one appears to have noticed that this lumber-cutting is taking place.

However, unknown to the men on the ground, a satellite has risen on the southern horizon. Racing around the Earth at 17,000 miles per hour (27,000 kilometers per hour), it reaches a point from which to observe the activity in the rainforest within three minutes. Not even the banks of mist hanging over the hilly landscape are of use in hiding the illegal loggers, and no one feels the radar waves as they flash at light speed from the satellite to the ground, returning echoes with incontrovertible proof of the

Les arbres semblent s'élancer jusqu'au ciel tandis qu'un épais sous-bois de broussailles, de fougères et de lianes rend la forêt vierge impénétrable. La chaleur et l'humidité de l'air sont à la limite du supportable. Les grillons grésillent et les singes hurlent dans le lointain. Des lambeaux de nuage vacillent à la cime des arbres. Quand le vrombissement d'une tronçonneuse vient troubler ce paradis quelque part dans la forêt amazonienne. Quelques minutes plus tard, un magnifique acajou s'écrase sur le sol. Des dizaines d'autres arbres tomberont à leur tour au cours des prochaines heures. Les ouvriers de l'exploitation forestière se sentent à l'abri des regards. Le village le plus proche se situe à des dizaines de kilomètres. Personne ne semble remarquer le déboisement.

Pourtant, un satellite se lève à l'horizon, à l'insu des bûcherons. Il gravite autour de la Terre à la vitesse de 27 000 kilomètres par heure, et en l'espace de trois minutes, le voici aux premières loges pour enregistrer ce qui se trame dans la forêt tropicale. Les trafiquants de bois ne peuvent même pas compter sur les nappes de brume accrochées aux collines alentour. Nul ne perçoit les ondes radar envoyées par le satellite à la vitesse de la lumière vers le sol, et dont

Los árboles parecen elevarse casi hasta el cielo, al tiempo que un denso sotobosque de maleza, helechos, lianas y musgo hace casi intransitable la selva virgen. El ambiente cálido, húmedo y pesado apenas se puede soportar. Los grillos cantan, los monos chillan a lo lejos. Alrededor de las cimas de los árboles flamean jirones de nubes. Súbitamente, el chirriar de las motosierras en algún lugar de la cuenca del Amazonas rompe el encanto del bosque húmedo tropical. Minutos después, un enorme tronco de caoba cae al suelo entre crujidos. En las horas siguientes se le van añadiendo otros muchos. Los trabajadores de la empresa maderera no se sienten observados. El lugar poblado más próximo está a muchos kilómetros de distancia. Nadie parece prestarle atención a la tala indiscriminada. Pero sin que los hombres que están en tierra firme se den cuenta, por el sur, en el horizonte, aparece un satélite. Gira alrededor de la Tierra a la increíble velocidad de 27.000 km por hora. En solo tres minutos ya está apostado en el lugar idóneo para documentar lo que se está tramando en el interior de la selva. Ni siquiera los vapores brumosos que flotan sobre el ondulado terreno les sirven de nada a los saqueadores de árboles. Nadie percibe las ondas de radar, rápidas como la luz, que el

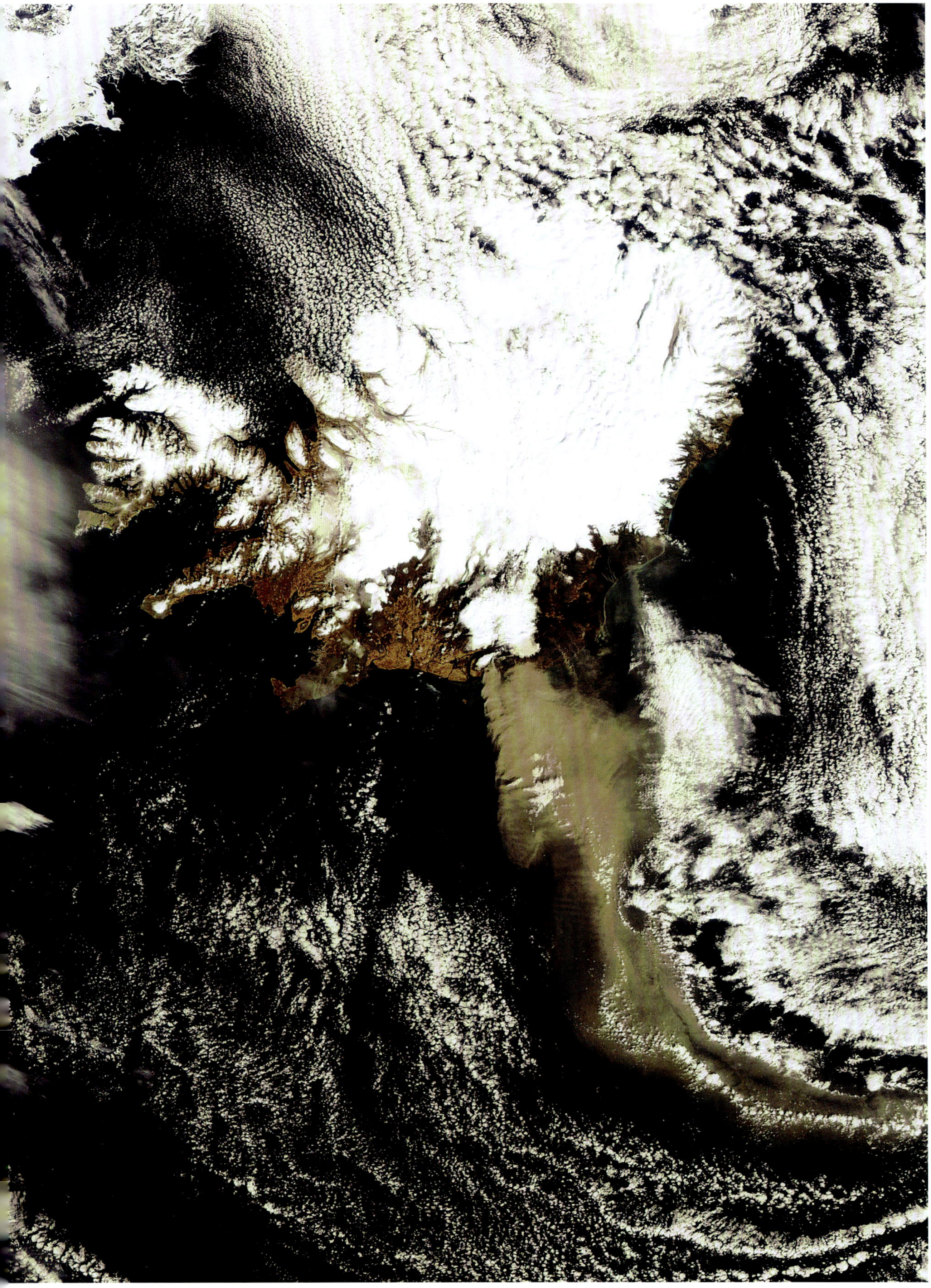

Links und Seite 60: Beim Ausbruch des Vulkans Eyjafjallajökull auf Island im April 2010 wehte die Aschewolke direkt auf Europa zu (Envisat)

Left and page 60: When Iceland's Eyjafjallajökull volcano erupted in April 2010, a cloud of ash drifted directly over Europe (Envisat)

À gauche et page 60 : le volcan de cendres généré par l'éruption du volcan islandais Eyjafjallajökull en avril 2010 se dirigeant droit sur l'Europe (Envisat)

Izda. y página 60: Durante la erupción del volcán islandés Eyafjallajökull, en abril de 2010, la nube de cenizas fue arrastrada por toda Europa (Envisat)

Rechte Seite: Die Asche stoppte den Flugverkehr in Europa. Südostengland mit Kondensstreifen (links) und ohne Flugverkehr (rechts, Envisat)

Right: The ash grounded air traffic all over Europe. Vapor trails over southeastern England (near right) and during the flight ban (far right, Envisat)

Page de droite : le nuage de cendres a paralysé le trafic aérien en Europe. Le ciel du Sud-est de l'Angleterre avec traînées de condensation (à gauche) et sans trafic aérien (à droite, Envisat)

Página dcha.: Las cenizas paralizaron el tráfico aéreo de Europa. Sudeste de Inglaterra con estelas (izda.) y sin tráfico aéreo (dcha., Envisat)

eine trostlose abgeholzte Fläche, wie das Radarbild zeigt.

Der Satellit entschwindet so schnell, wie er gekommen ist. Aber er wird die Abholzung des Regenwaldes immer wieder genau im Blick haben – alle einhundert Minuten zieht er über die Tropen hinweg.

Zweiter Weihnachtstag 2004. Nach einem verheerenden Erdbeben im Indischen Ozean rasen Tsunamiwellen durch den Indischen Ozean und zerstören große Küstenabschnitte in Indonesien, Thailand, Indien, Sri Lanka, Myanmar, Somalia, Malaysia auf den Malediven und anderen Staaten.

Die Hilfskräfte sind völlig überfordert. Über die Lage in Touristenzentren mit guter Infrastruktur weiß man ziemlich schnell Bescheid. Doch wie ist es den Menschen in abgelegenen Gebieten ergangen?

Schon wenige Stunden nach der Katastrophe gibt es die ersten Bilder aus dem All. Binnen weniger Tage gibt es präzise Informationen darüber, welche Gebiete wie stark betroffen und ob Straßen noch passierbar sind. Satelliten können den Opfern der Katastrophe nicht direkt helfen – aber sie helfen den Helfern. Ohne die technische Unterstützung aus dem All ließen sich die Hilfsaktionen kaum zielgerichtet durchführen.

deforestation. What only yesterday had been unspoiled jungle is now a bare and barren clearing, which shows up on the radar image.

The satellite disappears just as quickly as it arrived, but it will keep a close eye on the logging in the rainforest—it passes the tropics every 100 minutes.

December 26, 2004. After a devastating earthquake in the Indian Ocean, a succession of tsunamis roll across the high seas and destroy large sections of the coasts of Indonesia, Thailand, India, Sri Lanka, Myanmar, Malaysia, Somalia, Maldives, and other states.

The rescue services are completely overwhelmed. The situation in tourist centers with good infrastructure soon becomes clear, but how have people fared in the more remote areas?

Only a few hours after the catastrophe, the first images arrive from space and within a few days there is precise information about which areas are hardest hit and whether roads are still passable. Satellites are not able to aid disaster victims directly, but they can help the helpers—without technical support from space, it would be almost impossible to coordinate relief work effectively.

l'écho fournit les preuves irréfutables du déboisement en cours. Ce qui était hier encore une belle parcelle de forêt ne correspond plus aujourd'hui qu'à une triste surface défrichée sur l'image radar. Le satellite repart aussi vite qu'il est venu, mais il continuera à observer avec attention le déboisement de la forêt tropicale, toutes les cent minutes, lors de son passage au-dessus des tropiques.

26 décembre 2004. Suite à un terrible tremblement de terre, une gigantesque vague de tsunami déferle sur l'océan Indien, détruisant sur son passage de vastes régions côtières en Indonésie, en Thaïlande, en Inde, et au Sri Lanka.

Les équipes de secours sont débordées. L'état des lieux dans les centres touristiques disposant d'une bonne infrastructure est rapidement établi. Mais qu'en est-il des personnes vivant dans des zones reculées ?

Les premières images sont envoyées quelques heures seulement après la catastrophe. Quelques jours plus tard, des informations précises sur l'ampleur des dégâts et l'état des routes sont disponibles. Les satellites ne peuvent certes venir en aide aux victimes des catastrophes, mais ils contribuent aux secours, et sans cette assistance technique, il serait difficile de cibler les interventions d'urgence.

satélite envía al suelo y cuyos ecos son pruebas irrefutables de la deforestación. Lo que anteayer era todavía una completa arboleda, hoy es una desolada superficie talada, como muestra la imagen de radar. El satélite desaparece tan rápido como llegó. Pero cada cien minutos, lo que tarda en volver a sobrevolar los trópicos, seguirá concentrando su mirada en la deforestación del bosque húmedo.

Navidad de 2004, día de San Esteban. Tras un atroz terremoto en las profundidades del océano, un tsunami atraviesa a toda velocidad el océano Índico y destruye grandes zonas costeras en Indonesia, Tailandia, India, Sri Lanka y otros países. Los equipos de socorro están desbordados. La normalidad se restablece en los centros turísticos, por su buena infraestructura. Pero, ¿cómo les ha ido a las personas que viven en regiones aisladas? Sólo unas pocas horas después de la catástrofe ya se dispone de las primeras imágenes desde el espacio. En unos pocos días se tiene información precisa sobre cuáles son las regiones más afectadas y si las carreteras están transitables. Los satélites no pueden ayudar a las víctimas de la catástrofe, pero sí a los socorristas. Sin el apoyo técnico desde el espacio, no se hubiese podido realizar el socorro más urgente.

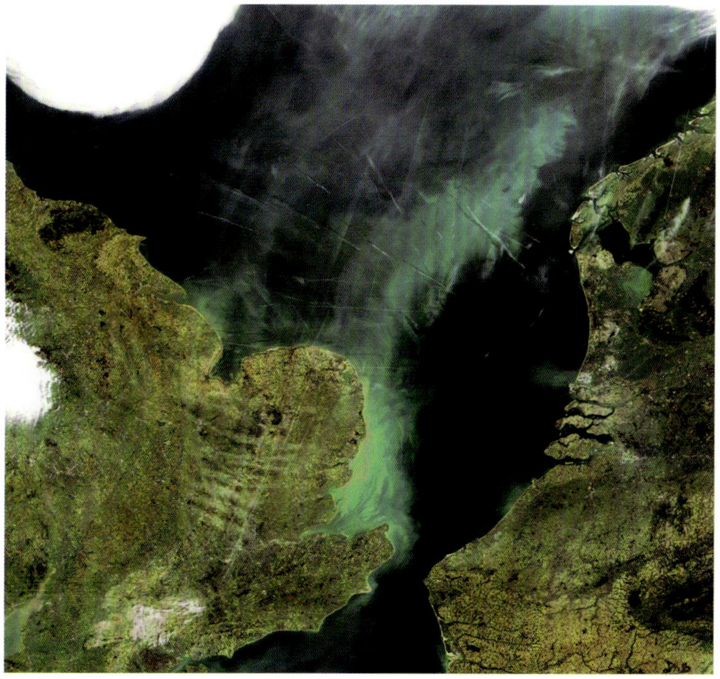

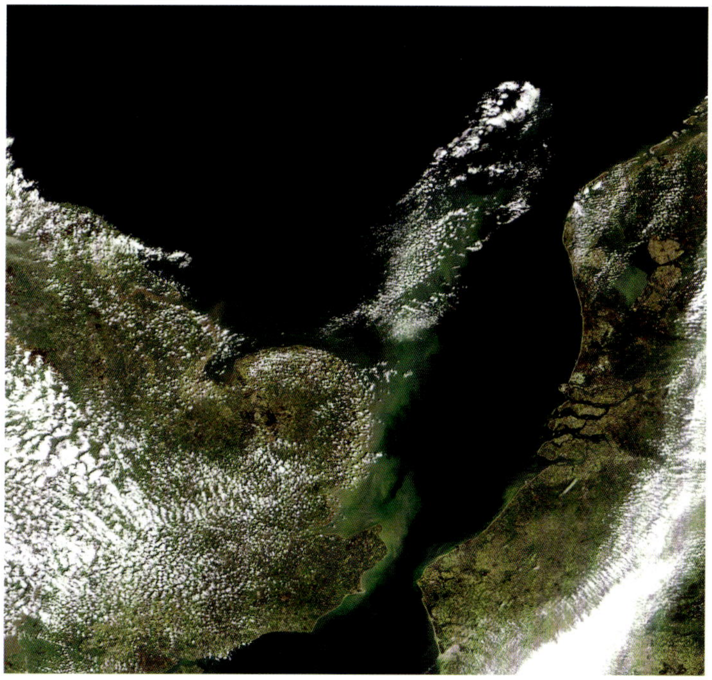

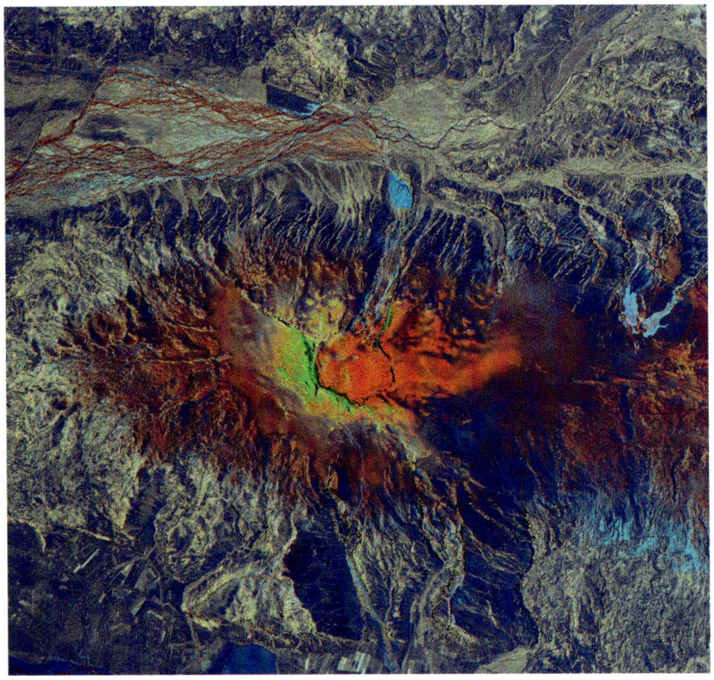

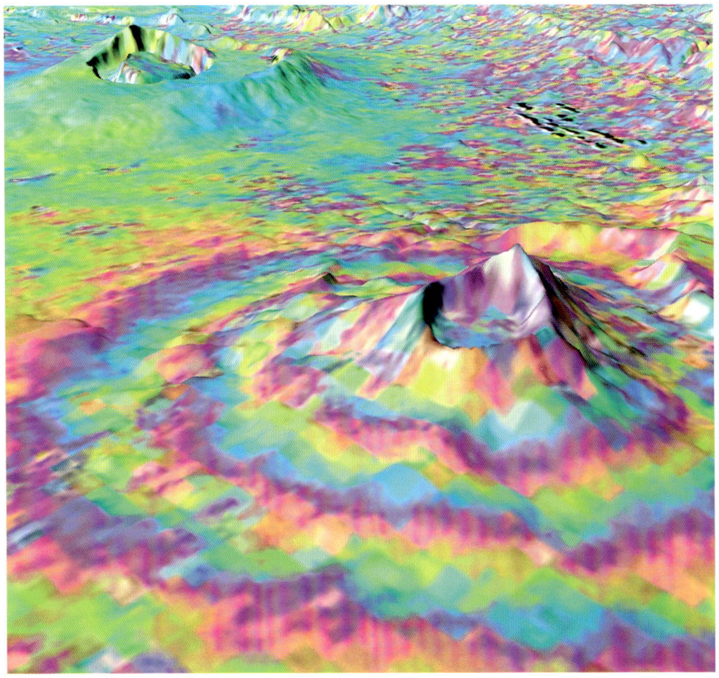

Logenplatz im Orbit

April 2010, Golf von Mexiko. Eine Bohrinsel explodiert und geht unter. Aus einer offenen Ölquelle am Meeresgrund sprudeln Unmengen des schwarzen Rohstoffs. Während die Betreiberfirma von einer kleinen Leckage schwadroniert und das Ausmaß des Unglücks herunterspielen will, zeigen die Satelliten aus dem All die traurige Wahrheit: Riesige Ölteppiche schwappen auf der Meeresoberfläche. Der vermeintlich nicht so dramatische Zwischenfall entpuppt sich als eine der größten Umweltkatastrophen.

Juli 2010, Pakistan. Der Himmel öffnet alle Schleusen. Wochenlang stürzt der Monsunregen auf die hilflosen Menschen herab. Es gibt kaum ein Entrinnen. Weite Teile des Landes werden überflutet. Von dieser Katastrophe, fernab beliebter Touristenzentren, nimmt zunächst kaum jemand Notiz. Reportagen vom Ort des Geschehens scheinen tragische Einzelfälle zu dokumentieren.

Hilfskräfte und Journalisten können oder dürfen viele betroffene Regionen nicht bereisen. Doch dann dokumentieren die himmlischen Augen das ganze Ausmaß der Katastrophe: Wo sich früher der mächtige Indus aus dem Karakorum-Gebirge seinen Weg zum Ozean bahnte, ist jetzt eine

A ringside seat in orbit

April 2010, the Gulf of Mexico. A drilling rig explodes and sinks, and an open oil well on the ocean floor begins to spill out vast amounts of black crude oil. Although the operating company equivocates about a small leak and tries to play down the full extent of the disaster, satellites in space record the sad truth: giant oil slicks have spread across the surface of the sea and the supposedly not so dramatic little accident reveals itself to be one of the worst environmental disasters ever to take place.

July 2010, Pakistan. The heavens open, allowing monsoon rains to fall onto the helpless people below for weeks on end. It is almost impossible to escape, and large parts of the country are flooded. Taking place far from popular tourist areas, the disaster is almost ignored, and reports from the locality seem to reflect only scattered individual tragedies.

Aid workers and journalists are either prohibited or unable to visit the regions affected, but the eye in the sky records the full extent of the disaster: a giant lake has appeared where the mighty Indus once forged its path from Karakoram

Aux premières loges

Avril 2010, Golfe du Mexique. Une plate-forme de forage explose et sombre dans les flots. D'énormes quantités de pétrole jaillissent du gouffre béant au fond de la mer. Tandis que la compagnie pétrolière prétend qu'il s'agit d'une fuite mineure et tente de minimiser l'ampleur du désastre, les satellites révèlent la triste réalité : une gigantesque marée noire se répand à la surface de la mer. Le prétendu incident marginal apparaît alors comme l'une des plus grandes catastrophes écologiques de l'histoire.

Juillet 2010, Pakistan. Le ciel ouvre ses vannes. Les pluies de mousson s'abattent sur la population impuissante. Impossible de fuir. La majeure partie du pays est sous les eaux. À l'écart des sites touristiques prisés, cette catastrophe passe tout d'abord inaperçue. Les reportages réalisés sur place semblent relater quelques destins tragiques isolés.

De nombreuses régions touchées sont inaccessibles aux équipes de secours et aux journalistes. C'est alors que les yeux du ciel dévoilent l'ampleur de la catastrophe : le lit de l'Indus, puissant fleuve qui serpente entre les montagnes du Karakoram

Atalaya orbital

Abril de 2010, golfo de México. Una plataforma petrolífera explota y se va a pique. De un pozo de petróleo abierto en el fondo del mar, surgen a borbotones enormes cantidades de crudo negro. Mientras la empresa explotadora habla de una pequeña fuga, pretendiendo minimizar la magnitud de la catástrofe, los satélites muestran desde el espacio la espantosa realidad. Enormes manchas de petróleo derramado en la superficie del mar. Aquel incidente, presuntamente no tan dramático, se revela como uno de los mayores desastres ecológicos de la zona.

Julio de 2010, Pakistán. El cielo abre sus compuertas. Desde hace semanas, las lluvias del monzón se precipitan sobre la desamparada población. Apenas hay escapatoria. Se inundan amplias zonas del país. Al principio, casi nadie se hace eco de esta catástrofe, tan lejana de los centros turísticos populares. Los reportajes en el lugar de los acontecimientos parecen documentar trágicos casos aislados. Los equipos de socorro y los periodistas no pueden viajar por muchas regiones afectadas o, simplemente, no tienen permiso. Pero los ojos

Linke Seite, links: Per Radar blicken die Forscher während des Ausbruchs in den Krater des Eyjafjallajökull (TerraSAR-X)

Far left: Scientists used radar to monitor Eyjafjallajökull's crater during the eruption (TerraSAR-X)

Page de gauche, à gauche : à l'aide de radars, les scientifiques inspectent le cratère de l'Eyjafjallajökull pendant l'éruption (TerraSAR-X)

Página izda., izda.: Durante la erupción, los investigadores observaban el cráter del Eyjafjallajökull por radar (TerraSAR-X)

Linke Seite, rechts: Der Vulkan Longonot in Kenia hat sich von 2004 bis 2006 um 9 cm aufgebläht; Radarmessung (Envisat)

Near left: Radar measurements revealed that Mount Longonot, a volcano in Kenya, expanded by 3.5 inches (9 cm) between 2004 and 2006 (Envisat)

Page de gauche, à droite : le volcan Longonot, au Kenya, a connu un gonflement de 9 cm entre 2004 et 2006. Mesure radar (Envisat)

Página izda., dcha.: Entre 2004 y 2006, el volcán Longonot de Kenia ha crecido unos 9 cm; medición de radar (Envisat)

Unten: Es muss nicht immer Island sein: Aschewolke des Ätna auf Sizilien im November 2006 (Aqua)

Below: It's not always Iceland—a cloud of ash from Mount Etna (Sicily) in November 2006 (Aqua)

Ci-dessous : ce n'est pas toujours la faute de l'Islande : nuage de cendres libéré par l'Etna en Sicile en novembre 2006 (Aqua)

Abajo: No siempre tiene que ser Islandia: cenizas del Etna sobre Sicilia en noviembre de 2006 (Aqua)

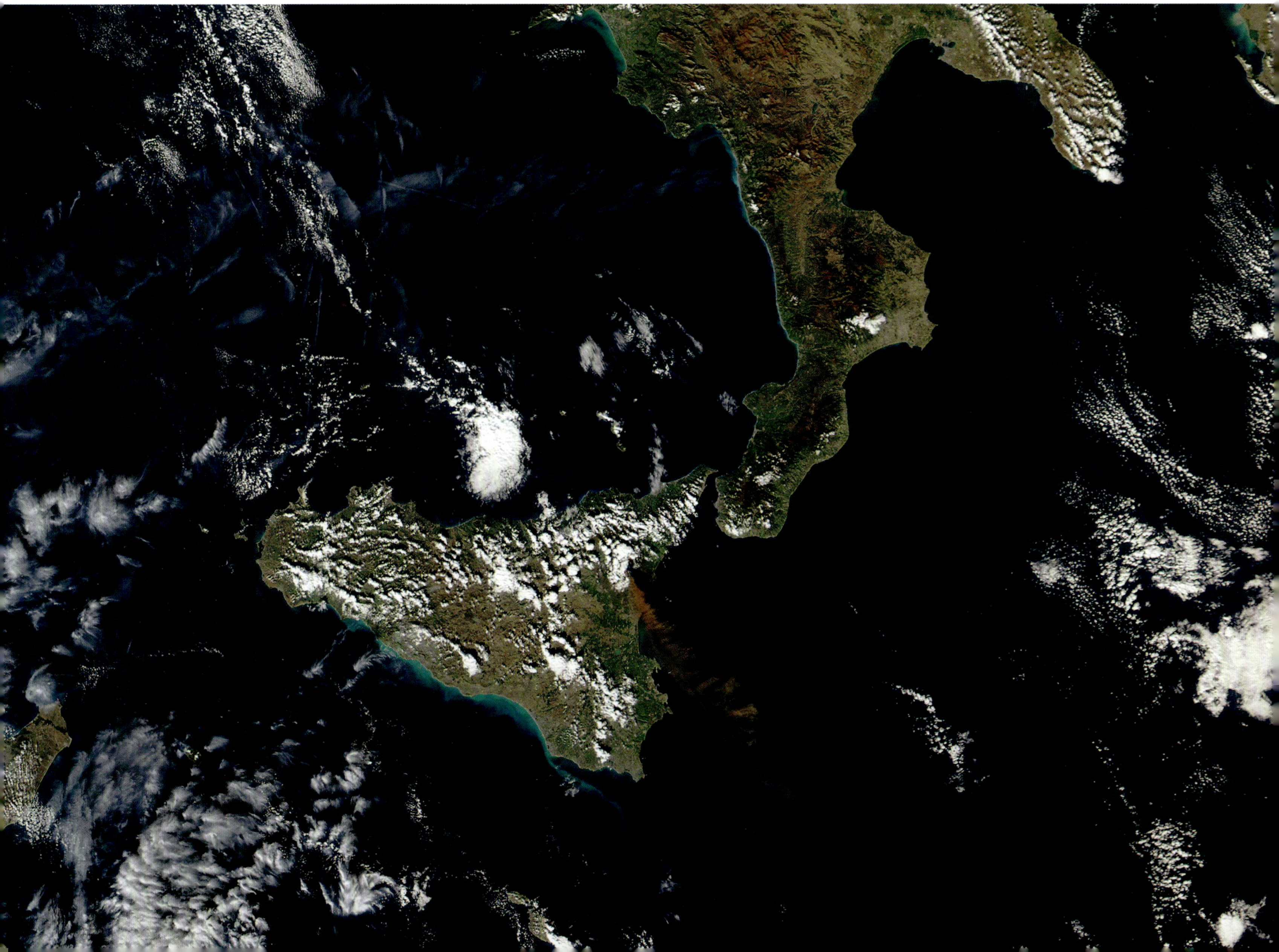

riesige Seenlandschaft entstanden. Mehrmals täglich ziehen Erdbeobachtungssatelliten über Pakistan hinweg und machen Bilder im sichtbaren Licht, im infraroten und per Radar. Aus dem Fluss ist eine Wasserwüste so groß wie Italien geworden. Die Beobachter in etwa 800 Kilometern Höhe können den Menschen am Boden keine Unterstützung zukommen lassen – aber sie sorgen dafür, dass alle das Ausmaß der Katastrophe erkennen können.

Sommer 2010, Nordpolarmeer. Der Eispanzer im kalten Wasser jenseits von Grönland und Spitzbergen wird dünn. Riesige Eisschollen brechen, driften ab und geben große Bereiche des polaren Ozeans frei.

Einst waren die Forscher schon froh, überhaupt sehen zu können, wo noch Treibeis das Meer bedeckt.

Jetzt kreuzt mehr als ein Dutzend Mal am Tag ein Satellit über diesem Gebiet, der per Radar genau vermisst, wie weit die Eisschollen aus dem Meer herausragen – und zwar bis auf zwei Zentimeter genau. Damit lässt sich exakt bestimmen, wie dick das Eis wirklich ist. Eisberge befinden sich bekanntlich größtenteils unter Wasser. Nur sehr wenig schwimmt über der Meeresoberfläche, so wie ein Eiswürfel im Fruchtsaft kaum zu sehen ist.

to the ocean. Earth observation satellites make several passes over Pakistan every day, creating images using visible light, infrared, and radar. The river has become a watery wilderness the size of Italy. These observers are unable to help the people on the ground from their position 500 miles (800 kilometers) above, but they can reveal the full magnitude of the disaster.

Summer 2010, the Arctic Ocean. The ice sheet beyond Greenland, Spitsbergen, Russia, Alaska, and Canada is getting thin. Giant ice floes break off, drift away, and expose wide expanses of the sea around the pole.

Scientists once thought themselves lucky even to glimpse drift ice covering the sea.

Now a satellite crosses the area more than a dozen times a day, using radar to measure precisely—to an accuracy of less than an inch (2.5 centimeters)—how far the icebergs stick up out of the sea. From this, the real thickness of the ice can be calculated—as is well known, most of an iceberg lies underwater with very little visible above the surface, just as an ice cube is almost completely submerged in a glass of juice.

avant de se jeter dans l'océan, n'est plus qu'une vaste étendue d'eau. Les satellites d'observation de la Terre survolent plusieurs fois par jour le Pakistan qu'ils photographient en lumière visible, en infrarouge et par radar. Le fleuve s'est transformé en un désert aquatique aussi grand que l'Italie. Du haut de leur huit cents kilomètres d'altitude, nos témoins oculaires sont incapables d'aider les populations au sol, mais ils permettent au reste du monde de mesurer l'ampleur du désastre.

Été 2010, océan glacial Arctique. La banquise qui s'étend au large du Groenland et du Spitzberg rétrécit. D'énormes blocs de glace se détachent, libérant de vastes zones dans l'océan Arctique.

Les chercheurs s'estimaient jadis heureux s'ils pouvaient repérer l'emplacement des glaces flottantes.

Aujourd'hui, un satellite parcourt cette région une douzaine de fois par jour. Il mesure à deux centimètres près la partie émergente des blocs de glace à l'aide d'un radar, ce qui permet de déterminer leur épaisseur avec exactitude. Comme chacun sait, la partie immergée des icebergs est bien plus importante que la partie visible. Ils surnagent à peine, comme un glaçon dans un jus de fruit.

celestes ilustran la dimensión completa de la catástrofe: el cauce del río Indo se ha convertido en una gigantesca extensión de agua del tamaño de Italia. Los observadores que están a 800 km de altura no pueden hacer llegar la ayuda a la gente, pero se ocupan de que todos reconozcan la magnitud de la catástrofe.

Verano de 2010, océano Ártico. El casquete de hielo, en las frías aguas de más allá de Groenlandia y Spitzbergen, cada vez es más fino. Se rompen enormes témpanos de hielo que navegan a la deriva por el océano polar. Antiguamente, los investigadores ya se daban por satisfechos si podían conocer hasta dónde permanecía cubierta la superficie del mar por los hielos flotantes. Ahora, un satélite cruza por encima de esta región más de una docena de veces al día y mide exactamente por radar cuánto sobresalen los témpanos del mar, con una precisión de hasta dos centímetros. Así se puede determinar exactamente el grosor real del hielo. Como es sabido, la mayor parte de la masa de los icebergs se esconde bajo el agua. Tan solo sobresale un poco por encima de la superficie del mar, lo mismo que un cubito de hielo apenas sobresale del refresco.

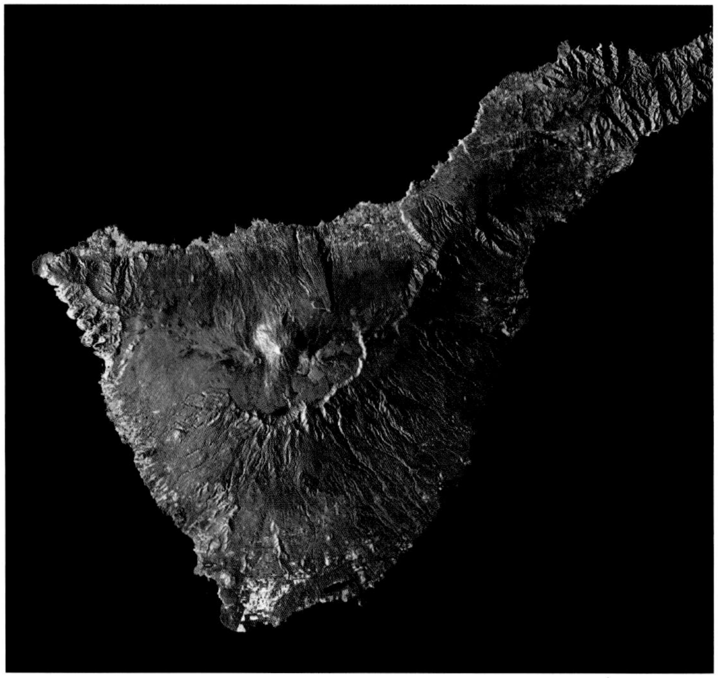

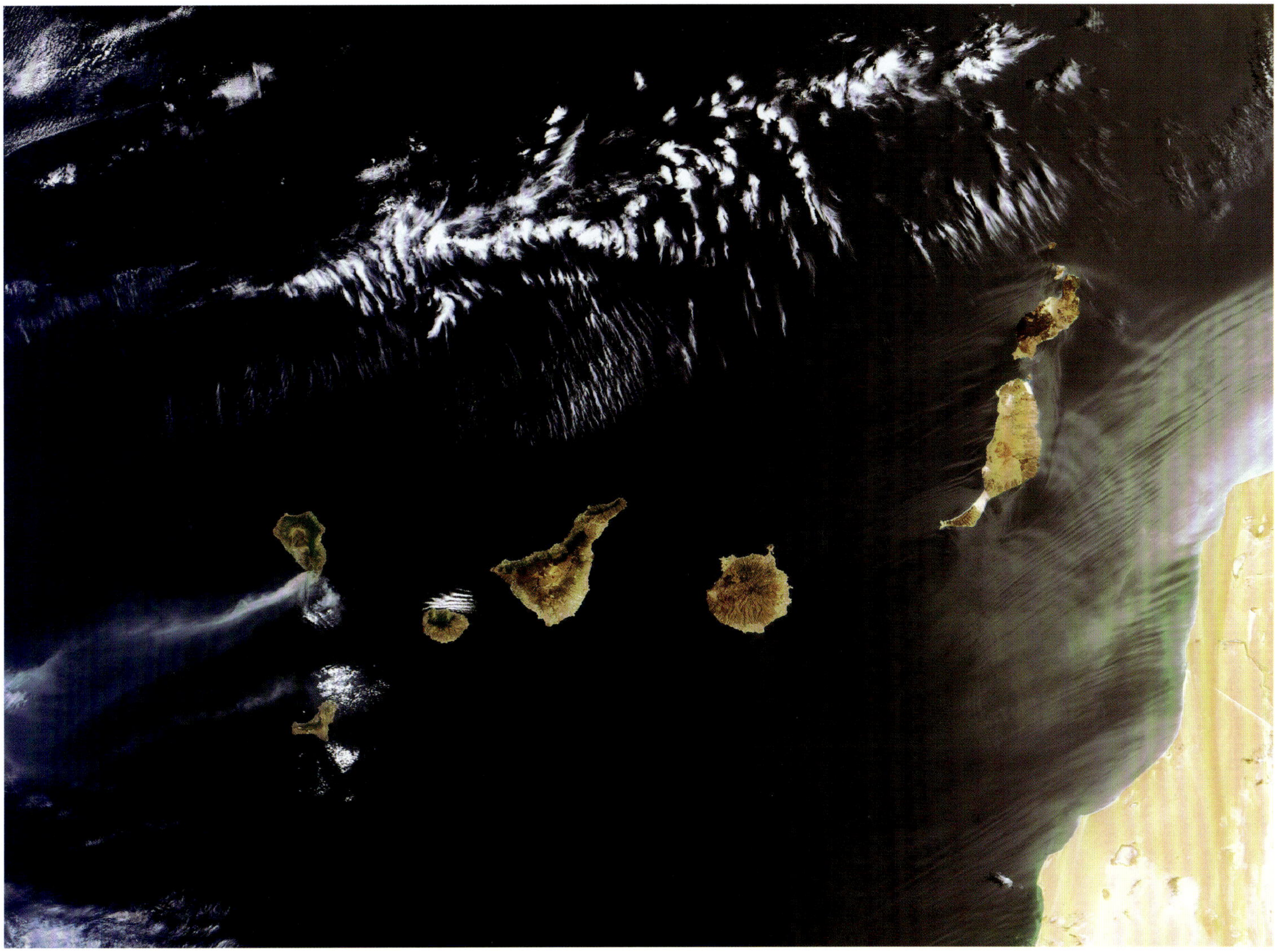

Linke Seite: Die Radaraufnahme der Kanareninsel Teneriffa zeigt eine interessante Geologie (TerraSAR-X)

Left: A radar image of Tenerife in the Canary Islands revealing some interesting geology (TerraSAR-X)

Page de gauche : l'image radar de Tenerife, dans l'archipel des Canaries, révèle la structure géologique de l'île (TerraSAR-X)

Página izda.: La toma de radar de la isla canaria de Tenerife muestra una interesante geología (TerraSAR-X)

Oben: Nicht der Vulkan, der Wald brennt: Rauchfahne auf der Kanareninsel La Palma (Envisat)

Above: This plume of smoke rising from La Palma in the Canary Islands is a forest fire, not a volcano (Envisat)

Ci-dessus : ce n'est pas un volcan, mais la forêt qui brûle. Nuage de fumée sur l'île de Palma, Canaries (Envisat)

Arriba: No arde el volcán, es el bosque: columna de humo en la isla canaria de La Palma (Envisat)

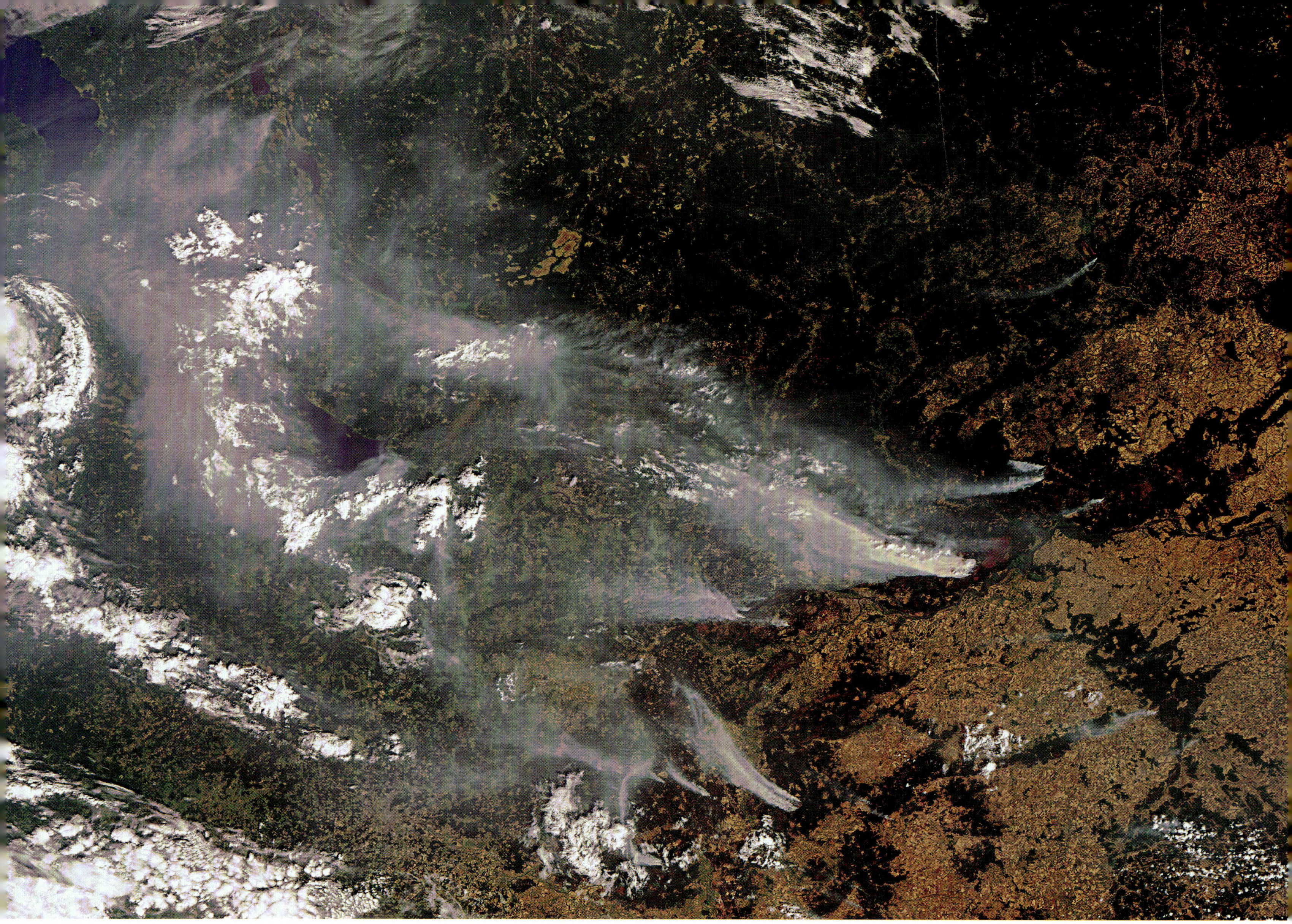

Im Sommer 2010 kam es in Russland zu lang anhaltenden Waldbränden (Envisat)

In Russia, devastating forest fires burned throughout summer 2010 (Envisat)

Des feux de forêt persistants se sont déclarés en Russie au cours de l'été 2010 (Envisat)

Incendios forestales continuos en el verano de 2010 en Rusia (Envisat)

Rechte Seite, links: 2007 brannten in Griechenland wochenlang die Wälder (Aqua)

Near right: It was the turn of Greece to suffer forest fires in 2007 (Aqua)

Page de droite, à gauche : en 2007, les forêts ont brûlé en Grèce durant plusieurs semaines (Aqua)

Página dcha., izda.: En 2007, los bosques ardieron durante semanas en Grecia (Aqua)

Rechte Seite, rechts: Spätere Aufnahme der verbrannten Gebiete (Envisat)

Far right: A later photograph of the burnt areas (Envisat)

Page de droite, à droite : image ultérieure des régions dévastées par le feu (Envisat)

Página dcha., dcha.: Fotografía posterior de las zonas quemadas (Envisat)

Etwas ist immer zu sehen

Dichte Rauchschwaden wabern über die Ebenen südlich von Moskau. Wald- und Buschbrände im extrem heißen Sommer 2010 gefährden Mensch und Natur und machen das Leben in der Metropole zur Qual. Optische Bilder aus 800 Kilometern Höhe zeigen riesige Rauchfahnen, die der Wind Hunderte Kilometer weit über das Land weht. Infrarotkameras erfassen die Wärmestrahlung der Brandherde. Andere Instrumente registrieren präzise, wie weit sich die Schadstoffe in der Atmosphäre verteilen. Selbst in Bereichen, die dem bloßen Auge als rauchfrei erscheinen, ist die Luft voller giftiger Partikel – gemessen nicht vor Ort am Boden oder in der Luft, sondern im Überflug aus dem All.

Die Sonne brennt vom strahlend blauen Himmel. Kein Wölkchen trübt die Sommerfreude. Doch der Sonnenbrand kommt schneller als noch vor einigen Jahren – Schuld ist der abnehmende Schutzschild der Atmosphäre gegen die gefährliche Ultraviolettstrahlung der Sonne.

Das natürliche Sonnenschutzmittel ist das Gas Ozon. Es leuchtet nicht, und es bildet keine Wolken. Kein Mensch hat es je mit bloßem Auge

There's always something to observe

Thick coils of smoke wreath the plains to the south of Moscow. Forest and bush fires in the extremely hot summer of 2010 endanger the lives of humans and animals, and make life in the metropolis a torment. Images captured from a height of 500 miles (800 kilometers) show giant plumes of smoke being blown for hundreds of miles (hundreds of kilometers) across the country and infrared cameras reveal how much heat the fires are radiating. Other instruments make a precise record of how far the pollutants are spreading in the atmosphere. Even in those areas which seem smoke-free to the naked eye, the air is full of toxic particles, and all this is measured not on the ground or in the air, but during a flypast in space.

The Sun is beating down from a brilliant blue sky and there is not a cloud to spoil the joys of summer, but sunburn arrives a little more quickly than a few years ago: a result of the atmosphere's diminishing protection against dangerous ultraviolet rays.

The gas ozone is a natural defense against the effects of the Sun; it casts no shadow, forms no clouds,

Il y a toujours quelque chose à voir

D'épais nuages de fumée s'élèvent sur la plaine au sud de Moscou. Dans l'extrême chaleur de l'été 2010, les feux de tourbières affectent la nature et la population, faisant de la capitale russe un enfer irrespirable. Les images satellite montrent des traînées de fumées emportées par le vent à des centaines de kilomètres à la ronde. Les caméras infrarouge mesurent le rayonnement thermique émis par les incendies, tandis que d'autres instruments examinent la propagation des gaz toxiques dans l'atmosphère, démontrant que même au-dessus des zones qui semblent épargnées par la fumée, l'air est saturé de particules toxiques.

Le soleil brille dans un ciel sans nuage. Rien ne semble contrarier la gaieté estivale. Les coups de soleil apparaissent pourtant plus rapidement depuis quelques années, en raison de l'affaiblissement du bouclier qui, dans l'atmosphère, nous protège du rayonnement ultraviolet.

La protection solaire naturelle, c'est l'ozone. Nul n'a jamais pu observer ce gaz invisible à l'œil nu, mais sa présence dans l'atmosphère est indispensable à la vie sur Terre.

Siempre hay algo a la vista

Sobre las llanuras del sur de Moscú se elevan densas nubes de humo. Los incendios forestales del extremadamente cálido verano de 2010 ponen en peligro a hombres y naturaleza y convierten en un tormento la vida en la metrópoli. Las imágenes ópticas tomadas a 800 km de altura muestran gigantescas columnas de humo que el viento arrastra a cientos de kilómetros de distancia sobre el país. Las cámaras infrarrojas captan las emisiones térmicas de los focos del incendio. Otros instrumentos registran con precisión el alcance de la dispersión de las sustancias tóxicas en la atmósfera. Incluso en aquellas áreas en las que a simple vista parece que no hay humo, el aire está lleno de partículas tóxicas; partículas que no se miden in situ, en el suelo o en el aire, sino desde el espacio.

El sol resplandece en la inmensidad del cielo azul. Pero las quemaduras solares se presentan mucho antes que en años anteriores. La culpable es la disminución de la pantalla protectora de la atmósfera contra la peligrosa radiación ultravioleta del Sol. El elemento protector natural es el gas ozono. No brilla ni forma nubes. Ninguna persona lo ha podido ver. Pero su presen-

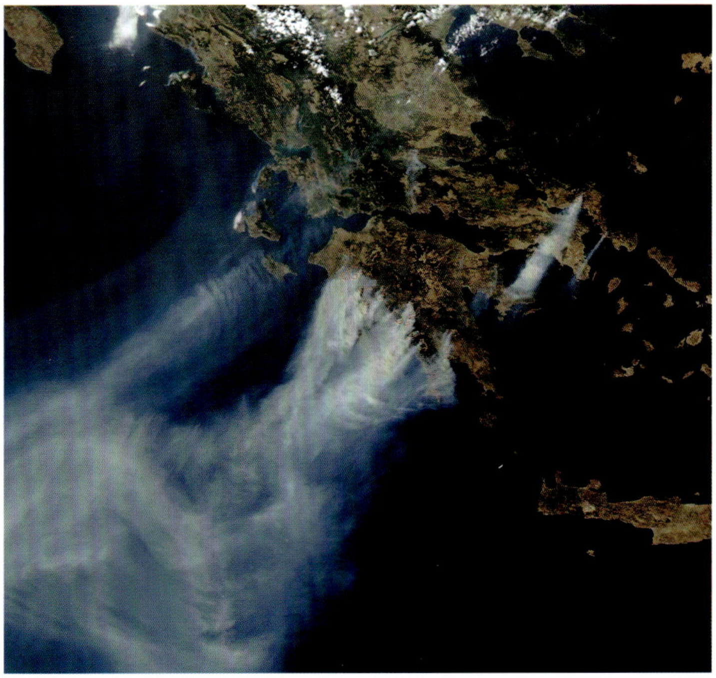

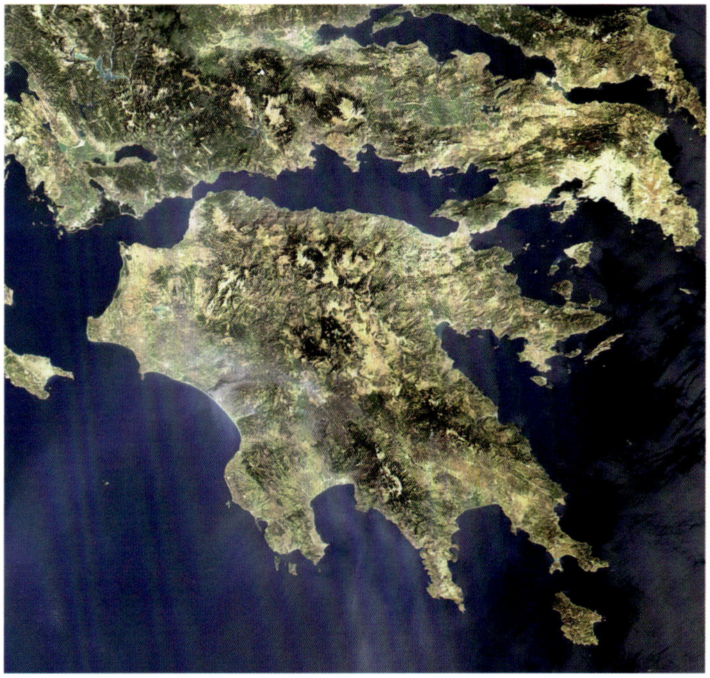

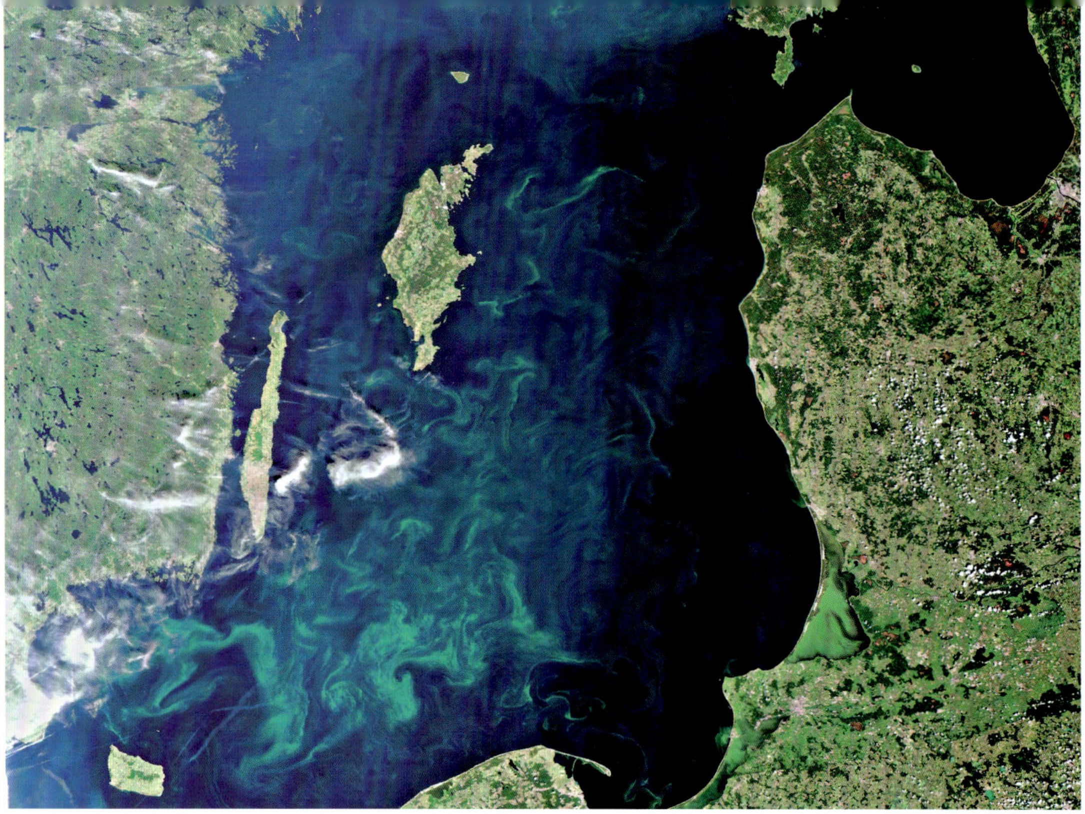

gesehen. Aber seine Beimischung in der Atmosphäre ist für das Leben auf der Erdoberfläche existenziell wichtig. Denn die Ultraviolettstrahlung schädigt Zellen und verändert das Erbgut. Mit speziellen Instrumenten lässt sich der Ozongehalt in der Atmosphäre messen. Das geht auch vom Erdboden oder Ballon aus. Den großen Überblick liefern aber nur Satellitenmessungen.

Der NASA-Satellit Nimbus-7, gestartet im Herbst 1978, spielte eine große Rolle bei der Entdeckung des Ozonlochs über der Antarktis. Mitte der achtziger Jahre sorgten Artikel in Fachpublikationen über einen dramatischen Rückgang des Ozongehalts der oberen Atmosphäre für großes Aufsehen. Zum ersten Mal bekamen wissenschaftliche Daten eines Erdbeobachtungssatelliten große öffentliche Aufmerksamkeit. Zeitweise waren nicht weniger als zehn Satelliten im Einsatz, um den Ozongehalt in der Atmosphäre zu überwachen. Das Ozonloch ist ein Musterbeispiel dafür, was Erdbeobachtung zu leisten imstande ist – und dass die dabei gewonnenen Erkenntnisse politische Folgen haben können: Zwanzig Jahre lang hat man das Ozonloch beobachtet, und nun sieht es so aus, als ob der Ozongehalt in der Atmosphäre wieder zunimmt und der Schutz vor der Ultraviolett-

and is invisible to the naked eye, yet its presence in the atmosphere is of crucial importance for the existence of life on Earth—ultraviolet rays damage cells and influence genetics. The ozone content of the atmosphere can be measured with special instruments, and, even though this is possible at surface level or from a balloon, only satellite readings can give the entire perspective.

The NASA satellite Nimbus 7, launched in the fall of 1978, played a significant role in the discovery of the hole in the ozone layer above the Antarctic, and specialist publications reporting a dramatic decline in the ozone levels in the upper atmosphere caused a great stir in the mid-1980s. It was the first time that scientific data from an Earth observation satellite had attracted considerable public attention and there were, at times, no fewer than ten satellites in use to monitor ozone levels in the atmosphere. The hole in the ozone layer is a classic example of what Earth observation can achieve, and also of what political consequences may ensue: the hole has been observed for 20 years, and, as a result of concerted action by Earth's inhabitants it now seems that atmospheric ozone levels are rising again and our defenses against

Sans cette pellicule protectrice, les rayons ultraviolets détruisent les cellules et modifient les gènes. La concentration de l'ozone dans l'atmosphère est mesurée à l'aide d'instruments spécifiques qui peuvent être actionnés au sol ou dans une montgolfière. Mais seuls les satellites fournissent une vision globale de sa répartition.

Le satellite Nimbus-7 de la NASA, lancé à l'automne 1978, a joué un rôle décisif dans la découverte du trou d'ozone au-dessus de l'Antarctique. Au milieu des années 1980, les publications spécialisées ont lancé un pavé dans la mare en révélant la diminution alarmante de la couche d'ozone stratosphérique. C'était la première fois que l'opinion publique s'intéressait aux données scientifiques fournies par un satellite d'observation de la Terre. Pas moins de dix satellites veillaient alors à tour de rôle sur la couche d'ozone. Le trou d'ozone est la parfaite illustration de la contribution apportée par l'observation de la Terre, et de son impact politique : vingt années ont été consacrées à l'étude de la couche d'ozone, et il semble aujourd'hui que la concentration de l'ozone dans l'atmosphère augmente à nouveau, renforçant progressivement la protection de la Terre contre

cia en la atmósfera es extremadamente importante para la vida en la superficie de la Tierra porque, sin su protección, la radiación ultravioleta dañaría las células y modificaría el patrimonio genético. La concentración de ozono en la atmósfera se puede medir con instrumentos especiales, tanto desde el suelo terrestre como desde globos aerostáticos. Pero la visión más amplia la proporcionan las mediciones de los satélites.

El satélite de la NASA Nimbus 7, lanzado en otoño de 1978, desempeñó un importante papel en el descubrimiento del agujero de ozono sobre la Antártida. A mediados de la década de 1980, las publicaciones especializadas procuraban llamar la atención sobre un dramático retroceso de la concentración de ozono en la atmósfera superior. Por primera vez, datos científicos de una observación terrestre mediante satélites conseguían gran repercusión pública. En ocasiones, se llegaron a utilizar no menos de diez satélites para controlar la concentración de ozono en la atmósfera. El agujero de ozono es un ejemplo típico del rendimiento que puede prestar la observación terrestre y de que los conocimientos así adquiridos pueden tener consecuencias políticas: se ha estado observando el agujero de ozono durante veinte años y ahora parece que la concentra-

Kein Plankton, sondern Eis – große Teile des Bottnischen und Finnischen Meerbusens waren Mitte März 2010 zugefroren (Envisat)

Ice, not plankton—large parts of the Gulf of Bothnia and Finnish coastal waters froze over in mid-March 2010 (Envisat)

Non pas du plancton mais de la glace. Les eaux des golfes de Botnie et de Finlande étaient en grande partie gelées à la mi-mars 2010 (Envisat)

No es plancton, sino hielo. Gran parte de los golfos de Botnia y Finlandia estaban congelados a mediados de marzo de 2010 (Envisat)

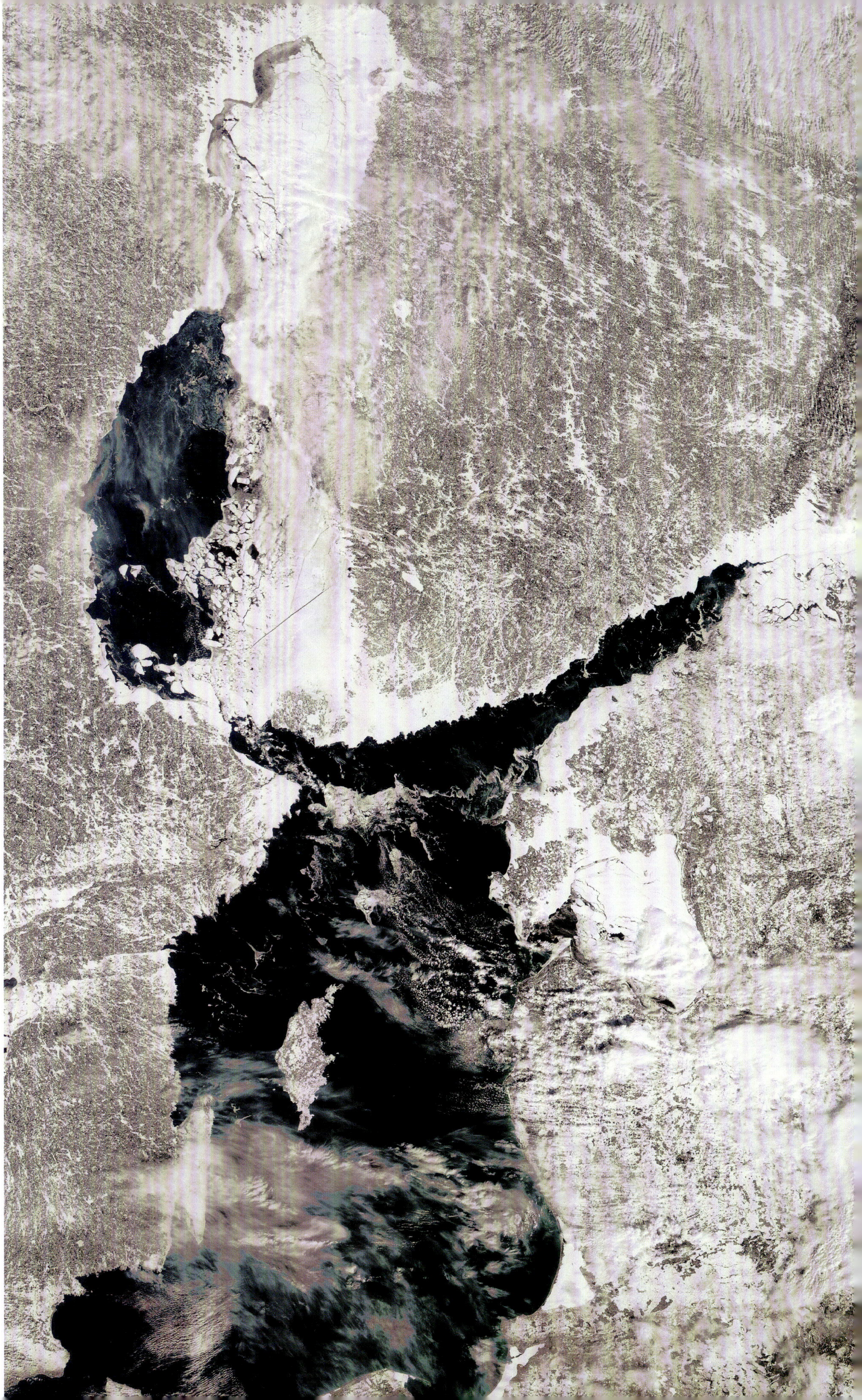

Linke Seite: Wie ein Gemälde der Natur – Planktonblüte in der zentralen Ostsee im Sommer 2010 (Envisat)

Left: Nature's paintbox—plankton growth in the middle of the Baltic Sea in summer 2010 (Envisat)

Page de gauche : telle une peinture naturelle, la floraison planctonique au centre de la Baltique à l'été 2010 (Envisat)

Página izda: Como un cuadro natural: floración de plancton en el mar Báltico central, en el verano de 2010 (Envisat)

Unten: Der Satellit im Blick: Ein Ingenieur arbeitet im Reinraum an CryoSat-2

Below: A satellite close-up: an engineer working on CryoSat-2 in the "clean room"

Ci-dessous : contrôle du satellite. Un ingénieur s'affairant sur CryoSat-2 dans la salle blanche

Abajo: Satélite en perspectiva: un ingeniero trabajando en la sala blanca del CryoSat-2

Rechts: Das Eis im Blick: CryoSat-2 im Einsatz über den Polargebieten (Montage)

Right: Keeping an eye on the ice; a montage of CryoSat-2 over the North Pole

À droite : contrôle de la glace. CryoSat-2 en mission au-dessus des régions polaires (photomontage)

Dcha.: El hielo en perspectiva: CryoSat-2 operando sobre las regiones polares (montaje)

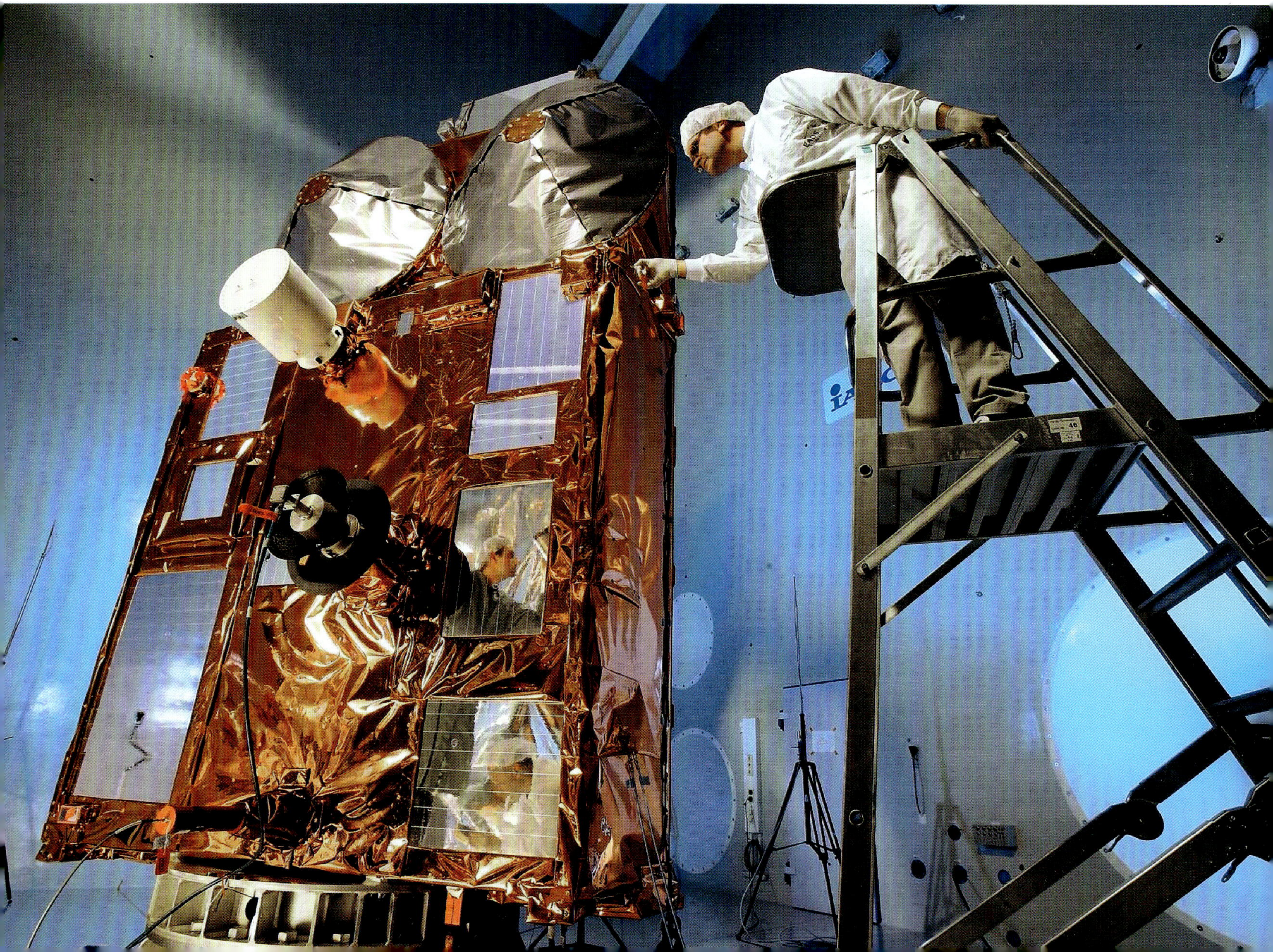

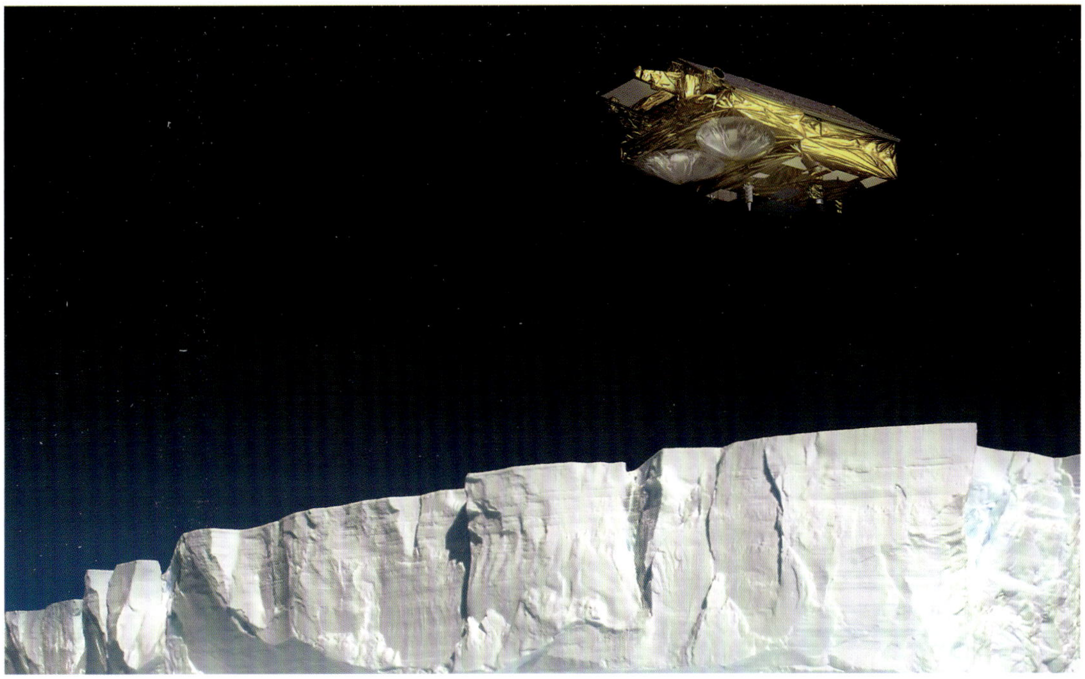

strahlung allmählich wieder besser wird. Solche Ergebnisse erfreuen die Forscher ganz besonders.

Viele Kleine statt eines Großen

Dies sind nur einige wenige Beispiele, welche Rolle die Daten spielen, die die Erdbeobachtungssatelliten zum Boden funken. Nicht immer führen solche Daten zu den erforderlichen politischen Konsequenzen. Aber dafür können die Instrumente nichts, die die Messungen machen. Die Satelliten beobachten die Erde aus großer Höhe. Doch beeinflussen lässt sich die Erde praktisch nur am Boden. Die Messinstrumente und Kameras an Bord der Satelliten sind wie Trainer, die am Spielfeldrand zusehen, was auf dem Platz passiert. Ihre Daten haben empfehlenden Charakter. Direkt ins Spielgeschehen eingreifen können Envisat, Landsat, TerraSAR-X und all die anderen Satelliten aber nicht. Spötter mögen ergänzen, dass die Satelliten offenbar ohne Unterlass das Spielfeld umkreisen, weil sie in Ruhe das Elend nicht mehr ertragen, das sie dort zu sehen bekommen. Doch Satelliten müssen stets in Bewegung sein. Blieben sie stehen, stürzten sie unweigerlich auf die Erde.

ultraviolet rays are slowly improving. Results like this are particularly pleasing for scientists.

Lots of little ones instead of one big one

These are just a few examples of the role played by the data transmitted from Earth observation satellites to the ground. Such information does not always result in the necessary political action, but that is not the fault of the instruments taking the readings. Satellites observe the Earth from a great height, but in practice, changes can only be made on the surface. The sensors and cameras on board the satellites are like coaches watching the action from the edge of the field, and although their data can influence tactics, Envisat, Landsat, TerraSAR-X, and all the other satellites cannot intervene directly in play. Some comedians might add that the satellites are clearly milling round the field as they cannot bear to sit still and watch the suffering there, but satellites have to stay in motion; were they to stop, they would fall straight to Earth.

The ranks of these coaches in the sky have been swelling for some years now, and the engineers and

le rayonnement ultraviolet. Des résultats assurément réjouissants pour les scientifiques.

Privilégier les petites missions

Il ne s'agit ici que de quelques exemples du rôle que jouent les données transmises par les satellites d'observation de la Terre. Qu'elles ne parviennent pas toujours à susciter les décisions politiques nécessaires, ne tient en aucun cas à la performance des instruments scientifiques. Les satellites observent la Terre à haute altitude, alors que seules les actions menées sur le terrain sont susceptibles d'influencer la planète. Les caméras et autres appareils embarqués sur les satellites sont comme l'entraîneur qui suit son équipe depuis le banc de touche. Envisat, Landsat, TerraSAR-X et leurs confrères sont des conseillers incapables d'intervenir dans le jeu. Les persifleurs ajouteront que les satellites ne cessent de s'agiter au bord du terrain, parce qu'ils ne supportent plus la misère qu'ils observent. En réalité, les satellites sont contraints au mouvement perpétuel, sans quoi ils s'écraseraient sur la Terre.

ción de este gas en la atmósfera vuelve a crecer y, con él, la protección contra la radiación ultravioleta va mejorando poco a poco. Estos resultados animan a los investigadores.

Muchos pequeños, en vez de uno grande

Estos son sólo algunos ejemplos de la repercusión que tienen los datos que se transmiten desde los satélites de observación terrestre. Estos datos no siempre tienen las respuestas políticas necesarias. Pero eso es algo que los instrumentos que hacen las mediciones no pueden controlar. Los satélites observan la Tierra a gran altura. Pero la influencia real sólo se puede ejercer desde tierra firme. Los instrumentos de medición y las cámaras a bordo de los satélites son como los entrenadores que observan desde el borde del terreno de juego lo que ocurre en la cancha. Sus datos tienen carácter de recomendación. Los Envisat, Landsat, TerraSAR-X y todos los demás satélites no pueden intervenir directamente en el juego. A modo de broma, se podría añadir que los satélites giran sin cesar alrededor del terreno de juego porque no pueden soportar con tranquilidad las desgracias que ven allí. En los últimos años, el equipo de

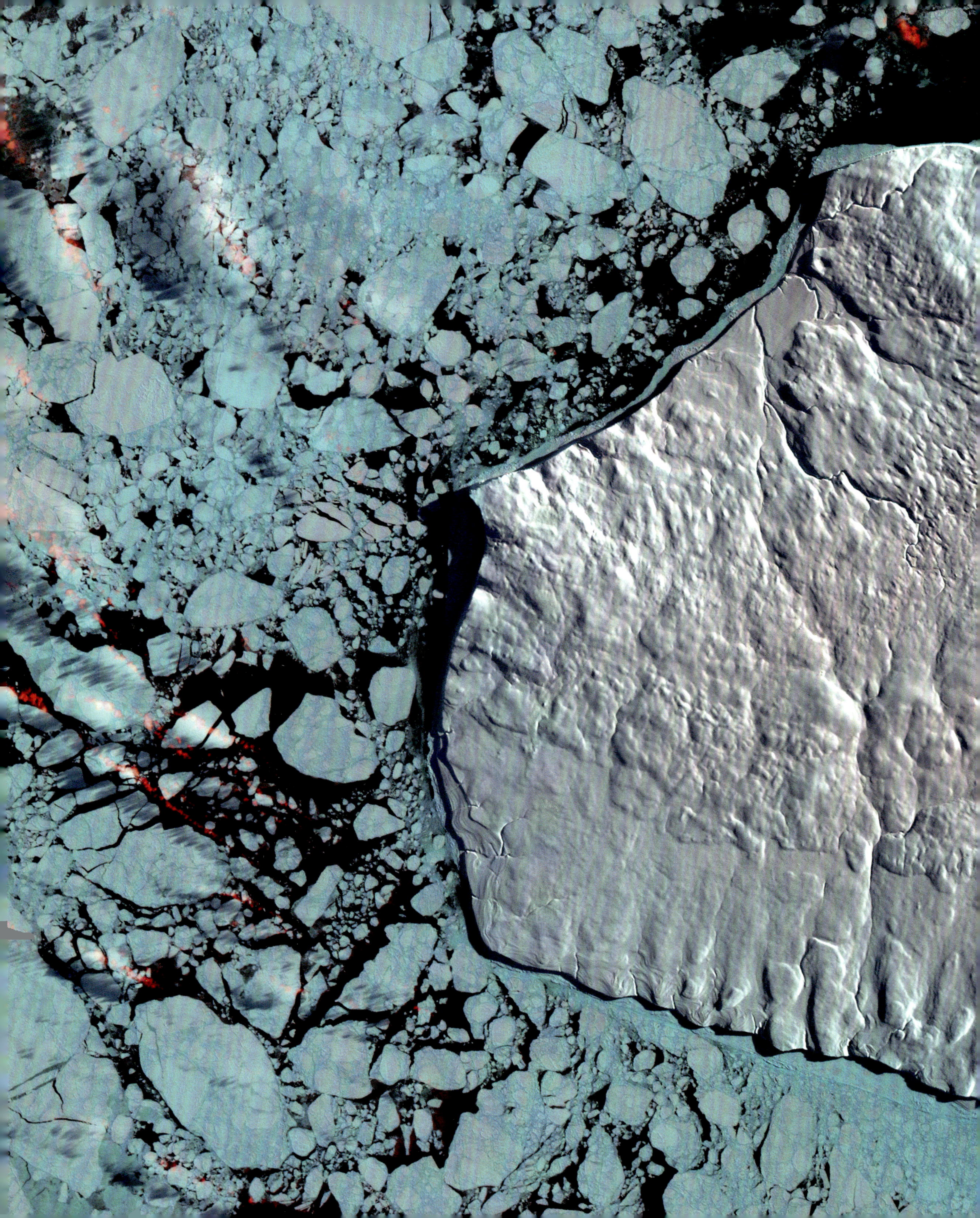

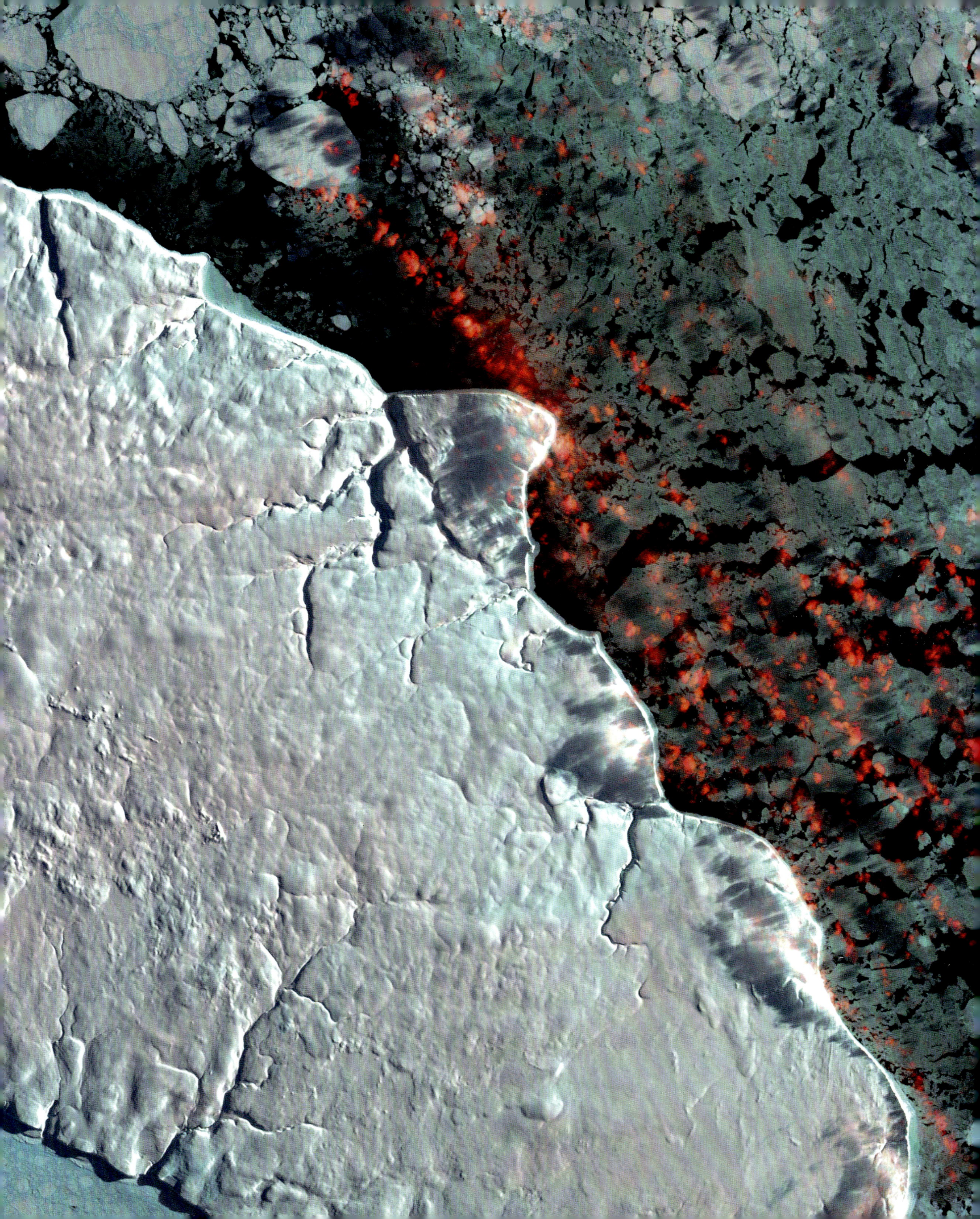

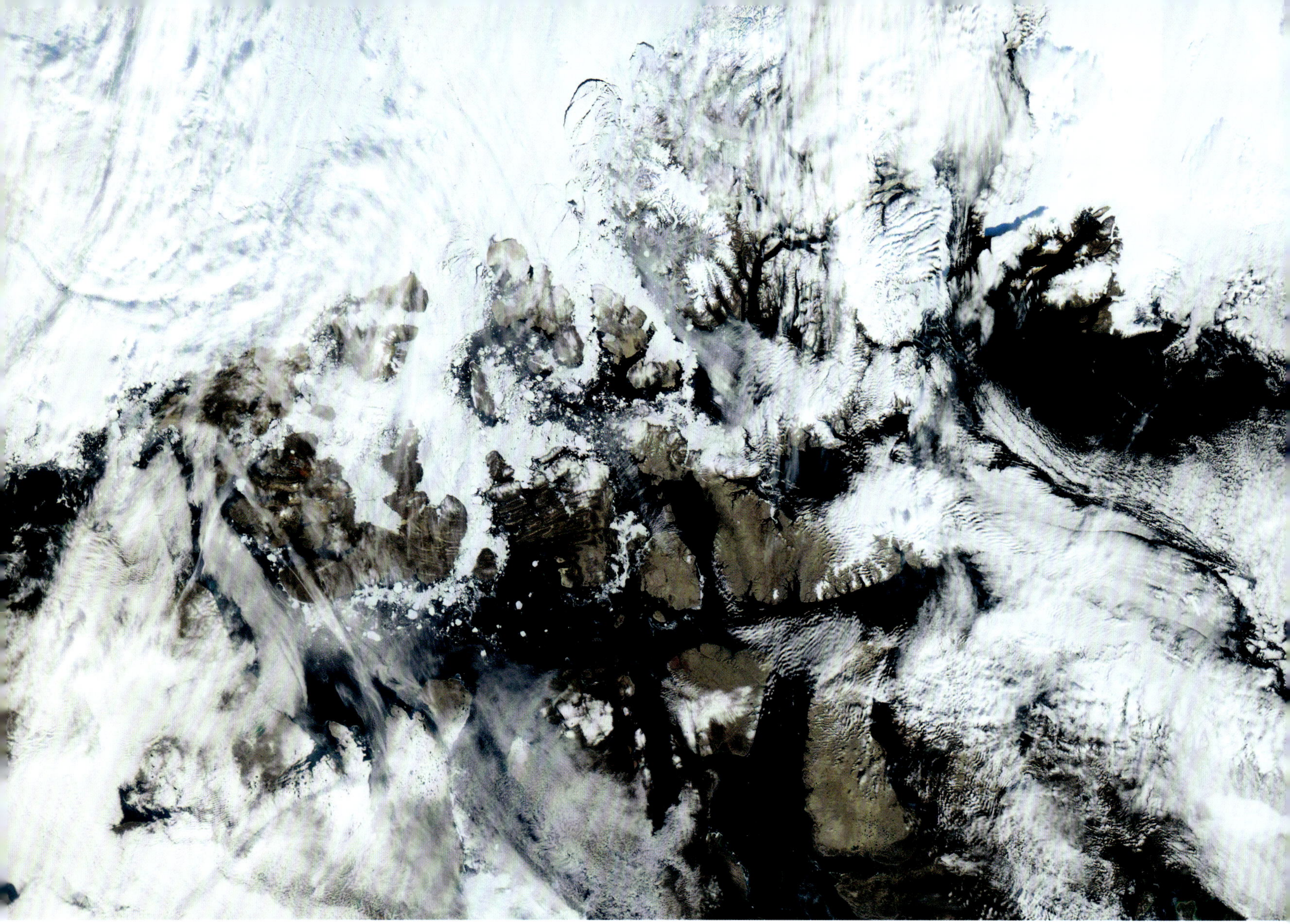

Seite 74/75: Die kanadische Insel Akpatok ist von Eisschollen umgeben (Landsat)

Pages 74–5: Canada's Akpatok Island is surrounded by pack ice (Landsat)

Pages 74 et 75 : l'île canadienne d'Akpatok est entourée de glaces flottantes (Landsat)

Páginas 74/75: Isla canadiense de Akpatok rodeada por el hielo (Landsat)

Oben: Ein Teil der legendären Nordwestpassage ist frei von Eis (Aqua)

Above: The legendary Northwest Passage is only partly free of ice (Aqua)

Ci-dessus : une section de la fameuse route maritime du Nord est libre de glace (Aqua)

Arriba: Una parte del legendario paso del Noroeste está libre de hielo (Aqua)

Rechte Seite: Teile Spitzbergens in einem Gewirr aus Eis und Wolken (Envisat)

Right: An area of Spitsbergen swathed in ice and clouds (Envisat)

Page de droite : territoires de l'île de Spitzberg dans un lacis de glace et de nuages (Envisat)

Página dcha.: Parte de las Spitzbergen entre una maraña de hielo y nubes (Envisat)

Seit einigen Jahren wird der himmlische Trainerstab immer größer. Denn bei Ingenieuren und Raumfahrtmanagern hat ein Umdenken stattgefunden. Heute werden eher kleine Satelliten gebaut, die klar bestimmte Aufgaben übernehmen, im Gegensatz zu Satelliten wie Envisat, der noch mehr als zehn Instrumente mit sich trägt. Das bedeutet nicht, dass die Forscher nichts mehr mit zehn Instrumenten anfangen könnten. Aber Satelliten mit wenigen Instrumenten an Bord sind kleiner und leichter, und sie kosten weniger Geld. Sie lassen sich somit sehr viel preiswerter mit kleinen Raketen starten. Zudem ist das Risiko beim Start nicht mehr so groß wie früher. Nicht auszudenken, Envisat wäre beim Start verloren gegangen. Die Erdbeobachtung nicht nur in Europa hätte fast ein Jahrzehnt lang nur mit halber Kraft erfolgen können. Geht eine kleine Mission beim Start verloren, ist Ersatz relativ schnell und vergleichsweise preiswert zu beschaffen.

Atomraketen für die Erde

Die kleinen, leichten, spezialisierten Satelliten starten oft mit Raketen, deren Verwendung noch vor zwei Jahrzehnten völlig undenkbar schien. Rockot heißt die Rakete, die von der

space program managers have rethought their approach. Nowadays, they prefer to build smaller satellites with very specific functions, as opposed to Envisat, which carries more than ten different types of instrument. This isn't supposed to suggest that modern researchers wouldn't know what to do with ten such instruments, merely that satellites with only a few sensors on board are lighter, more compact, and cost less money. They can be launched using smaller and cheaper rockets, and the risks associated with lift-off are also greatly reduced. Losing Envisat during the launch doesn't bear thinking about—Earth observation in Europe and beyond would have been running at half-power for almost a decade. If a small mission is lost due to an accident at launch, it is relatively easy and quick to provide a replacement.

Atomic rockets for the Earth

Small, light, specialized satellites are often launched using rockets whose use for this purpose would have been unthinkable even two decades ago. The rocket developed in Bremen by Eurockot, a joint

Les entraîneurs célestes se multiplient depuis quelques années en raison de la nouvelle approche adoptée par les ingénieurs spatiaux. Ces derniers préfèrent aujourd'hui construire de petits satellites spécialisés plutôt que des satellites du type d'Envisat équipés d'une dizaine d'instruments. Non pas que les chercheurs ne sachent que faire de dix instruments, mais les satellites moins chargés sont plus petits, plus légers et moins onéreux, et peuvent ainsi être lancés à moindre coût, avec des risques réduits d'échec au lancement. Imaginez qu'Envisat ait été détruit lors de son lancement. Les capacités d'observation de la Terre auraient pendant près d'une décennie été réduites de moitié. Si un petit satellite est détruit au lancement, il peut être remplacé relativement plus rapidement et à moindre frais.

Des missiles nucléaires pour la Terre

Les petits satellites spécialisés sont placés en orbite par des lanceurs encore inimaginables il y a vingt ans. Rockot, le lanceur proposé par la société allemande Eurockot, une joint-venture entre Airbus et

entrenadores celestes se ha hecho mucho más grande porque los ingenieros y los gestores espaciales han reconsiderado su posición. Actualmente, se prefiere construir pequeños satélites que asuman misiones claramente definidas, en lugar de satélites como el Envisat que incluyen más de 10 instrumentos diferentes. Esto no significa que los investigadores no vayan a emprender proyectos de estas características, pero los satélites con pocos instrumentos a bordo son más pequeños y ligeros y cuestan mucho menos dinero. Además, el riesgo en el lanzamiento ya no es tan grande como antes. Resulta impensable que el Envisat se hubiera podido perder en el lanzamiento. Por el contrario, si se perdiese una misión pequeña por un accidente en el despegue, la sustitución sería mucho más rápida y comparativamente mucho más barata.

Misiles nucleares para la Tierra

Estos pequeños, ligeros y especializados satélites se lanzan frecuentemente con cohetes cuya utilización todavía era impensable hace dos décadas. La empresa Eurockot de Bremen, una alianza comercial

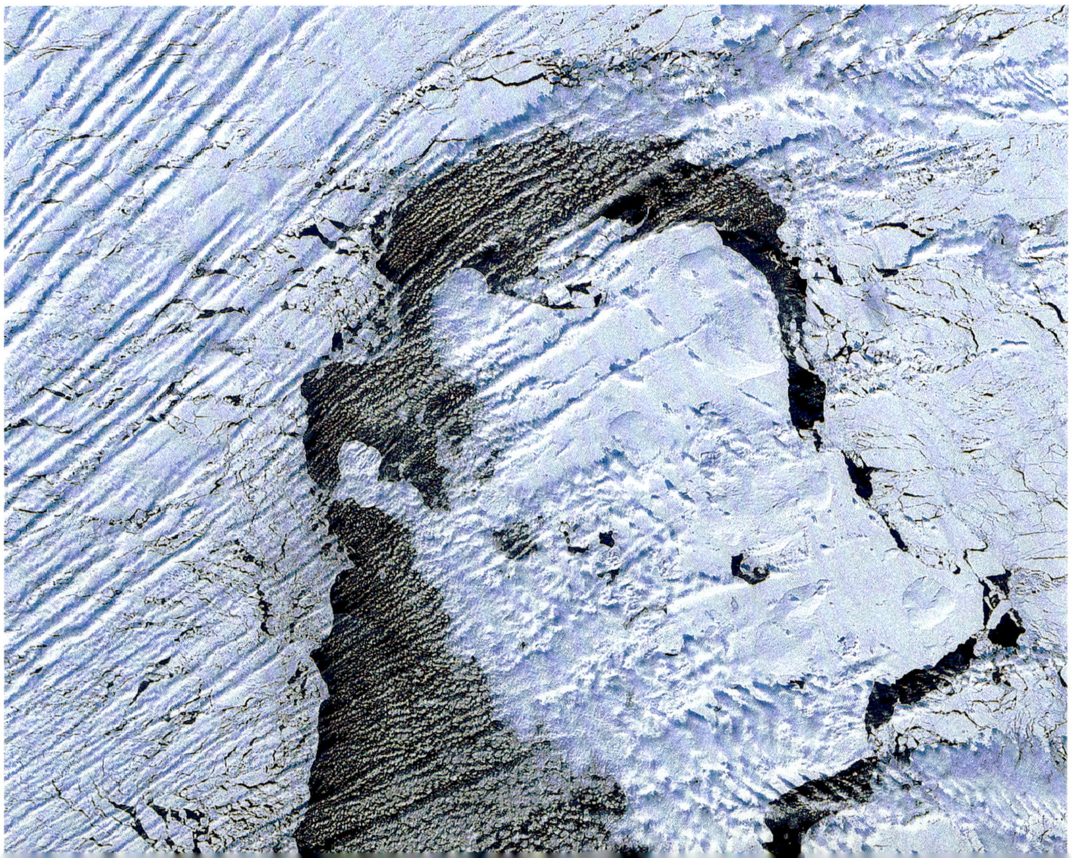

Bremer Firma Eurockot zur Verfügung gestellt wird, einem Joint Venture von Airbus und der russischen Firma Khrunichev. Sie ist eine Rakete mit ganz besonderer Vergangenheit: *Das ist eine konvertierte SS-19, also eine ehemalige interkontinentale Atomrakete,* erklärt York Viertel, Direktor von Eurockot. Ihm ist die Freude über das Erstaunen seines Gesprächspartners anzumerken. *Nach den Abrüstungsverträgen mit den Amerikanern müssen diese Raketen zerstört werden. Wir haben für die Zerstörung einen wirtschaftlicheren Weg gefunden, indem wir eine Oberstufe weiterentwickelt haben. Jetzt können wir zwei Tonnen in eine niedrige Umlaufbahn bringen.* Ein friedlicher Start ins All ist eine elegante Methode, eine frühere Schreckenswaffe zu vernichten. Auch der Startplatz hat es in sich: Plesetsk, 800 Kilometer nördlich von Moskau. Dieser Weltraumbahnhof ist längst nicht so bekannt wie Baikonur, von wo aus Russland seine Astronauten ins All schickt – völlig zu Unrecht: Denn Plesetsk ist der Startplatz mit den weltweit meisten Raketenstarts. Etwa 2000-mal hoben dort Raketen Richtung Umlaufbahn ab.

Mehr als vierzig Prozent aller Raketenstarts auf dieser Erde haben in Plesetsk stattgefunden. Allerdings waren das hauptsächlich militärische Missionen, die oft im Verborgenen realisiert wurden. Somit erklärt sich, dass Plesetsk nur wenigen Experten ein Begriff ist. Es ist eine Geschichte wie aus dem Kitschroman: Raketen, die einst Atomsprengköpfe nach Chicago tragen sollten, bringen nun westliche Forschungssatelliten in die Umlaufbahn. Ausgangspunkt ist ein Startgelände, das zu Zeiten des kalten Kriegs geradezu eine Frontstellung hatte. Heute sind die Gegner von einst Partner – es geht nicht mehr darum, die Erde zu zerstören, sondern sie zu beobachten und zu bewahren. Schwerter zu Flugscharen.

Wasser an Land

Mit einer dieser Raketen ist im November 2009 auch SMOS gestartet,

venture between Airbus and the Russian firm of Khrunichev, is known as the "Rockot" and has a very strange history. *It's a converted SS-19—in other words, an old intercontinental ballistic missile,* explains York Viertel, Eurockot's director; his pleasure at his interlocutor's amazement is evident. *These rockets were to be destroyed after the disarmament treaties with the Americans, but we found a more economical way of disposing of them by developing a top stage for the missile and we can now deliver a payload of 2 tons [2 tonnes] into a low Earth orbit.*

A peaceful launch into space is an elegant new use of a former deterrent weapon. The launch site—Plesetsk, 500 miles (800 kilometers) north of Moscow—is also no slouch. The space station is not as well known as Baikonur, where the Russians send their cosmonauts into space, but this is quite unfair; Plesetsk holds the world record for the number of launches, having sent more than 2000 rockets into orbit to date.

More than 40 percent of all the rocket launches from Earth have taken place in Plesetsk, although these were mainly military operations and were often conducted in secret; this may explain why Plesetsk is known only to a few experts.

It is almost like a far-fetched spy story: rockets once intended to rain atomic warheads down on Chicago are now carrying western research satellites into orbit and the launchpad used was more or less a frontline position during the Cold War.

The old enemies are now partners and are more concerned with observing and preserving the Earth than destroying it—swords have been beaten into plowshares.

Water on land

One such rocket was used to launch SMOS, the second of Europe's specialized Earth observation devices. The satellite, which was developed in Italy under the aegis of Thales Alenia, monitors properties of the

l'entreprise russe Krounitchev, a une histoire particulière : *Il s'agit d'un SS-19 reconverti, autrement dit d'un ancien missile balistique intercontinental,* explique York Viertel, directeur d'Eurockot, que réjouit visiblement la surprise de son interlocuteur. *Les accords de désarmement conclus avec les États-Unis prévoient le démantèlement de ces missiles. En développant un étage supérieur, nous avons trouvé une solution plus rentable que la simple destruction. Désormais nous sommes en mesure de placer en orbite basse une charge utile de deux tonnes.* Le lancement paisible dans l'espace est en effet une méthode raffinée de destruction d'anciennes armes de dissuasion. La base de lancement de Plessetsk, à huit cents kilomètres au nord de Moscou, n'est pas banale non plus. Ce cosmodrome est, à tort, beaucoup moins connu que Baïkonour, la base qui a vu le départ des cosmonautes russes dans l'espace. Plessetsk est en effet la base de lancement la plus active au monde : quelque 2 000 lanceurs ont décollé de ses pas de tir.

Plessetsk a accueilli plus de quarante pour cent des lancements effectués dans le monde, des missions toutefois essentiellement militaires réalisées dans le plus grand secret, ce qui explique pourquoi Plessetsk n'est connu que de quelques experts. Une histoire qui semble tirée d'un roman des fusées construites pour lancer des ogives nucléaires sur les États-Unis, emportent désormais les satellites scientifiques occidentaux dans l'espace, à partir d'une base de lancement autrefois haut lieu de la guerre froide. Les vieux ennemis sont devenus partenaires, et leur but n'est plus de détruire la Terre, mais de l'observer pour mieux la protéger. De leurs épées, ils ont forgé des socs, et de leurs missiles des vaisseaux pacifiques.

D'eau et de sel

C'est l'une de ces fusées qui, en novembre 2009, a placé SMOS, la deuxième mission européenne d'exploration de la Terre, sur orbite.

entre Airbus y la empresa rusa Khrunichev, suministra unos cohetes llamados Rockot. Se trata de cohetes con un pasado muy particular: *Es un SS-19 reconvertido; es decir, un antiguo misil nuclear intercontinental,* explica York Viertel, director de Eurockot. Su rostro muestra una expresión divertida ante el asombro de su interlocutor. *Según los tratados de desarme con los estadounidenses, estos cohetes tendrían que haber sido destruidos. Nosotros hemos encontrado una alternativa más económica a la destrucción desarrollándolos a un nivel superior.* Un lanzamiento pacífico al espacio es un método elegante para destruir una antigua arma disuasoria. También incluye el lugar del lanzamiento: Plesetsk, a 800 km al norte de Moscú. Este cosmódromo no es ni mucho menos tan conocido como Baikonur, lugar desde donde Rusia envía sus astronautas al espacio, lo que no deja de ser injusto, porque Plesetsk es el lugar de lanzamiento que más despegues de cohetes acumula en todo el mundo. Desde allí ya se han puesto en órbita unos 2.000 cohetes.

En Plesetsk se han llevado a cabo más del cuarenta por ciento de todos los lanzamientos de cohetes desde la Tierra. Sin embargo, la mayoría eran misiones militares que con frecuencia se realizaban en secreto. De ese modo, se explica que sólo algunos expertos lo conozcan. Es casi una historia de culebrón: los cohetes que deberían haber sembrado Chicago de cabezas atómicas, ponen ahora en órbita los satélites de investigación occidentales. El punto de partida es un terreno de lanzamiento que, en tiempos de la Guerra Fría, era un frente real. Los adversarios de antaño son socios en la actualidad y ya no se trata de destruir la Tierra, sino de observarla y protegerla. «De las espadas se forjarán arados...»... de los misiles, flotillas espaciales pacíficas.

El agua del suelo

Con uno de esos cohetes, se lanzó en noviembre de 2009 el segundo de los satélites europeos especializados en exploración terrestre, el SMOS. Este

Die vereisten Queen-Elizabeth-Inseln im Norden Kanadas. Rechts oben die Nordwestspitze Grönlands (Envisat)

The Queen Elizabeth Islands in northern Canada, surrounded by ice. The northwestern tip of Greenland can be seen top right (Envisat)

Les îles glacées de la Reine-Élisabeth dans le nord du Canada. En haut à droite, la pointe nord-ouest du Groenland (Envisat)

Islas de la Reina Isabel, al norte de Canadá, cubiertas de hielo. Arriba a la derecha, extremo noroeste de Groenlandia (Envisat)

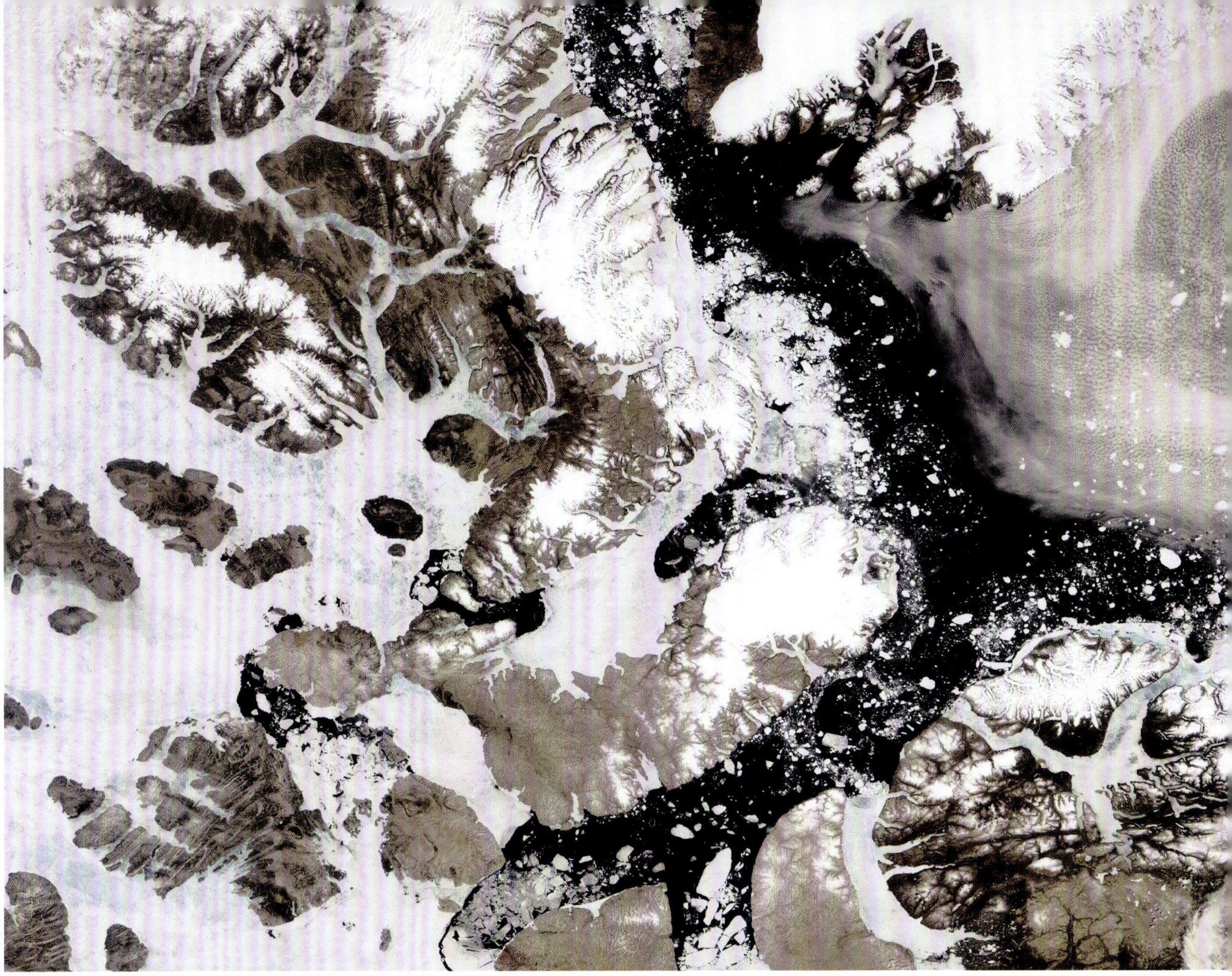

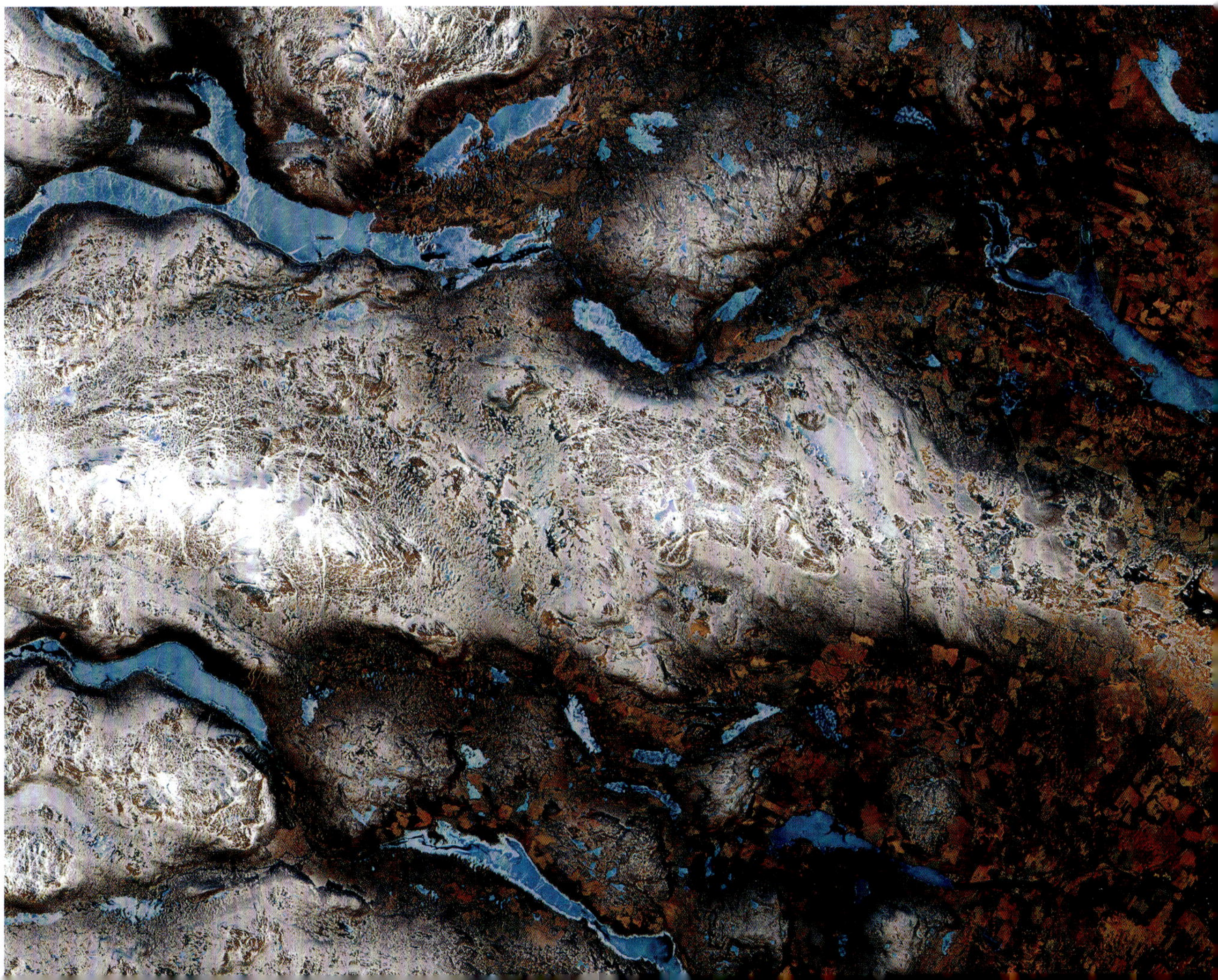

Erster Schnee auf der herbstlich verfärbten Landschaft: In Nordschweden hält der Winter Einzug (Spot)

The first snows obscure the colors of fall in northern Sweden; winter is coming (Spot)

Premières neiges sur un paysage automnal. L'hiver s'installe dans le nord de la Suède (Spot)

Primeras nieves sobre el colorido paisaje otoñal: el invierno comienza en el norte de Suecia (Spot)

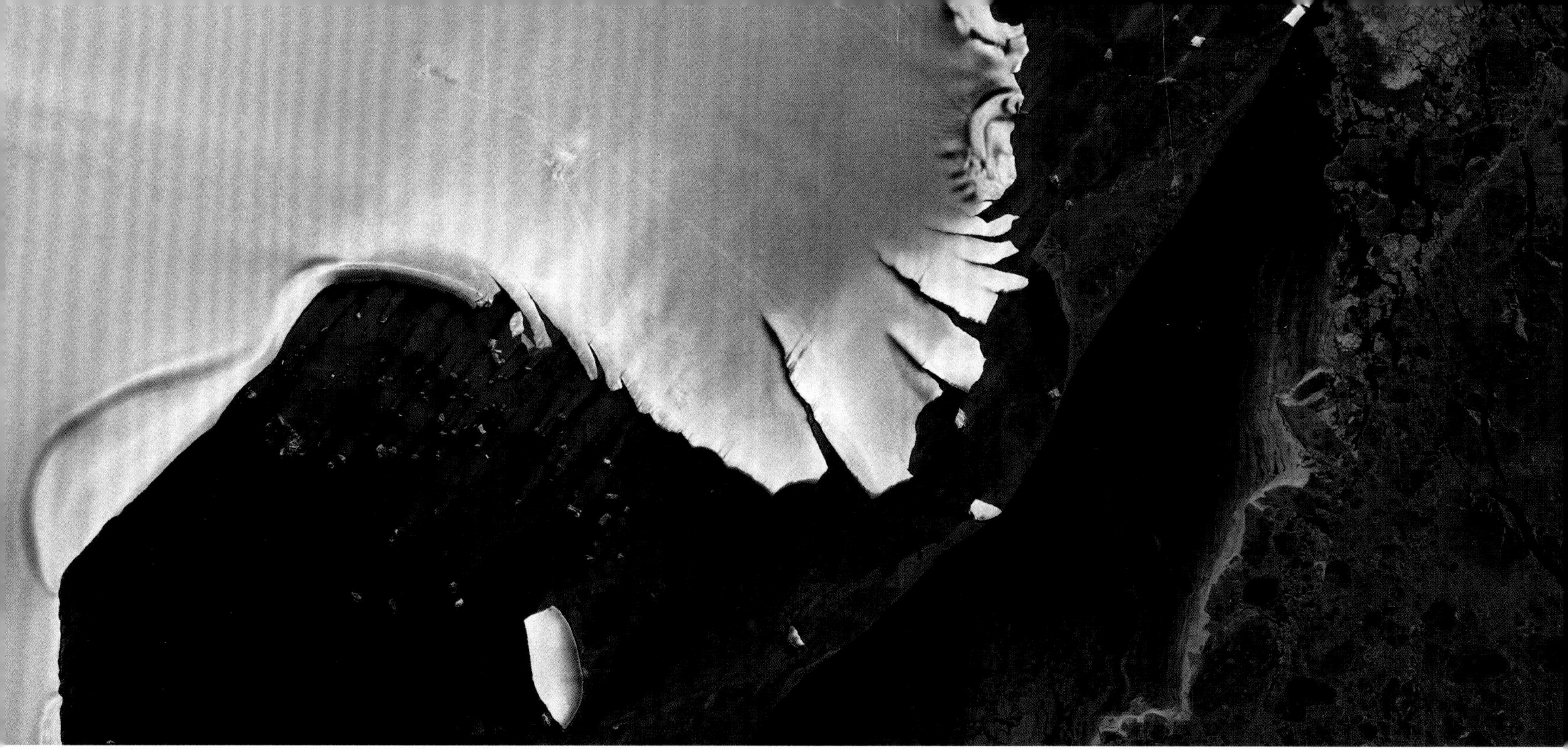

das zweite Exemplar von Europas spezialisierten Erderkundungssatelliten. Der Satellit, der unter Federführung von Thales Alenia in Italien gebaut wurde, beobachtet Eigenschaften der Erde, die Laien auf den ersten Blick völlig uninteressant erscheinen. SMOS steht für Soil Moisture and Ocean Salinity, also Bodenfeuchte und Salzgehalt der Meere. Susanne Mecklenburg, die am ESA-Forschungszentrum in Frascati bei Rom die Mission leitet, ist es gewohnt, dass bei diesen Themen manch einer die Stirn runzelt. Zweifel am Sinn einer solchen Mission weiß sie umgehend zu zerstreuen: *Bodenfeuchte ist eine der Größen, die sehr, sehr wichtig sind für den Wasserkreislauf der Erde. Im Moment gibt es nur In-Situ-Messungen. Das heißt, man nimmt Bodenproben und bestimmt daraus, wie viel Wasser sich im Boden befindet.* Erstmals misst jetzt ein Satellit weltweit, wie viel Wasser im Boden steckt. Wetterradar und Satellitenbilder hatten zuvor nur gezeigt, wo es regnet. Was genau mit dem Wasser passiert, ob es im Boden versickert oder schnell wieder verdampft, ließ sich nicht erkennen. Dabei spielt der Wasserkreislauf eine äußerst wichtige Rolle bei Klimamodellen und Wettervorhersagen. Zum Beispiel wird je nach Bodenfeuchte unterschiedlich viel Sonnen-

Earth which might seem to be of no interest to a layperson: SMOS stands for Soil Moisture and Ocean Salinity, and Susanne Mecklenburg, who directs the mission from the ESA Research Center in Frascati near Rome, is used to the puzzled looks that often greet the concept. She immediately clears up any doubts about the point of such an enterprise: *Soil moisture is a big deal and of enormous importance to the water cycle on Earth. At the moment, we are restricted to* in situ *readings, in other words, we take soil samples and find out how much moisture there is in the ground.*

For the first time, a satellite is now measuring how much moisture there is in the soil worldwide. Weather radar and satellite images had previously only shown where it was raining, and exactly what was happening to the water, whether it was draining into the soil or evaporating, couldn't be determined. The water cycle plays an extremely important role in climate modeling and weather forecasts; more or less Sun is required to evaporate water depending on soil moisture levels, for example.

Data such as this also enables meteorologists to issue warnings

Ce satellite construit sous la houlette de Thales Alenia en Italie observe certaines propriétés de la Terre dont l'intérêt échappe au premier abord aux yeux du profane. SMOS (Soil Moisture and Ocean Salinity) mesure l'humidité des sols et la salinité des océans. Susanne Mecklenburg, qui dirige la mission depuis le centre de recherche spatiale de l'ESA à Frascati, près de Rome, est souvent confrontée à l'incompréhension de ses interlocuteurs. Mais elle balaie rapidement les doutes quant à l'utilité de telles missions : *l'humidité des sols est un paramètre extrêmement important pour l'étude du cycle de l'eau sur la Terre. À l'heure actuelle, nous disposons uniquement de mesures terrestres. Nous prélevons des échantillons de sol à partir desquels nous déterminons la teneur en eau des terres.* C'est la première fois qu'un satellite mesure l'humidité des sols à l'échelle du globe. Les radars météorologiques et les images satellites se sont contentés jusqu'ici d'indiquer les zones de pluie. Impossible de savoir cependant si l'eau s'infiltrait dans les sols ou s'évaporait rapidement. Pourtant, les échanges d'eau affectent directement le temps et les phénomènes climatiques. L'humidité des sols détermine, par exemple, l'intensité

satélite observa características de la Tierra que pueden parecer poco interesantes a los profanos. SMOS es el acrónimo de Soil Moisture and Ocean Salinity, es decir, humedad del suelo y salinidad de los océanos. Susanne Mecklenburg, que dirige la misión en el centro de investigación de la ESA en Frascati, cerca de Roma, está acostumbrada a que más de uno arrugue la frente ante estos temas. No obstante, consigue disipar inmediatamente las dudas sobre este tipo de misiones: *La humedad del suelo es una de las magnitudes más importantes para el ciclo del agua de la Tierra. De momento, sólo se dispone de mediciones in situ. Es decir, se toman muestras de suelo y se determina la cantidad de agua que se encuentra en el mismo.* Ahora, por primera vez, un satélite mide la cantidad de agua presente en el suelo. Hasta ahora, el radar meteorológico y las imágenes de satélite únicamente nos mostraban donde llovía. Lo que ocurría después con el agua, si era absorbida por el suelo o se evaporaba rápidamente, no se podía detectar. Sin embargo, el ciclo del agua juega un papel extremadamente importante en los modelos climáticos y en los pronósticos meteorológicos.

Además, los meteorólogos necesitan este tipo de datos para poder advertir mejor ante situaciones meteorológicas

Die Antarktis im Bereich der Neumeyer-Station (linke Seite) und des Königin-Maud-Lands (rechts, TerraSAR-X)

The Antarctic around the Neumayer Station (left) and Queen Maud Land (right, TerraSAR-X)

L'Antarctique à proximité de la station polaire Neumeyer (page de gauche) et de la Terre de la Reine-Maud (à droite, TerraSAR-X)

La Antártida en la región de la base Neumeyer (página izda.) y de la Tierra de la reina Maud (dcha., TerraSAR-X)

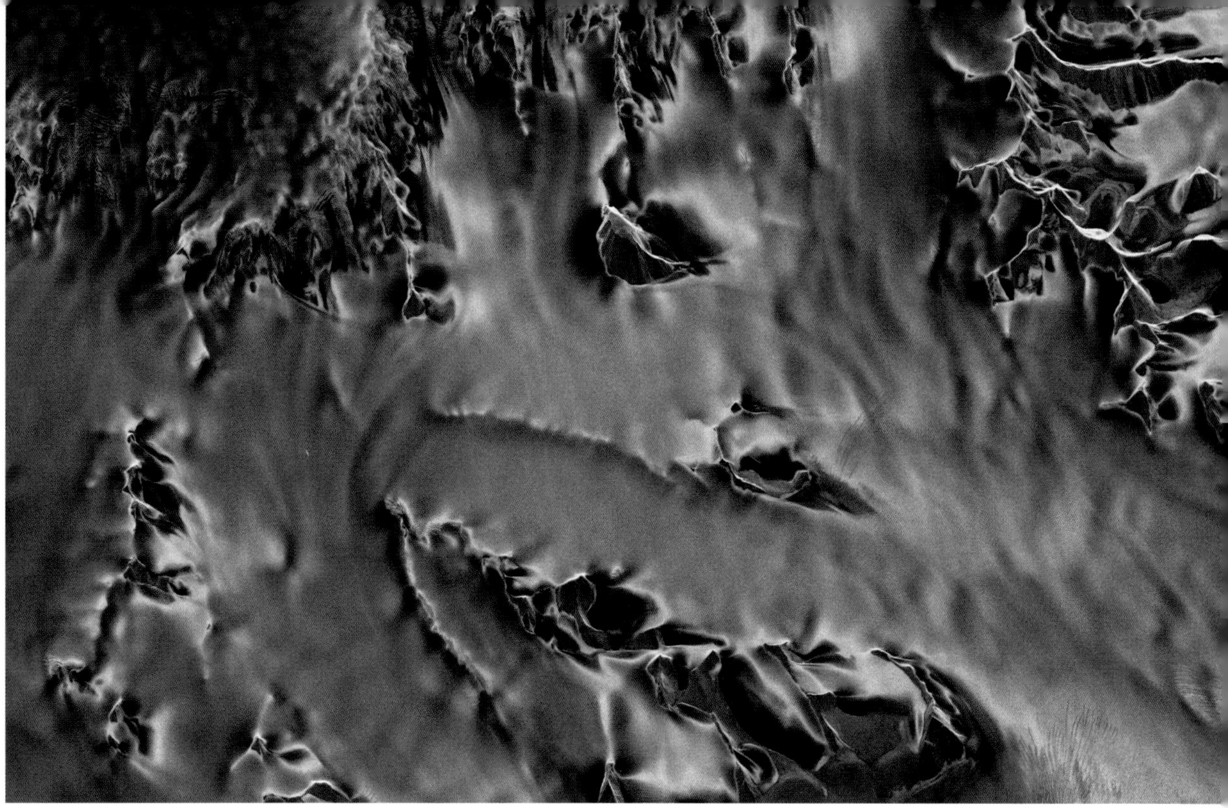

strahlung für das Verdampfen des Wassers gebraucht.

Zudem brauchen die Meteorologen solche Daten, um besser vor extremen Wettersituationen warnen zu können, etwa Überschwemmungen. *Die natürliche Bodenfeuchte liegt zwischen null und fünfzig Prozent,* erläutert Susanne Mecklenburg, Hydrologin und Atmosphärenwissenschaftlerin. *Alles, was darüber liegt, bedeutet, dass der Boden im Prinzip gesättigt ist. Dann würde Regenwasser auf der Oberfläche abfließen.* SMOS zeigt den Forschern also, wo der Boden noch Wasser aufnehmen kann und wo nicht.

Das Instrument an Bord, das die Feuchte im Boden gleichsam »fühlt«, testet ganz nebenbei auch noch den »Geschmack« des Meerwassers. *Wir messen auch noch den Salzgehalt der Ozeane, der ebenfalls sehr bedeutend für den Wasserkreislauf ist. Auch vom Salzgehalt gibt es bisher keine globalen Messungen.* Die Dichte des Meerwassers hängt vor allem von seinem Salzgehalt ab. Schweres salzreiches Wasser sinkt ab und steuert so die weltweite Zirkulation in den Meeren, der wir zum Beispiel den warmen Golfstrom und damit das gemäßigte Klima in Europa verdanken. Wie salzhaltig das Meerwasser ist, bestimmt ein komplexes Wechselspiel von Niederschlag und Verdunstung.

about extreme weather events such as floods. *Natural soil moisture lies between zero and 50 percent,* explains Susanne Mecklenburg, who worked as a hydrologist and atmospheric scientist before joining the ESA. *Everything above that generally means that the soil is saturated, in which case rainwater would flow off the surface.* SMOS thus shows scientists where the ground can still absorb water and where it can't.

The functions of the onboard instrument are to "feel" the moisture in the soil and to "taste" seawater. *We also record ocean salinity, which is also extremely important for the water cycle. Salt levels have still not been measured globally.*

The density of seawater is primarily determined by its salinity—heavy, salty water sinks, promoting the worldwide circulation of ocean currents, and this brings us the warm Gulf Stream and Europe's temperate climate, for example. The salt content of seawater determines a complex alternation of precipitation and evaporation.

Researchers intend to use SMOS to investigate these phenomena more closely and incorporate them into weather and climate modeling for the first time.

du rayonnement nécessaire à l'évaporation de l'eau.

Ces données permettent par ailleurs aux météorologues de mieux appréhender les événements climatiques extrêmes tels que les inondations. *L'humidité naturelle des sols est comprise entre zéro et cinquante pour cent,* explique Susanne Mecklenburg, hydrologue et spécialiste en sciences de l'atmosphère. *Un taux supérieur indique que le sol est saturé. L'eau de pluie s'écoulerait donc à sa surface.* SMOS révèle ainsi aux scientifiques la capacité des sols à absorber l'eau.

L'instrument scientifique qui « ressent » pour ainsi dire l'humi-dité des terres, teste par la même occasion le « goût » de l'eau de mer. *Nous mesurons également la salinité des océans, un autre élément essentiel du cycle de l'eau, pour lequel il n'existe pas encore de données globales.* La densité de l'eau de mer dépend avant tout de la salinité. L'eau très salée, plus lourde, plonge et régit ainsi la circulation globale des courants marins tels que le Gulf Stream auquel nous devons le climat tempéré européen. La salinité des mers est le résultat d'une alternance complexe entre précipitation et évaporation. SMOS permettra aux scientifiques d'analyser ces phénomène à l'échelle

extremas, como las inundaciones. *La humedad natural del suelo se encuentra entre el cero y el cincuenta por ciento,* aclara Susanne Mecklenburg, hidróloga y científica atmosférica que ahora trabaja en la ESA. *Todo lo que supere ese porcentaje significa, en principio, que el suelo está saturado. Entonces, el agua de lluvia se deslizará por la superficie.* Por tanto, SMOS muestra a los investigadores los lugares en los que el suelo todavía puede admitir agua y en los que eso no es posible.

El instrumento de a bordo que, en cierto modo, «siente» la humedad del suelo, examina al mismo tiempo el «sabor» del agua de mar. *También medimos la salinidad de los océanos, que igualmente es muy significativa para el ciclo hidrológico. Hasta ahora tampoco se disponía de mediciones globales de salinidad.* La densidad del agua de mar depende, sobre todo, de su contenido en sal. El agua rica en sal, más pesada, se hunde y gobierna así la circulación mundial de los mares. La salinidad del agua de mar viene determinada por un complejo juego de intercambio de precipitaciones y evaporación. Con el SMOS, los investigadores pretenden analizar estos fenómenos con absoluta precisión en todo el mundo y mencionarlos por primera vez en los modelos meteorológicos y climáticos.

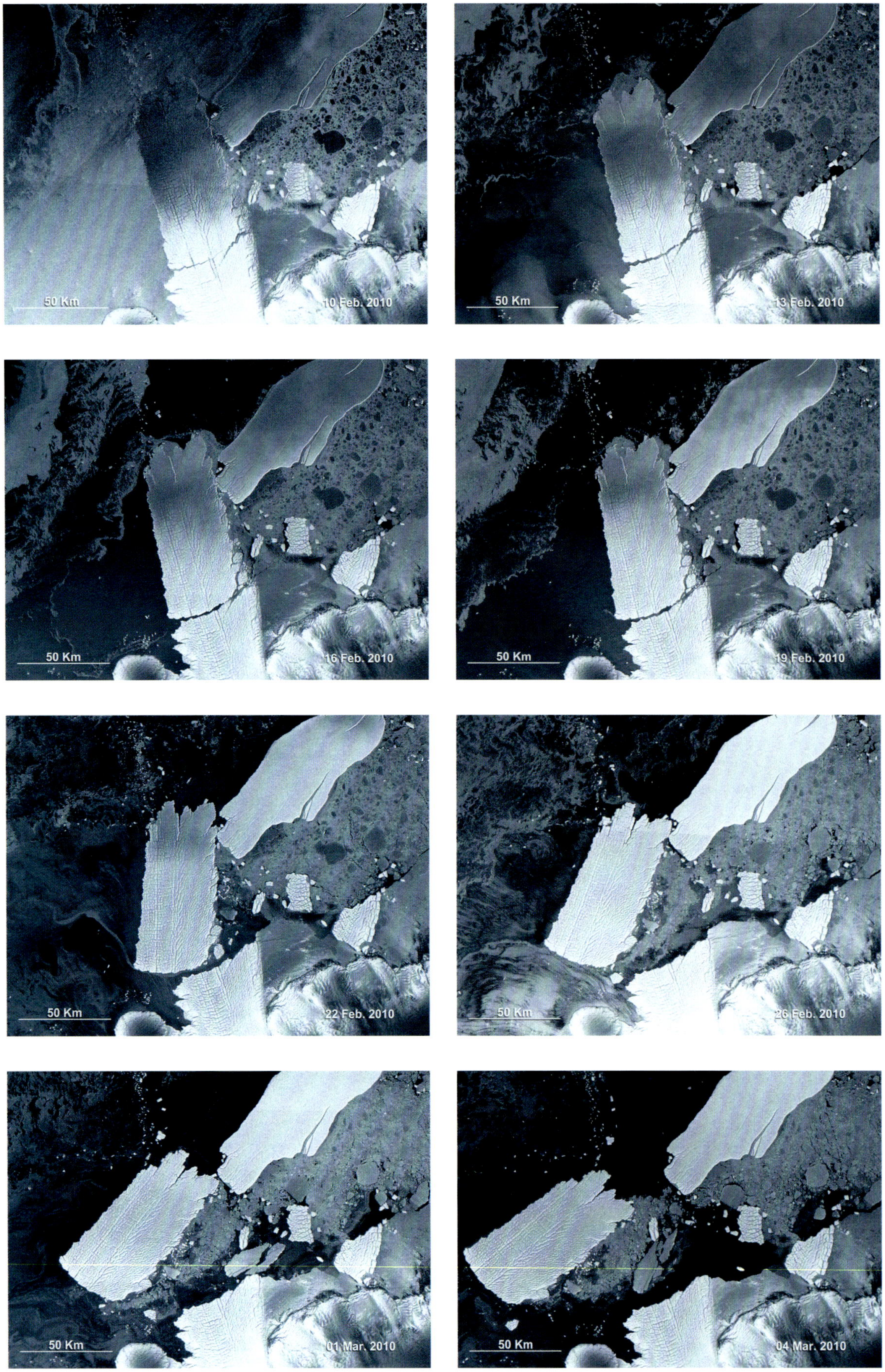

Halb zog es ihn, halb sank er hin: Der etwa 100 km lange Eisberg brach im Februar 2010 in der Antarktis ab und driftete binnen drei Wochen deutlich ab (Envisat)

Not waving but drowning: this 60-mile-long (100-km) iceberg broke off in February 2010 and was well adrift in less than three weeks (Envisat)

À vos marques, prêts, partez : en février 2010, un iceberg d'environ 100 km de long s'est détaché de l'Antarctique avant de partir à la dérive en l'espace de trois semaines (Envisat)

Entre tocado y hundido: un iceberg de casi 100 km de largo se desprende de la Antártida en febrero de 2010 y se desplaza a la deriva durante tres semanas (Envisat)

Mit SMOS wollen die Forscher nun diese Phänomene weltweit präzise untersuchen und erstmals in die Wetter- und Klimamodelle einfließen lassen.

Die Mission kostet gut 300 Millionen Euro und wird mindestens bis 2017 dauern – eine Verlängerung ist wahrscheinlich.

Über die Berge fallen

Eisbedeckung, Vegetation, Wolkenbildung, Gasverteilung in der Atmosphäre – die Satelliten in der Umlaufbahn machen Fotografien, sie nehmen Spektren auf, sie schicken Radar- oder Laserpulse nach unten. Der Satellit GOCE arbeitete auf höchst kuriose Weise. Er beobachtete auch die Erde, aber ganz ohne Kamera. Er fühlte die sich von Ort zu Ort minimal verändernde Anziehungskraft der Erde. Seit März 2009 spürt GOCE der Gravitation nach, 2013 war die Mission planmäßig beendet. Schon nach einem Jahr Messzeit war klar, dass der Satellit stellenweise ein ganz neues Bild der Erde zeichnet: *Bisher standen uns oft nur relativ schlechte Messungen auf der Erde zur Verfügung,* erklärt Reiner Rummel, Geodät an der TU München. *Jetzt sehen wir mit GOCE deutliche Unterschiede vor allem in Südamerika, in Afrika, im Himalaya, im Gebiet um Indonesien und in der Antarktis.* Diese Gebiete sind tektonisch aktiv und für Geophysiker von großer Bedeutung. Der Satellit registrierte, was wo im Innern der Erde passiert. Faustregel: Wo viel Anziehung ist, muss auch viel Materie sein. Das bezieht sich keineswegs nur auf Gestein. GOCE zeigte den Forschern auch genau, was in den Weltmeeren passiert. *Wir können die Meeresströmungen kartieren. Das ist mit dieser räumlichen Auflösung erstmals vom Weltraum aus möglich,* freut sich Reiner Rummel. *Wir sehen die Erhebungen der mittleren Ozeanoberfläche, die nur ein bis zwei Meter hoch sind. Sie zeigen uns an, wie die Ozeane fließen und wie schnell sie fließen.* Führt etwa der Golfstrom zu einer kleinen Erhöhung der Meeresoberfläche, so spürte GOCE die verstärkte Anziehungskraft des Wassers. Was zunächst als ein etwas

The mission, which cost more than 300 million euros, will continue at least until 2017.

Falling on the mountains

Ice-cover, vegetation, cloud formation, gas distribution in the atmosphere—all these are revealed as the orbiting satellites take photographs, make spectroscopic observations, and beam down radar and laser pulses. The GOCE (Gravity Field and Steady-State Ocean Circulation Explorer) satellite worked in a very strange way, observing the Earth without using any cameras; instead, it registered the minimal variations in the Earth's pull from place to place, and researched gravity from March 2009 to 2013, when the mission finished.

After only a year's results, it had become clear that in some places the satellite was producing an entirely new picture of the Earth: *Until then, we had only rather low-quality measurements made on the surface,* explains Reiner Rummel, a geometer at Munich's Technical University. *GOCE is now allowing us to see significant differences, particularly in South America, Africa, the Himalayas, the area around Indonesia, and the Antarctic.* These regions of tectonic activity are extremely important to geophysicists, and the satellite recorded everything happening beneath the Earth's surface. As a rule of thumb, wherever there is more powerful attraction, there must be more matter, and this is not just true of rocks—GOCE also showed scientists what is going on in the oceans. *We are able to chart the ocean currents; the perspective from space has made this possible for the first time,* grins Reiner Rummel. *We can see peaks in the middle of the ocean surface; they are only 3–6 feet high [1–2 meters] but they showed us where and how fast the oceans are flowing.* If the Gulf Stream caused the surface of the ocean to rise slightly, GOCE would register the increased attraction of the water. What might at first seemed to be a

planétaire et de les intégrer aux modèles climatiques et météorologiques.

Cette mission dotée d'un budget de 300 millions d'euros doit s'étaler au moins jusqu'en 2017.

Palper le relief

Pour étudier les phénomènes terrestres, les satellites scientifiques prennent généralement des photos, réalisent des images spectrales et émettent des ondes radar ou laser. La méthode du satellite GOCE était bien plus étrange. Il observait la Terre certes, mais sans caméra. Il a cartographié les variations minimales du champ gravitationnel de la Terre de mars 2009 jusqu'à la fin programmée de la mission en 2013. Un an après le début de la mission, le satellite présentait déjà une toute nouvelle image de la Terre : *Jusqu'ici, nous ne disposions souvent que de données peu précises,* explique Reiner Rummel, géodésien à l'Université technique de Munich. *Grâce à GOCE, nous constatons des variations considérables, notamment en Amérique du sud, en Afrique, dans l'Himalaya, autour de l'Indonésie, et en Antarctique.* Ces régions connaissent une activité tectonique primordiale aux yeux des géophysiciens, c'est pourquoi le satellite enregistrait les phénomènes survenant à l'intérieur de la Terre. Pour simplifier : plus la gravitation est importante, plus la matière est dense, une règle qui s'applique à tous les corps. GOCE observait les phénomènes sous-marins avec la même précision. *Nous sommes en mesure de cartographier les courants océaniques avec une résolution spatiale sans précédent,* se félicite Reiner Rummel. *Nous distinguons les variations du relief au milieu de l'océan, qui sont inférieures à deux mètres, mais nous informent sur la direction et la vitesse des courants marins.* Si le Gulf Stream soulève légèrement la surface océanique, GOCE notait une augmentation de la force gravitationnelle de l'eau. Ce qui, au premier abord, semblait la folle expérience de physiciens enthousiastes, s'est révélé être

La misión cuesta más de 300 millones de euros y durará al menos hasta 2017.

Más allá de las montañas

La cubierta de hielo, la vegetación, la formación de nubes o la distribución de los gases en la atmósfera son fenómenos fotografiados por los satélites en órbita, que también capturan espectros o envían pulsos de radar o láser. Pero el satélite GOCE trabajaba de una manera sumamente curiosa. Observaba la Tierra, sin cámara, y detectaba variaciones mínimas de la fuerza de gravedad en diferentes lugares de la Tierra desde marzo de 2009 hasta 2013, año en que finalizó su misión. Tras un año de mediciones, estaba claro que el satélite podía dibujar una imagen completamente nueva de la Tierra: *Hasta ahora solo disponíamos de unas mediciones poco precisas tomadas desde la Tierra,* aclara Reiner Rummel, geómetra en la Universidad Politécnica de Múnich. *Ahora, con GOCE, vemos claras diferencias sobre todo en Sudamérica, en África, en el Himalaya, en la región en torno a Indonesia y en la Antártida.* Estas regiones son tectónicamente activas y muy importantes para los geofísicos, y el satélite registraba lo que ocurre en el interior de la Tierra. La regla general es que donde hay mucha gravedad también tiene que haber mucha materia. Esto no se refiere solo a la roca. GOCE también mostraba detalladamente a los investigadores lo que pasa en los océanos. *Podemos cartografiar las corrientes marinas. Es la primera vez que resulta posible desde el espacio con esta resolución tridimensional,* celebra Reiner Rummel. *Observamos las elevaciones de la superficie media del océano, que son de uno a dos metros de alto. Nos indican cómo circulan los océanos y lo rápido que lo hacen.* Si la corriente del Golfo provocaba una pequeña elevación de la superficie del mar, GOCE percibía un aumento de la fuerza de gravedad del agua. Lo que en principio podía parecer un experimento estrambótico de físicos entusiastas, se reveló rápidamente como

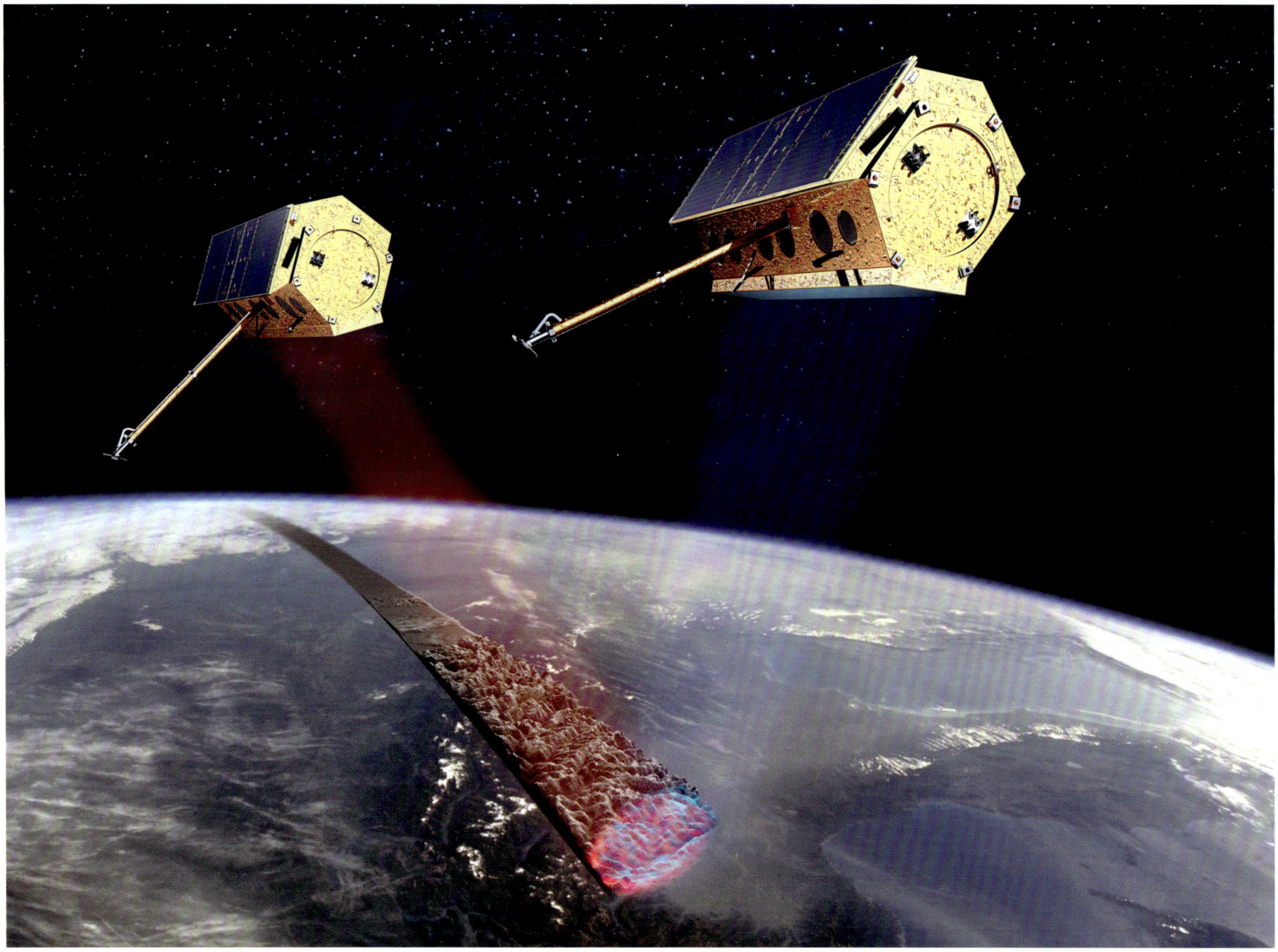

Oben: Zu zweit sieht man mehr – TerraSAR-X und TanDEM-X machen 3D-Bilder der Erde

Above: Two eyes are better than one—TerraSAR-X and TanDEM-X take 3-D pictures of the Earth

Ci-dessus : deux yeux voient plus qu'un seul. TerraSAR-X et TanDEM-X réalisent des images 3D de la Terre

Arriba: Dos ven más que uno. TerraSAR-X y TanDEM-X toman imágenes 3D de la Tierra

Rechte Seite: Erste Aufnahme des Radarsatelliten TanDEM-X – eine Bucht von Madagaskar

Right: The first image created by the TanDEM-X radar satellite—a bay in Madagascar

Page de droite : première image fournie par le satellite radar TanDEM-X, une baie de Madagascar

Página dcha.: Primera imagen del satélite de radar TanDEM-X: una bahía de Madagascar

schräges Experiment von begeisterten Physikern aussehen mag, entpuppte sich schnell als wichtiges Werkzeug für Wetter- und Klimaforscher. Denn die genaue Kenntnis der großen Ozeanströmungen ist unerlässlich, um das Klimageschehen auf der Erde besser zu verstehen.

Der Boom der Erdbeobachtung geht munter weiter. Die nächsten Satelliten sind schon lange in Bau. Zu SMOS, dem Eissatelliten Cryosat-2 und den Magnetfeldsonden Swarm gesellen sich in den kommenden Jahren mindestens zwei weitere fliegende Spezialisten: ADM-Aeolus wird die Dynamik unserer Atmosphäre erfassen, also welche Stoffe wo in der Luft vorkommen und wie sie sich mit Wind und Wetter verteilen. Als im April 2010 die Aschewolke eines Vulkans auf Island den Flugverkehr in fast ganz Europa lahm gelegt hat, hätte ADM-Aeolus wertvolle Dienste leisten können.

Dieser Satellit wird dreidimensional vermessen, wo welche Partikel durch die Atmosphäre schweben. Mit seinen Daten können künftig die Flugsicherheitsbehörden sehr viel fundierter beurteilen, ob ein Flugverbot notwendig ist. Heute basieren solche Entscheidungen vor allem auf recht groben Modellannahmen. Künftig werden stets aktuelle Messdaten zur Verfügung stehen.

slightly off-the-wall experiment by enthusiastic physicists soon revealed itself as an important tool for weather and climate researchers. Precise know-ledge of the great ocean currents is essential if we are better to understand the way the Earth's climate works.

This boom in Earth observation is continuing unabated and the next satellites have been under construction for some time. Three more orbiting specialists will be joining SMOS, GOCE, and the ice satellite CryoSat-2 in the next few years: ADM-Aeolus will monitor the dynamics of the atmosphere—in other words, which substances occur in the air and how they are distributed by wind and weather. ADM-Aeolus would have done sterling work in April 2010, when a cloud of ash from the volcano Eyjafjallajökull in Iceland grounded almost all European air traffic for several days.

The satellite will make 3-D observations of the particles floating in the atmosphere, and the air safety authorities will be able to use its data to decide with more confidence if a flight ban is required. Such decisions are currently made on the basis of rather crude simulations, but real-time data will be available in the future.

un instrument précieux pour la science climatique. La connaissance exacte des grands courants océaniques est en effet indispensable pour mieux comprendre les phénomènes climatiques terrestres.

L'exploration de la Terre se poursuit avec ferveur. Les prochains satellites sont d'ores et déjà en construction. Au moins deux nouveaux spécialistes viendront rejoindre au cours des prochaines années SMOS, le satellite CryoSat-2 en mission glaciaire et les sondes d'étude du champ magnétique de Swarm. ADM-Aeolus observera la dynamique de l'atmosphère terrestre, c'est-à-dire la composition de l'air et la circulation des vents. Lorsqu'en avril 2010 le nuage de cendres issu d'un volcan islandais a paralysé l'ensemble du trafic aérien en Europe, ADM-Aeolus aurait pu apporter une aide précieuse.

Ce satellite livrera des analyses tridimensionnelles des particules en suspension dans l'atmosphère. Grâce à ces données, les autorités de contrôle du trafic aérien seront en mesure de déterminer avec plus d'exactitude la nécessité d'interdire le trafic aérien. Ces décisions se fondent actuellement sur des calculs très approximatifs. À l'avenir, des mesures actualisées seront disponibles en permanence.

una herramienta importante para los investigadores del tiempo y del clima. Porque para entender mejor los acontecimientos climáticos de la Tierra, es indispensable un conocimiento exacto de las corrientes oceánicas.

El auge de la observación terrestre prosigue con una gran vitalidad. Ya hace tiempo que se están construyendo nuevos satélites. A los SMOS, el satélite CryoSat-2 para el estudio del hielo y las sondas para la investigación del campo magnético de Swarm, se unirán en los próximos años al menos otros dos especialistas volantes. Uno de ellos, el ADM-Aeolus, estudiará la dinámica de nuestra atmósfera, es decir, las sustancias que se encuentran en el aire y su sistema de distribución con el viento y el tiempo. Cuando en abril de 2010, la nube de cenizas de un volcán islandés paralizó el tráfico aéreo en casi toda Europa, el ADM-Aeolus podría haber prestado un valioso servicio. Este satélite medirá tridimensionalmente dónde y qué partículas están en suspensión en la atmósfera. En el futuro, con los datos que proporcione, las autoridades de seguridad aérea podrán valorar, con mayor fundamento, si es necesaria una prohibición de vuelo. Actualmente, estas decisiones se basan en suposiciones referidas a modelos aproximados.

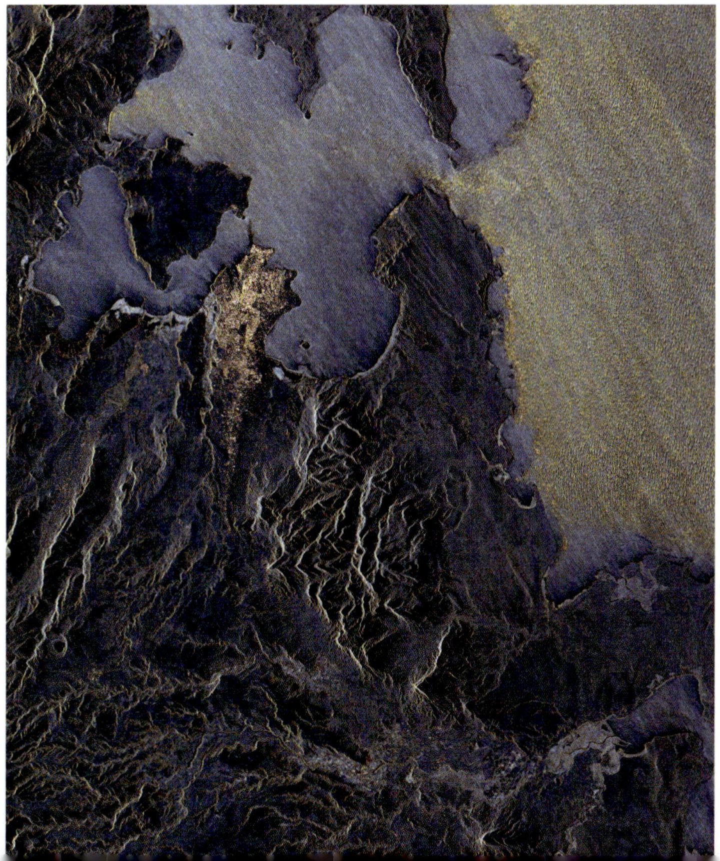

Seite 86/87: Berge, Eis und Seen in einer Andenregion in Argentinien (MIR)

Pages 86–7: Mountains, ice, and lakes in a region of the Andes in Argentina (MIR)

Pages 86 et 87 : monts, lacs et glaces d'une région andine en Argentine (MIR)

Páginas 86/87: Montañas, hielo y lagos en una región andina de Argentina (MIR)

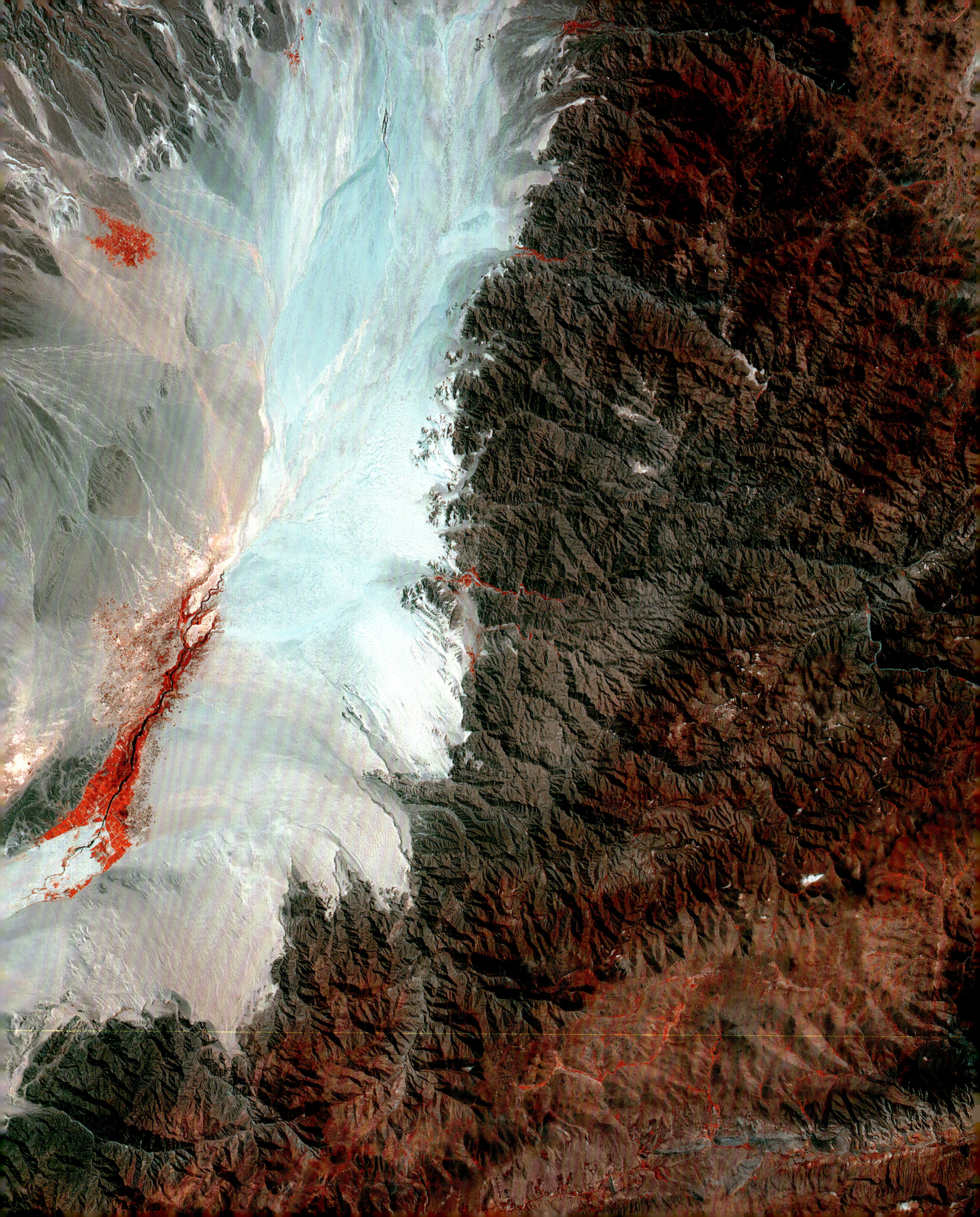

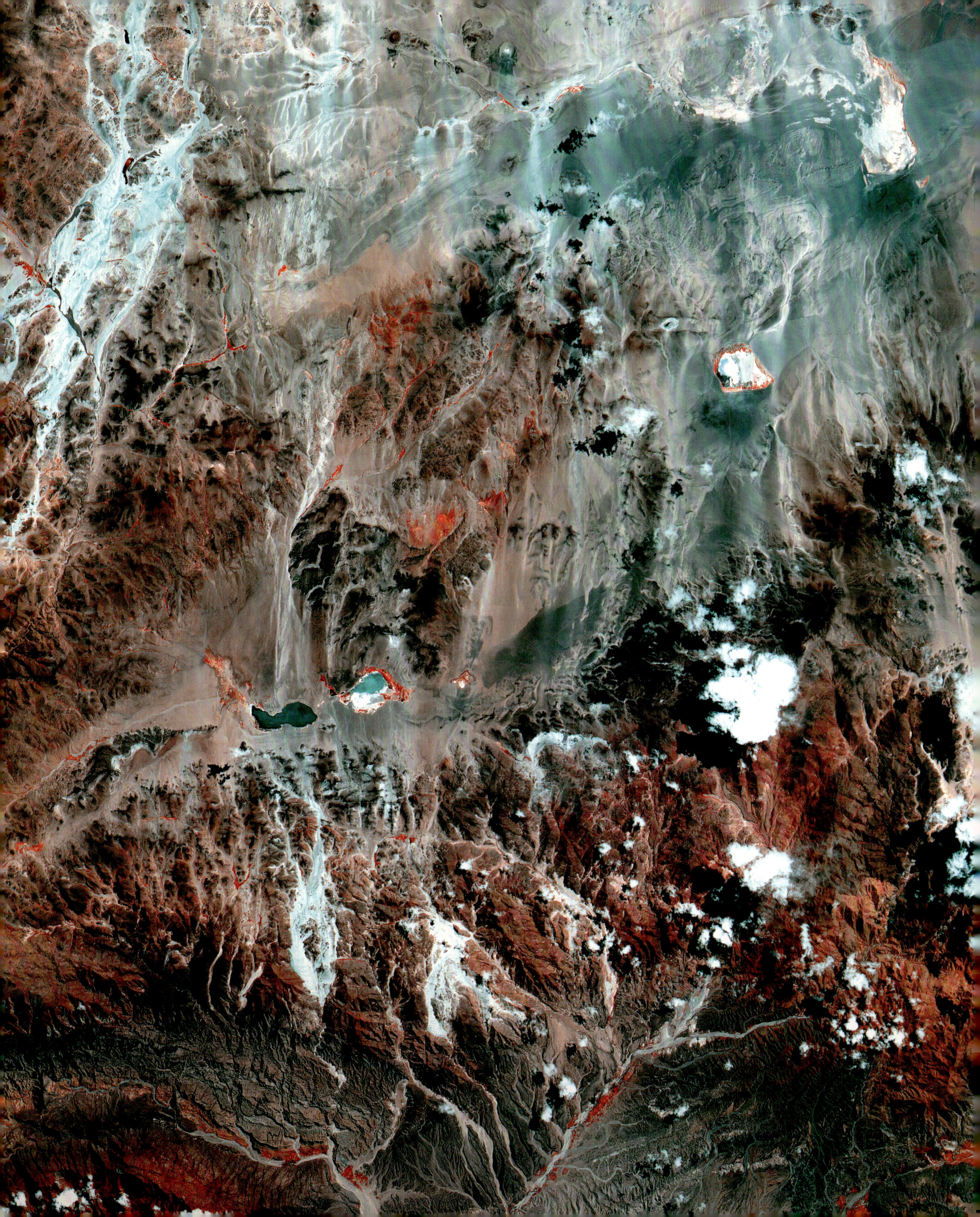

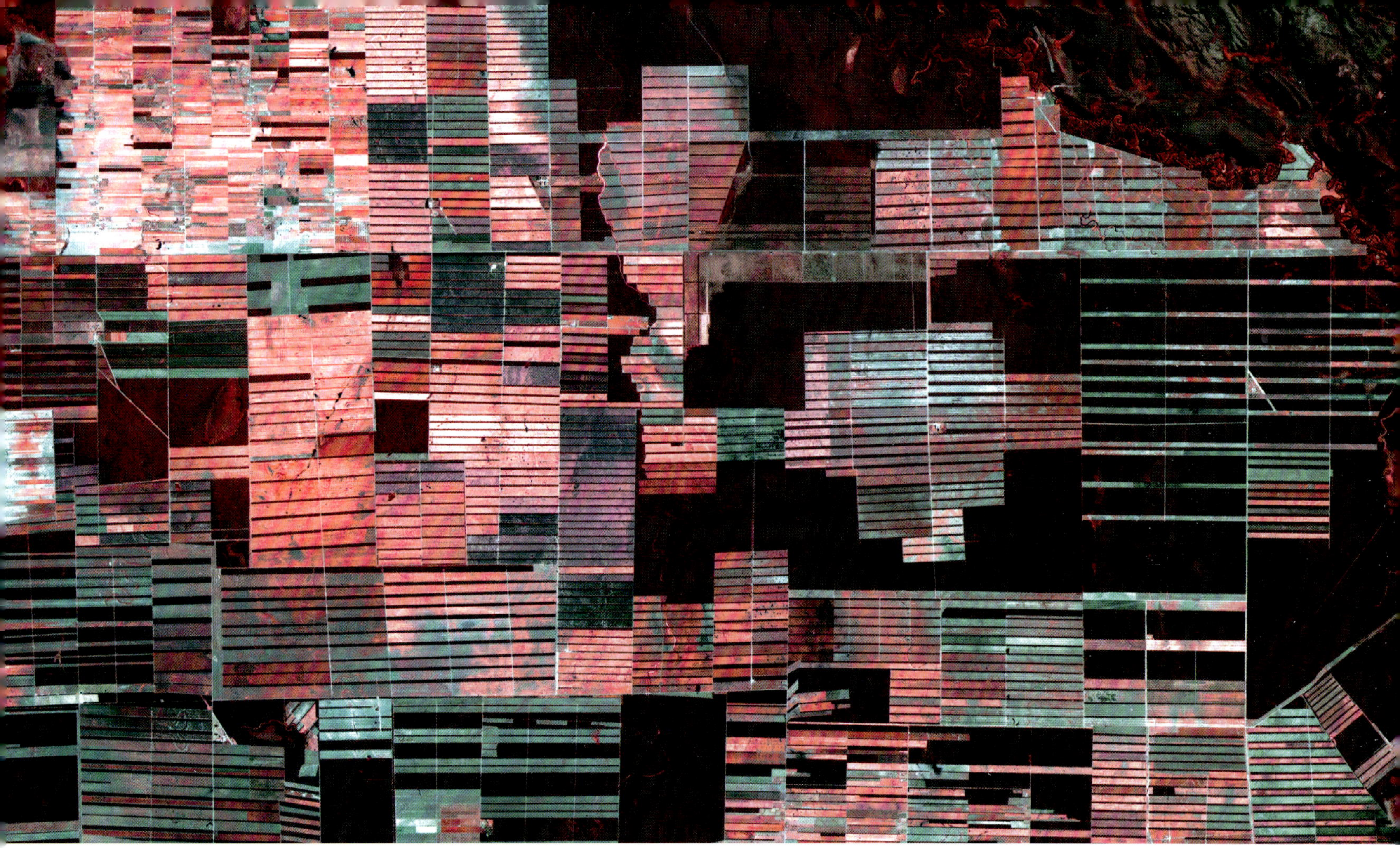
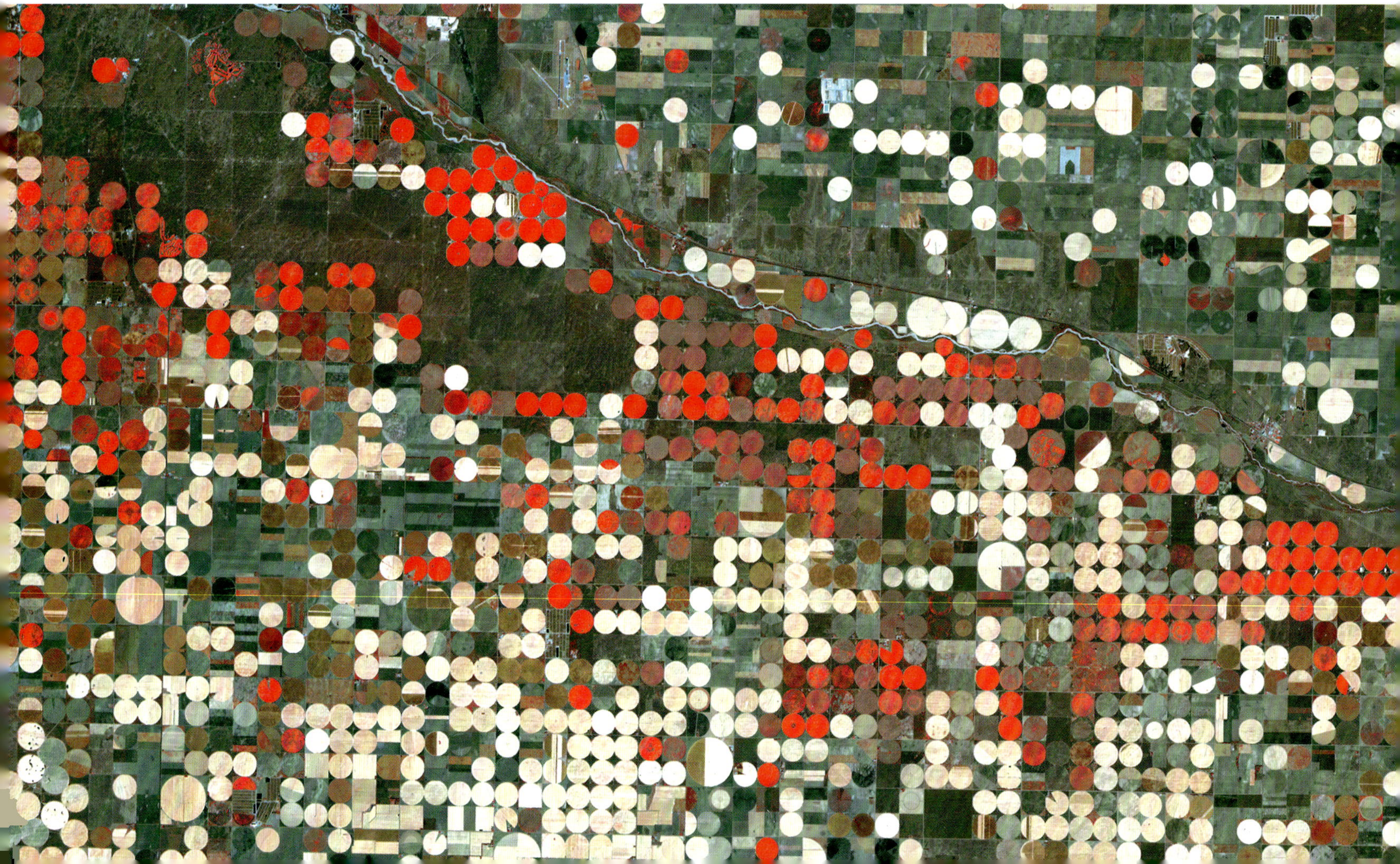

Luft kann man sehen

Swarm ist die Mission, die seit 2013 das Magnetfeld der Erde untersucht. Drei baugleiche Satelliten verteilen sich rund um unseren Planeten und ermöglichen so eine räumliche Vermessung des Magnetfeldes der Erde. Schließlich soll EarthCARE, eine Mission, die gemeinsam mit Japan realisiert wird, genauer untersuchen, wie die Wolkenbildung in der Atmosphäre, der Gehalt an Schwebteilchen und der Strahlungshaushalt unserer Erde zusammenhängen. Bis heute gibt es nur vage Vorstellungen davon, wann und wie sich Wolken bilden. Für die Forscher bleibt viel zu tun. Die Erde ist in vielem noch immer ein geheimnisvoller Körper.

Im Rahmen des Copernicus-Programms von ESA und Europäischer Union werden mindestens fünf Sentinel-Satelliten gestartet. Sentinel ist das englische Wort für Wache, im Sinne von »über etwas wachen«. Sentinel-1 ist ein fliegendes Radargerät, das Landmassen und Ozeane überwachen soll. Der zweite Satellit der Serie wird hochaufgelöste und extrem detailreiche Aufnahmen im Optischen und nahen Infrarot machen. Sentinel-3 besteht aus zwei Satelliten, die jeweils binnen zwei Tagen die gesamte Erde fotografisch

You can see air

The objective of the ESA's Swarm mission has been to explore the Earth's magnetic field since 2013. Three identical satellites will be arranged in different polar orbits around our planet, enabling spatial measurement of the geomagnetic field. EarthCARE, a joint mission undertaken with Japan, should facilitate a closer examination of cloud formation, particle distribution in the atmosphere, and the Earth's radiation budget. Until now, there have only been vague theories about how clouds are formed, and the scientists still have much ground to cover—in many respects, the Earth is still an enigmatic body.

At least five Sentinel satellites are to be launched as part of the ESA and European Union's Copernicus program. As the name perhaps implies, Sentinel-1 is a flying radar station, monitoring the land masses and oceans. Sentinel-2 will take extremely detailed, high-resolution photographs using visible light and the low infrared spectrum. Sentinel-3 consists of two satellites, each of which will photograph the entire world every two days. The instruments aboard Sentinel-4 will examine chemical processes in the atmosphere

L'air rendu visible

Swarm est une mission qui étudie le champ magnétique terrestre depuis 2013. Elle prévoit le lancement d'une constellation de trois satellites chargés de cartographier le champ magnétique de la Terre. Enfin, la mission euro-japonaise EarthCARE étudiera les nuages, les microparticules en suspension dans l'atmosphère et leur influence sur le rayonnement atmosphérique. Nous n'avons aujourd'hui qu'une vague idée de la manière dont se forment les nuages. Les scientifiques ont donc encore fort à faire pour percer les nombreux mystères de la planète.

L'ESA et l'Union européenne prévoient de lancer dans le cadre du programme Copernicus au moins cinq satellites sentinelles chargés de veiller sur la Terre. Sentinelle-1 est équipé d'un radar surveillant les masses terrestres et les océans. Le deuxième satellite de la série fournira des images optiques et proche infrarouge haute résolution, riches en détails. Sentinelle-3 est composé de deux satellites capables de photographier l'ensemble de la Terre en deux jours. Sentinelle-4 examinera pour sa part les processus chimiques dans l'atmosphère terrestre et fournira des données à haute résolution

Se puede ver el aire

La misión de satélites Swarm analiza el campo magnético de la Tierra desde 2013. Se colocarán tres satélites iguales alrededor de nuestro planeta, que permitirán una medición tridimensional del campo magnético terrestre. Finalmente, EarthCARE, una misión que se realiza en colaboración con Japón, abordará las interacciones entre la formación de nubes en la atmósfera, el contenido de partículas en suspensión y el balance de radiación de nuestra Tierra. Hasta la actualidad solo existe una vaga idea de cuándo y cómo se forman las nubes. Los investigadores tienen todavía mucho por hacer. La Tierra aún es, en muchos aspectos, un objeto celeste misterioso.

En el marco del programa conjunto Copérnico de la ESA y de la Unión Europea, se lanzarán al menos cinco satélites Sentinel, palabra que proviene del inglés y significa centinela, el que vigila. Sentinel-1 es un equipo volante de radar que vigilará las masas de tierra y los océanos. El segundo satélite de la serie, Sentinel-2, hará fotografías ópticas y cercanas al infrarrojo, de alta resolución y extremadamente detalladas. Sentinel-3 constará, a su vez, de dos satélites que pueden fotografiar toda la Tierra

Linke Seite, oben: Einst unberührter Regenwald in Bolivien, jetzt intensiv genutzt für Felder und Weiden (Landsat)

Top: Once unspoiled rainforest in Bolivia has now been turned over to intensive agriculture and pasture land (Landsat)

Page de gauche, en haut : jadis forêt vierge de Bolivie, désormais transformée en exploitation agricole et pâturages (Landsat)

Página izda., arriba: Antes selva virgen en Bolivia, ahora usado intensivamente como campo y pasto (Landsat)

Linke Seite, unten: Spezielle Bewässerungssysteme führen zu kreisrunden Feldern, hier in den USA (Landsat)

Bottom: Some irrigation systems create circular fields, such as here in the USA (Landsat)

Page de gauche, en bas : des systèmes d'irrigation spéciaux arrosent des champs circulaires, ici aux États-Unis (Landsat)

Página izda., abajo: Sistemas especiales de riego que condicionan la forma redonda de los campos, en EE. UU. (Landsat)

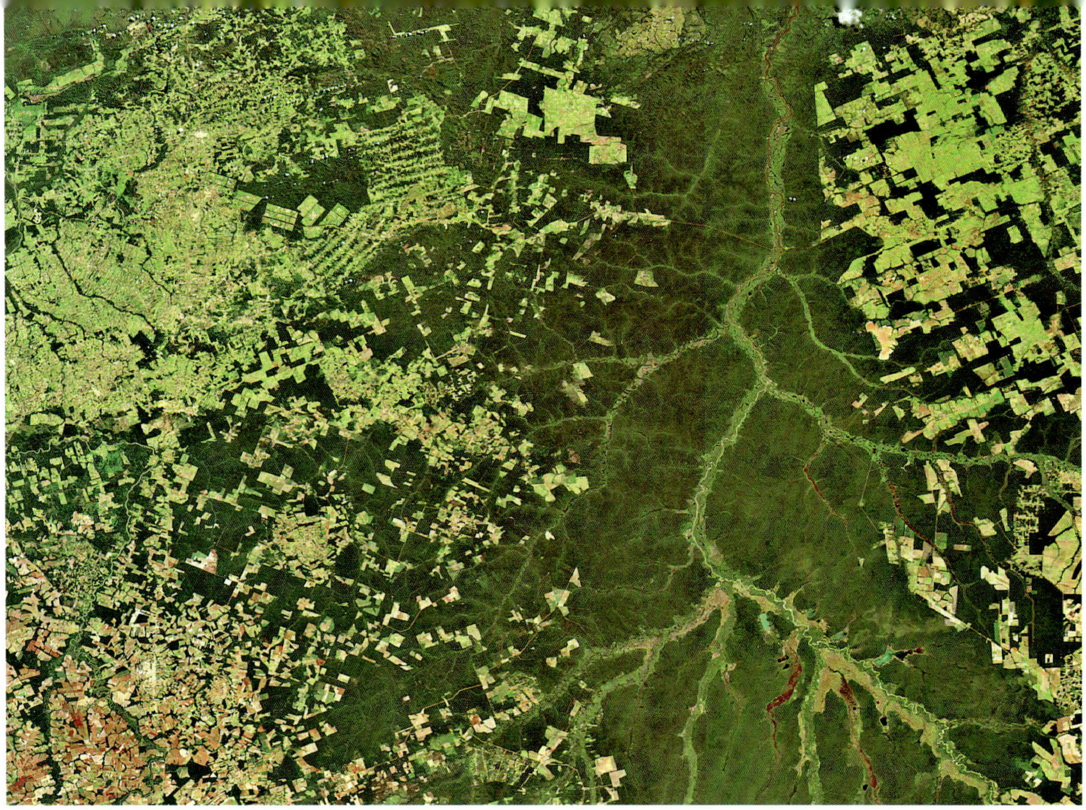

erfassen. Sentinel-4 ist ein Instrument, das die chemischen Prozesse in der Erdatmosphäre mit hoher zeitlicher und räumlicher Auflösung untersucht. Es wird Teil eines Meteosat-Satelliten der dritten Generation. Diese modernsten Meteosats werden gerade unter Federführung von Deutschland und Frankreich gebaut. Sentinel-5 schließlich soll von einer niedrigen Umlaufbahn aus die Vorgänge in der irdischen Lufthülle detailliert vermessen.

Die USA fallen zurück

Europa ist in der Erdbeobachtung weltweit führend. Aber natürlich ist der alte Kontinent nicht allein im All. Auch die US-amerikanische NASA betreibt etliche Satelliten, die sich speziell der Erforschung der Erde widmen. Das »Earth Observing System«, so der Name des Programms, begann mit den ersten Starts 1997. Missionen wie Terra, der legendäre Landsat-7, Jason-1 oder CloudSat machen hochauflösende Bilder der Erdober fläche, untersuchen die Meere und verfolgen die Vorgänge in der Atmosphäre. Doch in den USA war für einige Jahre die Erdbeobachtung politisch nicht mehr ganz oben auf der Agenda. So sind viele NASA-

at great temporal and spatial resolutions. All these are part of the third generation of Meteosats and the most modern of these satellites are being built under the aegis of Germany and France. Sentinel-5, the last in the series, will adopt a low Earth orbit to make detailed measurements of atmospheric processes.

The USA is lagging behind

Europe leads the world in Earth observation, but the Old World is not alone in space. The USA's NASA maintains numerous specially dedicated satellites for Earth research, and the "Earth Observing System," as the program is called, began with its first launch in 1997. Missions such as Terra-SAR, the legendary Landsat-7, Jason-1, and CloudSat have provided high-resolution images of the Earth's surface, investigated the sea, and tracked changes in the atmosphere. For several years, however, Earth observation was low on the political agenda in the USA, and many NASA missions in the field are relatively outdated in comparison with their European equivalents and US scientists often observe the satellite projects of their European colleagues with envy.

spatiale et temporelle. Cet instrument sera intégré à un satellite Météosat de troisième génération, actuellement en cours de construction sous l'égide de la France et de l'Allemagne. Enfin, Sentinelle-5 mesurera avec précision les phénomènes atmosphériques depuis une orbite basse.

Les États-Unis distancés

Si l'Europe a pris la tête dans le domaine de l'observation de la Terre, elle n'est pas seule dans l'espace. La NASA, l'agence spatiale américaine, exploite également quantité de satellites voués à l'exploration de la Terre dans le cadre du programme « Earth Observing System » lancé en 1997. Les missions telles que Terra, le fameux Landsat-7, Jason-1 ou CloudSat fournissent des images haute résolution de la surface terrestre, étudient les mers et observent les phénomènes atmosphériques. Cependant, durant plusieurs années, l'observation de la Terre semblait avoir disparu du programme politique américain. C'est pourquoi, comparées aux missions européennes, les missions de la NASA paraissent souvent dépassées. Les scientifiques américains suivent ainsi

en un plazo de dos días. Sentinel-4 será un instrumento que analizará los procesos químicos de la atmósfera terrestre con una alta resolución temporal y tridimensional. Finalmente, Sentinel-5 deberá medir, desde una órbita baja, los procesos en la atmósfera terrestre.

Estados Unidos se queda atrás

Europa es el líder mundial en la observación terrestre, pero no está sola en el espacio. También la NASA estadounidense opera algunos satélites que se dedican a la investigación de la Tierra. El «Earth Observing System», que es como se llama el programa, comenzó sus primeros lanzamientos en 1997. Misiones como Terra, la legendaria Landsat-7, Jason-1 o CloudSat, toman imágenes de alta resolución de la superficie terrestre, analizan los mares y observan los fenómenos atmosféricos. Pero, durante algunos años, la observación de la Tierra dejó de ser un asunto político de primer orden en Estados Unidos. Así, en comparación con sus equivalentes europeas, muchas misiones de la NASA en este sector se han quedado muy anticuadas. A menudo, los científicos estadounidenses contemplan envidiosos los proyectos de

Linke Seite: Am Rio Xingu in Brasilien fressen sich die Felder immer weiter in den Regenwald (Envisat)

Left: Crop cultivation is eating away more and more of the rainforest surrounding the Rio Xingu in Brazil (Envisat)

Page de gauche : sur les rives du Rio Xingu, au Brésil, les champs empiètent de plus en plus sur la forêt tropicale (Envisat)

Página izda.: En el río Xingú de Brasil los campos van devorando la selva tropical (Envisat)

Unten: In Indonesien wurden große Flächen (oben im Bild) für den Palmenanbau gerodet (Envisat)

Below: Vast areas of Indonesia (visible at the top of the image) have been cleared to cultivate palm trees (Envisat)

Ci-dessous : en Indonésie, de vastes surfaces ont été défrichées au profit de la culture du palmier (Envisat)

Abajo: En Indonesia se talan grandes áreas (parte superior de la imagen) para cultivar palmeras (Envisat)

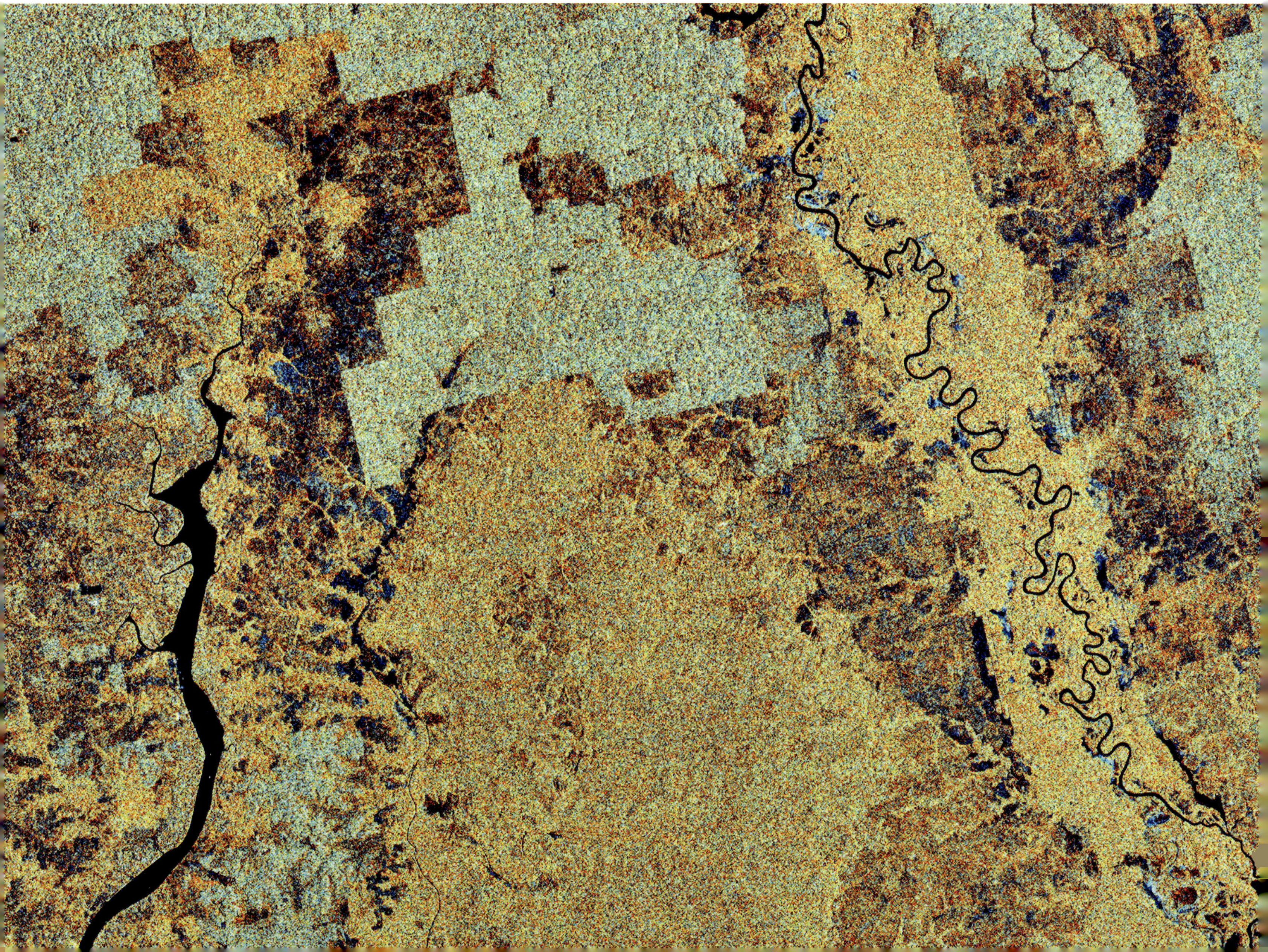

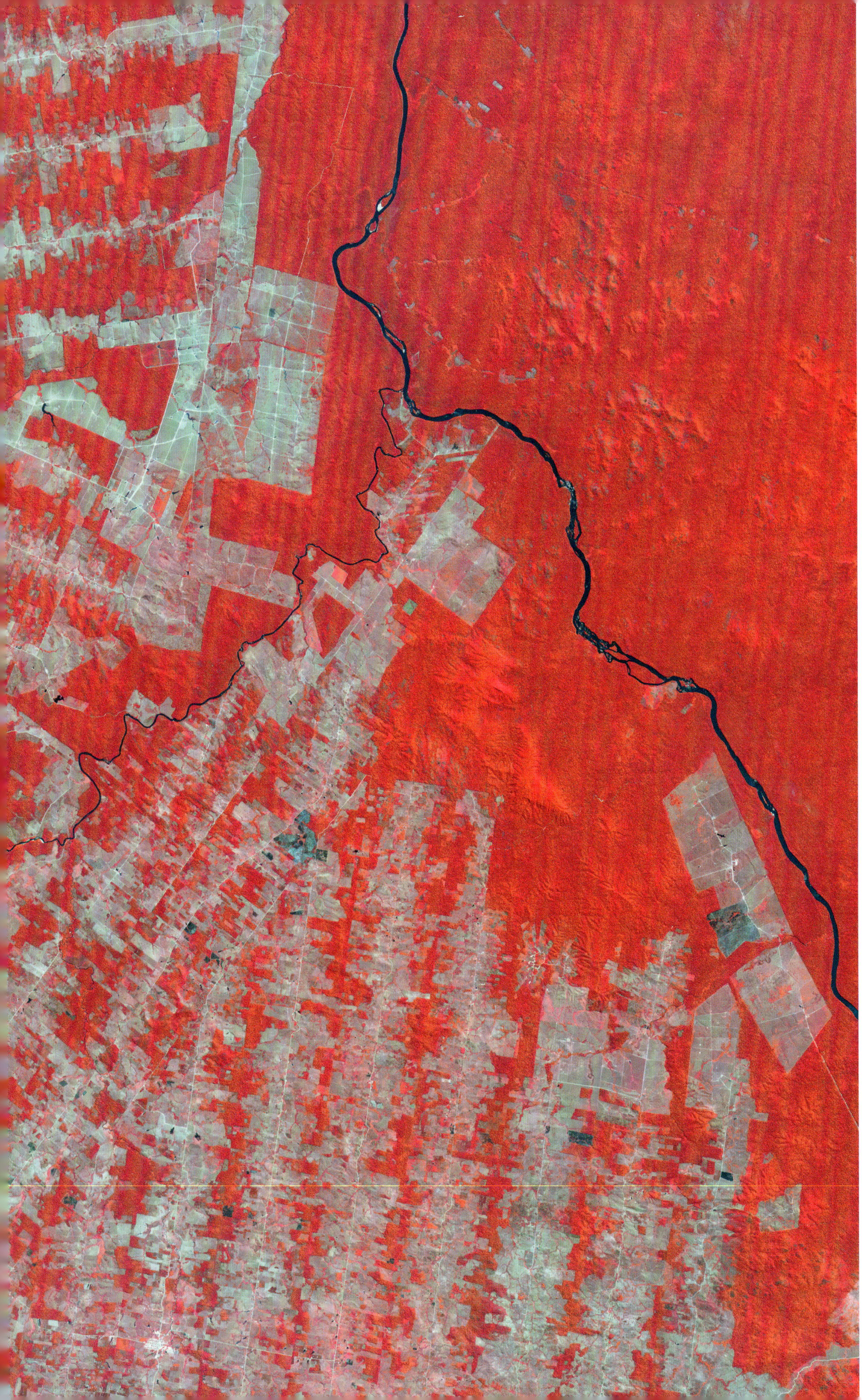

Die Rodung des Regenwalds schreitet in Brasilien immer weiter voran. Rot zeigt Bereiche intakter Vegetation (Terra)

More and more of the Brazilian rainforest is being cleared. Areas of untouched vegetation show up red (Terra)

Dans la forêt tropicale du Brésil, le déboisement gagne sans cesse du terrain. Les zones rouges représentent les zones de végétation encore intacte (Terra)

La tala de la selva tropical no se detiene en Brasil. El rojo muestra las zonas de vegetación intacta (Terra)

Rechte Seite: Eine Gegend im brasilianischen Staat Mato Grosse 1992 (links) und 2006 (rechts). Vom Regenwald blieb nicht viel übrig (Terra)

Right: An area in the Brazilian state of Mato Grosso in 1992 (near right) and 2006 (far right). Little rainforest now remains (Terra)

Page de droite : une région de l'état brésilien du Mato Grosso en 1992 (à gauche) et en 2006 (à droite). Il ne reste plus grand-chose de la forêt amazonienne (Terra)

Página dcha.: Una comarca en el estado brasileño de Mato Grosso, en 1992 (izda.) y 2006 (dcha.). No queda mucho de la selva tropical (Terra)

Missionen in diesem Bereich verglichen mit den europäischen Gegenstücken stark veraltet. US-Wissenschaftler blicken oft neidvoll auf die Satellitenprojekte der europäischen Kollegen. Doch im Sinne der Forschung arbeiten an den Daten meist internationale Teams. In der Regel kommt es zu einem regen Datenaustausch. Großes Pech hatte die NASA 2009, als ihr OCO, das Orbiting Carbon Observatory, beim Start verunglückte und in das Südpolarmeer stürzte. OCO hätte aus der Umlaufbahn Vorkommen und Verteilung unter anderem von Kohlendioxid in der Atmosphäre studieren sollen. Kohlendioxid gilt als eine der Hauptursachen für den vom Menschen verursachten Wandel des Erdklimas. Immerhin hat die US-Regierung schnell beschlossen, OCO ein zweites Mal zu bauen. Mitte 2014 ist OCO-2 erfolgreich gestartet.

Andere Nationen, die gezielt die Erde beobachten, sind Russland, China, Indien, Brasilien und einige andere Staaten. Vor allem Indien verfolgt ein sehr ehrgeiziges Programm, um den Subkontinent aus dem All genauestens zu beobachten, Karten zu erstellen, Wetterdaten zu sammeln und Ressourcen nutzbar zu machen. Auch in Europa betreiben einige Länder, zusätzlich zu ihrem Engage-

As far as research is concerned, however, the data generated is worked upon by international teams and the results are regularly shared. NASA suffered a great misfortune in 2009 when its OCO, the Orbiting Carbon Observatory, misfired on launch and crashed into the Antarctic Ocean. Once in orbit, OCO was to have studied the occurrence and distribution of various compounds, including carbon dioxide, in the atmosphere. Carbon dioxide is thought to be one of the chief agents in the climate change that Man has caused on Earth. The US government quickly decided to construct a replacement OCO. Baptized as OCO-2, the second launch was a success in the middle of 2014.

Other nations making Earth observations include Russia, China, India, Brazil, and a few other states. India in particular is pursuing an extremely ambitious program of observing the subcontinent from space, making maps, collecting weather data, and opening up resources. In addition to their contribution to the ESA, some European countries are maintaining other satellites—Germany has launched radar missions known as TerraSAR-X and TanDEM-X, for example, and Italy has a whole constellation of

souvent avec convoitise les projets de satellite de leurs confrères européens. Les données sont toutefois traitées par des équipes internationales dans l'intérêt de la recherche, au cours d'échanges généralement intensifs. Précisons que la NASA a été victime de malchance en 2009 : son Orbiting Carbon Observatory (OCO) a été détruit au cours de son lancement avant de s'abîmer dans l'océan Antarctique. OCO devait mesurer la concentration et la répartition du dioxyde de carbone dans l'atmosphère terrestre, ce gaz considéré comme l'un des principaux facteurs du changement climatique d'origine humaine. Toujours est-il que le gouvernement américain a décidé de construire un nouvel exemplaire d'OCO. Baptisé OCO-2, il a été lancé avec succès au milieu de l'année 2014.

Parmi les autres nations qui observent la Terre avec attention figurent notamment la Russie, la Chine, l'Inde et le Brésil. L'Inde a ainsi lancé un programme extrêmement ambitieux visant à surveiller et cartographier le sous-continent avec précision, collecter des données climatiques et valoriser les ressources du pays. Certains états européens exploitent indépendamment de leur participation aux programmes de l'ESA

satélites de sus colegas europeos. Pero si nos atenemos a la investigación, la mayor parte de las veces los datos son tratados por equipos internacionales. Por regla general, se suele producir un intenso intercambio de datos. La NASA tuvo muy mala suerte en 2009 cuando su OCO, el Orbiting Carbon Observatory, sufrió un accidente en el despegue y acabó cayendo al océano Antártico. OCO tendría que haber estudiado, desde la órbita, la presencia y distribución en la atmósfera de, entre otros, dióxido de carbono. Este gas está considerado una de las causas principales del cambio climático antropogénico de la Tierra. De todos modos, el gobierno estadounidense ha decidido construir un segundo OCO. Bautizado como OCO-2, se lanzó con éxito a mediados de 2014.

Otras naciones que dedican recursos a la observación de la Tierra son Rusia, China, India, Brasil y algunos otros estados. La India, en particular, sigue un programa muy ambicioso para poder observar el subcontinente desde el espacio con la máxima precisión, elaborar mapas, recoger datos meteorológicos y aprovechar los recursos. También en Europa, algunos países gestionan otros satélites al margen de su compromiso con la ESA. Alemania está presente con las misio-

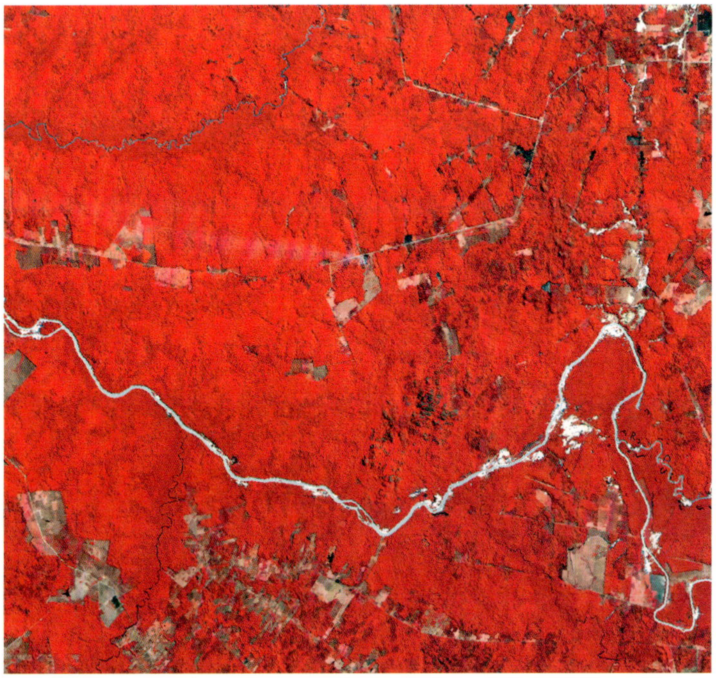

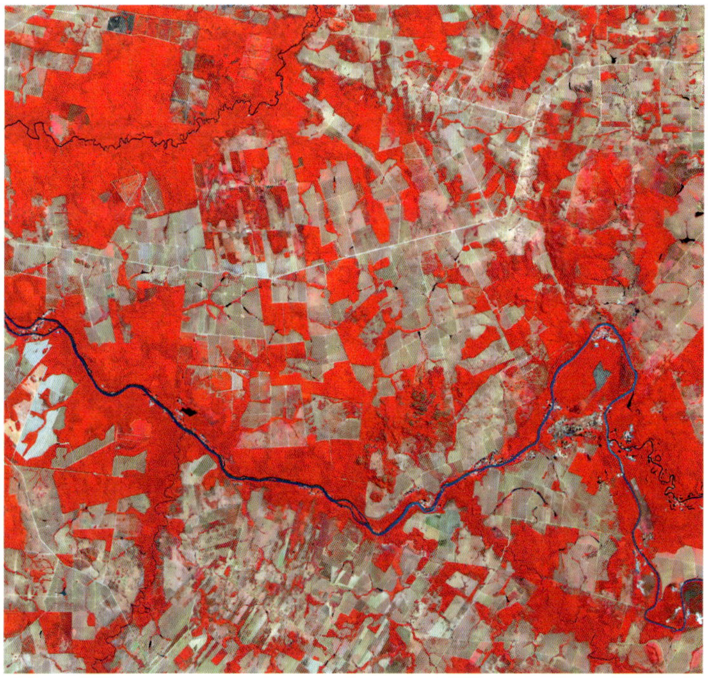

ment in der ESA, weitere Satelliten. Deutschland ist mit den Radarmissionen TerraSAR-X und TanDEM-X vertreten. Italien verfügt über eine Konstellation von Kleinsatelliten, die den Mittelmeerraum per Radar im Blick haben. Ziel ist sowohl die zivile Erdbeobachtung als auch die militärische Überwachung. Die Satelliten der französischen Spot-Serie können praktisch jeden Tag jeden beliebigen Punkt auf der Erde aufnehmen. Dieser kommerzielle Dienst steht allen Interessenten offen.

Die Beobachtung der Erde erfolgt heute auf äußerst vielfältige Weise. Dutzende Satelliten haben unseren Heimatplaneten im Blick. Zum Einsatz kommen ganz unterschiedliche Techniken, je nachdem, was die Satelliten erforschen sollen. Auch die Umlaufbahnen der himmlischen Späher werden der jeweiligen Aufgabe optimal angepasst. Beim Blick nach unten auf die Erde haben die Forscher genau die Entwicklung nachvollzogen, die die Astronomen beim Blick hinauf von der Erde durchgemacht haben. Ursprünglich gab es nur Beobachtungen im sichtbaren Licht. Bei der Erforschung des Kosmos kamen schnell neue »Fenster« hinzu: Radiowellen, Röntgenlicht, Ultraviolett- und Infrarotstrahlung. Erst die Kombination all dieser Daten ergibt

smaller satellites monitoring the Mediterranean with radar.

The aim of this activity is probably both civilian Earth observation and military surveillance. The satellites in France's Spot series can photograph practically any point on Earth every day, and this has been made available to all as a commercial service.

Earth observation is now carried out in all kinds of ways and dozens of satellites are watching our home planet. A wide variety of techniques is deployed, depending on what the satellites are intended to explore. Even the orbits adopted by the eyes in the sky are adapted to each specific task. Scientists looking down on Earth have followed the same learning curve as the astronomers who look out from it. Observations were originally made using only visible light, but as the cosmos was probed, other avenues of exploration opened up: radio waves, X-rays, and the ultraviolet and infrared spectra. A combination of all this data provides the most comprehensive image of the processes going on in the cosmos, and the same is true for Earth observation: the first weather satellites began by using onboard television cameras transmitting grainy black-and-white pictures.

quelques satellites supplémentaires. Ainsi, l'Allemagne dispose des satellites radar TerraSAR-X et TanDEM-X, tandis que l'Italie surveille l'espace méditerranéen à l'aide d'une constellation de minisatellites à vocation civile et militaire. Les satellites français Spot sont capables d'acquérir chaque jour une image de n'importe quel point du globe, un service commercial à la disposition de tous.

À l'heure actuelle, les méthodes d'observation de la Terre sont multiples. Des dizaines de satellites scrutent notre planète, à l'aide de techniques diverses adaptées à leur mission, sur une orbite appropriée. L'observation de la Terre a suivi la même évolution que l'astronomie. Les scientifiques ont commencé par des observations en lumière visible, puis l'exploration de l'univers a ouvert de nouvelles « fenêtres » : ondes radio, rayons X, rayonnements ultraviolet et infrarouge. Seule l'association de toutes ces données permet d'obtenir une image aussi complète que possible des phénomènes cosmiques. Il en va de même pour l'observation de la Terre. Les premiers satellites météorologiques, équipés de caméras de télévision, livraient des images floues en noir et blanc. Plus tard, des instruments plus perfectionnés capables de dé-

nes de radar TerraSAR-X y TanDEM-X. Italia dispone de una constelación de pequeños satélites que mantienen en observación por radar la cuenca del Mediterráneo. El objetivo es tanto la observación de la Tierra como la vigilancia militar. Los satélites franceses de la serie Spot pueden tomar imágenes, prácticamente en cualquier momento, del punto que se desee.

En la actualidad, la observación de la Tierra se realiza de maneras extremadamente diversas. Docenas de satélites someten a nuestro planeta a una minuciosa observación. Las técnicas que se utilizan varían mucho, dependiendo de aquello que los satélites deban investigar. Las órbitas de los observadores celestes también se deben adaptar a la correspondiente tarea. En la observación de la Tierra, los investigadores actuales han seguido exactamente el mismo camino que tomaron los antiguos astrónomos. Al principio, sólo había observaciones con luz visible. En la investigación del cosmos se fueron añadiendo nuevas «ventanas»: ondas de radio, rayos X, radiación ultravioleta y rayos infrarrojos. Únicamente la combinación de todos esos sistemas hace posible una imagen lo más amplia posible de los procesos cósmicos. Lo mismo sirve para la observación de la Tierra. Los primeros

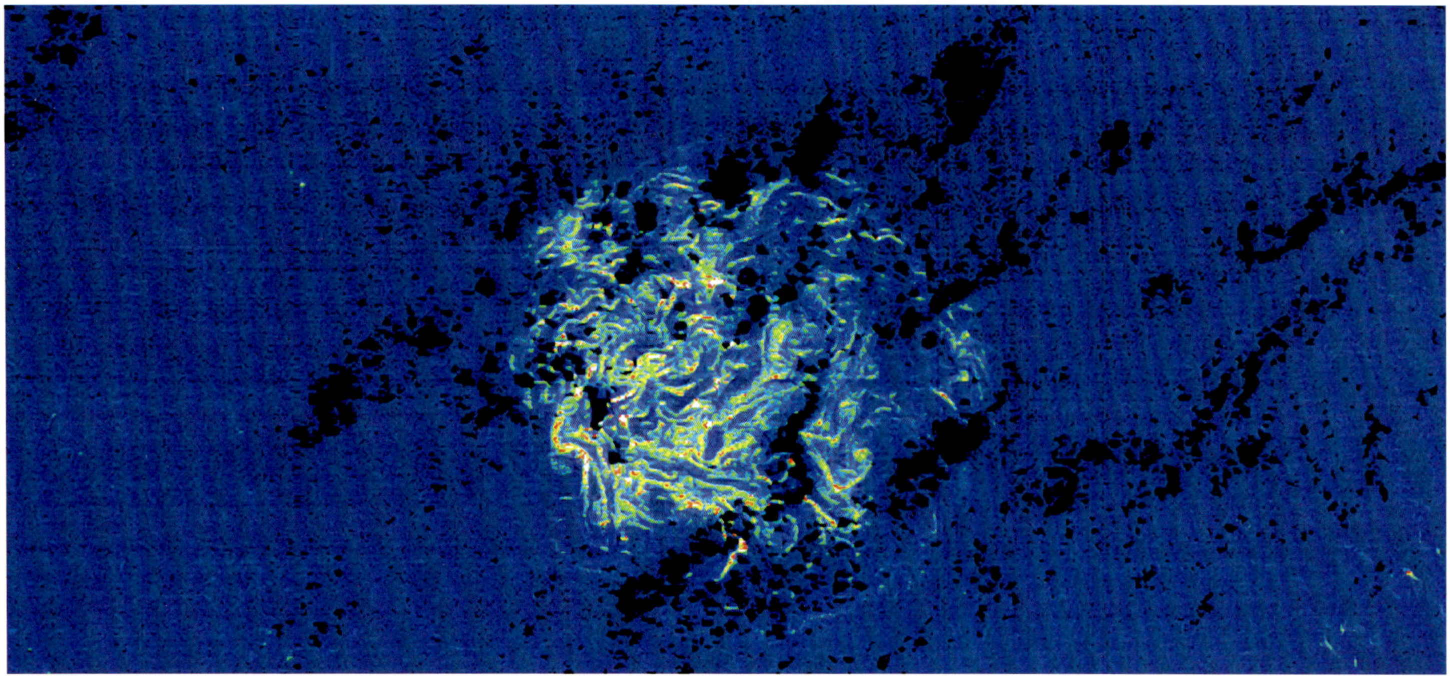

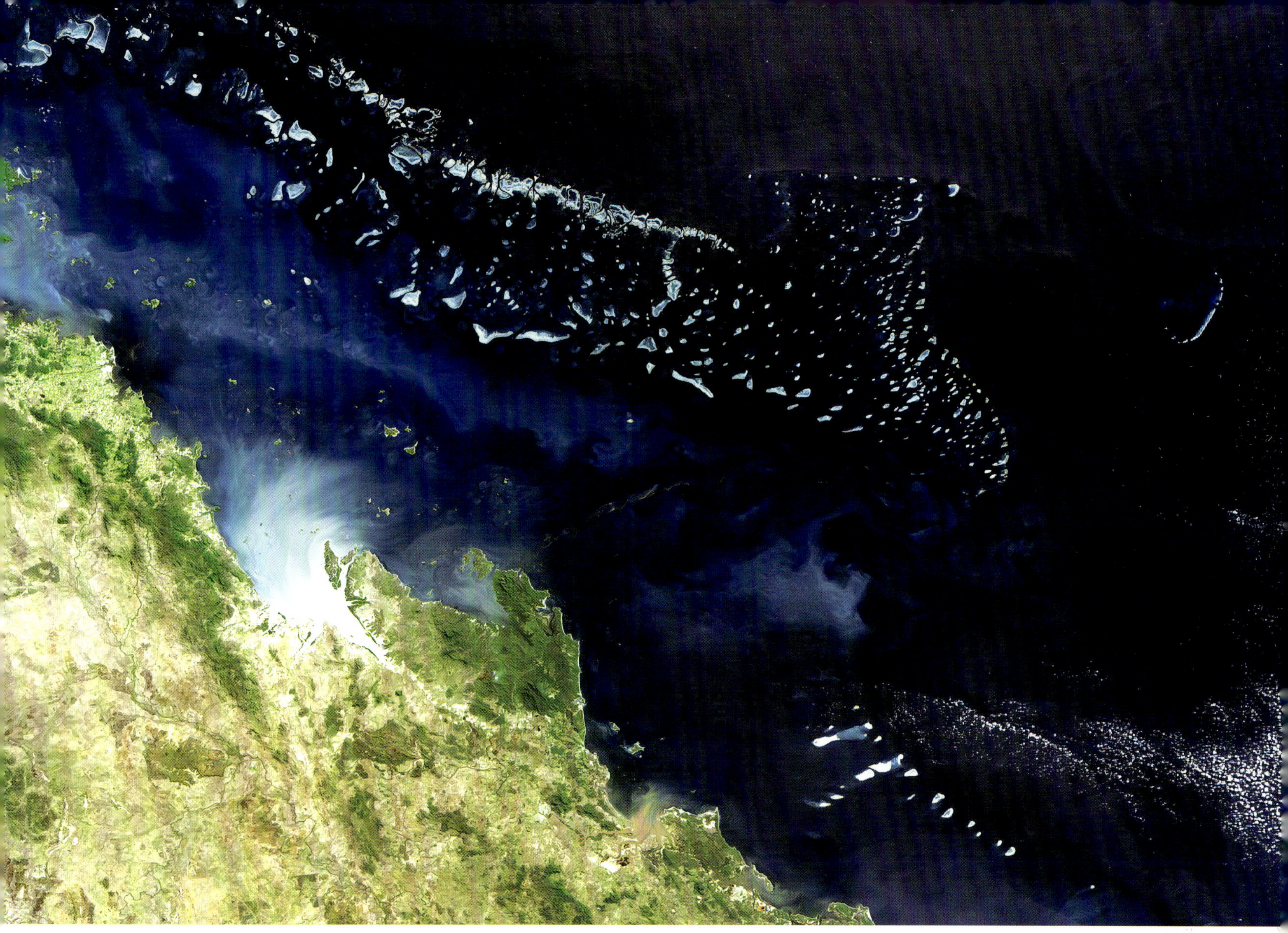

Linke Seite: Ein 45 km großes Knäuel von treibendem Seetang im Golf von Mexiko (Envisat)

Left: A 28-mile-wide (45-km) tangle of seaweed in the Gulf of Mexico (Envisat)

Page de gauche : masse de varech de 45 km flottant dans le golfe du Mexique (Envisat)

Página izda.: Aglomeración de 45 km de algas marinas a la deriva, en el golfo de México (Envisat)

Oben: Die Küste Australiens mit Korallenriffen und eingeschwemmten Sedimenten (Envisat)

Above: Coral reefs and sedimentary deposits along the Australian coast (Envisat)

Ci-dessus : littoral australien présentant des récifs coralliens et dépôts de sédiments (Envisat)

Arriba: Costa australiana con arrecifes de coral y sedimentos arrastrados (Envisat)

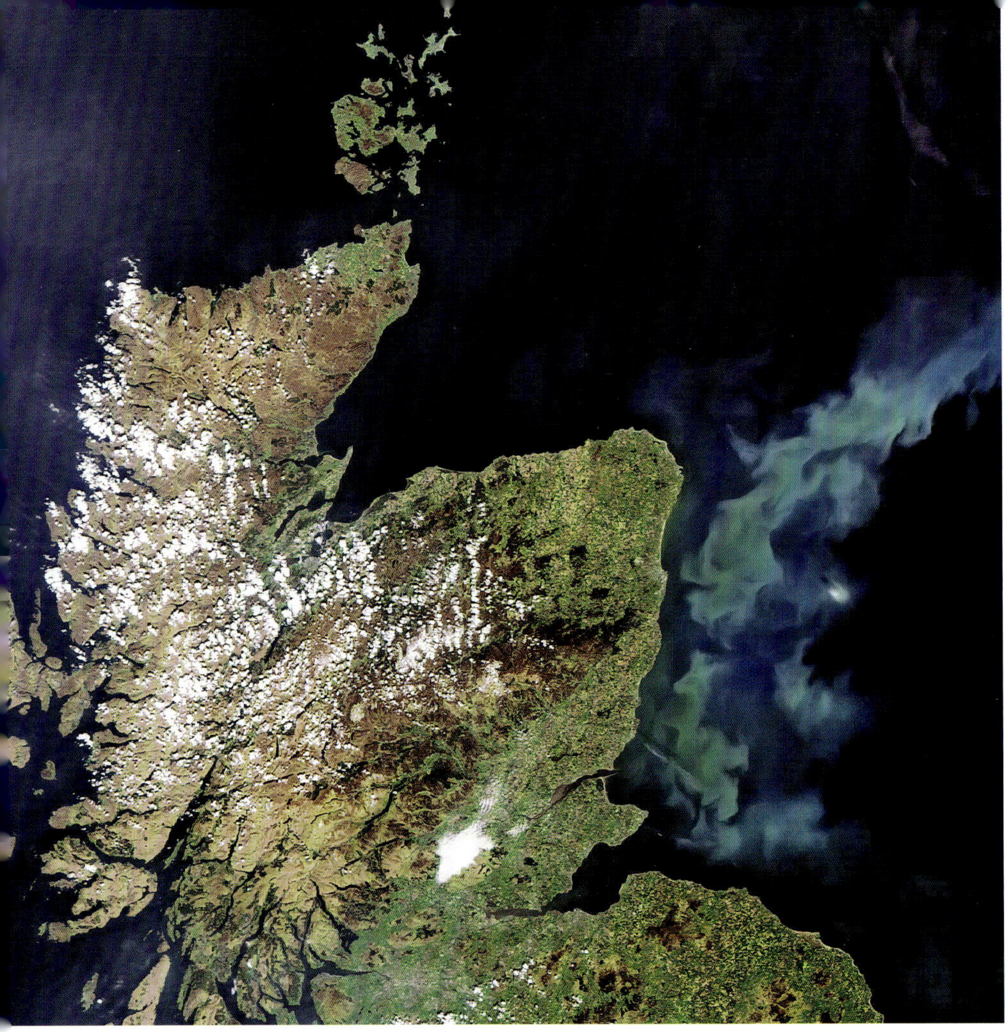

Plankton-Blüte in der Nordsee vor Schottland, aufgenommen im Sommer 2008 (Envisat)

Algae thrive in the North Sea off Scotland, photographed in summer 2008 (Envisat)

Image de floraison planctonique sur la côte écossaise dans la mer du Nord, datant de l'été 2008 (Envisat)

Floración de plancton en el Mar del Norte, ante Escocia, tomada en el verano de 2008 (Envisat)

Rechte Seite: Das niederländische Wattenmeer mit der Insel Texel (im Bild links unten, Spot)

Right: The Dutch island of Texel (bottom left in the image) in the Wadden Sea (Spot)

Page de droite : île Texel (en bas à gauche de l'image) dans la mer des Wadden au large des Pays-Bas (Spot)

Página dcha.: Marismas neerlandesas con la isla Texel (en la ilustración, abajo a la izda., Spot)

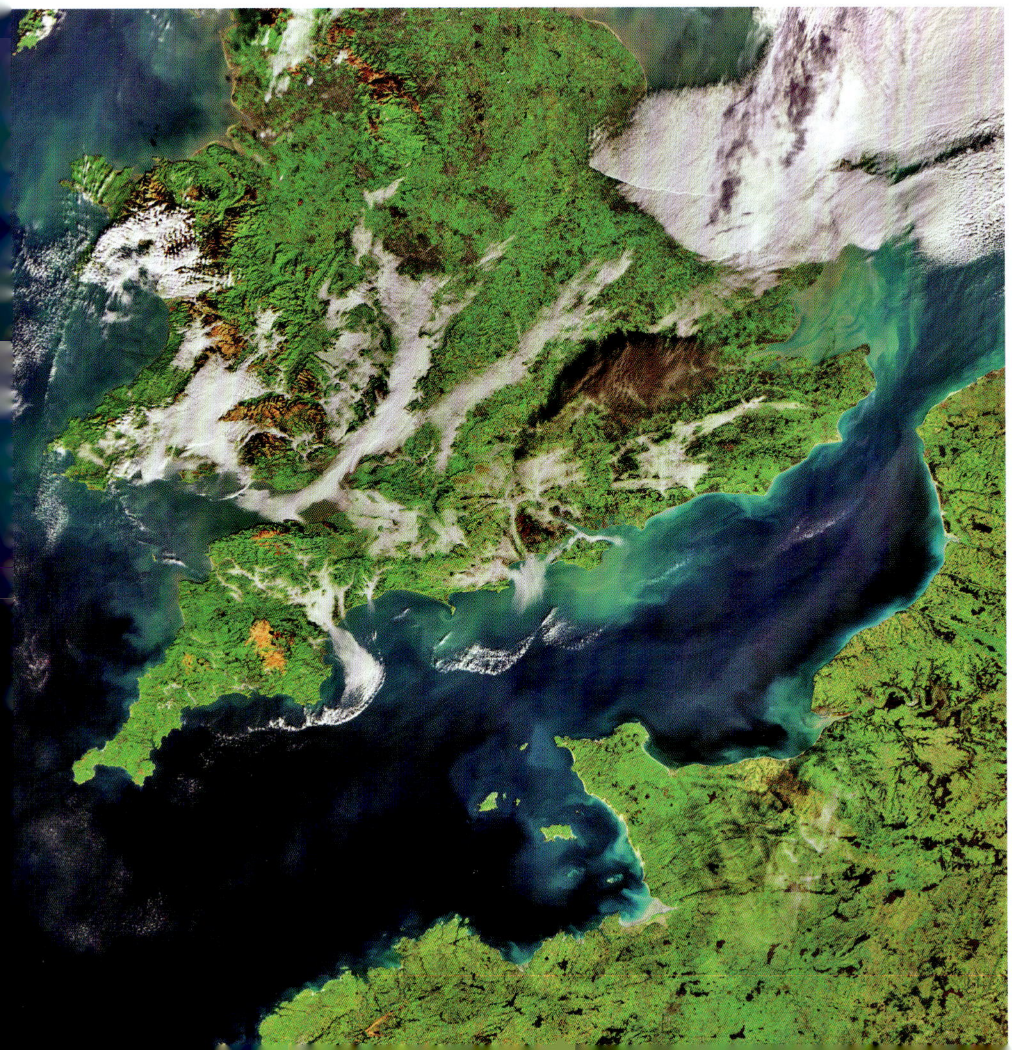

Rauchwolke bei London nach Explosion eines Öllagers im Dezember 2005. Es war der schlimmste Brand in Europa zu Friedenszeiten (Envisat)

A cloud of smoke covers London after an explosion at an oil storage facility in December 2005. The resulting fire was the worst in Europe during peace time (Envisat)

Nuage de fumée près de Londres suite à l'explosion d'un dépôt pétrolier en décembre 2005. Le plus grand incendie en Europe depuis la dernière guerre (Envisat)

Nube de humo cerca de Londres tras la explosión de un depósito de petróleo en diciembre de 2005. Fue el peor incendio europeo en tiempos de paz (Envisat)

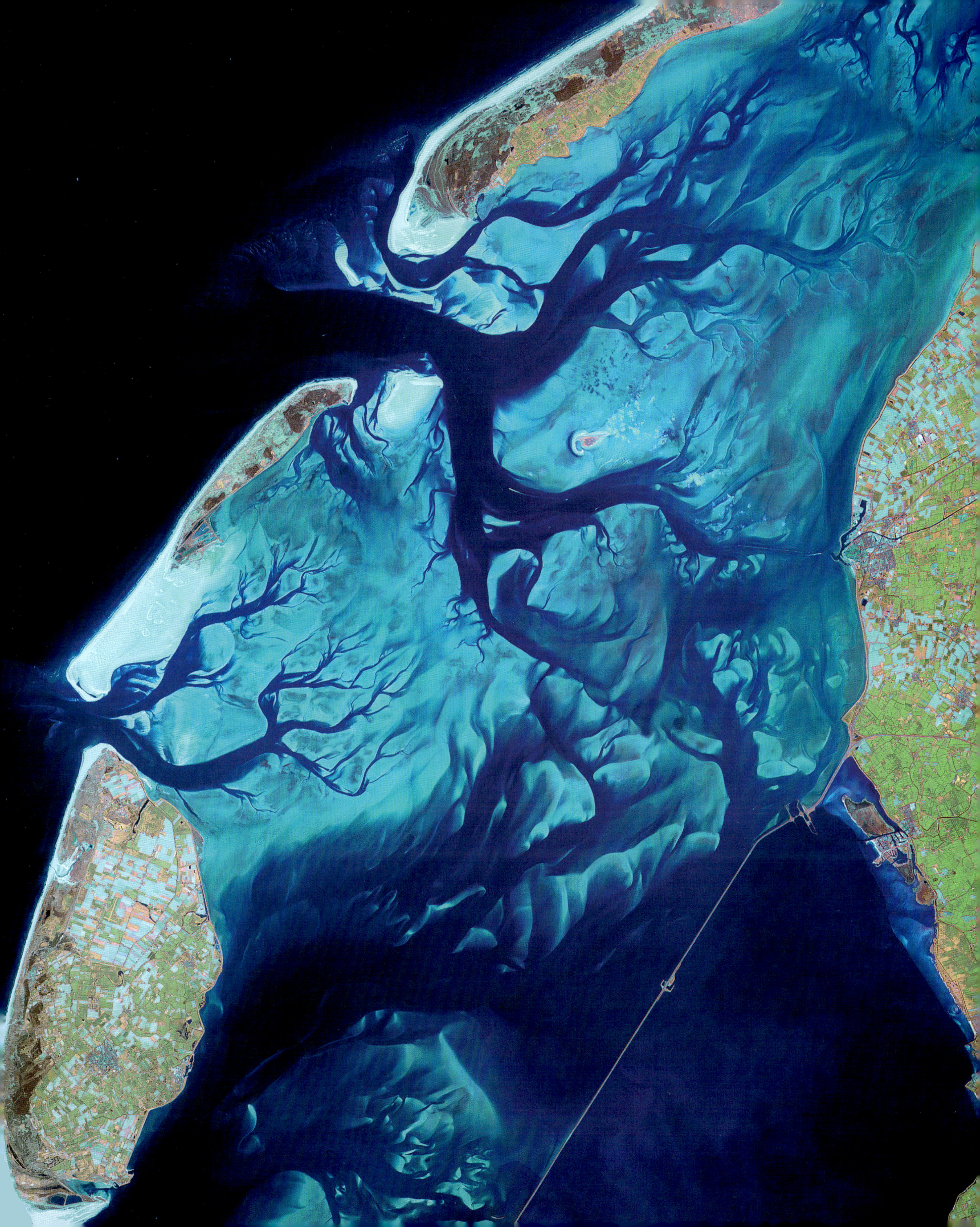

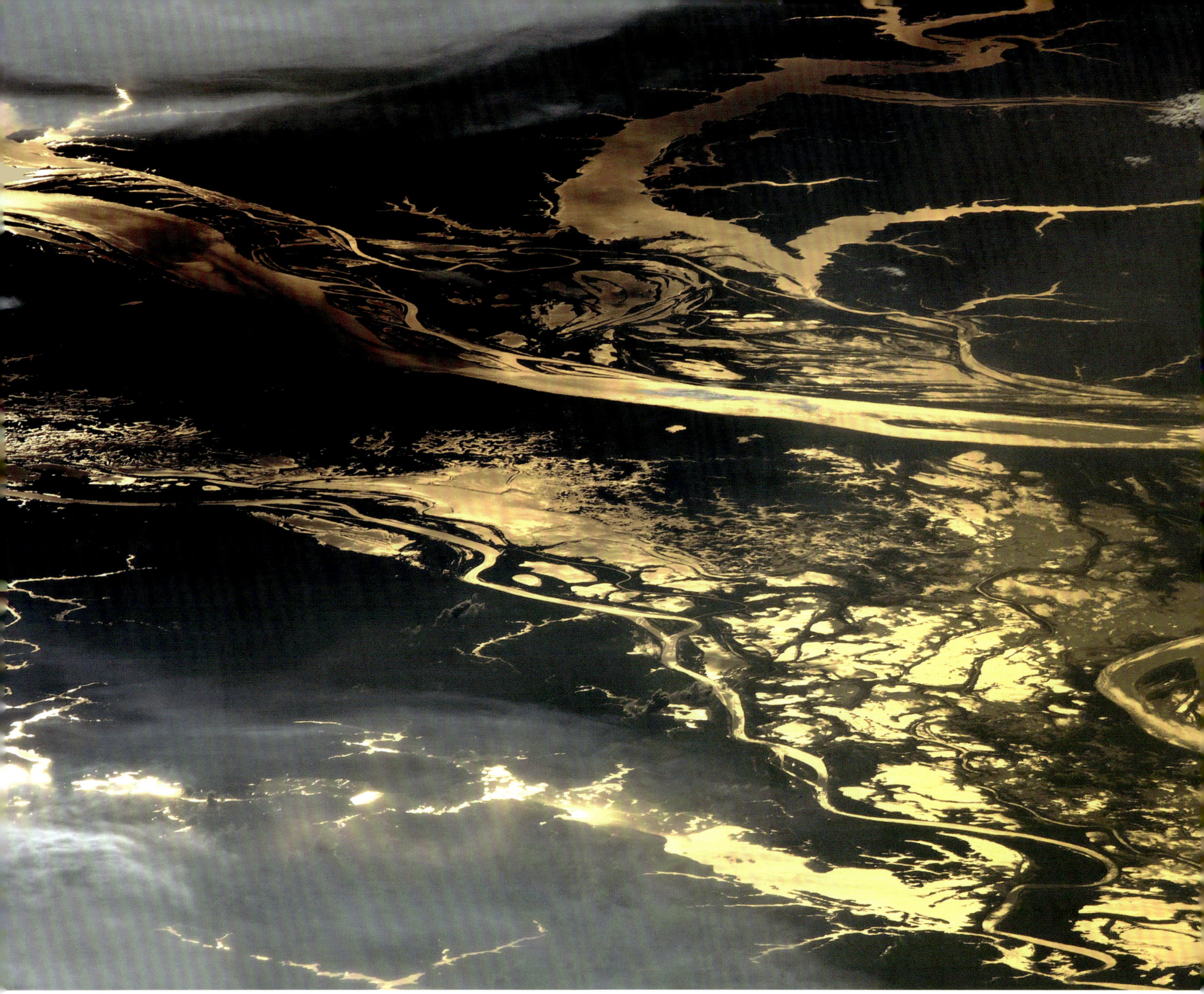

Oben: Die Sonne reflektiert im Wasser des weit verzweigten Amazonas (Raumstation)

Above: The convolutions of the Amazon, glinting in the sun (Space Station)

Ci-dessus : reflet du soleil dans les méandres du long fleuve Amazone (station spatiale)

Arriba: El sol reflejándose en las aguas de las ramificaciones del Amazonas (estación espacial)

Rechte Seite: Plankton-Blüte vor der Südküste Islands im Sommer 2005 (Envisat)

Right: Algae thrive off the southern coast of Iceland in summer 2005 (Envisat)

Page de droite : floraison planctonique sur le littoral du sud de l'Islande en été 2005 (Envisat)

Página dcha.: Floración de plancton frente al sur de Islandia, en el verano de 2005 (Envisat)

ein möglichst umfassendes Bild der Vorgänge im Kosmos. Gleiches gilt für die Erdbeobachtung. Die ersten Wettersatelliten haben mit Fernsehkameras an Bord begonnen und unscharfe Schwarzweißbilder geliefert. Später kamen bessere Instrumente und Detektoren in anderen Wellenlängenbereichen hinzu, etwa im ultravioletten oder infraroten. Manche Satelliten beleuchten die Erde aktiv mittels Radarwellen. Andere lassen einfach Kräfte und Felder auf sich wirken.

Wer wissen will, was in der Andromedagalaxie vor sich geht, muss Daten von Röntgen- und Ultraviolettteleskopen sowie von optischen Instrumenten und Beobachtungen im Infrarot- und Radiobereich zusammenfügen. Wer wissen will, was auf der Erdoberfläche geschieht, muss Satelliten nutzen, die mit allen Sinnen, die zur Verfügung stehen, sehen, fühlen, schmecken und riechen. Und wie in den Tiefen des Universums so gilt auch bei der blauen Kugel, die wir bewohnen: Je genauer die Forscher hinsehen, desto faszinierender erweist sich das Studienobjekt. Aber mit neuen Daten tauchen auch immer neue Rätsel auf. Forschung ist nie fertig. Das gilt für Welteninseln in Milliarden Lichtjahren Entfernung genauso wie für unsere kleine Erde.

Better instruments and sensors followed, detecting other wavelengths such as ultraviolet and infrared. Some satellites actively send beams to the Earth in the form of radar waves, while others merely allow forces and fields to act upon them.

Anyone wishing to find out what is happening in the Andromeda Galaxy has to compile data from X-ray and ultraviolet detectors as well as from optical instruments, infrared sensors, and radio receivers. Anyone wishing to know what is happening on the Earth's surface has recourse to satellites using almost every sense—sight, touch, taste, and smell. And it is the same in the depths of the universe as it is on the blue planet we inhabit: the more closely the scientists look, the more fascinating the object of study turns out to be.

New data turns up fresh puzzles, and research will never come to an end; this is as true of galaxies millions of light years away, as it is of our own little Earth.

tecter différentes longueurs d'onde, notamment dans les domaines ultraviolet et infrarouge, ont pris la relève. Certains satellites envoient des ondes radar vers la Terre. D'autres enregistrent les effets des forces et champs divers.

Quiconque s'intéresse à ce qui se passe dans la galaxie d'Andromède doit collecter les données provenant des télescopes ultraviolets et à rayons X, ainsi que des instruments optiques et des détecteurs radio et infrarouge. Quiconque s'intéresse aux phénomènes terrestres doit avoir recours à des satellites capables d'utiliser tous les sens dont ils disposent pour regarder, goûter et sentir. Et sur la planète bleue que nous habitons, tout comme dans les profondeurs de l'univers, la fascination augmente à mesure que s'affine l'observation scientifique. Néanmoins, chaque découverte nouvelle est porteuse d'un nouveau mystère. Qu'elle s'intéresse aux astres gravitant à des milliards d'années lumière ou à notre si précieuse Terre, la recherche n'est jamais terminée.

satélites meteorológicos comenzaron con cámaras de televisión a bordo y suministraban borrosas imágenes en blanco y negro. Más tarde, se fueron añadiendo mejores instrumentos y detectores en otras longitudes de onda, por ejemplo, en ultravioleta o infrarrojo. Algunos satélites examinan activamente la Tierra mediante ondas de radar. Otros, simplemente detectan las fuerzas y los campos.

Para saber lo que ocurre en la galaxia de Andrómeda, se deben reunir datos de los telescopios de rayos X y ultravioleta, así como de instrumentos ópticos y observaciones en la gama infrarroja y de radio. Para saber lo que sucede en la superficie de la Tierra se deben utilizar satélites que estén disponibles para ver, sentir, gustar y oler. Lo que sirve para las profundidades del universo también sirve para la esfera azul en la que vivimos: cuanto mayor es la precisión de los investigadores, tanto más fascinante resulta el objeto de estudio. Pero a medida que van apareciendo nuevos datos, también se multiplican los enigmas. La investigación no ha concluido, en absoluto. Y esto vale tanto para las galaxias a millones de años luz de distancia como para nuestra pequeña Tierra.

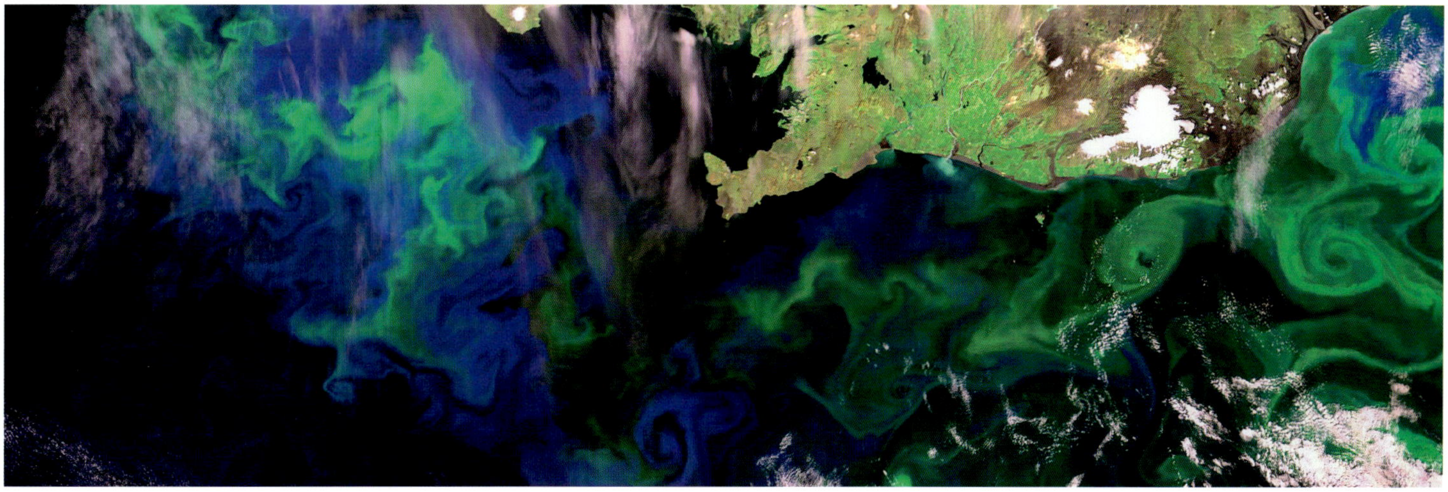

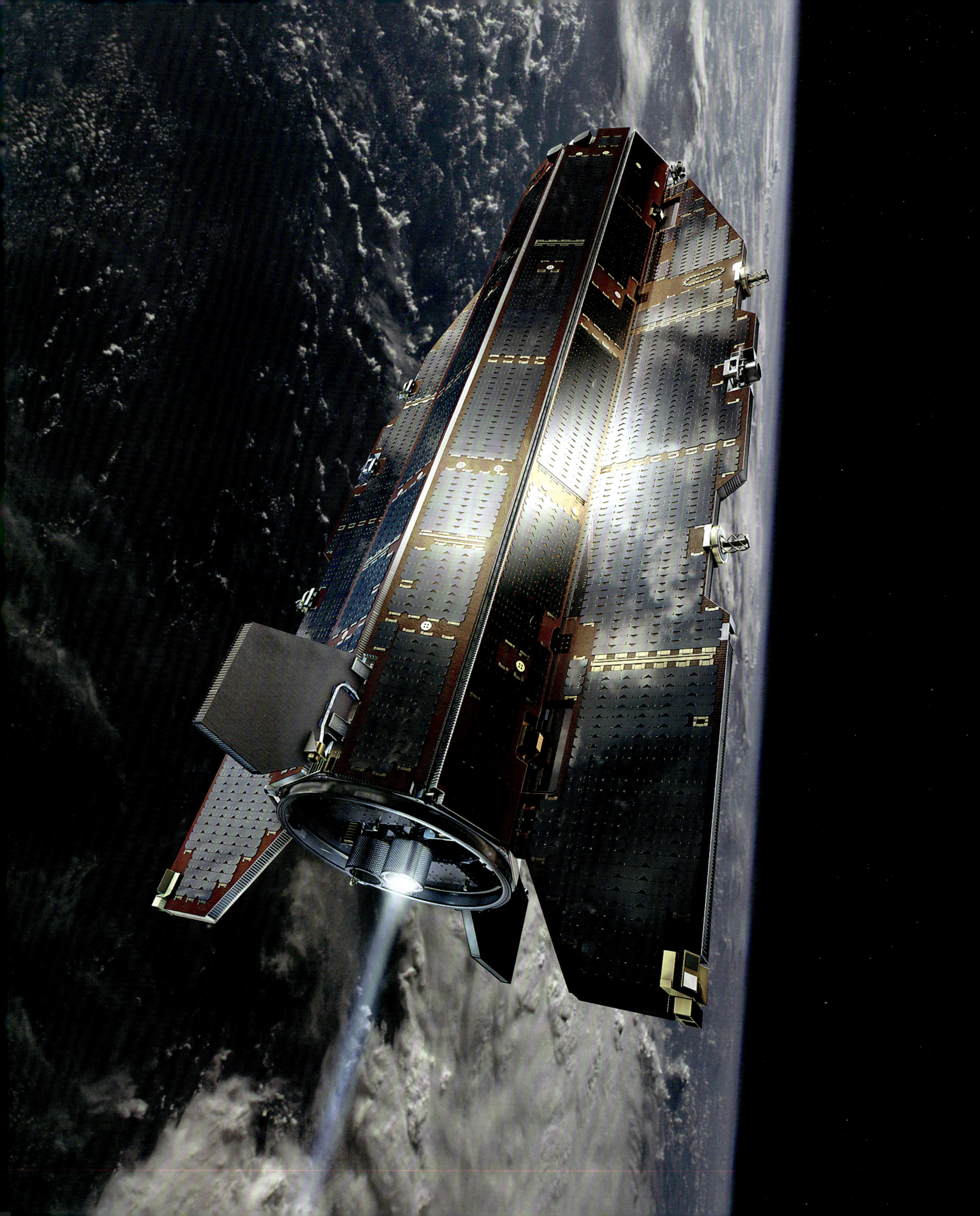

Sehen, Fühlen, Riechen: Die himmlischen Sinne
Seeing, feeling, smelling: senses in the sky
Voir, toucher, sentir : les sens célestes
Ver, sentir, oler: los sentidos celestes

Der Legende nach soll Isaac Newton einst beim Anblick eines fallenden Apfels die geniale Idee der Schwerkraft gekommen sein. Newtons Nachfolger sind mehr als drei Jahrhunderte später nicht viel weiter. Sie lassen immer noch fallen, heute aber High-tech-Äpfel. Diese fallen nicht vom Baum, sondern im Innern des GOCE-Satelliten um die Erde herum, erklärt Reiner Rummel vom Institut für Astronomische Geodäsie der Technischen Universität München: *In unserem Satelliten schweben sechs Testkörper aus einer Platin-Legierung. Die haben etwa einen halben Meter Abstand voneinander und werden daher durch die Massen auf der Erde minimal unterschiedlich angezogen.* GOCE kreiste viereinhalb Jahr lang um die Erde und spürte deren Anziehungskraft. Die ist nicht überall exakt gleich. Sie schwankt minimal, je nach Beschaffenheit der Erdoberfläche und der oberen Schichten des Erdmantels.

Flog GOCE beispielsweise über den Himalaya, so spürte der Platinwürfel weiter vorne das Gebirge etwas früher als der Platinwürfel weiter hinten. Diesen Effekt haben Rummel und seine Kollegen gemessen. *Jede Veränderung in der Massenverteilung unseres Erdsystems ist in der Schwerkraft messbar,* betont Rummel. *Wir spüren mit GOCE also Berge und Täler, Ozeane, Verände-*

Legend has it that the idea of gravity was inspired in Isaac Newton when he saw an apple fall, but Newton's successors have not come much further in three centuries. The apples that are falling now are high-tech, and they don't grow on trees—they are orbiting the Earth within the GOCE satellite, explains Reiner Rummel from the Institute for Astronomical Geodesy at Munich's Technical University. *We have six test masses made of a platinum alloy suspended in our satellite. They are about 18 inches [46 centimeters] apart, so there is a tiny difference in the attraction exerted on them by solid bodies on Earth.* During its four and a half years orbiting the Earth, GOCE detected variations in the gravitational pull, which is not uniform and varies by a tiny degree according to the composition of the surface and the upper layers of the Earth's crust.

If GOCE flies over the Himalayas, for example, the platinum cube toward the front senses the mountains a little earlier than the platinum cube toward the back, and this effect was what Rummel and his colleagues measured. *Any change in the distribution of mass in our planetary system can be measured using gravity,* reminds Rummel. *With GOCE,*

La légende raconte que c'est la vue d'une pomme tombant sur le sol qui a inspiré à Isaac Newton l'intuition géniale de la gravitation universelle. Plus de trois siècles plus tard, ses successeurs ne sont guère plus avancés. Ils continuent à observer la chute de pommes désormais technologiques. Elles ne tombent plus des arbres, mais flottent à l'intérieur du satellite GOCE, comme l'explique Reiner Rummel de l'Institut de géodésie astronomique de l'Université technique de Munich : *Six petits cubes en alliage de platine sont suspendus dans notre satellite. Ils sont distants d'environ cinquante centimètres et enregistrent par conséquent d'infimes écarts de gravité.* Pendant ses quatre années et demie d'orbite autour de la Terre, GOCE a enregistré les variations de la force d'attraction de la planète dues aux irrégularités de la surface terrestre et à la structure du manteau supérieur.

Quand GOCE survolait l'Himalaya, par exemple, le dé de platine situé à l'avant percevait la montagne légèrement plus tôt que son confrère placé à l'arrière. Ce sont ces différences que Reiner Rummel et son équipe ont étudiées. *Chaque variation de la distribution des masses sur notre planète se répercute sur les valeurs de gravité mesurées*, souligne Reiner Rummel.

Cuenta la leyenda que a Isaac Newton se le ocurrió la genial idea de la fuerza de la gravedad al contemplar cómo caía una manzana. Trescientos años después, los seguidores de Newton no han ido mucho más allá. Siguen viendo caer manzanas, pero de alta tecnología. Estas no caen de los árboles al suelo, sino en el interior del satélite GOCE mientras gira alrededor de la Tierra, aclara Reiner Rummel, del Instituto de Geodesia Astronómica de la Universidad Politécnica de Múnich: *En nuestro satélite están suspendidos seis chivatos hechos con una aleación de platino. Entre ellos hay aproximadamente medio metro de distancia*
y por eso son atraídos por las masas de la Tierra con unas diferencias mínimas. Durante sus cuatro años y medio de actividad alrededor de la Tierra, GOCE registró las variaciones de la fuerza de gravedad, que fluctúa mínimamente dependiendo de la calidad de la superficie terrestre y de los estratos del manto superior.

Si GOCE volaba, por ejemplo, sobre el Himalaya, el cubo de platino más adelantado percibía la montaña un poco antes que el cubo de platino que está más atrás. Este efecto era lo que medían Rummel y sus colegas. *Cualquier modificación en la distribución de masas de nuestro sistema te-*

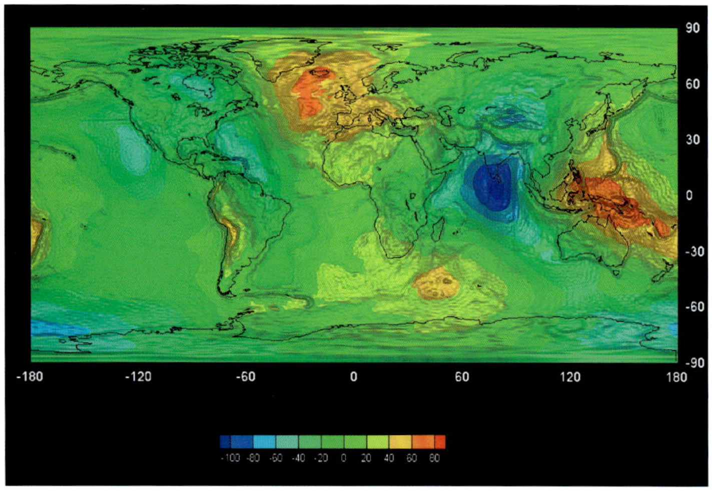

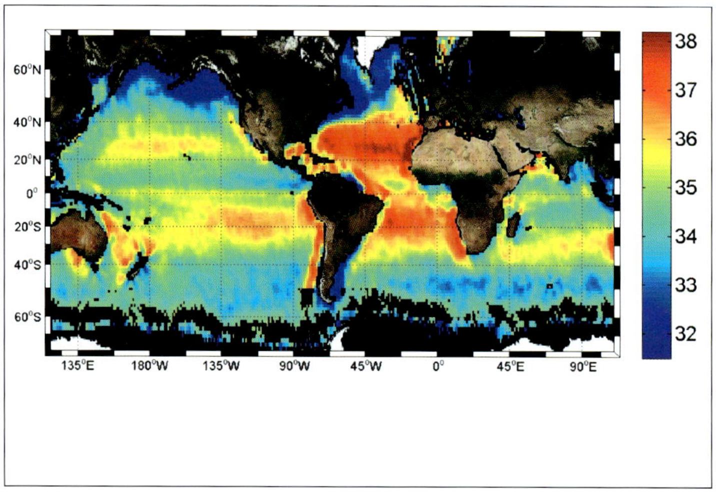

rungen des Meeresspiegels, abschmelzende Eiskappen und sogar das fallende Laub, wenn man sehr, sehr genau misst.

Vor allem die globalen Strömungen in den Weltmeeren haben es den Forschern angetan. Sie verraten sich durch leichte »Flutberge« mitten im Ozean: Dort ist der Wasserstand ein bis zwei Meter höher als in anderen Bereichen. Was Forschern auf einem Schiff vor Ort gar nicht auffiele, misst GOCE aus der Umlaufbahn. Denn die zusätzlichen Wassermassen erhöhen ein wenig die Anziehungskraft auf den Satelliten, was die empfindlichen Instrumente an Bord registrieren.

Wahnsinn, Wunder, GOCE

Der Satellit, der im Wesentlichen bei Thales Alenia in Turin und bei Airbus (damals noch Astrium) in Friedrichshafen gebaut wurde, erfüllte höchste Qualitätskriterien. Manche Ingenieure räumen hinter vorgehaltener Hand ein, sie seien »fast wahnsinnig« geworden, welche Anforderungen an GOCE gestellt wurden: Der Satellit bestand aus speziellen Kohlefasern, die sich praktisch nicht ausdehnen oder zusammenziehen, egal ob der Satellit im heißen Sonnenlicht fliegt oder im kalten Erdschatten. An Bord gab es keine Motoren mit

we can sense mountains and valleys, oceans, changes in sea level, melting ice caps, and even trees shedding their leaves, if we measure very, very precisely.

The scientists are particularly interested in global ocean currents. These reveal themselves by causing small "humps" in the middle of the sea, where the water level is 3–6 feet (1–2 meters) higher than elsewhere. GOCE can measure from orbit what would usually escape scientists in the locality—the additional mass of water slightly increases the gravitational force acting on the satellite and this is recorded by the sensitive instruments on board.

Madness, miracles, GOCE

The satellite, whose fuselage was constructed by Thales Alenia in Turin, Italy, and by Airbus (formerly Astrium), in Friedrichshafen, was built to the highest standards. Many of the engineers would have quietly admitted that the specifications for GOCE had almost droven them mad: the satellite was made of special carbon fibers which to all intents and purposes neither contract nor expand, whether they are burning in sunlight or in the chill of the

Nous percevons avec GOCE les montagnes et les vallées, les océans et les variations du niveau de la mer, la fonte des glaces et même les feuilles qui tombent des arbres, quand nos mesures sont assez fines.

Mais c'est la circulation des masses d'eau dans les océans qui passionnent le plus les chercheurs. Les courants marins produisent des ondulations à la surface des océans, créant par endroit des « bosses » de près de deux mètres. GOCE a repéré depuis l'espace des phénomènes imperceptibles par l'équipage d'un bateau. Les masses d'eau supplémentaires provoquent en effet une légère augmentation de la force gravitationnelle exercée sur le satellite, que percevaient ses instruments ultrasensibles.

GOCE : de la folie au miracle

Ce satellite construit pour l'essentiel par Thales Alenia et Airbus (anciennement Astrium), répondait aux critères de qualité les plus exigeants. Plusieurs ingénieurs avouent discrètement avoir « frisé la folie » au vu des exigences imposées à GOCE. En effet, ce satellite était constitué de fibres de carbone spéciales qui ne se dilatent et ne se rétractent pas, que l'appareil évolue sous la brûlante

rrestre se puede medir en la fuerza de gravedad, destaca Reiner Rummel. *Así, con GOCE detectamos montañas y valles, océanos, modificaciones del nivel del mar, casquetes helados derritiéndose e, incluso, la caída de las hojas cuando se mide con gran exactitud.*

Lo que más ha impresionado a los investigadores han sido las corrientes oceánicas. Éstas se delatan por ligeras «montañas de marea» en medio del océano. Allí el nivel del agua es entre uno y dos metros mayor que en otras zonas. Aquello que nunca llamaría la atención a los investigadores en un barco, GOCE lo mide desde la órbita.

Locura, milagro, GOCE

Este satélite, que fue fabricado esencialmente por Thales Alenia en Turín y por Airbus (antiguamente Astrium), en Friedrichshafen, cumplía con los más altos criterios de calidad. Algunos ingenieros reconocen extraoficialmente que «casi se vuelven locos» ante las exigencias de GOCE: el satélite estaba hecho con fibras de carbono especiales que prácticamente no se dilatan ni se contraen, independientemente de que el satélite vuele bajo la ardiente luz del Sol o a la fría sombra de la Tierra. A bordo no existían motores con partes móviles.

Himalaya mit Mt. Everest (links der Mitte); Vegetation erscheint rot (Envisat)

The Himalayas with Mount Everest (slightly to the left in the middle of the image); vegetation is shown in red (Envisat)

Himalaya et mont Everest (à gauche du centre) ; les zones rouges indiquent la végétation (Envisat)

Himalaya, con el Everest (izda. centro); la vegetación aparece en rojo (Envisat)

Linke Seite: Neue Daten über das Schwerefeld der Erde (linkes Bild, GOCE) und den Salzgehalt der Meere (rechtes Bild, SMOS)
Seite 100: ESA-Satellit GOCE vermisst das Schwerefeld der Erde
Seite 104/105: Die Wand des Himalaya, 750 km sind im Bild zu sehen, erhebt sich aus der dunstigen Ebene Indiens

Left: New data about the Earth's gravitational field (GOCE, near right) and the salinity of the oceans (SMOS, far right)
Page 100: The ESA's GOCE satellite measures the Earth's gravitational field
Pages 104–105: The Himalayas, 460 miles (750 km) of which can be seen in the image, rise like a wall from the misty plains of India

Page de gauche : nouvelles données du champ gravitationnel de la Terre (image de gauche, GOCE) et de la salinité des océans (image de droite, SMOS)
Page 100 : le satellite GOCE de l'ESA mesure le champ gravitationnel de la Terre
Pages 104 et 105 : la chaîne de l'Himalaya, dont 750 km apparaissent sur l'image, s'élève au-dessus de la plaine brumeuse de l'Inde

Página izda.: Nuevos datos sobre el campo gravitatorio (ilustración izda., GOCE) y la salinidad de los mares (ilustración dcha., SMOS)
Página 100: El satélite GOCE de la ESA mide el campo gravitatorio de la Tierra
Páginas 104/105: En la imagen se pueden ver 750 km del muro del Himalaya levantándose sobre las brumosas llanuras de la India

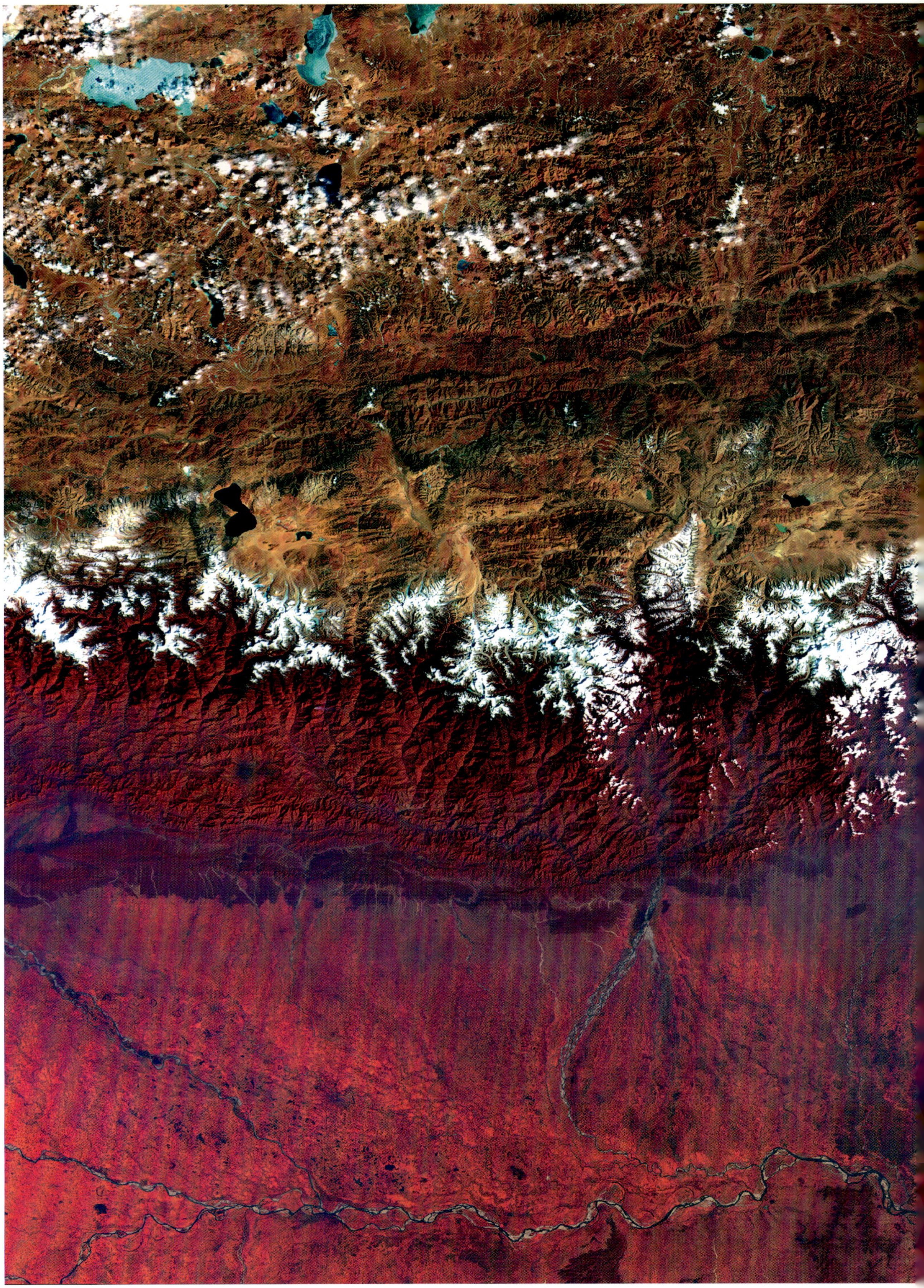

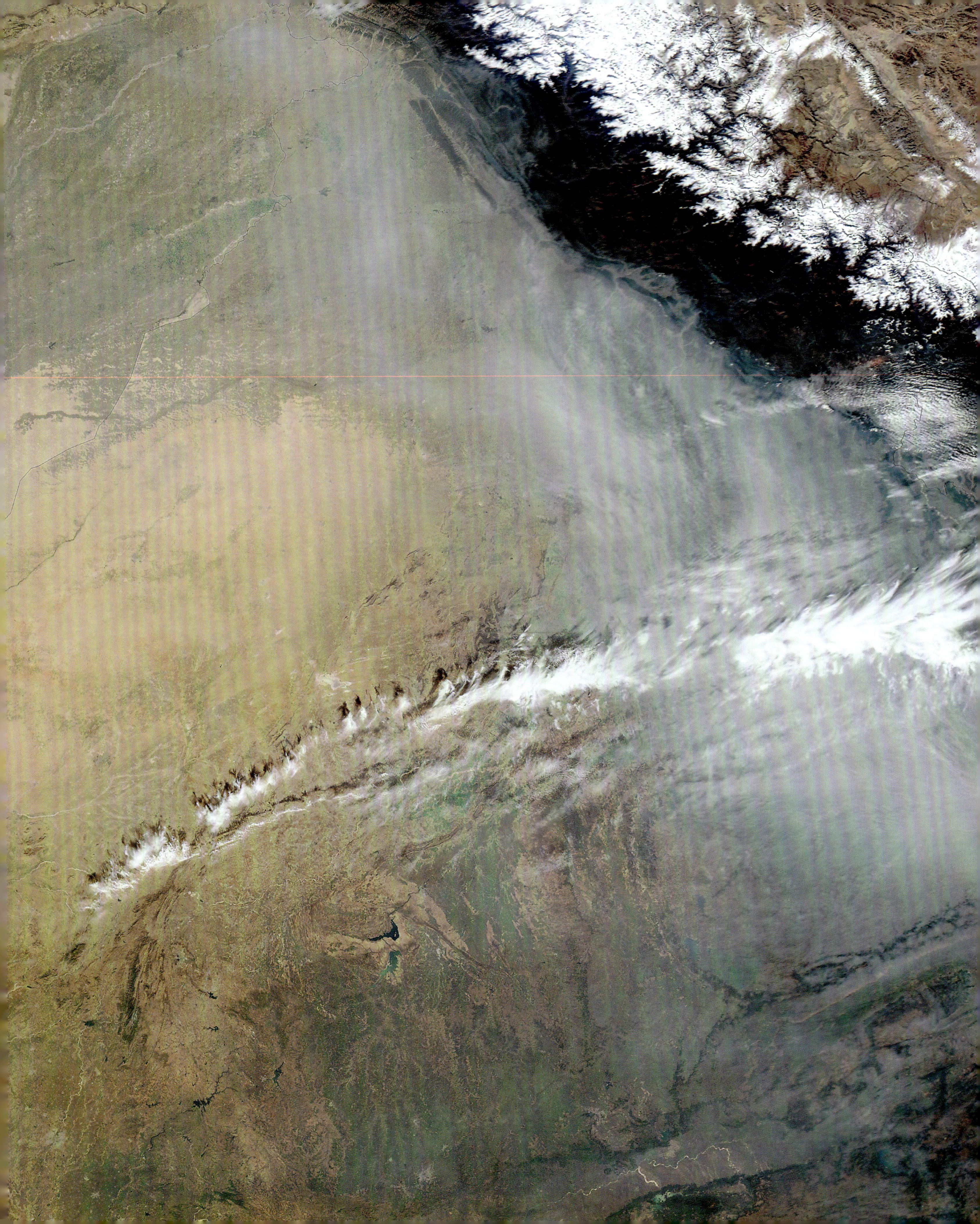

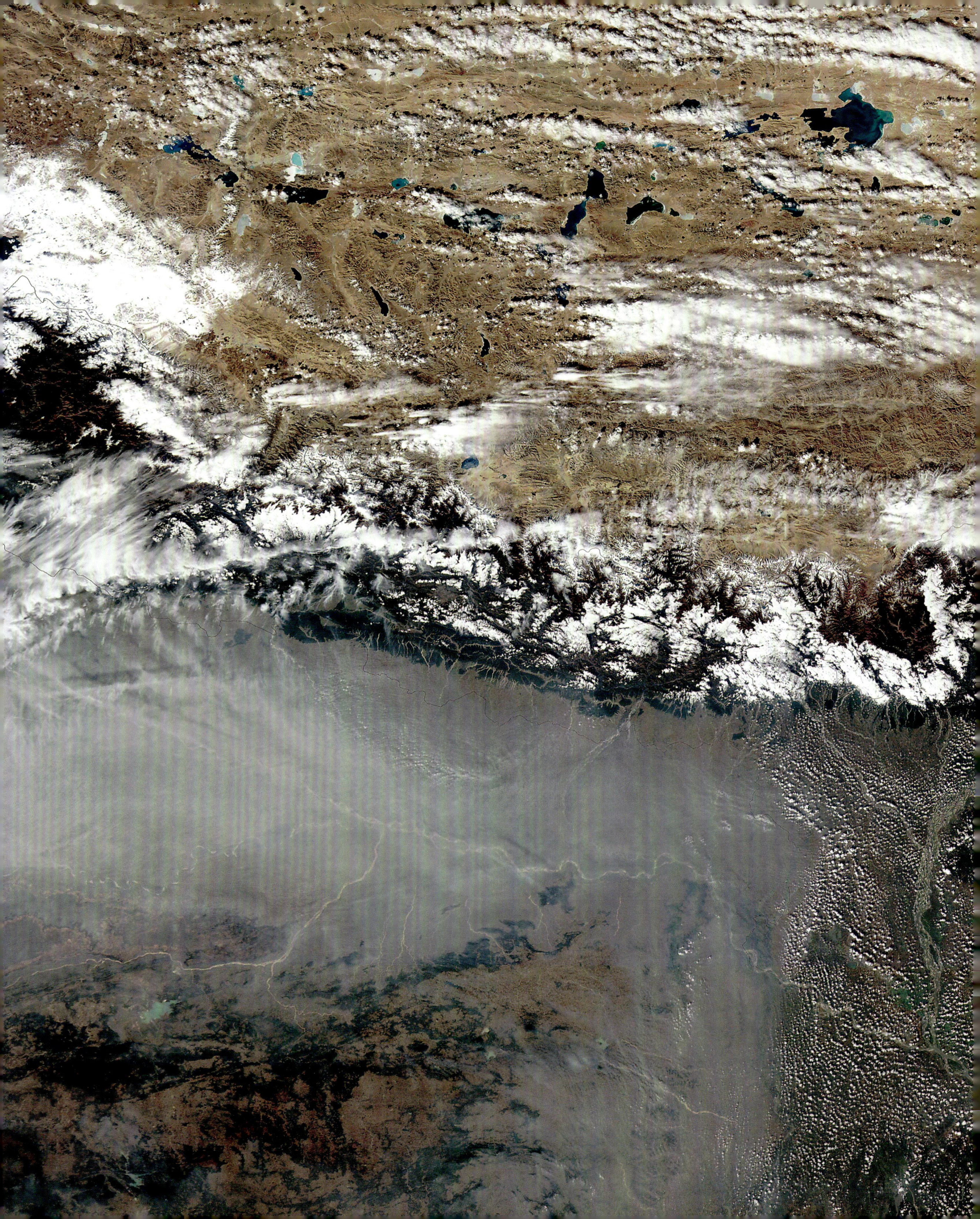

beweglichen Teilen. Jede Veränderung oder Erschütterung des Satelliten hätte die Messungen erheblich gestört. Lohn der Mühe war eine schier unglaubliche Präzision.

Stellen Sie sich eine Schneeflocke vor, die auf einen Supertanker fällt, sagt Mark Drinkwater, GOCE-Projektwissenschaftler bei der ESA. *Dann spürt der Supertanker den ›Einschlag‹ der Schneeflocke und fährt hinterher etwas langsamer weiter. Natürlich ist dieser Effekt winzig klein. Aber die GOCE-Instrumente sind so empfindlich, dass sie das Auftreffen der Schneeflocke auf den Tanker noch registrieren würden.* Der britische Forscher strahlt, wenn er die Messgenauigkeit seines Satelliten mit diesem verblüffenden Beispiel verdeutlicht. Nach dem Start im Frühjahr 2009 hatte der Satellit bewiesen, wie scharf sein Sinn für die Schwerkraft ist. Vor dem Flug hatten manche Raumfahrt-Offizielle im vertraulichen Gespräch Skepsis geäußert. Wenn GOCE seine geplante Präzision tatsächlich erreiche, wäre das »ein Wunder«. Die gibt es bekanntlich immer wieder.

Um die Erdanziehung so gut wie möglich zu vermessen, flog GOCE auf einer extrem niedrigen Bahn in nur 250 Kilometern Höhe. Zum Vergleich: Die Internationale Raumstation ISS befindet sich gut 150 Kilometer

Earth's shadow. The onboard motors have no moving parts didn't exist. Any change or vibration in the satellite would have made measurements impossible. The reward for all this trouble and effort was almost unimaginable precision.

Imagine a snowflake falling on a supertanker, says Mark Drinkwater, a GOCE project scientist at the ESA. *The supertanker feels the snowflake's 'impact' and is subsequently slowed down a little. The effect is of course extremely slight, but the GOCE's instruments are so sensitive that they would record a supertanker's encounter with a snowflake.* The British scientist beams at the example he has chosen to explain the accuracy of the satellite's measuring devices. The satellite demonstrated just how sensitive it was to gravity since it launched in spring 2009. Before the satellite entered orbit, some space officials had privately voiced their skepticism—it would be "a miracle" if GOCE ever achieved its predicted accuracy. The miracles have been happening ever since.

To measure the Earth's attraction as accurately as possible, GOCE followed an extremely low orbit at a height of 150 miles (240 kilometers). By comparison, the International Space Station (ISS) is a good

lumière du soleil, ou dans l'ombre glacée de la Terre. Son moteur ne contenait aucune pièce mobile, afin d'éviter toute vibration. Le fruit de tous ces efforts était une précision absolument incroyable.

Imaginez un flocon de neige tombant sur un supertanker, explique Mark Drinkwater, un scientifique travaillant pour l'ESA au projet GOCE. *Le supertanker perçoit l'impact du flocon de neige qui freine légèrement sa progression. Bien sûr, cet effet est infime, mais les instruments de GOCE sont tellement sensibles qu'ils pourraient enregistrer le choc du flocon sur le pétrolier.* Le visage du chercheur britannique s'illumine lorsqu'il décrit la précision de son satellite à l'aide de cette comparaison saisissante. Depuis son lancement au printemps 2009, le satellite avait démontré son sens aiguisé de la gravité. Avant son départ, quelques représentants de la branche spatiale s'étaient montrés sceptiques, affirmant qu'atteindre une telle précision relèverait du « miracle ». Mais comme chacun sait, un miracle est toujours possible.

Pour cartographier avec une très haute précision le champ de pesanteur de la Terre, GOCE évoluait sur une orbite très basse située à 250 kilomètres d'altitude. À titre de comparaison, la station spatiale internatio-

Cualquier modificación o sacudida del satélite habría hecho imposible las mediciones. La recompensa por el esfuerzo era una precisión increíble.

Imagínese un copo de nieve que cae sobre un superpetrolero, dice Mark Drinkwater, científico responsable del proyecto GOCE en la ESA. *Entonces, el superpetrolero siente el «impacto» del copo de nieve y continúa su camino un poco más lento. Naturalmente, este efecto es diminuto. Pero los instrumentos de* GOCE *son tan sensibles que incluso podrían registrar el impacto del copo de nieve en el barco.* El investigador británico sonríe complacido mientras explica la precisión de las mediciones de su satélite a través de este sorprendente ejemplo. Tras el lanzamiento, en la primavera de 2009, el satélite demostró lo rigurosa que era su sensibilidad para la fuerza de la gravedad. Antes del vuelo, algunas fuentes espaciales habían manifestado su escepticismo. Si GOCE consiguiese la precisión proyectada, sería «un milagro». Pero, como sabemos, milagros ocurren todos los días.

Para medir la gravedad terrestre con la mayor exactitud posible, GOCE volaba en una órbita extremadamente baja, a sólo 250 km de altura. La estación espacial internacional ISS, por ejemplo, se encuentra 150 km más arriba. A la altura de vuelo de GOCE

Rechts: Oberflächentemperatur des Wassers am 14. April 2010; *S. 107:* Chlorophyllgehalt der Meere 2009/2010 (beide Envisat)

Near right: Ocean surface temperatures on April 14, 2010. *Far right:* Ocean chlorophyll levels 2009/2010 (both Envisat)

À droite : température de surface de l'eau le 14 avril 2010 *Page 107 :* teneur en chlorophylle des océans en 2009/2010 (Envisat)

Dcha.: Temperatura superficial del agua el 14.04.2010; *p. 107:* Contenido en clorofila de los océanos 2009/2010 (ambas, Envisat)

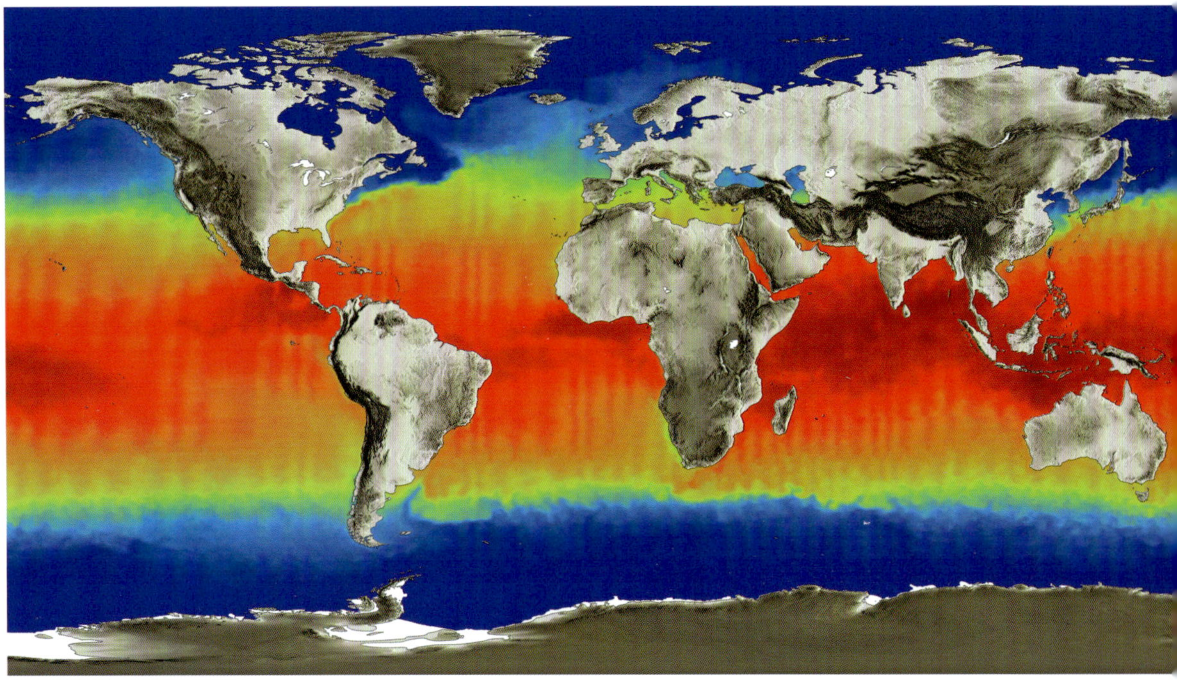

höher. Auf der Flughöhe von GOCE herrschte noch kein perfektes Vakuum des Weltalls. Zwar ist die Atmosphäre dort bereits extrem dünn, aber die Reste reichten aus, um GOCE ständig zu bremsen. Dieser »Gegenwind« hätte die Messungen unmöglich gemacht. Daher untersuchte der Satellit zehnmal pro Sekunde, wie sehr ihn die wenigen Luftteilchen da oben abbremsen. Diesen Effekt hat er umgehend mit seinem Ionentriebwerk ausgeglichen, das ständig geladene Xenon-Teilchen ausstieß.

Auch dieser Aspekt der Mission bereitete den Ingenieuren manch schlaflose Nacht. Denn das Triebwerk durfte nicht immer mal wieder einen kleinen »Feuerstoß« abgeben, sondern es musste perfekt gleichmäßig brennen. Jeder »Ruck« hätte den zu messenden Effekt völlig überlagert. Die meisten Satelliten sind von recht klobiger Statur und sehen eher aus wie fliegende Kühlschränke. GOCE dagegen hatte eine sehr windschnittige Form. So minimiert man die durch den Gegenwind beeinflussten Störeffekte. Bei Thales Alenia in Turin sprechen die Ingenieure gerne vom Ferrari unter den Satelliten. GOCE hatte auch etwa die Länge eines Rennwagens, aber einen Ferrari hat er mühelos abgehängt: Mit 28.000 Km pro Stunde jagte der Satellit mit den Platinwürfeln um die Erde.

60 miles (100 kilometers) higher. The perfect vacuum of space still didn't exist at the GOCE's altitude and, although the atmosphere is extremely thin up there, what remains was still enough to act as a continual brake on GOCE. Such a "headwind" made it measurement impossible, and so the satellite checked ten times a second how much these few air particles were slowing it down and immediately compensated for their effect with its ion drive, which emited a constant stream of charged xenon particles.

This aspect of the mission was also to cause the engineers a few sleepless nights, as the drive could not undergo a "spurt" every so often but had to burn entirely consistently. Any jolt would have completely overshadowed the phenomena they were trying to measure. Most satellites have an angular appearance, somewhat resembling flying refrigerators, but GOCE had an extremely aerodynamic profile. This helped to minimize the disturbance caused by headwind—at Thales Alenia in Turin, they called it the Ferrari of satellites. GOCE is indeed about the length of a racing car, but it would leave a Ferrari standing; the satellite and its platinum cubes orbited the Earth at 17,400 miles per hour (28,000 kilometers per hour).

nale ISS se trouve 150 kilomètres plus haut. À l'altitude de GOCE le vide parfait de l'univers ne régnait pas encore. La densité de l'atmosphère était certes faible, mais l'atmosphère résiduelle qui subsistait suffisait à freiner le satellite. L'effet de ce « vent contraire » aurait rendu toute mesure caduque si GOCE n'avait pas enregistré dix fois par seconde la décélération provoquée par ces quelques particules d'air. Il compensait immédiatement cet effet en actionnant un moteur à propulsion ionique qui éjectait des particules de xénon chargées.

Les ingénieurs ont passé d'innombrables nuits blanches à plancher sur cet aspect de la mission. La combustion du moteur devait être parfaitement régulière, sans quoi chaque secousse se serait superposée à l'effet mesuré. La plupart des satellites ont une forme assez grossière et ressemblent plutôt à des frigidaires volants. GOCE était en revanche doté d'un profil aérodynamique qui minimisait la résistance due à la friction de l'air. Chez Thales Alenia à Turin les ingénieurs parlent volontiers de la Ferrari de l'espace. S'il était aussi long qu'une voiture de course, GOCE distançait toutefois sans effort le meilleur de ces bolides, en filant autour de la Terre à la vitesse de 28 000 kilomètres par heure.

aún no existía un vacío perfecto en el espacio. Aunque la atmósfera está allí muy enrarecida, sus restos eran suficientes para frenar permanentemente el avance de GOCE. Este «viento en contra» hacía imposible cualquier tipo de medición. Por eso, el satélite analizó diez veces por segundo en qué medida las pocas partículas de aire de allá arriba le frenaban. De inmediato compensó ese efecto con su motor iónico, que expulsaba continuamente partículas de xenón cargadas.

También este aspecto de la misión hizo pasar más de una noche en vela a los ingenieros. El motor no podía ir produciendo continuamente pequeñas «ráfagas», sino que debía funcionar con una regularidad perfecta. Cada «arranque» debía interferir sobre el efecto a medir. La mayor parte de los satélites tienen una constitución compacta y más bien parecen frigoríficos volantes. GOCE, por el contrario, tenía una forma muy aerodinámica. Así se minimizaban los efectos perturbadores ocasionados por el viento en contra. En Thales Alenia, en Turín, los ingenieros hablaban de él encantados y le llamaban el «Ferrari de los satélites». GOCE tenía aproximadamente la longitud de un coche de carreras, pero dejaba atrás sin esfuerzo a un Ferrari, pues el satélite giraba alrededor de la Tierra a la increíble

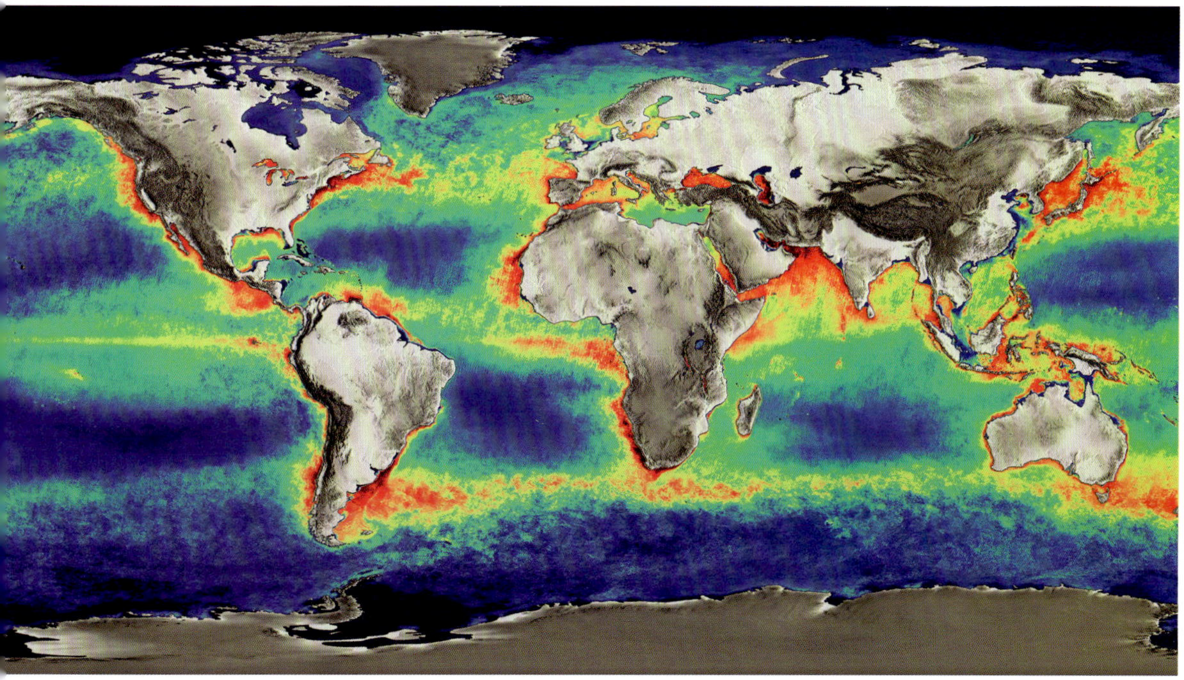

Seite 108/109: Je röter, desto schmutziger: Envisat misst Stickstoffdioxid im April 2010. China fällt besonders auf

Pages 108–9: The redder the area, the greater the pollution—Envisat's measurements of nitrogen dioxide in April 2010. China is particularly striking

Pages 108 et 109 : plus c'est rouge, plus c'est sale. Mesure du dioxyde d'azote par Envisat en avril 2010. Les données de la Chine sont frappantes

Páginas 108/109: Cuanto más rojizo, tanto más sucio: dióxido de nitrógeno medido por Envisat en abril de 2010. Destaca China

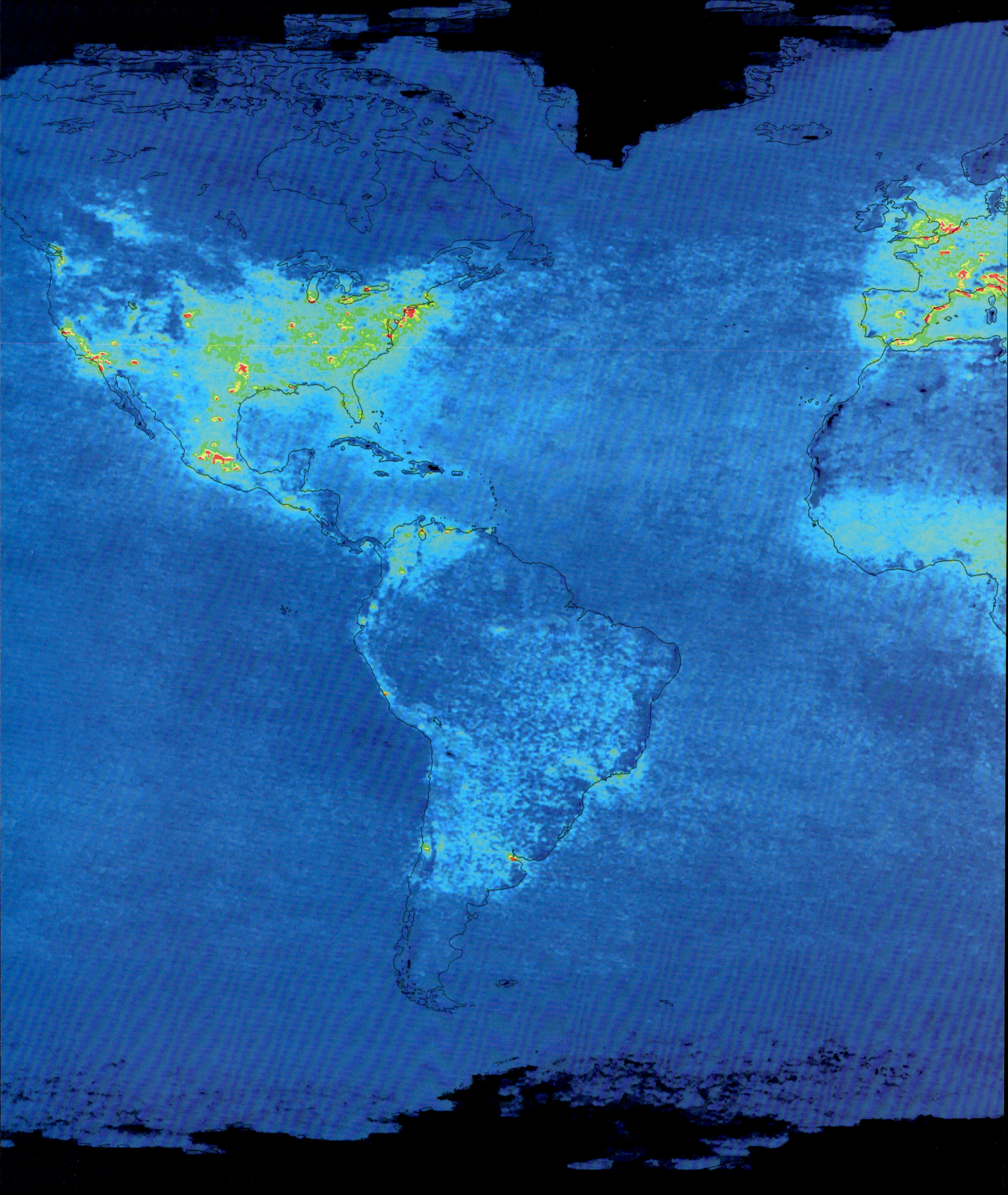

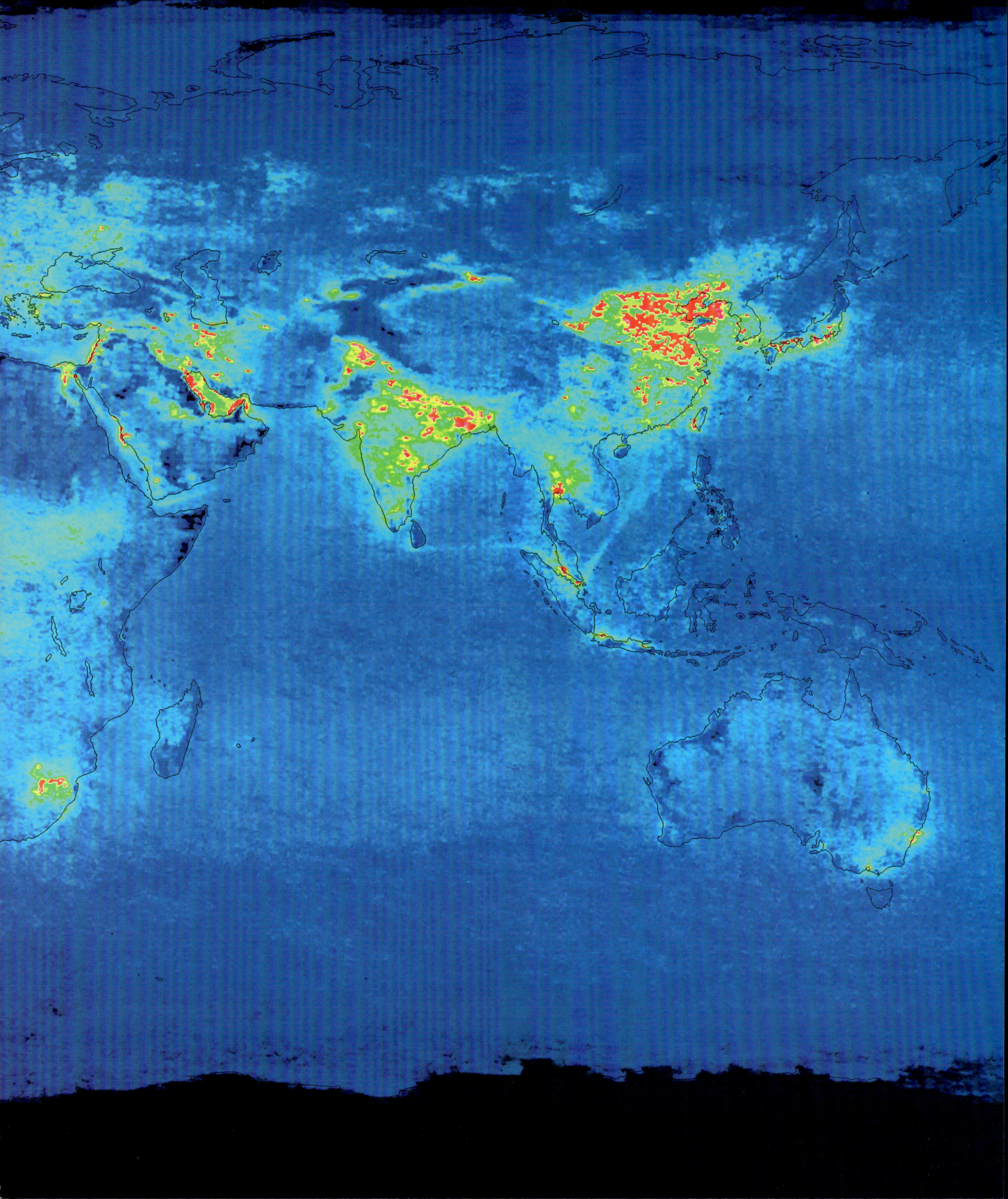

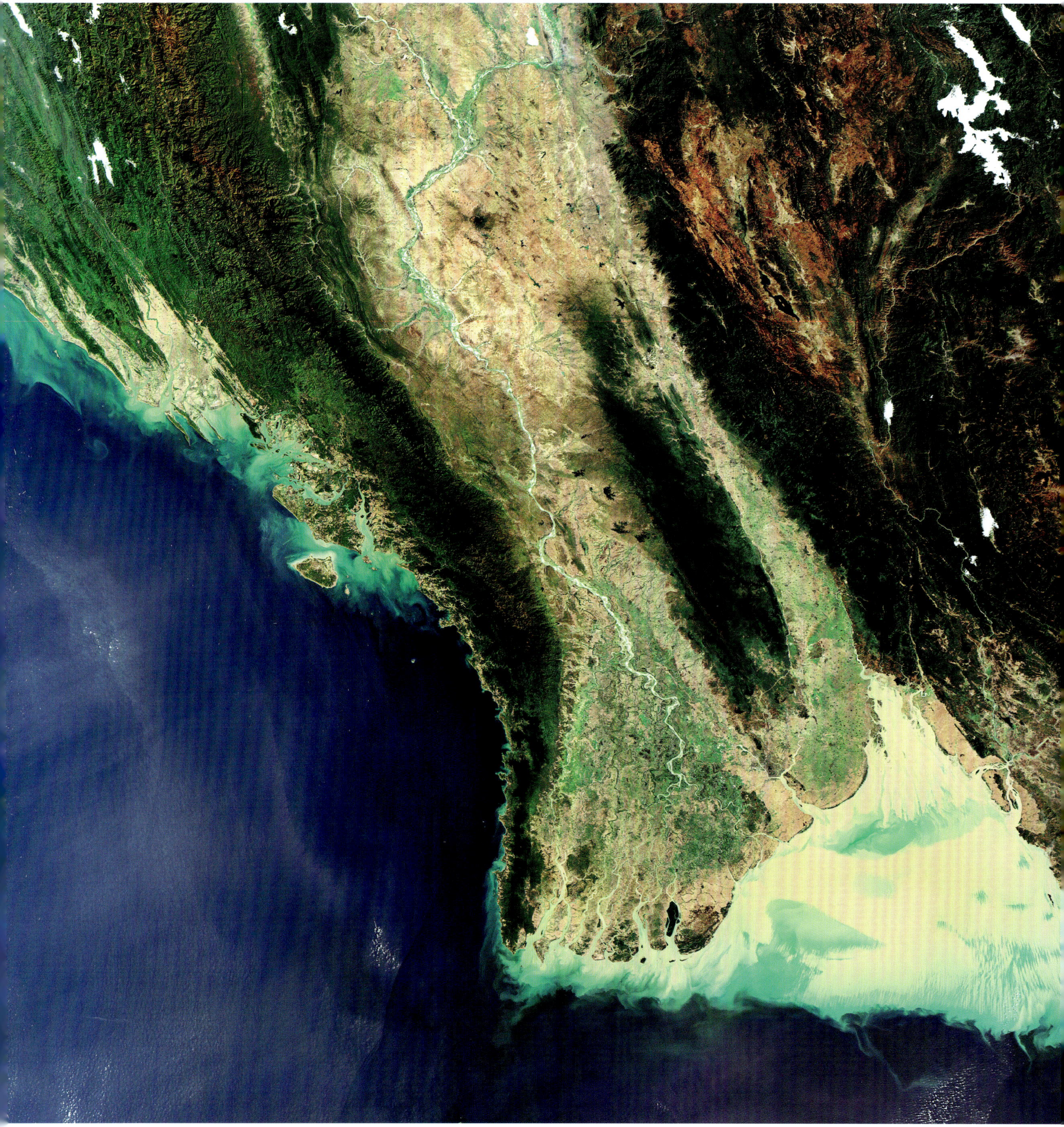

Wunschzettel sind keine Baupläne

GOCE ist ein besonders prägnantes Beispiel dafür, wie faszinierende Forschungsideen und schier unfassbare Ingenieurskunst ineinandergreifen. Weil alle Beteiligten dem Satelliten ein buchstäblich außerirdisches Gespür für Schwerkraft mit auf den Weg gegeben haben, beobachtete er unsere Erde auf so ungewöhnliche Weise. Aber auch »klassische« Erdbeobachtungssatelliten, die Fotos aus der Umlaufbahn machen, sind technische Meisterwerke.

An einem tüftelt gerade das Team um Heinz Sontag. Der gelernte Laserphysiker leitet bei Airbus in Friedrichshafen die Entwicklung der beiden Satelliten von Sentinel-2, dem zweiten Projekt des großen Erdbeobachtungsprogramms Copernicus, das die Europäische Union und die ESA gemeinsam realisieren. *Wir wollen Daten in dreizehn spektralen Kanälen liefern, im sichtbaren Licht und im nahen Infrarotbereich.* Die Satelliten, die frühestens 2017 im Einsatz sein werden, sollen nicht einfach nur Bilder der Erdoberfläche machen. Die Aufnahmen erfolgen sozusagen stets durch dreizehn verschiedene Filter bei unterschied-lichen Wellenlängen. Das ist kein Luxus, sondern unerläss-

Wish lists are not blueprints

GOCE is a particularly good example of how fascinating research ideas and almost unimaginable engineering skill can work together. The satellite was able to observe the Earth in such an unusual way because the workers on the project endowed the satellite with a literally extraterrestrial sensitivity to gravity, but even "classic" Earth observation satellites, taking photographs from orbit, are technical masterpieces.

Heinz Sontag's team is working on one right now. Having trained as a laser physicist, he now works for Airbus in Friedrichshafen, Germany, directing the development of both Sentinel-2 rockets, the second project in the great Copernicus Earth observation program, a joint venture between the European Union and the ESA. *We wanted to provide data in 13 spectral bands, in visible light, and in the low infrared spectrum.* The satellites, which are to be launched in 2017 will not just take pictures of the Earth's surface. The images will be captured through 13 different filters at different wavelengths, but this is not just self-indulgence—the filters are required to extract as much information as

Transformer les désirs en réalité

GOCE est le résultat remarquable de l'association d'une fascinante idée de recherche et d'un extraordinaire talent en ingénierie. C'est parce que les équipes impliquées ont doté ce satellite d'une faculté littéralement extraterrestre de sentir la pesanteur, qu'il percevait notre planète d'une si étrange manière. Toutefois, les satellites « classiques » d'observation de la Terre, photographes célestes de la planète, n'en sont pas moins des chefs-d'œuvre techniques.

L'équipe de Heinz Sontag s'ingénie justement à la construction de l'un d'eux. Ce spécialiste de la physique du laser dirige pour Airbus le développement des deux satellites de Sentinelle-2, le deuxième projet mené de front par l'Union européenne et l'ESA dans le cadre du programme Copernicus. *Nous voulons acquérir des images sur treize canaux spectraux, en lumière visible et dans le domaine du proche infrarouge.* Ces satellites, qui entreront en activité à partir de 2017, ne se contenteront pas de photographier la surface terrestre. Les images seront pour ainsi dire acquises à travers treize filtres distincts sur différentes longueurs d'onde. Ce qui n'est pas un luxe, mais une néces-

velocidad de 28.000 km por hora.

Der Flussverlauf des Irrawaddy in Myanmar; Envisat-Aufnahme im sichtbaren Licht

The course of the Irrawaddy in Myanmar; Envisat image using visible light

Cours du fleuve Ayeyarwady en Birmanie ; image acquise par Envisat en lumière visible

Curso del río Irawadi, en Birmania; fotografía del Envisat con luz visible

Immer wieder von Überflutungen heimgesucht: das Irrawaddy-Delta (Radaraufnahme, Envisat)

Plagued by flooding: the Irrawaddy Delta (radar image, Envisat)

Le delta de l'Ayeyarwady est fréquemment touché par les inondations (image radar, Envisat)

Continuamente devastado por las inundaciones: el delta del Irawadi (toma de radar, Envisat)

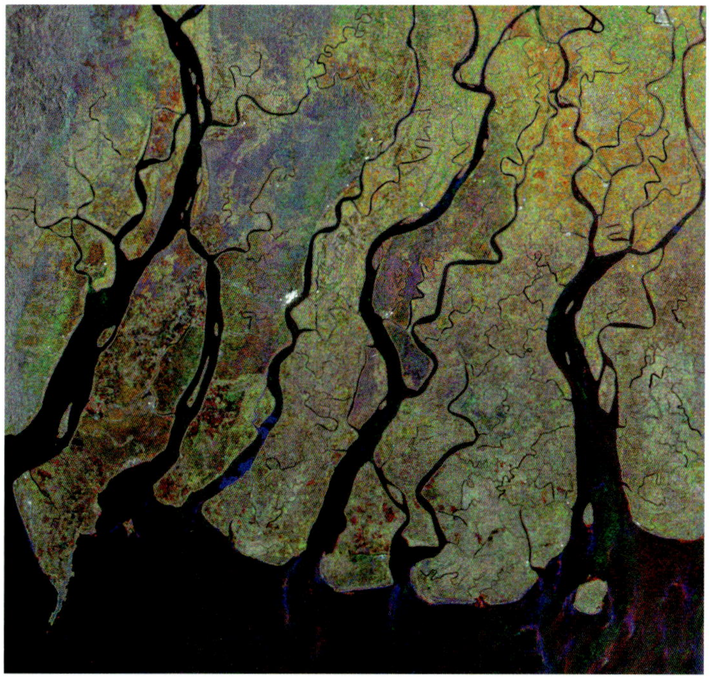

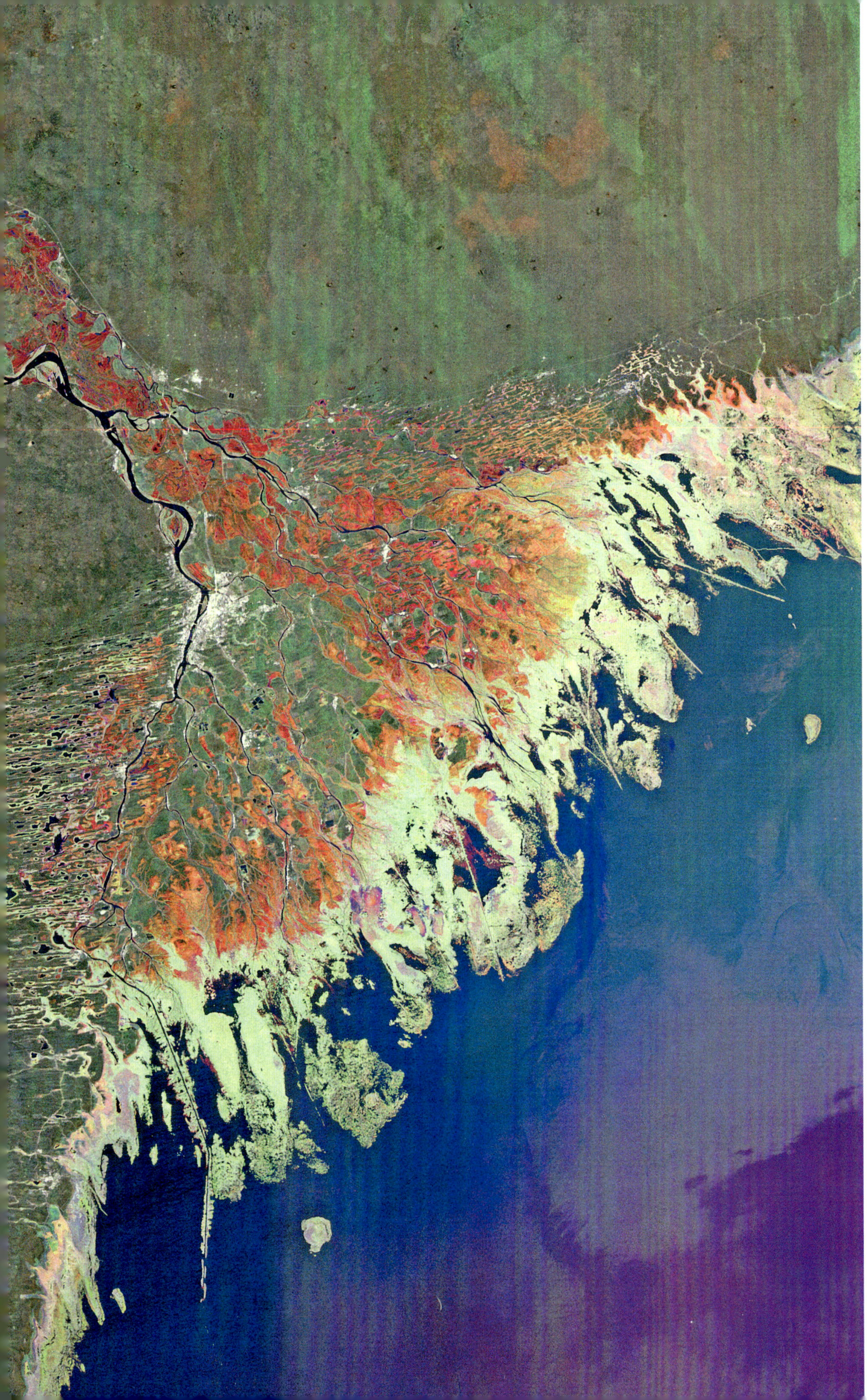

Die Wolga, der längste Fluss Europas, mündet in einem 200 km breiten Delta ins Kaspische Meer. Der weiße Fleck am Flussufer ist die Stadt Astrachan (Radaraufnahme, Envisat)

The Volga, Europe's longest river, flows into the Caspian Sea through a delta 125 miles (200 km) wide. The white area on the riverbank is the city of Astrakhan (radar image, Envisat)

La Volga, le plus long fleuve d'Europe, se jette dans la mer Caspienne en formant un delta de 200 km de large. La tache blanche sur les rives du fleuve est la ville d'Astrakhan (image radar, Envisat)

El Volga, el río más largo de Europa, desemboca en el mar Caspio por un delta de 200 km de ancho. La mancha blanca a orillas del río es la ciudad de Astracán (toma de radar, Envisat)

lich, um aus den Fotografien möglichst umfassende Informationen zu gewinnen. Denn erst aus den gefilterten Aufnahmen lässt sich viel lernen über Vegetation, geologische Strukturen, Bebauung, Gewässer etc.

So eine Mission beginnt stets damit, dass die beteiligten Wissenschaftler, zum Beispiel Meereskundler, Klimaforscher, Regionalplaner oder Landwirtschaftsexperten, erklären, wofür sie den Satelliten verwenden wollen, wie groß das Blickfeld sein sollte, welche Auflösung die Bilder haben müssen usw. Die zuständige Raumfahrtagentur, etwa die ESA, legt daraufhin die Basisanforderungen der Mission fest. Heinz Sontag und sein Team müssen dann oft aus kühnen Träumen harte Fakten werden lassen: *Wir brechen diese Spezifikationen herunter auf die Bestandteile des Satelliten und auf die Instrumente an Bord.*

Klein, leicht, stark

Die Kunst liegt darin, den Satelliten so zu bauen, dass er möglichst klein und leicht ist und so wenig elektrische Leistung wie möglich braucht. Zugleich soll er aber ein Maximum an Wissenschaft ermöglichen. Ein typischer Erdbeobachtungssatellit ist inklusive Grundgerüst, Bordcomputer, Antennen, Messinstrumente, Triebwerke etc. nur etwa so groß wie ein VW Golf. Meist wiegt er kaum eine Tonne und verbraucht weit weniger Strom als eine Kaffeemaschine.

Wer meint, in den Weiten des Alls und bei einer Flughöhe von etwa 800 Kilometern komme es auf die Bahnlage nicht ganz so genau an, irrt gewaltig. *Bei so einer Mission ist vor allem die Geolokalisierung anspruchsvoll,* erklärt Heinz Sontag. *Man will ja genau wissen, wo sich das, was man aufgenommen hat, auf der Erde befindet. Dafür braucht man präzise Kenntnis, wo der Satellit sich gerade befindet und unter welchem Winkel er zur Erde schaut.* Kurz gesagt: Sollen die Bilder nachher auf einige Meter genau sein, dann müssen die Ingenieure auch auf einige Meter genau wissen, wo sich der Satellit auf der Umlaufbahn befindet – und das,

possible from the photographs. Only after filtering do the images reveal a great deal about vegetation, geological structures, building density, and bodies of water and the like.

A mission like this always begins with a statement from the scientists involved—be they oceanographers, climatologists, regional planners, or agronomists—regarding the use to which they will put the satellite, how large its scope should be, and the required resolution for the images, and so on. The relevant space agency—the ESA for example—will then decide the basic requirements for the mission and Heinz Sontag and his team have to create cold hard facts out of these bold dreams: *We break down the specifications between the satellite and the instruments on board.*

Small, light, strong

The trick is to build as small and light a satellite as is feasible and which uses as little electric power as it can. At the same time, it should facilitate as much science as possible. Including its fuselage, onboard computer, antennae, measuring instruments, and thrust system, a typical Earth observation satellite is only about as large as a VW Golf, usually weighs less than a ton, and requires far less electrical current than a coffee maker.

Anyone thinking that adopting an orbit at an altitude of about 500 miles (800 kilometers) out in space is not a precise business can think again. *Geo-localization is extremely problematic with such missions,* explains Heinz Sontag. *You have to know exactly where on Earth what you have photographed is located, and for that you have to know exactly where the satellite is at each moment and at what angle it is looking down on the Earth.* In short: if the images are to be accurate to within a few feet (a meter or so), the engineers need to know to within a few feet (a meter) at exactly what point on its orbit the satellite was standing—even as it races round the Earth at

sité pour obtenir le maximum d'informations. Ces images filtrées permettront d'étudier la végétation, les structures géologiques, l'urbanisation, les eaux, etc.

Au début d'une telle mission, tous les scientifiques impliqués, dont océanographes, spécialistes du climat, planificateurs régionaux et experts agricoles, décrivent ce qu'ils souhaitent faire du satellite, le champ de vision souhaité, la résolution requise pour les images, etc. L'ESA définit ensuite le cahier des charges, puis Heinz Sontag et son équipe s'attellent à traduire dans les faits des rêves souvent audacieux : *Nous déclinons ces spécifications au niveau du satellite et de ses instruments.*

Petit mais robuste

Le génie consiste à construire un satellite petit et léger qui fonctionne avec une faible puissance électrique, tout en fournissant un maximum de données scientifiques. En général, un satellite d'observation de la Terre incluant plate-forme, ordinateur de bord, antennes, instruments scientifiques, moteurs, etc. n'est pas plus grand qu'une Golf. Il pèse à peine une tonne et consomme nettement moins d'énergie qu'une machine à café.

Il serait erroné de croire qu'à 800 kilomètres d'altitude, définir précisément la position orbitale est accessoire. *Dans ce genre de mission, la géolocalisation est particulièrement délicate,* explique Heinz Sontag. *Il s'agit de déterminer la position exacte sur la Terre des phénomènes enregistrés. Pour ce faire, il est indispensable de connaître avec précision la position du satellite et l'angle sous lequel il observe la Terre.* En un mot, pour obtenir des images précises au mètre près, les ingénieurs doivent connaître au mètre près la position orbitale du satellite, alors que ce dernier gravite autour de la Terre à la vitesse de huit kilomètres par seconde. Avis aux amateurs !

Les satellites Sentinelle-2 génèrent un tel volume d'informations que les

Las listas de deseos no son proyectos de construcción

GOCE es un magnífico ejemplo de cómo fascinantes ideas de investigación pueden ir de la mano con una ingeniería casi inconcebible. El satélite observaba nuestra Tierra de un modo tan insólito porque todos los participantes le dotaron de un olfato, literalmente extraterrestre, para la fuerza de la gravedad. Pero también los satélites de observación terrestre «clásicos», que hacen fotos en órbita, son obras maestras de la técnica.

Precisamente, en uno de ellos trabaja el equipo de Heinz Sontag. El físico especializado en láser dirige en Airbus, Friedrichshafen, el desarrollo de los dos satélites de Sentinel-2, el segundo proyecto del gran programa de observación terrestre Copérnico que la Unión Europea y la ESA llevan a cabo en común. *Queremos suministrar datos en trece canales espectrales de luz visible e infrarrojo cercano.* Estos satélites, que se lanzarán a partir de 2017, no deben limitarse a tomar imágenes de la superficie terrestre. Las fotografías se efectúan siempre, por así decirlo, a través de trece diferentes filtros con distintas longitudes de onda. Esto, más que un lujo, es algo indispensable para obtener la más amplia información posible de esas fotografías, ya que sólo las tomas filtradas permiten adquirir muchos conocimientos sobre vegetación, estructuras geológicas, edificación, aguas, etc.

Así, una misión comienza siempre con aquello que los científicos implicados aclaran sobre el empleo que le quieren dar al satélite, lo extenso que debería ser el campo visual, la resolución que deberían tener las imágenes, etc. La agencia espacial competente, por ejemplo la ESA, establece entonces los requisitos básicos de la misión. Heinz Sontag y su equipo tienen que convertir esos sueños, frecuentemente temerarios, en hechos indiscutibles: *Trasladamos estas especificaciones a los satélites y a los instrumentos de a bordo.*

Pequeño, ligero, robusto

während er mit acht Kilometern pro Sekunde um die Erde rast. Viel Spaß beim Nachmachen!

Die Sentinel-2-Satelliten produzieren so große Datenmengen, dass die Rohdaten bereits an Bord ein wenig verarbeitet und auf ein Viertel komprimiert zur Erde gefunkt werden. Anders wäre das wegen der begrenzten Übertragungskapazitäten gar nicht möglich. All diese Aspekte müssen die Ingenieure von Anfang an bedenken. Sonst droht die Mission zu scheitern.

Heinz Sontag ist ein alter Hase im Feld der Erdbeobachtung. Er war bereits in den neunziger Jahren beim Bau von Envisat dabei und hat die wichtigste Kamera an Bord betreut. Jetzt muss er sich selbst übertreffen. Sentinel-2 soll beim Fotografieren der Erde etwa dreimal besser sein als die bisherigen Missionen und die erfolgreiche Arbeit der US-amerikanischen Landsat- und der französischen Spot-Satelliten fortsetzen. In einigen Jahren werden die ersten Bilder zeigen, ob das große Ziel erreicht wird.

Planet in der Radarfalle

Spektakuläre Aufnahmen der Erdoberfläche, etwa von schroffen Vulkanbergen, endlosen Sanddünen in der

5 miles (8 kilometers) a second. You try doing that!

The Sentinel-2 satellites produce such vast amounts of information that the raw data is provisionally processed on board, compressed to a quarter of its size, and then transmitted to Earth. Limited upload capacity makes it impossible to do this any other way, and the engineers have to incorporate all these considerations from the very beginning—otherwise the mission may fail.

Heinz Sontag is no novice where Earth observation is concerned. He was involved in the construction of Envisat during the 1990s, overseeing the most important onboard camera, but now he has to excel himself. Sentinel-2 will continue the successful work of the US Landsat and the French Spot satellites, but is intended to be three times better at photographing the Earth than any preceding mission. In a few years, the first pictures will show if this great goal has been achieved.

A planet in a speed trap

Spectacular images of the Earth's surface—of steep volcanic slopes, perhaps or endless expanses of sand

données brutes doivent être traitées à bord puis comprimées avant d'être transmises vers la Terre. Le succès de la mission repose sur la prise en compte en amont de tous ces aspects par les ingénieurs.

Heinz Sontag est un vieux routier de l'observation de la Terre. Il a dans les années 1990 participé à la construction d'Envisat, supervisant la réalisation des principales caméras embarquées. Aujourd'hui, il lui faut se surpasser. Sentinelle-2 doit être trois fois plus performant que les missions précédentes, et assurer la relève des satellites français Spot et de leur confrère américain Landsat. Dans quelques années, les premières images révèleront si cet objectif ambitieux a été atteint.

La planète flashée au radar

Cônes volcaniques aux versants escarpés, immenses dunes de sable, volutes singulières des nuages sur l'océan, bras méandreux des deltas fluviaux : les images de la surface terrestre sont époustouflantes. La photographie présente toutefois un inconvénient de taille. La lumière du soleil est indispensable à de telles prises de vue, par conséquent,

De lo que se trata, entonces, es de construir el satélite de tal modo que sea lo más pequeño y ligero posible y así necesite la menor cantidad de potencia eléctrica. Pero, al mismo tiempo, debe incluir todo lo necesario para su misión. Un típico satélite de observación terrestre, incluyendo estructura básica, ordenador de a bordo, antenas, instrumentos de medición, propulsores, etc., no es más grande que un VW Golf. Normalmente, apenas llega a pesar una tonelada y consume menos electricidad que una máquina de café.

Quien piense que en la vastedad del espacio y a una altura de vuelo de unos 800 km tampoco es tan importante la situación de la órbita, se equivoca por completo. *En una misión de este tipo, la geolocalización es uno de los asuntos más críticos*, aclara Heinz Sontag. *Se quiere saber dónde se encuentra en la Tierra lo que se ha observado desde el satélite. Para eso se necesita un conocimiento preciso de la situación exacta de éste y del ángulo con el que mira hacia la Tierra.* Resumiendo: si las imágenes deben tener una precisión de pocos metros, entonces los ingenieros deben saber con exactitud dónde se encuentra el satélite en la órbita y todo ello mientras vuela a 8 km por segundo. ¡Una labor titánica!

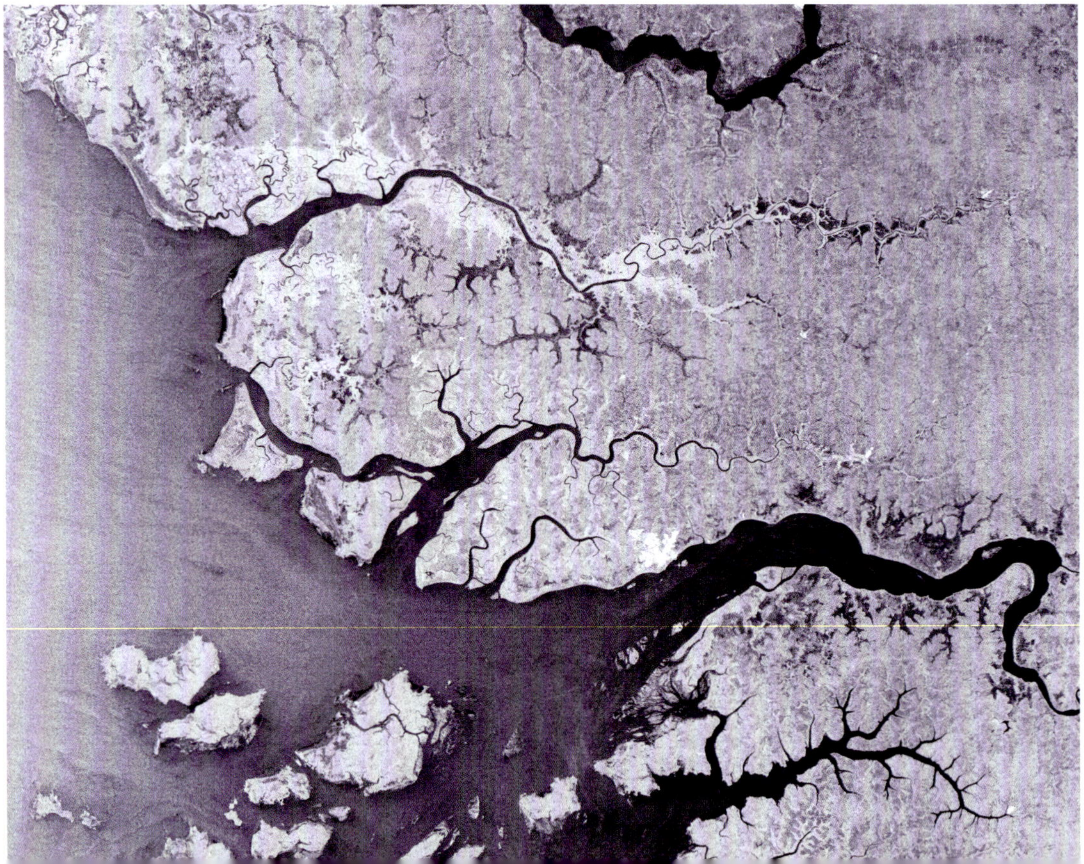

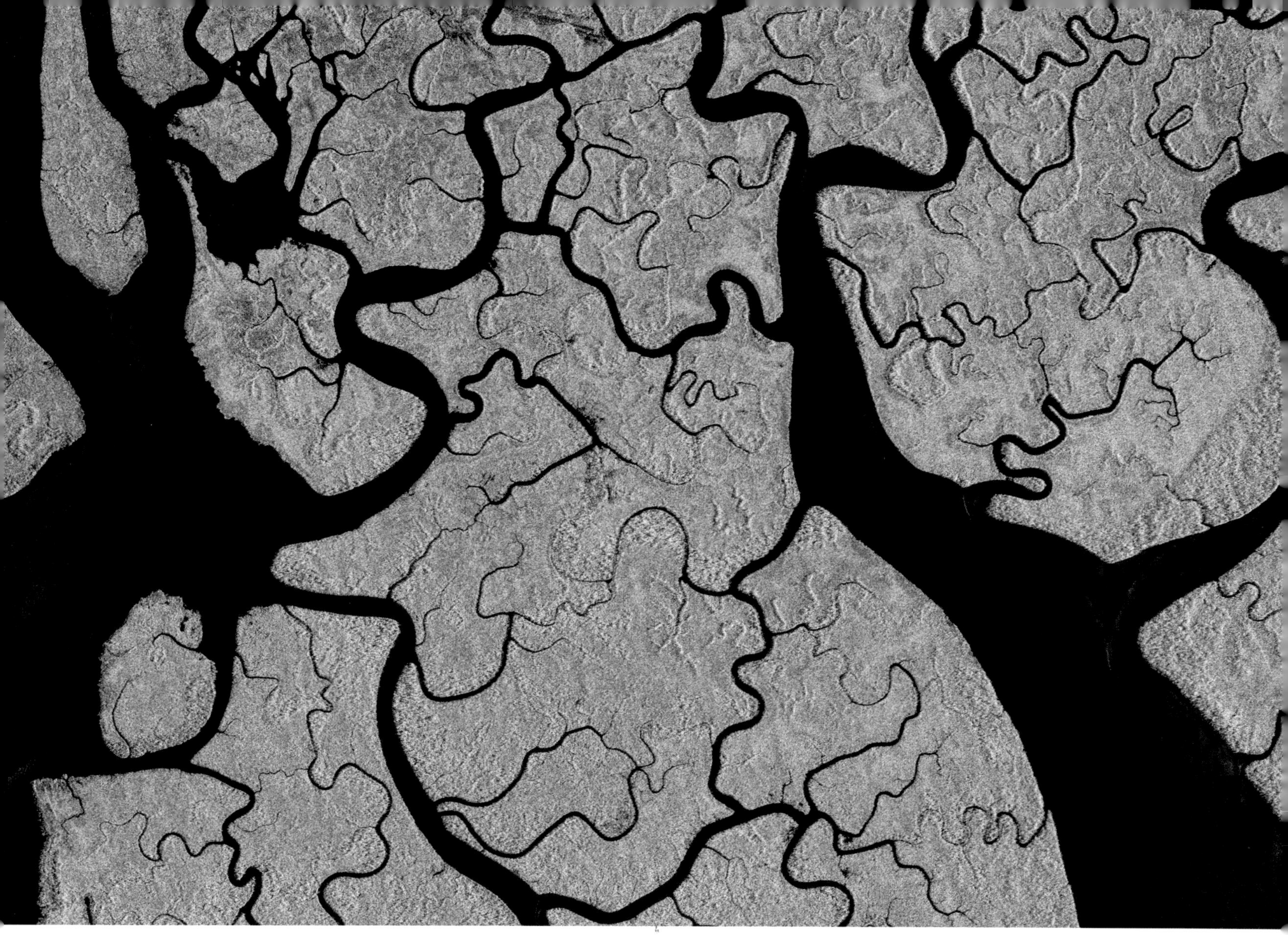

Linke Seite: Guinea-Bissau in einer Radaraufnahme. Der Bildausschnitt ist 190 km breit (Envisat)

Left: A radar image of Guinea-Bissau. The area shown is 120 miles (190 km) wide (Envisat)

Page de gauche : image radar de la Guinée-Bissau couvrant une largeur de 190 km (Envisat)

Página izda.: Toma de radar de Guinea-Bissau. El encuadre es de 190 km de ancho (Envisat)

Oben: Radarbild von Mangroven in Kamerun. Der Bildausschnitt ist 15 km hoch (TerraSAR-X)

Above: A radar image of mangroves in Cameroon. The detail shown is 9 miles (15 km) from top to bottom (TerraSAR-X)

Ci-dessus : image radar de la mangrove du Cameroun couvrant une hauteur de 15 km (TerraSAR-X)

Arriba: Imagen de radar de los manglares de Camerún. El encuadre es a 15 km de altura (TerraSAR-X)

Unten: Rotterdam in einer Radaraufnahme von Envisat mit 12,5 m Auflösung

Below: An Envisat radar image of Rotterdam with a resolution of 41 ft (12.5 m)

Ci-dessous : image radar de Rotterdam acquise par Envisat avec une résolution de 12,5 m

Abajo: Rotterdam en una toma de radar del Envisat, con 12,5 m de resolución

Rechte Seite: Der Radar zeigt die Schiffe in der Straße von Gibraltar (TerraSAR-X)

Right: Boats in the Strait of Gibraltar show up on radar (TerraSAR-X)

Page de droite : le radar repère les navires dans le détroit de Gibraltar (TerraSAR-X)

Página dcha.: El radar muestra los barcos en el estrecho de Gibraltar (TerraSAR-X)

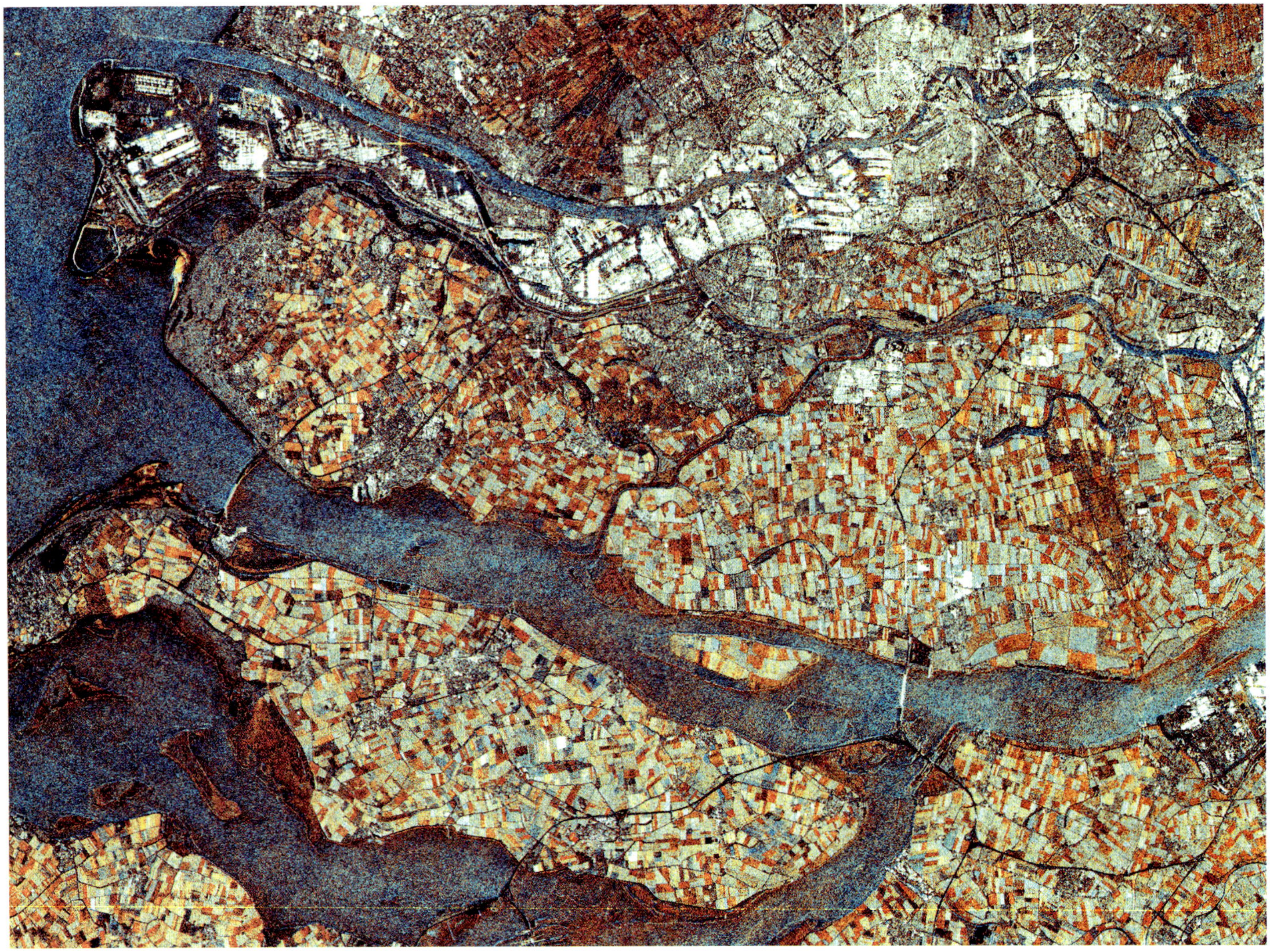

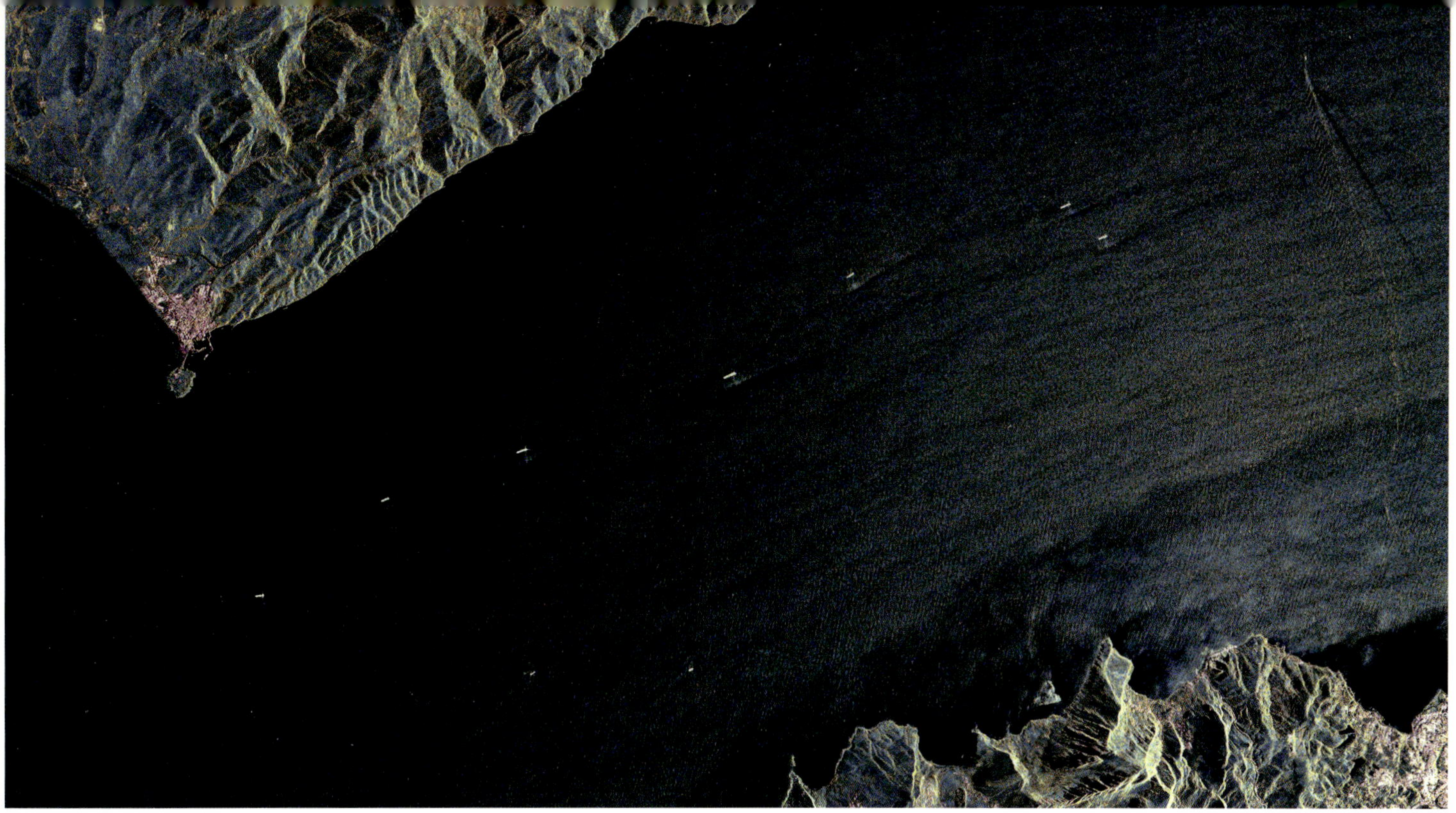

Wüste, skurrilen Wolkenwirbeln auf dem Ozean oder filigran verzweigten Flussdeltas, sind wunderschön. Doch beim Fotografieren gibt es eine gravierende Einschränkung. Solche Bilder gelingen nur bei Sonnenschein, also auf der Tagseite der Erde. Fliegen Satelliten wie Envisat oder später die von Sentinel-2 über die Nachtseite unseres Planeten, sind grandiose Farbfotos nicht möglich. Ein Blitzlicht haben die Satelliten natürlich nicht. Es würde aus so großer Höhe auch nichts bringen – jedenfalls nicht im sichtbaren Licht.

Im Radarbereich ist den Forschern und Ingenieuren der Sonnenstand völlig egal. *Wir beleuchten die Erdoberfläche selbst und nehmen die Bilder bei jeder Tageszeit und bei jedem Wetter auf,* betont schmunzelnd Ralf Zimmermann. Der gelernte Informatiker ist bei Airbus für die Radarmissionen TerraSAR-X und TanDEM-X zuständig. *Wir haben eine Radarantenne, die gleichzeitig Sender und Empfänger ist. Sie sendet Radarimpulse aus und misst hinterher das Echo. Auf diese Weise können wir jederzeit gute Bilder von der Erdoberfläche machen.* Die Auflösung beträgt im Idealfall etwa einen Meter.

TerraSAR-X ist bereits seit 2007 im Einsatz, TanDEM-X kam Mitte 2010 hinzu. Beide Satelliten fliegen zur

dunes in the desert, wisps of clouds floating over the ocean, or the dendritic patterns of river deltas—are extremely beautiful to look at, but capturing such images is subject to an important limitation: the pictures can be taken only in natural light, in other words, on the day side of the Earth. Marvelous color photography is just not possible when satellites like Envisat and later Sentinel-2 are passing the night side of our planet. There is no flash on board the satellites, of course, and it would have no effect at such a great height anyway—in visible light at least.

However, the amount of daylight is irrelevant to the researchers and engineers working in the radar bandwidth. *We send waves down to the surface itself and can capture images at any time of the day or night and in any weather,* grins Ralf Zimmermann, the Information Technology graduate who now leads the TerraSAR-X and TanDEM-X radar missions at Airbus. *We have a radar antenna which is both a transmitter and receiver. It sends out radar pulses and measures their echo afterwards, enabling us to take high-quality pictures of the Earth's surface.* At its best, the resolution is about 3 feet (1 meter).

quand les satellites tels qu'Envisat ou Sentinelle-2 survolent le côté de la Terre plongé dans l'obscurité, ils sont incapables de fournir de superbes photos couleurs. Ils ne sont pas équipés de flash, qui de toute façon ne serait à cette altitude d'aucune utilité, tout au moins en lumière visible.

Dans le domaine radar en revanche, chercheurs et ingénieurs n'ont que faire de la position du soleil. *Nous éclairons nous-mêmes la surface de la Terre que nous photographions à toute heure et par tous les temps*, précise dans un sourire Ralf Zimmermann. Cet informaticien est responsable des satellites radar TerraSAR-X et TanDEM-X chez Airbus. *Nous disposons d'une antenne qui fait simultanément office d'émetteur et de récepteur. Elle émet des impulsions radar et mesure leur écho. Nous obtenons avec cette méthode des images de bonne qualité.* Dans des conditions idéales, ces images ont une résolution d'un mètre.

En service opérationnel depuis 2007, TerraSAR-X a été rejoint par TanDEM-X à la mi-2010. Tous deux volent désormais en formation pour le plus grand plaisir de Ralf Zimmermann : *Grâce à ces deux satellites, nous réalisons la cartographie numérique du relief terrestre.*

Los satélites Sentinel-2 generan tal cantidad de datos que es necesario hacer un primer procesamiento a bordo de los datos primarios, para transmitirlos por radio a la Tierra comprimidos a un cuarto de su tamaño. De otro modo, la transmisión no sería posible. Los ingenieros tienen que pensar en todos estos detalles desde el principio. De lo contrario, la misión podría fracasar.

Heinz Sontag está curtido en el campo de la observación terrestre. En la década de 1990, ya participó en la construcción del Envisat, ocupándose de la cámara más importante de a bordo. Ahora se tiene que superar a sí mismo. Fotografiando la Tierra, Sentinel-2 tiene que ser casi tres veces mejor que las misiones anteriores y continuar el eficaz trabajo de los satélites Landsat estadounidenses y los Spot franceses. Dentro de unos pocos años, las primeras imágenes nos dirán si se pudo conseguir este gran objetivo.

Todo un planeta bajo el control del radar

Algunas imágenes espectaculares de la superficie terrestre, como las de interminables dunas de arena en el desierto, grotescos remolinos de nubes

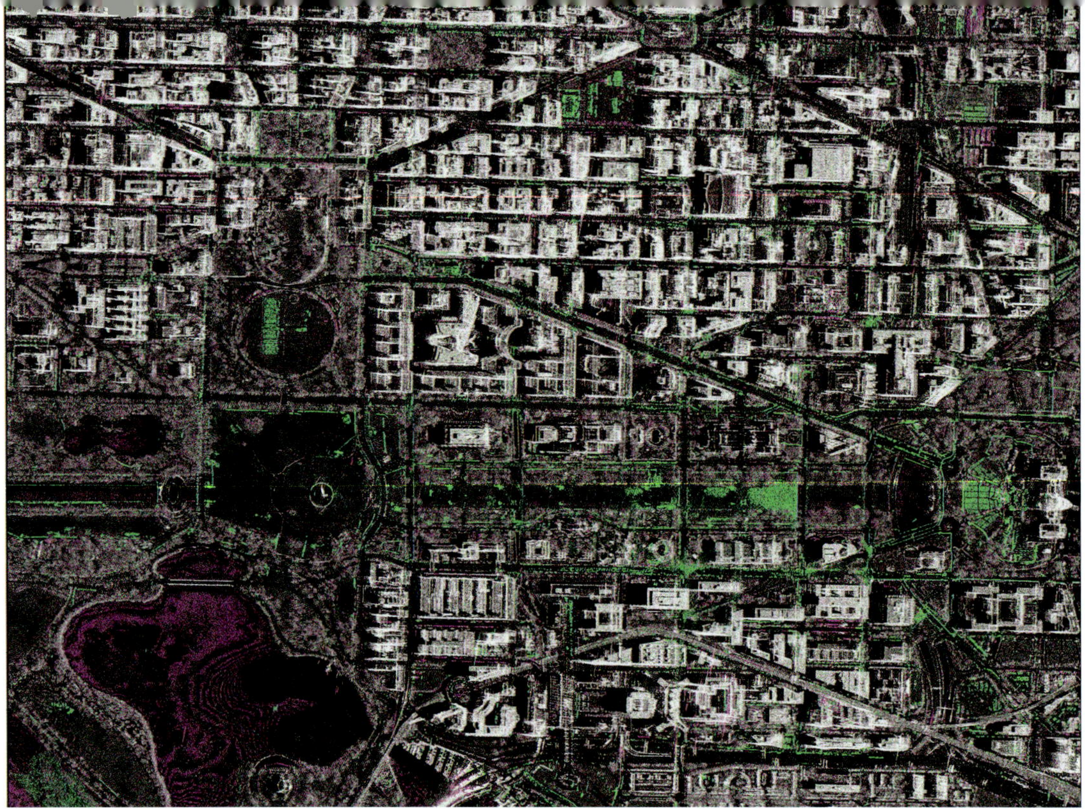

Freude von Ralf Zimmermann jetzt im Formationsflug: *Mit zwei Satelliten erstellen wir eine digitale Höhenkarte der ganzen Erde. Dafür brauchen wir einen Satelliten, der den Radarpuls sendet, und zwei, die ihn empfangen. Das kann man in etwa mit dem Stereo-Sehen vergleichen.* Erst zwei Satelliten ermöglichen gutes räumliches Sehen. Das magische Augenpaar im All nimmt nun erstmals die gesamte Erde in 3D auf – mit einer Genauigkeit von wenigen Metern.

Sogar einige Zentimeter Genauigkeit soll das Radarauge erreichen, das an Bord von CryoSat-2 um die Erde fliegt. Allerdings beobachtet dieser Satellit auch nicht die ganze Erde, sondern nur die Eismassen in den Polargebieten. *Als Kind bin ich im Winter immer auf jede Eisfläche gestiegen,* erinnert sich Duncan Wingham vom University College in London an seine ersten eisigen Erfahrungen. Etwa vier Jahrzehnte später steigt der Klimaphysiker wieder aufs Eis, dieses Mal aber viel professioneller: *Wir vermessen mit unserem Satelliten CryoSat-2 erstmals präzise, wo in den Polargebieten der Erde wie viel Eis vorkommt.*

Das Messprinzip von CryoSat-2 ist erstaunlich simpel: Der Satellit hat ein äußerst empfindliches Radargerät an Bord. Es sendet stets kurze Impulse nach unten und misst exakt,

TerraSAR-X has been in operation since 2007 and TanDEM-X was added in mid-2010. Ralf Zimmermann is delighted to report that the two satellites are now flying in formation: *We are using the two satellites to produce a digital relief map of the entire world, and for this we need one satellite to transmit the radar pulse and two to receive it. It's a little bit like stereo vision.* Seeing in depth like this is only possible with two satellites, and this magic pair of eyes in the sky is now recording the entire Earth in 3-D to an accuracy of within a few feet (a meter or so).

The radar receiver flying round the Earth on board CryoSat-2 is expected to have an accuracy of a few inches (just centimeters). It is not observing the entire Earth, however, but is concentrating on the ice caps at the poles. *As a child, I always used to climb on any ice I found in winter,* recalls Duncan Wingham of University College, London, thinking back to his first icy experiences. Some four decades later, and now a climate scientist, Wingham is stepping back onto the ice in a rather more professional capacity: *We use our CryoSat-2 satellite to measure how much ice there is at the Earth's poles and how it is distributed.*

Un satellite émet des impulsions radar et tous deux reçoivent les retours. C'est la présence de deux satellites qui permet d'obtenir une vision en relief. Ces deux yeux magiques photographient pour la première fois la Terre en 3D, avec une précision de quelques mètres.

Le radar qui gravite autour de la Terre à bord de CryoSat-2 atteint un degré de précision encore plus élevé. Il est vrai que ce satellite n'observe pas la totalité du globe, mais uniquement les masses de glace des régions polaires. *Enfant, je ne pouvais m'empêcher l'hiver de marcher sur les surfaces gelées,* se souvient Duncan Wingham du University College de Londres. Près de quatre décennies plus tard, le physicien du climat remonte sur la glace, avec un objectif nettement plus scientifique cette fois : *Grâce au satellite CryoSat-2, nous mesurons l'épaisseur des glaces polaires avec une précision sans précédent.*

Le principe de mesure de CryoSat-2 est étonnant de simplicité : un radar ultrasensible émet une série d'impulsions très courtes en direction de la Terre, et enregistre avec exactitude l'écho renvoyé vers le satellite. La durée du signal renseigne les scientifiques sur la distance entre la glace et le satellite.

o ramificaciones afiligranadas de un delta, son maravillosas. Pero para fotografiarlas, existe una seria restricción. Estas imágenes únicamente pueden obtenerse con luz solar, es decir, en el lado diurno de la Tierra. Cuando los satélites, por ejemplo, el Envisat u otros posteriores como el Sentinel-2, vuelan sobre el lado nocturno de nuestro planeta, estas grandiosas fotos en color no son posibles. Naturalmente, los satélites no tienen flash. Desde una altura tan grande tampoco serviría para nada.

En el caso de la cobertura de radar, a los investigadores e ingenieros les da exactamente igual la posición del Sol. *Nosotros mismos examinamos la superficie terrestre y tomamos imágenes a cualquier hora del día y con cualquier tiempo,* afirma satisfecho Ralf Zimmermann. Este cualificado informático es el responsable en Airbus de las misiones de radar TerraSAR-X y TanDEM-X. *Tenemos una antena de radar que es, al mismo tiempo, emisor y receptor. Emite impulsos de radar y posteriormente mide su eco. De esta manera podemos obtener buenas imágenes del suelo terrestre.* La resolución asciende, en el mejor de los casos, a aproximadamente un metro.

TerraSAR-X está operativo desde 2007 y TanDEM-X se le unió a media-

Linke Seite: Da ist auch Obama drauf – die National Mall in Washington während der Amtseinführung 2009 (TerraSAR-X)

Left: Obama's in there somewhere – the National Mall in Washington as the US President took the oath of office in 2009 (TerraSAR-X)

Page de gauche : Obama aussi est quelque part sur la photo. Le National Mall à Washington lors de son investiture en 2009 (TerraSAR-X)

Página izda.: Aquí sale hasta Obama: el National Mall de Washington durante el acto de investidura en 2009 (TerraSAR-X)

Unten: Radarbild von Hamburg mit Flughafen. Eine höhere Auflösung wurde von den Behörden nicht freigegeben (TerraSAR-X)

Below: A radar image of Hamburg, showing the airport. Higher resolution images are restricted by the authorities (TerraSAR-X)

Ci-dessous : image radar de Hambourg et son aéroport. Les autorités se sont opposées à la publication d'une image de résolution plus haute (TerraSAR-X)

Abajo: Imagen de radar de Hamburgo, con su aeropuerto. Las autoridades no permitieron una mayor resolución (TerraSAR-X)

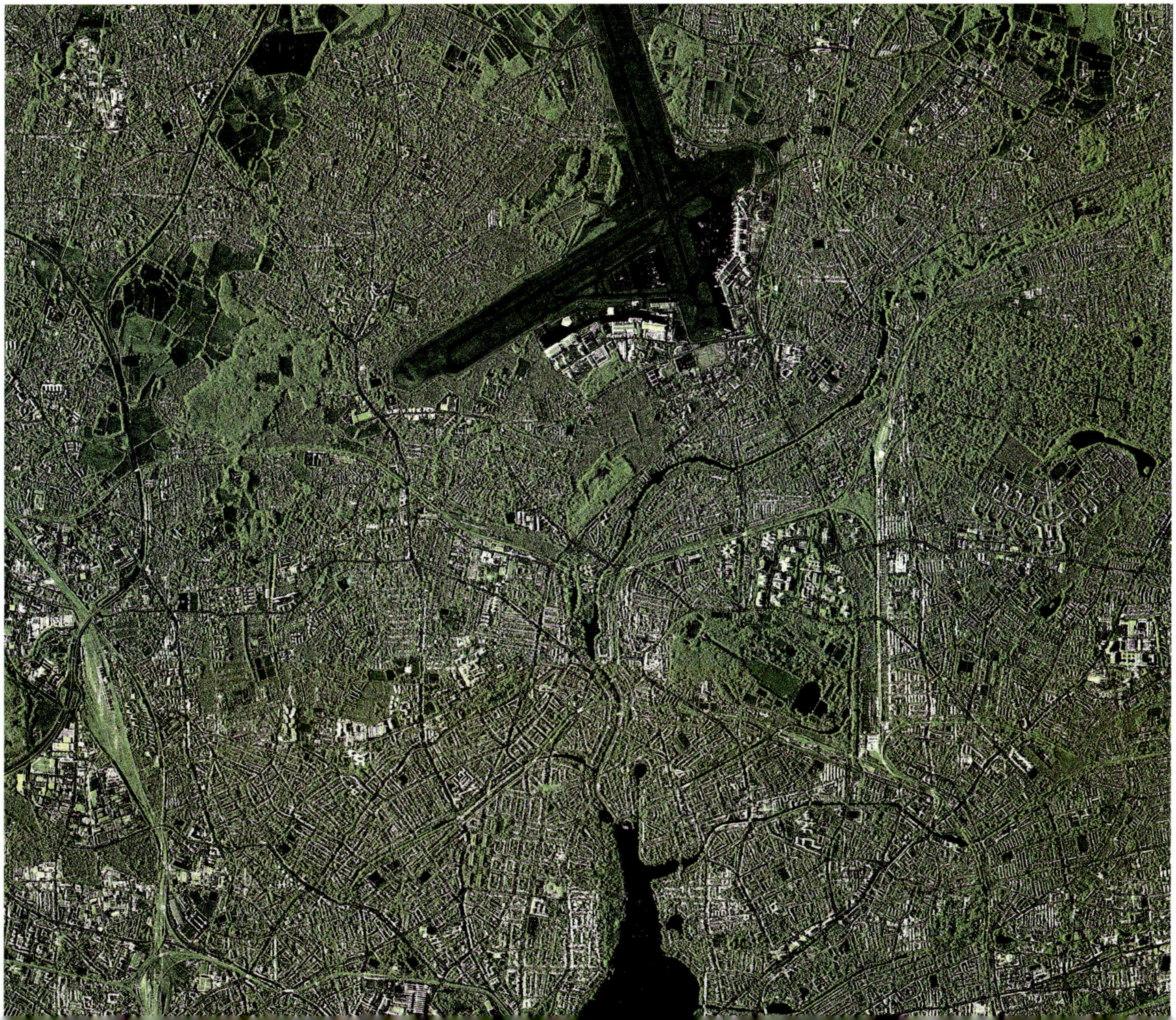

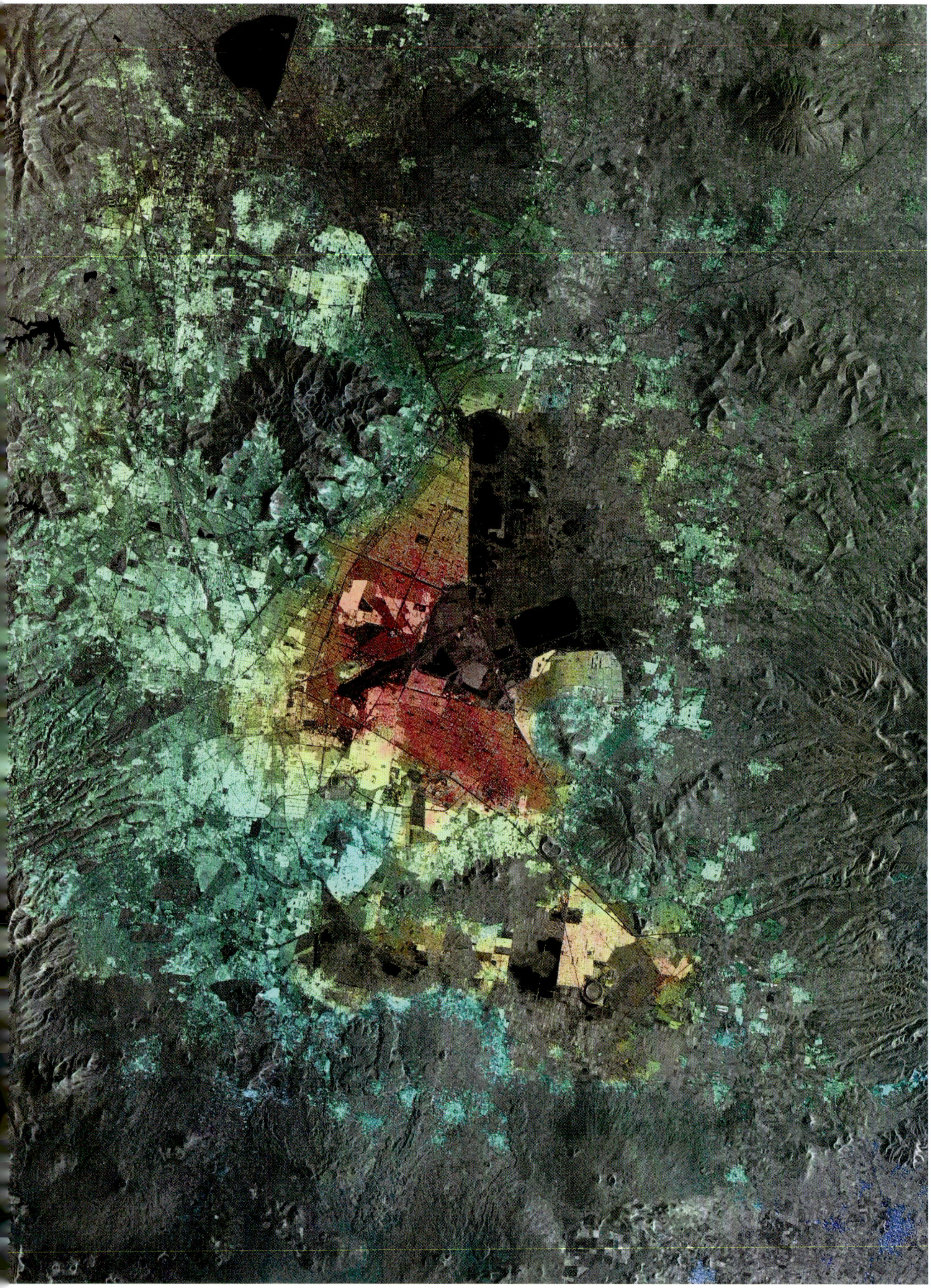

Radaraufnahmen zeigen, dass sich der Boden in Mexico City binnen vier Monaten um bis 10 cm gesenkt hat (rote Bereiche). Grund ist die Entnahme von Grundwasser (TerraSAR-X)

Radar images reveal that the ground in Mexico City subsided (red areas) by up to 4 inches (10 cm) in less than four months as a result of groundwater use (TerraSAR-X)

Les images radar indiquent que le sol de México s'est affaissé de 10 cm en quatre mois (zones rouges), en raison de ponctions dans les nappes phréatiques (TerraSAR-X)

Las tomas de radar muestran que el suelo de la ciudad de México se ha hundido 10 cm en cuatro meses (sectores en rojo). Se debe a la extracción de aguas subterráneas (TerraSAR-X)

Rechte Seite: Der kanadische Radarsat-2 macht Aufnahmen mit bis zu 3 m Auflösung (Zeichnung)

Right: Providing images with a resolution of up to 10 ft (3 m) is no problem for Canada's Radarsat-2 (illustration)

Page de droite : le satellite radar canadien Radarsat-2 permet l'acquisition d'images avec une résolution de 3 m (vue d'artiste)

Página dcha.: El Radarsat-2 canadiense toma imágenes de hasta 3 m de resolución (dibujo)

wann das Echo den Satelliten erreicht. Aus der Laufzeit des Signals bestimmen die Forscher, wie weit das Eis vom Satelliten entfernt ist. CryoSat-2 nimmt sozusagen ständig eine präzise Höhenmessung vor.

CryoSats Gespür für Schnee

Die Forscher erfahren so, wie dick der Eispanzer auf dem antarktischen Kontinent ist, aber auch, wie viel Eis im Nordpolarmeer treibt. Bisher konnten Satelliten nur mit Hilfe von Fotos zeigen, wo Eisschollen sind. Wie dick sie sind, also welche Mengen an Eis sie tatsächlich enthalten, blieb unklar. *Das abbildende Radarsystem hat mehr als die Hälfte der Baukosten für den Satelliten verschlungen*, erläutert Duncan Wingham. *Dafür ist es so gut geworden, dass wir damit auf zwei Zentimeter genau messen werden, wie weit die Eisschollen aus dem Wasser ragen. Dann lässt sich berechnen, wie viel Eis dort wirklich ist.*

Die Firma Airbus in Friedrichshafen, neben Thales Alenia der große Satellitenbauer in Europa, hat das Geld bekommen und die Qualität geliefert. Projektleiter war der Maschinenbauingenieur Klaus Köble. Gemeinsam mit etwa vierzig Mitarbeitern und weiteren neunzig Personen

CryoSat-2's measuring technique is astonishingly simple: there is an incredibly sensitive radar device on board which sends down short pulses and records precisely when the echo reaches the satellite. The time taken for the signal to return tells the scientists how far away the ice is from the satellite—in effect, CryoSat-2 is carrying out a continuous and precise measurement of altitude.

CryoSat's feeling for snow

Scientists can thus find out how thick the crust of ice on the continent of Antarctica is or how much ice is floating in the Arctic Ocean. Satellites had previously only been able to show where ice floes were with the aid of photographs. It was not possible to tell how thick they were—in other words, how much ice they actually contained. *The radar imaging system ate up more than half of the construction costs for the satellite,* explains Duncan Wingham. *But because of this, it is now so good that we can measure to the inch how far the ice floes jut out of the water. From that we can work out how much ice there really is.*

The firm of Airbus in Friedrichshafen, Germany, the largest satellite-

CryoSat ou le sens de la neige

Les scientifiques déterminent ainsi la hauteur de la calotte glaciaire de l'Antarctique ainsi que l'épaisseur de la banquise arctique. Jusqu'ici les satellites se contentaient d'indiquer la position des blocs de glace à l'aide de photos. Ils étaient incapables de mesurer leur épaisseur, autrement dit le volume de la glace flottante. *Le système radar a englouti plus de la moitié du budget de construction du satellite*, déclare Duncan Wingham. *Mais la qualité atteinte est telle qu'il nous permettra de mesurer à deux centimètres près la hauteur de la glace émergée, et donc de calculer le volume réel de glace.*

La société Airbus, le plus grand constructeur européen de satellites avec Thales Alenia, a obtenu les fonds nécessaires et livré la qualité requise. C'est Klaus Köble, ingénieur mécanique, qui a dirigé le projet. Il a développé, avec une équipe d'une quarantaine de personnes et les renforts de quatre-vingt-dix employés travaillant chez les fournisseurs, cette faculté particulière de CryoSat de sentir la glace. Les nouvelles technologies ont été affinées pendant des années, puis testées sur des maquettes. Tous ces efforts tendaient vers un but unique : placer sur orbite

dos de 2010. Para satisfacción de Ralf Zimmermann, ambos satélites vuelan ahora en formación: *Con dos satélites, elaboramos un mapa altimétrico digital de toda la Tierra. Para eso necesitamos un satélite que envíe el pulso de radar y dos que lo reciban. Se podría comparar, vagamente, con la vista estereoscópica.* Únicamente dos satélites hacen posible la vista tridimensional. Por primera vez, se fotografía en 3D toda la Tierra con una exactitud de unos pocos metros.

El radar que vuela alrededor de la Tierra a bordo del CryoSat-2 debe lograr una precisión de centímetros. Sin embargo, este satélite tampoco observa toda la Tierra, sino las masas de hielo de las regiones polares. *Cuando era niño, en invierno me subía a la primera superficie helada que veía*, recuerda Duncan Wingham, del University College de Londres. Casi cuatro décadas después, este físico especializado en climatología se sube de nuevo el hielo, pero esta vez de un modo mucho más profesional: *Con nuestro satélite CryoSat-2 medimos por primera vez con precisión cuánto hielo hay en las regiones polares de la Tierra.*

El principio de medición del CryoSat-2 es sorprendentemente simple: el satélite lleva a bordo un equipo de radar de extrema sensibilidad. Envía a

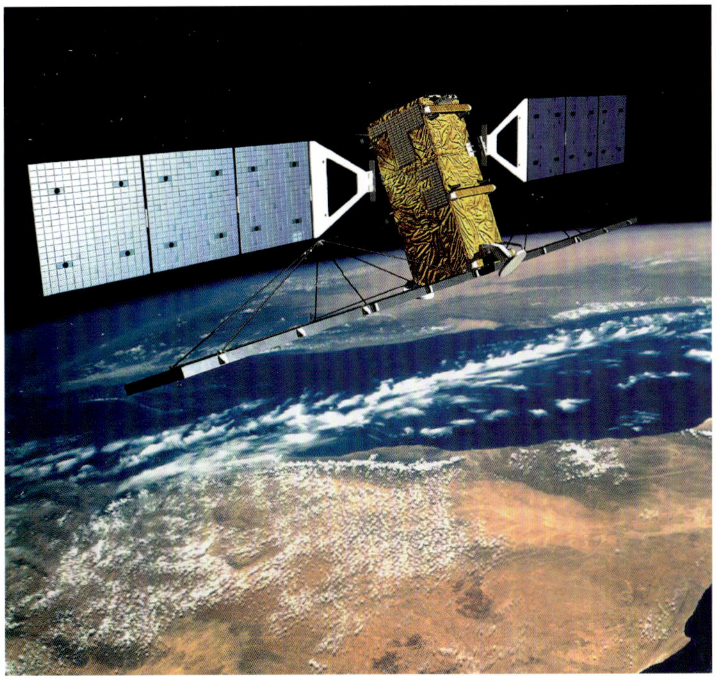

Seite 122: Radarbild von Eisschollen am Gletscher Vatnajökull auf Island (TerraSAR-X)

Page 122: A radar image of drift ice around the Vatnajökull Glacier in Iceland (TerraSAR-X)

Page 122 : image radar des blocs de glace flottant au pied du glacier Vatnajökull en Islande (TerraSAR-X)

Página 122: Imagen de radar de témpanos de hielo en el glaciar Vatnajökull, de Islandia (TerraSAR-X)

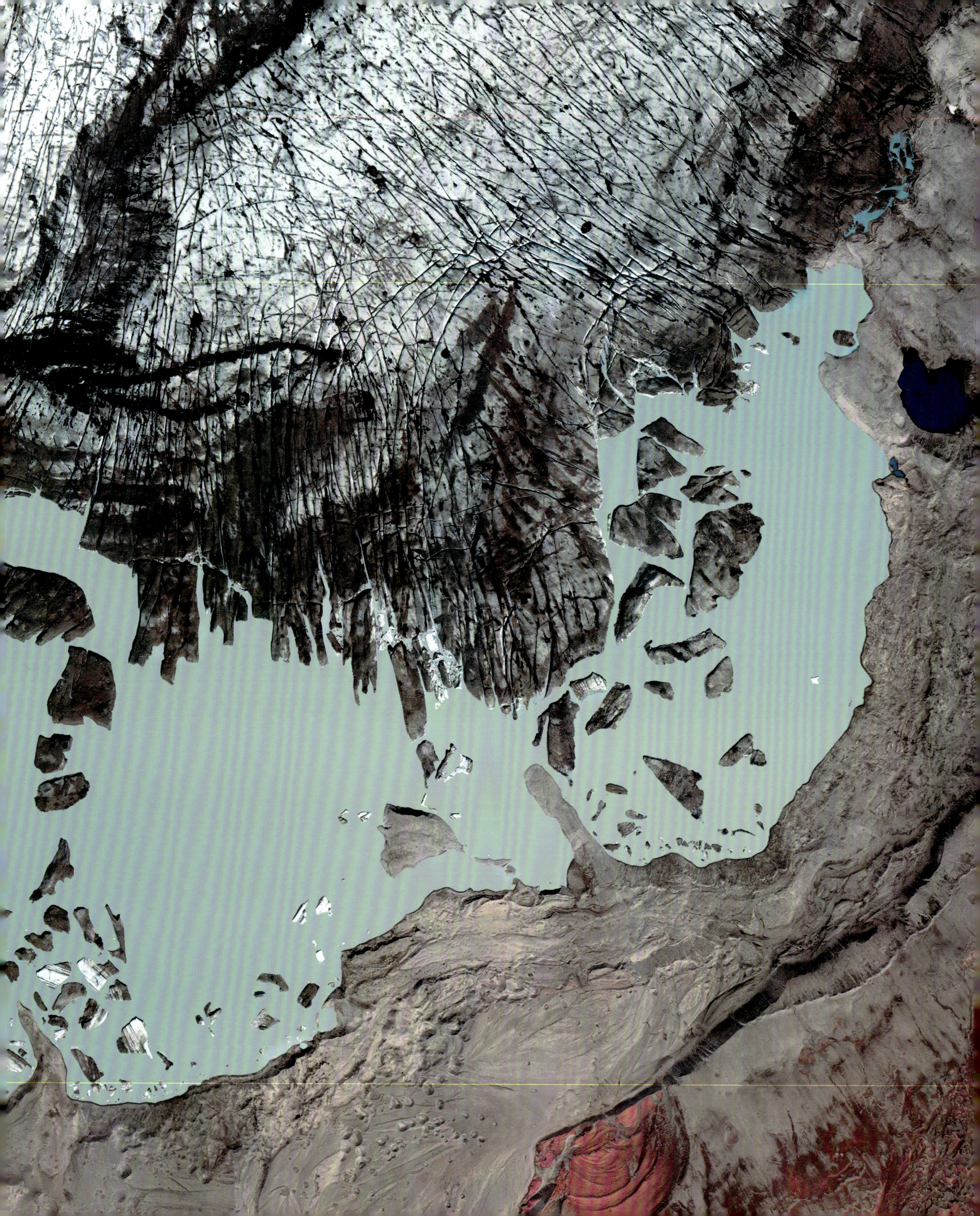

bei Zulieferern hat er CryoSats besonderes Gespür für Eis entwickelt. Über Jahre hinweg wurde an neuen Technologien gefeilt, die in Testmodellen erprobt wurden. Alle Arbeit galt einem Ziel: CryoSat ins All zu bringen und in die Lage zu versetzen, die ambitionierten Messungen durchzuführen.

Weltraummissionen sind nie Routine. Die Ingenieure testen jedes Mal aufs Neue ihre Grenzen und betreten bei wissenschaftlichen Projekten praktisch immer Neuland. Rückschläge bleiben da nicht aus, erinnert sich Klaus Köble: *Es passiert schon, dass solche Tests nicht erfolgreich sind. Allerdings sind wir in der Zwischenzeit in der Lage, viele Dinge am Computer zu simulieren, auch elektrische Schaltungen oder die Leistung des Radars. Das spart relativ viel Zeit und Geld.*

Doch Simulationen sind das eine, echte Tests von Prototypen das andere. Immer wieder müssen die Ingenieure zurück in Labor und Werkstatt, um Fehler auszubügeln. Im Oktober 2005 sollte der Satellit endlich starten. Doch sein Drang zum Eis war etwas zu stark. Schon zehn Minuten nach dem Abheben schlug er auf arktischen Eisschollen auf – die Rockot-Rakete, sonst sehr zuverlässig, hatte einen technischen Defekt. Bei den Ingenieuren in Friedrichshafen, die mit der Rakete nichts zu tun hatten, war das Entsetzen groß: *Das ist ein Schock für alle Beteiligten. Die Mitarbeiter haben viele Jahre sehr, sehr intensiv mit Herz und Seele an diesem Projekt gearbeitet. Und dann gibt es ein Unglück. Es mag vielleicht ein wenig übertrieben klingen: Aber es ist wie der Verlust eines guten Freundes.* So groß die Trauer war, so groß war einige Monate später die Erleichterung. Die ESA hatte sensationell schnell Mittel bereitgestellt, um CryoSat-2 zu bauen. Dessen Start im April 2010 glückte perfekt. Happyend!

Das Radarauge an Bord tastet nun buchstäblich die Eismassen ab. Die Pulse prallen auf die Eismassen, auf das Meer oder auf die eisfreie Erdoberfläche. Das Echo wird registriert – mit erstaunlicher Genauigkeit, sofern die Raumfahrtingenieure die Position des Satelliten präzise kennen. So erläutert Klaus Köble: *Ein Netz von gut*

builder in Europe after Italy's Thales Alenia, was given the money—and delivered the quality. Klaus Köble, a mechanical engineer, led the projekt with a team of 40 employees and a further 90 contractors in developing CryoSat particular feeling for ice. New technologies were refined over years and tested using experimental models. All this work had but one goal: getting CryoSat into space to enable it to carry out these ambitious measurements.

Space missions are never simply routine. Each time, the engineers are testing their limits and almost always breaking new ground with their scientific projects. There are sometimes setbacks, as Klaus Köble remembers: *Sometimes we find that these tests are unsuccessful. That said, we are now able to simulate many things on a computer, even electric switches or the performance of the radar. This saves a fair amount of time and money.*

Simulations are one thing, but actual tests of prototypes are quite another. The engineers have to return to the drawing board and the lab again and again to iron out errors. CryoSat was finally ready to be launched in October 2005, but its enthusiasm for ice was a little too pronounced—only ten minutes after take-off, it crashed into the ice floes of the Arctic; the Rockot rocket, usually so reliable, had developed a technical fault. The engineers at Friedrichshafen, who had had nothing to do with the rocket, were horrified: *This is a shock for everyone involved. Our colleagues pour their hearts and souls into this project for years, and then there is an accident. This might sound a little exaggerated, but it is like losing a good friend.* As great as this grief was, it was matched a few months later with relief: the ESA acted sensationally quickly in providing funds to build CryoSat-2, whose launch in April 2010 was a complete success. A story with a happy ending!

The onboard radar sensor literally feels its way across the ice sheets. The pulses bounce off the masses of ice, the sea, or the ice-free surface

un CryoSat capable de mener à bien les ambitieuses mesures visées.

Les missions spatiales ne connaissent pas la routine. Les ingénieurs ne cessent de repousser leurs limites et défrichent avec chaque projet de nouveaux domaines scientifiques. Dans ces conditions, le risque d'un revers de fortune est toujours présent, se souvient Klaus Köble : *Il arrive que les essais échouent. Nous sommes toutefois capables de simuler beaucoup de choses sur ordinateur, y compris les connexions électriques et les performances des radars. Ce qui permet d'économiser du temps et de l'argent.*

Malgré tout, simulations et tests de prototypes sont deux choses distinctes, et les ingénieurs sont souvent contraints de retourner à leur laboratoire pour corriger les erreurs. Le lancement du satellite avait été programmé en octobre 2005. Malheureusement, son attrait pour la glace étant sans doute trop fort, il s'écras sur la banquise arctique après seulement dix minutes de vol : une défaillance du lanceur Rockot, d'ordinaire pourtant si fiable. À Friedrichshafen, en Allemagne, le désarroi des ingénieurs, tributaires du lanceur, est à son comble. *Nous étions tous sous le choc. Les employés ont travaillé corps et âme à ce projet pendant de nombreuses années. Puis survient l'accident. Cela peut paraître excessif, mais c'était comme perdre un être cher.* Quelques mois plus tard, le soulagement est à la mesure de ce désespoir lorsque l'ESA débloque avec une rapidité extraordinaire les fonds nécessaires à la construction de CryoSat-2. Son lancement, en avril 2010, est un succès total.

Le radar qu'il emporte tâte littéralement les masses de glace. Ses impulsions percutent les couches de glaces marines et terrestres. L'écho qu'elles renvoient est ensuite enregistré avec une incroyable exactitude, à condition que les ingénieurs spatiaux connaissent précisément la position des satellites. Comme l'explique Klaus Köble : *Une cinquantaine de stations émettrices réparties aux quatre coins du globe envoient des signaux radio aux satellites dont l'écho permet de localiser avec précision le satellite*

tierra impulsos breves de forma continuada y mide exactamente el momento en que el eco alcanza el satélite. Dependiendo del tiempo que tarde la señal, los investigadores determinan lo lejos que está el hielo del satélite. Por así decirlo, CryoSat-2 realiza una altimetría de precisión.

El olfato de CryoSat para la nieve

Los investigadores averiguan así el grosor del casquete de hielo en el continente antártico, pero también cuánto hielo flota en el océano Ártico. Hasta ahora, los satélites sólo podían señalar, con la ayuda de fotografías, dónde se encontraban los témpanos. El grosor y la cantidad de hielo quedaban poco claros. *El sistema de acotación por radar se lleva más de la mitad de los costes de construcción del satélite*, explica Duncan Wingham. *Por eso es tan satisfactorio que con él se pueda medir, con una exactitud de hasta dos centímetros, cuánto sobresalen los témpanos del agua. Después, se puede calcular cuánto hielo hay allí realmente.*

La empresa Airbus de Friedrichshafen, la gran constructora de satélites europea, junto con Thales Alenia, recibió el dinero y entregó calidad. El director del proyecto fue el ingeniero mecánico industrial Klaus Köble. Junto con unos cuarenta colaboradores y otras noventa personas, incluidos los proveedores, él fue quien desarrolló ese especial olfato para el hielo que ha demostrado el CryoSat.

Las misiones espaciales nunca son rutinarias. Los ingenieros someten continuamente a prueba sus límites y abren nuevos horizontes casi en cada proyecto científico. Los reveses son inevitables, recuerda Klaus Köble: *Puede ocurrir que estas pruebas no tengan éxito. Sin embargo, por el momento estamos en condiciones de simular muchas cosas con ordenador, incluidos circuitos eléctricos o rendimiento del radar. Esto supone un relativo ahorro de tiempo y dinero.*

Pero las simulaciones son una cosa y las pruebas con prototipos, otra. Los ingenieros tienen que regresar continuamente al laboratorio y al

vierzig Sendestationen auf der Erde schickt seinerseits Pulse zum Satelliten, um aus den Echos die Lage des Satelliten im Orbit zu bestimmen. Die Auswertung dieser Daten liefert die Position des Satelliten auf seiner Bahn auf wenige Zentimeter genau. Damit sind die Forscher in der Lage, auch die vom Satelliten ausgesandten Radarpulse und deren Echos präzise auszuwerten. Das nahezu perfekte Radarauge sieht die Eismassen so genau wie kein anderes Instrument vor ihm.

Nass und salzig

Blicken Sie einmal aus dem Fenster, und betrachten Sie Nachbars Garten. Können Sie beim bloßen Hinsehen genau sagen, wie viel Wasser dort im Boden steckt? Nein? Dann sind Sie aber sicher in der Lage, beim nächsten Deichspaziergang allein beim Betrachten des Meeres den Salzgehalt des Wassers zu ermitteln. Auch nicht? Dann brauchen Sie die Hilfe von SMOS, einem der ESA-Satelliten im All. Er fliegt in 760 Kilometern Höhe um die Erde. An Bord ist nur ein einziges Messinstrument. Beim Überfliegen der Erde registriert es in langen Streifen die Abstrahlung der Erdoberfläche im Mikrowellenbereich. Aus solchen Daten lassen sich Bodenfeuchte und Salzgehalt der Meere bestimmen.

Faustregel: Je feuchter der Boden ist, desto geringer ist seine Abstrahlung. Je salziger das Wasser ist, desto schwächer strahlt die Wasseroberfläche. Dasselbe Instrument, dasselbe Messprinzip: SMOS kann auf diese Weise sowohl fühlen, wie feucht große Gebiete der Erde sind, als auch schmecken, wie viel Salz in einzelnen Meeresregionen vorhanden ist. Das ist kein Wunschdenken. Die Wissenschaftler testen die Mission ständig in bestimmten Gebieten der Erde, in denen sie aufgrund direkter Messungen vor Ort den Wassergehalt des Bodens und den Salzgehalt des Meeres kennen. SMOS hat die Tests locker bestanden.

Schwerkraftfühler, Hightechkameras, Radarsensoren und Salzschmecker – all das tummelt sich im Orbit und hat die Erde stets im Blick. Der Erfindungsreichtum der Wissenschaftler

of the Earth and the echo is recorded with astonishing precision, as long as the aerospace engineers know the exact position of the satellite. Klaus Köble explains: *A network made up of some 40 transmitters on Earth sends pulses to the satellite in order to determine its position in orbit from the echoes. Evaluation of this data narrows down the position of CryoSat-2 on its trajectory to within a few inches.* The scientists are then in a position to evaluate the radar pulses emitted from the satellite and the echoes it receives. This more or less perfect radar "eye" can make out the masses of ice beneath it better than any other instrument.

Wet and salty

Have a look out the window at the neighbor's garden. Can you tell how much water is in the ground just by looking at it? No? Perhaps you'll be able to see how much salt there is in the water next time you're at the seaside. Still no? You need the help of SMOS, the most recent ESA satellite to be launched into space. Orbiting at an altitude of 470 miles (760 kilometers), it has only a single measuring device on board. As it circles the Earth, it records microwave emissions from long tranches of the Earth's surface, and from this data soil moisture and ocean salinity can be calculated.

As a rule of thumb, the damper the ground, the lower its microwave signature, and the higher the salinity, the weaker the reflection from the sea's surface. Each is recorded by the same instrument using the same measuring technique, and SMOS can use it to feel how wet large regions of the Earth are and to taste how much salt is present in individual sea areas. This is no pipe dream. The scientists regularly test the probe in certain areas of the Earth where they already know the water content of the soil and the salinity of the sea from direct samples—SMOS passes the tests each time with flying colors.

Gravity detectors, high-tech cameras, radar arrays, and salt sensors—

en orbite. L'examen de ces données détermine sa position orbitale au centimètre près. Les scientifiques évaluent ensuite les impulsions radar émises par le satellite et leur écho. C'est ainsi que l'œil quasi parfait du radar observe les masses de glace avec une précision inégalée.

Humide et salée

Jetez un coup d'œil par la fenêtre et regardez le jardin de vos voisins. Pouvez-vous déterminer la quantité d'eau contenue dans le sol ? Non ? Alors vous saurez sans doute, lors de votre prochaine promenade sur la plage, évaluer la salinité de l'eau en observant la mer. Non plus ? Alors c'est SMOS qu'il vous faut, l'un des tout derniers satellites placés en orbite par l'ESA. Il gravite autour de la Terre à 760 kilomètres d'altitude, avec à son bord un unique instrument scientifique qui enregistre le rayonnement micro-onde émis par la surface de la Terre. Ces données permettent de déterminer l'humidité des sols et la salinité des océans.

Pour simplifier : plus le sol est humide, plus son rayonnement est faible. Plus l'eau est salée, plus le rayonnement de sa surface est réduit. Un instrument, un principe de mesure, qui autorisent SMOS à percevoir de la même manière le taux d'humidité de vastes régions continentales et la salinité des océans. Il ne s'agit pas d'une chimère. Les scientifiques ne cessent de tester ses performances dans diverses zones, où ils connaissent, par le biais de mesures *in situ*, l'humidité du sol et la salinité de la mer. SMOS réussit tous ces tests haut la main.

Détecteurs de gravité, caméras high-tech, capteurs radar et goûteurs de sel sont autant d'explorateurs qui s'ébattent en orbite en veillant constamment sur la Terre. L'inventivité des scientifiques est sans limite. Ainsi Envisat, le plus grand satellite polyvalent d'observation de la Terre, était équipé d'un « chasseur d'ombres ». Baptisé SCIAMACHY, il figurait parmi la dizaine d'instruments scientifiques du satellite. Il décomposait

taller para corregir los errores. El satélite despegó, finalmente, en octubre de 2005. Pero su ansia de hielo era demasiado fuerte. Diez minutos después de despegar, chocó contra los témpanos árticos porque el cohete Rockot, por lo general muy fiable, tenía un defecto técnico. La consternación de los ingenieros de Friedrichshafen, que nada tenían que ver con el cohete, fue muy grande: *Es una conmoción para todos los participantes. Los colaboradores se han entregado durante muchos años en cuerpo y alma a este proyecto y entonces ocurre una desgracia. Quizá pueda sonar un poco exagerado, pero es como la pérdida de un buen amigo.* Su tristeza fue grande, pero también fue muy grande el alivio que sintieron unos meses después. La ESA facilitó, excepcionalmente rápido, los medios para construir el CryoSat-2. Su despegue en abril de 2010 fue perfecto. ¡Final feliz!

Los ojos del radar de a bordo rastrean ahora las masas de hielo. Los pulsos chocan contra ellas, contra el mar o contra la superficie terrestre libre de hielo. Se registra el eco con una precisión asombrosa, siempre que los ingenieros espaciales conozcan con precisión la posición del satélite. Así lo explica Klaus Köble: *Una red de más de cuarenta estaciones emisoras en la Tierra envía pulsos al satélite para determinar con los ecos la posición de éste en la órbita. La evaluación de estos datos da la posición del satélite en su órbita con una precisión de centímetros.* Así, los investigadores también están en situación de analizar con precisión los pulsos de radar emitidos por el satélite y sus ecos.

Húmedo y salado

Mire por la ventana y contemple el jardín del vecino. ¿Podría decir a simple vista cuánta agua hay exactamente en el suelo? ¿No? Pero seguro que en su próximo paseo por el malecón usted estará en condiciones de calcular la salinidad del agua marina contemplando el mar. ¿Tampoco? Entonces necesita la ayuda de SMOS, uno de los más modernos satélites de la ESA. Vuela a 760 km de altura

Der Majn-Fluss schlängelt sich durch vereiste Gebiete im äußersten Osten Sibiriens. Der Bildausschnitt ist gut 30 km breit (Landsat)

The Majn River winds its way through the icy wastes of the furthest reaches of Eastern Siberia. The area shown is about 19 miles (30 km) wide (Landsat)

La rivière Majn serpente dans les régions glacées de la Sibérie extrême-orientale. L'image couvre une largeur de 30 km (Landsat)

El río Majn serpentea por las regiones heladas del extremo de Siberia oriental. El encuadre es de 30 km de ancho (Landsat)

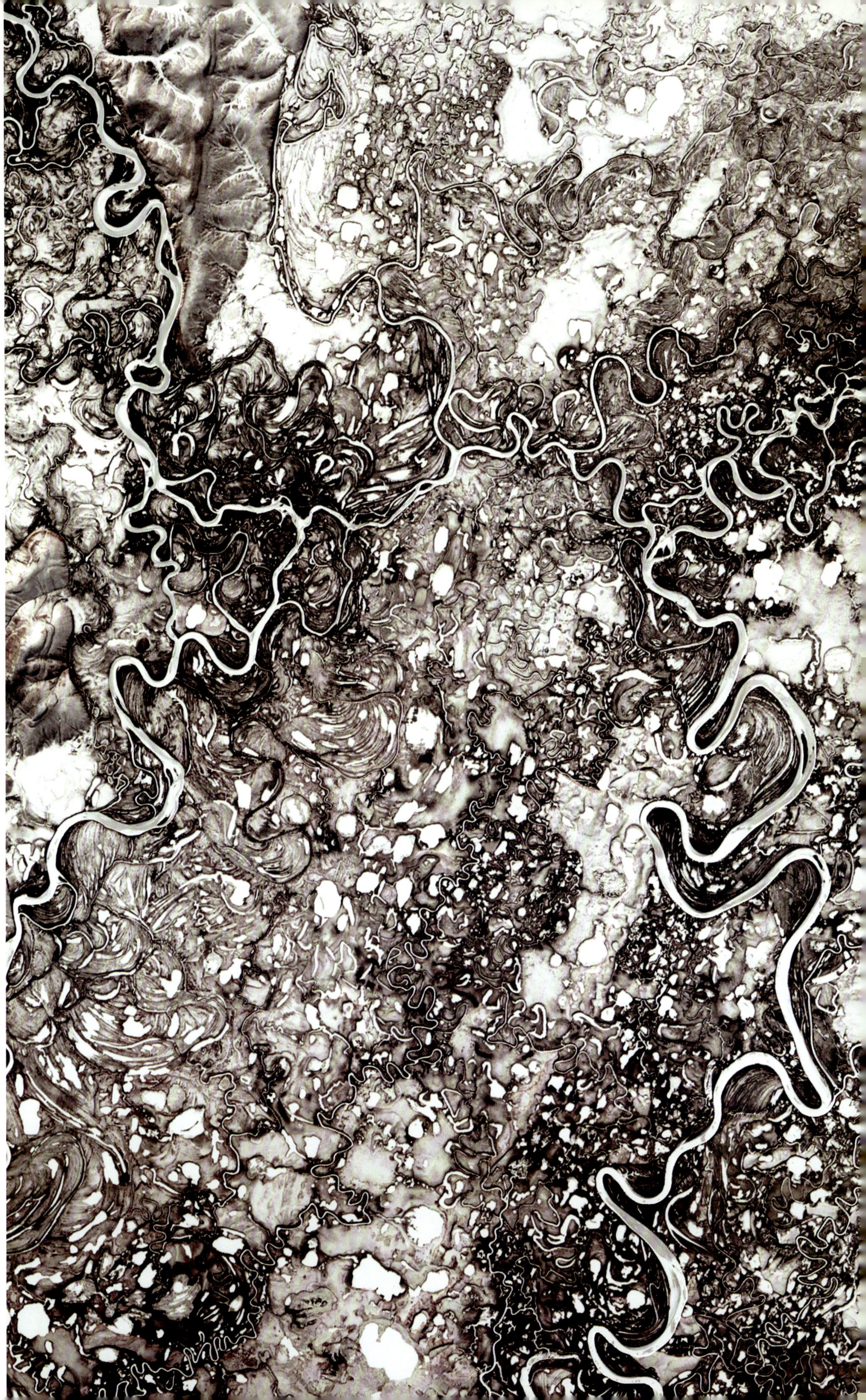

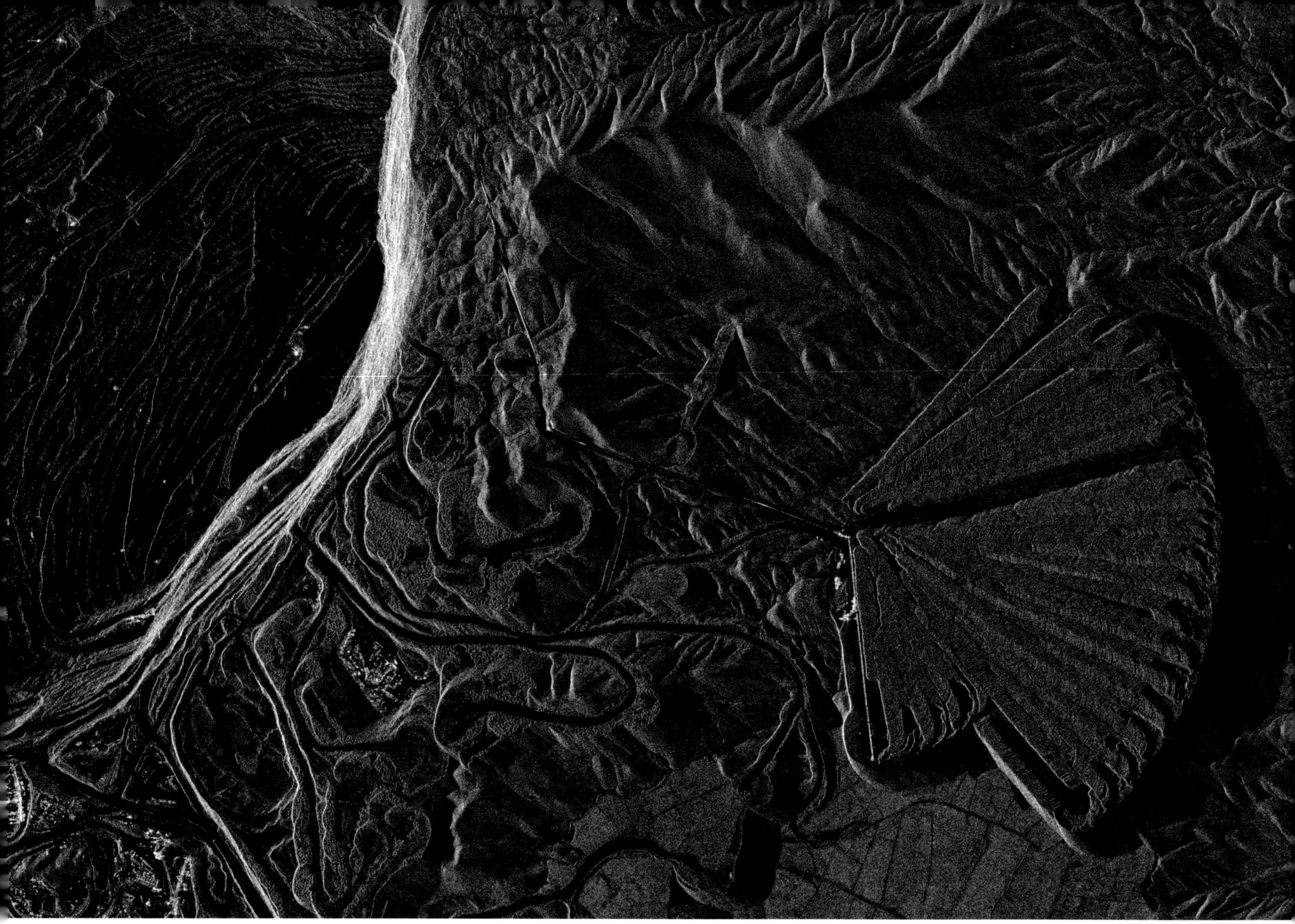

Oben: Das größte Loch der Erde, die Mine Chuquicamata in der Atacama-Wüste im Norden Chiles (TerraSAR-X)

Above: The biggest hole on Earth—the Chuquicamata mine in Northern Chile's Atacama Desert (TerraSAR-X)

Ci-dessus : la plus grande cavité de la Terre, la mine de Chuquicamata dans le désert de l'Atacama, au nord du Chili (TerraSAR-X)

Arriba: El mayor agujero de la Tierra, la mina Chuquicamata en el desierto de Atacama, al norte de Chile (TerraSAR-X)

Rechte Seite: Kein bewohnter Ort der Erde ist so isoliert wie die Osterinsel im Südostpazifik (TerraSAR-X)

Right: Easter Island in the South Pacific is the most remote inhabited area on Earth (TerraSAR-X)

Page de droite : aucun endroit habité sur Terre n'est plus isolé que l'île de Pâques, dans le sud-est du Pacifique (TerraSAR-X)

Página dcha.: Ningún otro lugar habitado de la Tierra está tan aislado como la isla de Pascua, en el sudeste del Pacífico (TerraSAR-X)

kennt keine Grenzen. So verfügte Envisat, der vielseitigste und größte aller Erdbeobachtungssatelliten, auch noch über einen Schattenboxer. SCIAMACHY, so das griechische Wort für diese Disziplin, war eines der gut zehn Messinstrumente auf dem Satelliten. Es zerlegte das Sonnenlicht, das vom Erdboden reflektiert wird oder direkt von der Quelle kommt und durch die irdische Lufthülle gelaufen ist, in seine einzelnen Wellenlängen.

Magische Schatten

So wie Astronomen das Licht ferner Galaxien in seine Bestandteile zerlegen, um die chemische Zusammensetzung von Objekten in Milliarden Lichtjahren Entfernung zu bestimmen, so zerlegten die Atmosphärenchemiker mit SCIAMACHY das empfangene Licht, um etwas über die Vorgänge in der Atmosphäre einige hundert Kilometer unter dem Satelliten zu lernen.

Läuft das Licht durch die irdische Lufthülle, so hinterlassen die chemischen Stoffe charakteristische Spuren im Spektrum. Bei bestimmten Wellenlängen scheint das Licht verschluckt. Diese Absorptionsmuster sind wie Fingerabdrücke. Sie verraten sofort, welche chemischen Stoffe wo und in welcher Konzentration in der Atmo-

all these are racing around in orbit, their eyes trained on the Earth, and the researchers' inventiveness knows no bounds. Envisat, for example, the most complex and the largest of all the Earth observation satellites, even had a "shadow boxer." The Greeks called this discipline sciamachy, and SCIAMACHY was also the name of one of the ten instruments on board the satellite. It analyzed and breaks down into its component wavelengths the sunlight reflected from the Earth's surface and the solar rays which arrive in the Earth's atmosphere directly from their source.

Magic shadows

In the same way that astronomers break down the light from distant galaxies into its components in order to determine the chemical composition of objects which are billions of light years away, the atmospheric chemists used the light captured by SCIAMACHY to learn more about the processes that are taking place in the atmosphere a few hundred miles (a few hundred kilometers) beneath the satellite.

When light passes through the Earth's atmosphere, chemical compounds leave characteristic traces

la lumière solaire reflétée par le sol, et le rayonnement solaire qui traverse l'atmosphère terrestre, en fonction de leurs différentes longueurs d'ondes. Le spectre ainsi représenté n'est peut-être pas aussi beau que la photo d'un paysage spectaculaire, mais il fournit bien plus d'informations.

Ombres magiques

Tout comme les astronomes décomposent la lumière émise par les galaxies lointaines, afin de connaître la composition chimique des corps situés à des milliards d'années lumière, les chimistes de l'atmosphère décomposaient la lumière captée par SCIAMACHY pour comprendre les processus atmosphériques observés quelques kilomètres sous le satellite.

Quand la lumière traverse l'atmosphère, les substances chimiques qu'elle recèle laissent des traces caractéristiques dans le spectre. Sur certaines longueurs d'onde, la lumière semble avoir été avalée. Ces structures d'absorption sont comme des empreintes digitales. Elles révèlent la présence et la concentration des substances chimiques dans l'atmosphère que SCIAMACHY chassait à l'aide de ces « ombres », détectant

alrededor de la Tierra. A bordo sólo lleva un instrumento de medición. Al sobrevolar la Tierra, registra en largas franjas la radiación de microondas emitida por la superficie terrestre. Con esos datos se puede determinar la humedad del suelo y la salinidad de los mares.

La regla general es que cuanto mayor sea la humedad del suelo, menor será su radiación. Cuanto más salada esté el agua, tanto más débil será la irradiación de su superficie. El mismo instrumento, el mismo principio de medición: de esta manera, SMOS puede sentir tanto la humedad de grandes zonas de la Tierra como saborear cuánta sal hay en determinadas regiones del mar. No se trata de una ilusión. Los científicos verifican continuamente la misión en regiones determinadas de la Tierra de las que conocen, por medición directa in situ, el contenido en agua del suelo y la salinidad del mar. SMOS ha superado las pruebas sin problemas.

Detectores de gravedad, cámaras de alta tecnología, sensores de radar y probadores de sal, todo en órbita y sin perder de vista a la Tierra. La inventiva de los científicos no tiene límites. Así, Envisat, el mayor y más polivalente de todos los satélites de observación terrestre, también disponía de un «boxeador de sombras».

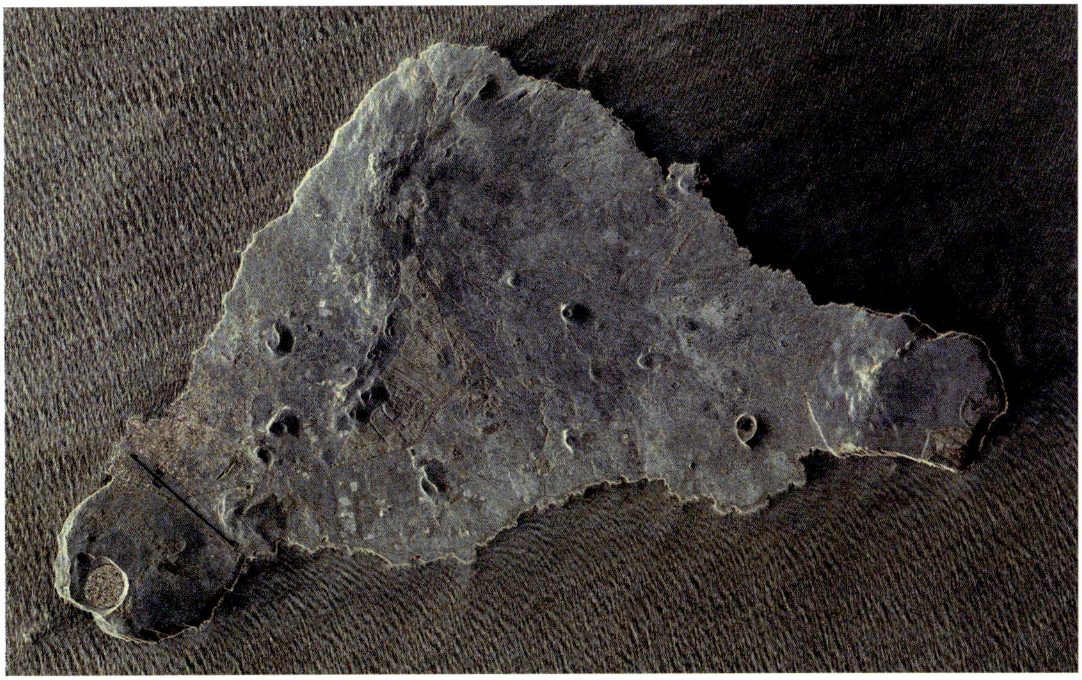

sphäre vorkommen. Strahlen ist Silber, Absorbieren ist Gold. Mit Hilfe der »Schatten« der chemischen Stoffe boxte sich SCIAMACHY durch die Atmosphäre und ermittelte mit hoher zeitlicher und räumlicher Auflösung, wo Spurengase wie etwa Ozon, Stickoxid, Methan oder Kohlendioxid vorkommen. Die Spektren, die SCIAMACHY an Bord von Envisat aufgenommen und zur Erde gefunkt hat, werteten Wissenschaftler in Deutschland, Belgien und den Niederlanden aus. Die Schattendaten aus den Spektren standen dann allen Forschern weltweit zur Verfügung.

Mit dem bloßen Auge sind Spektren in der Regel nicht zu erkennen. Nur bei einem Regenbogen sehen wir, wie das Sonnenlicht in seine Farbbestandteile zerlegt wird. Für den Menschen direkt gar nicht wahrnehmbar sind Magnetfelder. Wir spüren sie nicht, wir hören oder sehen sie nicht. Man kann sie nicht anfassen und auch nicht schmecken. Das heißt aber keineswegs, dass sie für die Erde oder für uns Menschen unwichtig wären. *Das Magnetfeld hat für uns alle eine besondere Rolle – denn ohne Magnetfeld gäbe es uns nicht.* Albert Zaglauer hat auch keinen Magnetsinn, aber er und sein Team haben bei Airbus die Instrumente gebaut, die Magnetfelder vermessen. Swarm heißt ein

in the spectrum; the light seems to be absorbed at certain wavelengths and this absorption pattern is like a fingerprint, immediately revealing which elements are present in the atmosphere and in what concentration. If transmission of light is described as silver, absorption is gold. With the aid of the "shadow" thrown by the compounds, SCIAMACHY "boxed" it way through the atmosphere, showing us with great temporal and spatial precision exactly where trace gases such as ozone, nitrogen oxide, methane, or carbon dioxide are found. The spectra recorded on board Envisat and transmitted to Earth by SCIAMACHY were evaluated by scientists from Germany, Belgium, and the Netherlands and then the "shadow" data were made available to researchers all over the world.

Magnetic fields are completely imperceptible to humans—we can't feel them, hear them, or see them, nor yet touch them or taste them, but that is certainly not to say that they are unimportant for the Earth or for humans. *The magnetic field plays a crucial role for all of us—we wouldn't exist without it.* Albert Zaglauer is not personally sensitive to magnetism, but he and his team at Airbus built instruments to measure magnetic fields. The trio of satellites

avec une haute résolution spatiale et temporelle la présence de gaz traces tels qu'ozone, oxyde d'azote, méthane et dioxyde de carbone. Les spectres enregistrés puis transmis vers la Terre par SCIAMACHY étaient analysés par des scientifiques basés en Allemagne, en Belgique et aux Pays-Bas, qui mettaient ensuite les informations collectées à la disposition des chercheurs du monde entier.

Les spectres sont généralement invisibles à l'œil nu. Seuls les arcs-en-ciel dévoilent à nos yeux les composantes colorées de la lumière solaire. Il en va de même des champs magnétiques, imperceptibles aux sens humains. Pourtant, ils ne sont pas insignifiants pour la Terre et ses habitants. *Le champ magnétique joue pour nous tous un rôle essentiel. Sans lui, nous n'existerions pas.* Albert Zaglauer n'a pas non plus le sens magnétique, mais avec son équipe il a construit au sein d'Airbus des instruments capables de mesurer les champs magnétiques. Swarm est le nom d'une constellation de trois satellites mis en orbite fin 2013.

Ces satellites magnétiques commandés par l'ESA appartiennent à la série des Earth Explorer, les petites vigies célestes hautement spécialisées. Le champ magnétique terrestre fait office de bouclier invi-

SCIAMACHY, como la palabra griega para esta disciplina, era uno de los 10 instrumentos de medición del satélite. Descomponía en sus longitudes de ondas individuales la luz solar reflejada por el suelo, o la que viene directamente de la fuente y ha pasado por la atmósfera terrestre. Un espectro así, aunque no tan bonito como una imagen en color, la mayoría de las veces dice mucho más que ella.

Sombras mágicas

Del mismo modo que los astrónomos descomponen la luz de galaxias lejanas en sus componentes para determinar la composición química de objetos que se encuentran a millardos de años luz de distancia, con SCIAMACHY los químicos atmosféricos descomponían la luz recibida para entender los procesos atmosféricos que suceden a unos cientos de kilómetros bajo el satélite.

Cuando la luz pasa por la atmósfera terrestre, las sustancias químicas dejan trazas características en el espectro. A determinadas longitudes de onda, la luz parece ser absorbida. Estos patrones de absorción son como las huellas dactilares. Delatan inmediatamente qué sustancias químicas

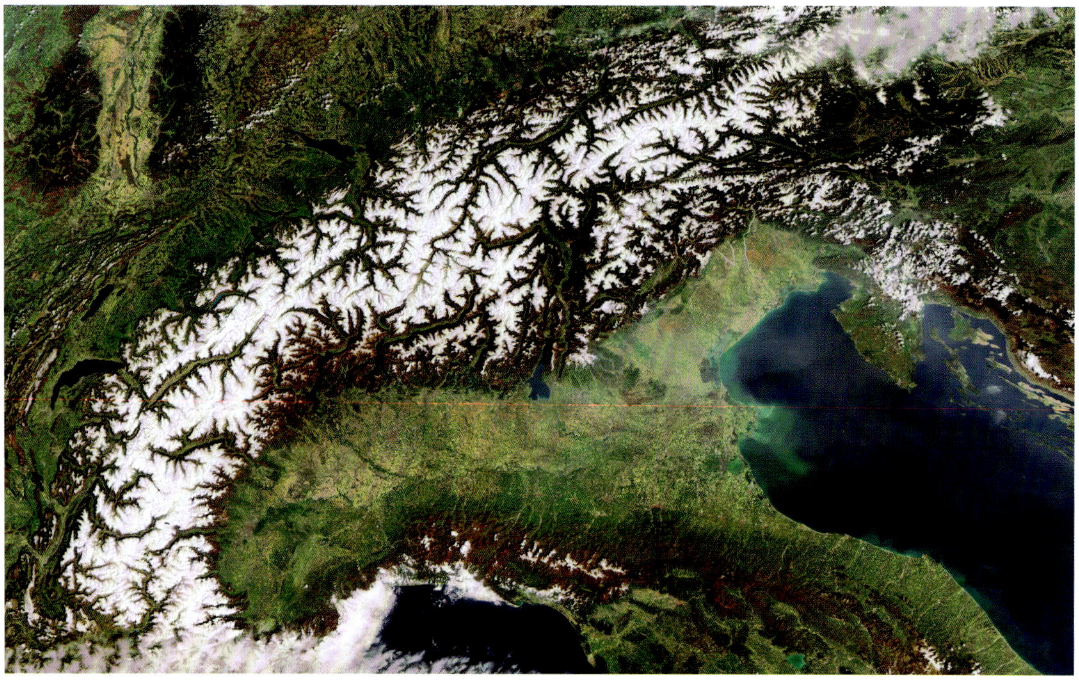

Links: Sonnige Alpen mit Bodensee und Genfer See.
Rechts: Am 19. Januar 2010 lag die Po-Ebene komplett im Nebel (Envisat)

Left: Lakes Constance and Geneva surrounded by sun-kissed Alps.
Right: The Po Valley was shrouded in mist on January 19, 2010 (Envisat)

À gauche : les Alpes ensoleillées autour du lac de Constance et du lac de Genève.
À droite : la plaine du Pô était entièrement dans le brouillard le 19 janvier 2010 (Envisat)

Izda.: Alpes soleados, con el lago de Constanza y el lago Lemán.
Dcha.: El 19.01.2010, la llanura padana estaba completamente cubierta de niebla (Envisat)

Satellitentrio, das Ende 2013 ins All gestartet ist.

Auftraggeber ist die ESA, und die Magnetsatelliten gehören in die Reihe der Earth Explorer, der recht kleinen, hoch spezialisierten Späher in der Umlaufbahn. Das irdische Magnetfeld ist wie ein unsichtbarer Schutzschirm. Es lenkt geladene energiereiche Teilchen ab, die aus den Tiefen des Alls und von der Sonne zu uns kommen. Solare Teilchenstürme quetschen und dehnen das Magnetfeld der Erde. Doch die gefährlichen Teilchen gelangen in der Regel nur in der Nähe der Pole in tiefere Atmosphärenschichten. Dort führen sie dann zu Polarlichtern, für die wiederum auch wir Menschen einen Sinn haben. Das farbenprächtige Spektakel der meist in Grün- und Rottönen schillernden Lichtvorhänge verzaubert alle Betrachter. Für die Wikinger waren sie Zeichen, dass die Götter in den Krieg zogen – und in gewisser Weise sind die Polarlichter ein Zeichen für den Kampf zwischen eindringenden Teilchen und dem tapfer kämpfenden Magnetfeld, das in vielem noch immer unerforscht ist.

Swarm soll das ändern. Für Albert Zaglauer und seine Kollegen bedeutet das viel Arbeit: *Wir wollen das natürliche Magnetfeld der Erde messen und nicht das durch die Menschheit künst-*

were launched into space at the end of 2013.

The satellites were commissioned by the ESA and form part of the Earth Explorers program, a series of extremely small and highly specialized orbiting Earth observers. The Earth's magnetic field is like an invisible protective force field, diverting the charged, energy-rich particles which come our way from the Sun or out of the depths of space. Solar particle storms buffet and stretch the Earth's magnetic field but, as a rule, the most dangerous particles only make it as far as the lower levels of the atmosphere near the poles, where they cause the Northern and Southern Lights that we humans find so captivating. This colorful spectacle, glittering curtains of usually green or red light, casts a spell on all who see it. For the Vikings, the lights were a sign that the gods were going to war, and in a way the polar lights are an indication of the battle raging between the invading particles and the valiant magnetic field defending us, a field about which we still know so little.

Swarm has changed this, although it meant a lot work for Albert Zaglauer, the leader of the Astrium team working on the construction of this ESA project in Friedrichshafen,

sible. Il dévie les particules chargées électriquement qui proviennent du soleil et des profondeurs de l'univers. Les tempêtes solaires compressent et dilatent le champ magnétique de la Terre, tandis que les particules nocives ne pénètrent généralement dans les couches inférieures de l'atmosphère qu'à proximité des pôles. Elles sont à l'origine des aurores polaires, qu'il nous est donné d'admirer. Le fabuleux spectacle de ces voiles de lumière irisée aux teintes vertes et rouges est un enchantement pour les yeux. Les Vikings y voyaient le signe du combat des dieux ; et d'une certaine manière, les aurores polaires témoignent de la valeureuse lutte du champ magnétique contre les particules tentant de s'infiltrer, un phénomène encore largement inexploré.

Swarm comble cette lacune. C'est un travail énorme pour Albert Zaglauer et son équipe : *Nous souhaitons mesurer le champ magnétique naturel de la Terre, et non le champ artificiel résultant de l'activité humaine. Tout courant électrique, tout métal possédant une signature magnétique génère un champ magnétique artificiel qui nous dérange.* Bien sûr, aucun satellite ne peut se passer entièrement d'électricité, sans quoi l'enregistrement et la transmis-

se encuentran en la atmósfera y en qué concentración. La plata refleja, el oro absorbe. Con la ayuda de las «sombras» de las sustancias químicas, SCIAMACHY «boxeaba» por la atmósfera y determinaba con una resolución temporal y tridimensionalmente alta donde existían gases traza, como ozono, óxido nítrico, metano o dióxido de carbono. Los espectros que SCIAMACHY registraba a bordo del Envisat y transmitía a la Tierra, eran analizados por científicos de Alemania, Bélgica y los Países Bajos. Los datos de las sombras de los espectros quedaban después a disposición de todos los investigadores del mundo.

Por regla general, los espectros no se pueden reconocer a simple vista. Solamente en un arco íris podemos ver cómo se descompone la luz solar en sus colores elementales. Para las personas no es posible la percepción directa de los campos magnéticos. No, no los podemos sentir, ver u oír. Pero esto no significa, en absoluto, que no tengan importancia para la Tierra o para nosotros, los seres humanos. *El campo magnético representa un papel que es esencial para nosotros, porque sin él, no existiríamos.* Albert Zaglauer tampoco es «magnetoconsciente», pero junto con su equipo construyó instrumentos en Airbus que miden los campos

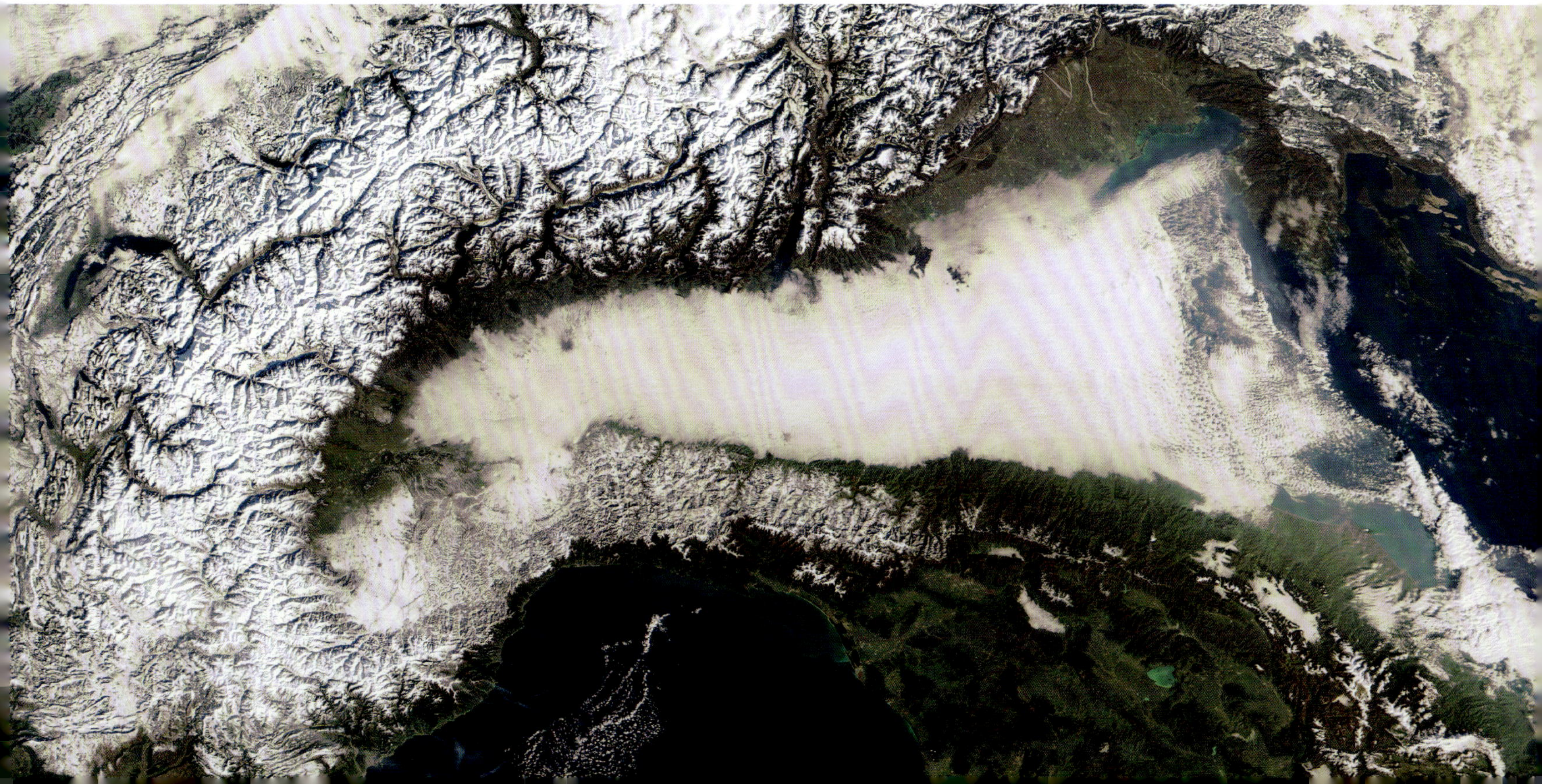

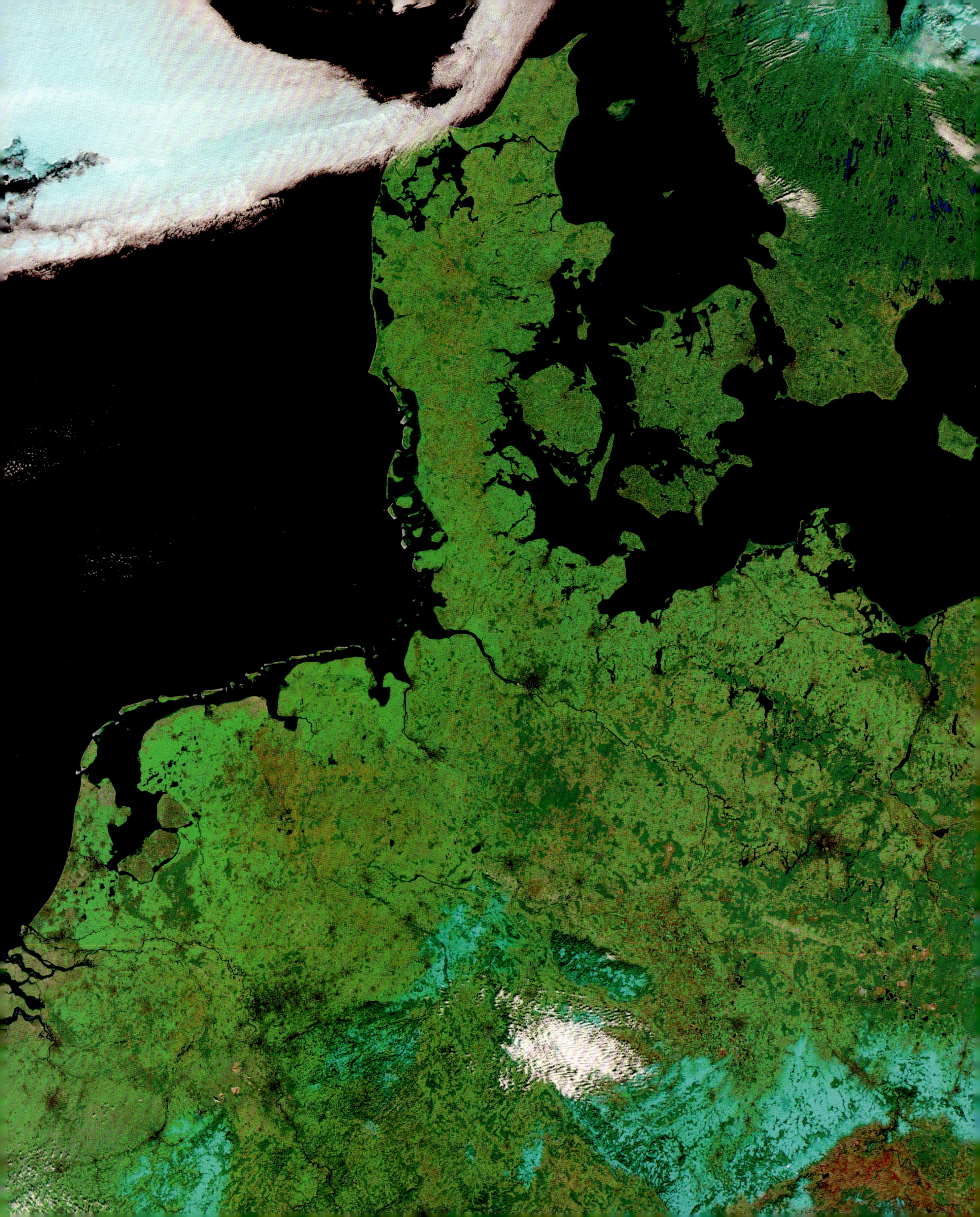

Linke Seite: Dänemark und Norddeutschland – wolkenlos (Envisat)
Rechts: Hurrikan, im sichtbaren Licht und in der Spektraldarstellung des Drucks an der Wolkenobergrenze (Envisat, 2003)

Left: Not a cloud in the sky over Denmark and northern Germany (Envisat)
Right: Representations of the pressure at a hurricane's cloud top (Envisat, 2003)

Page de gauche : ciel sans nuage au-dessus du Danemark et du nord de l'Allemagne (Envisat)
À droite : un cyclone et sa couverture nuageuse très dense (Envisat, 2003)

Página izda.: Dinamarca y norte de Alemania sin nubes (Envisat)
Dcha.: Huracán con presión en el límite superior de nubes (Envisat, 2003)

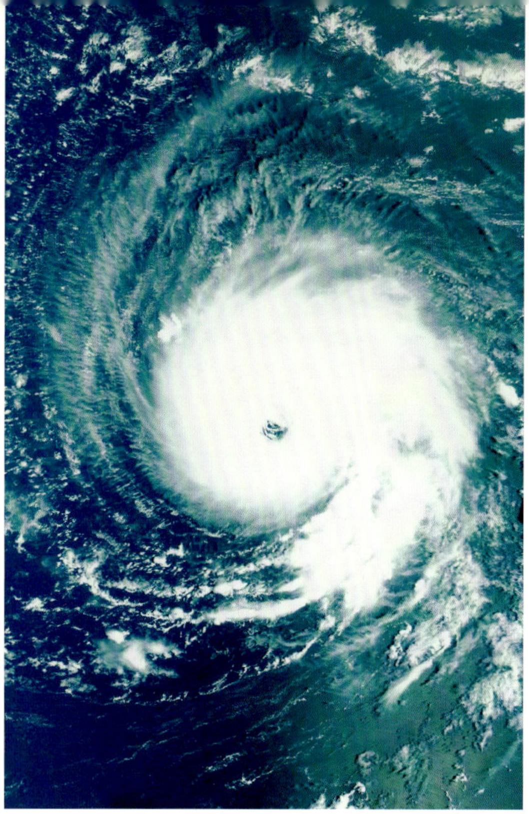

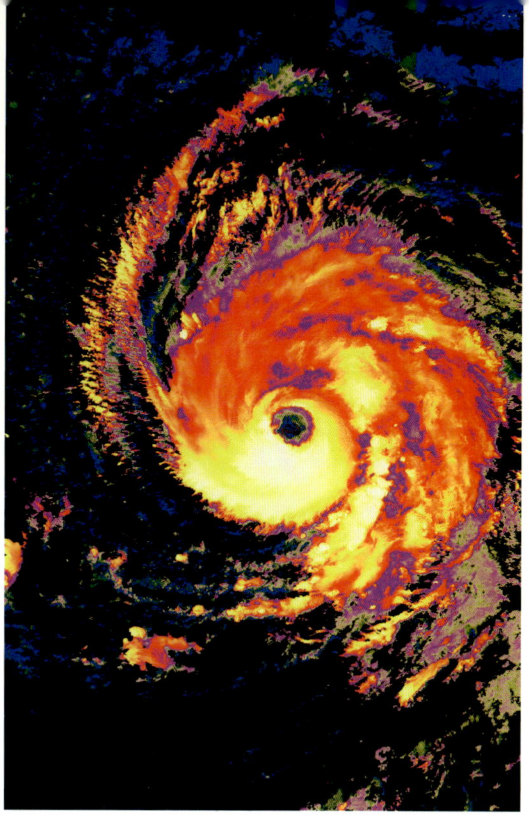

lich generierte Feld. Wann immer ein elektrischer Strom fließt oder ein Metall mit magnetischer Signatur verwendet wird, gibt es ein künstliches Feld. Das stört uns. Zwar kann kein Satellit ganz auf Strom verzichten, weil sich dann auch keine Daten speichern oder übertragen ließen. Aber das Ingenieursteam greift auf seine große Erfahrung mit früheren Magnetmissionen zurück, etwa der Sonnensonde Ulysses.

Große Teile des Satelliten wurden aus einer Kohlefaserstruktur gefertigt, die die Magnetfeldmessungen nicht beeinträchtigt. *Die Beschränkungen bei Masse, Volumen und Leistung waren für uns kein Thema, weil wir das gut im Griff hatten,* betont Albert Zaglauer. *Wir hätten uns dafür öfters mal reineres Material gewünscht. Wir hatten zum Beispiel ein Design in Keramik geplant. Das sah ganz gut aus, und wir waren davon ausgegangen, dass die Keramik magnetisch keine Signatur hat. Wir waren alle überrascht, als unsere Messungen gezeigt haben, dass von der Keramik doch ein magnetisches Störsignal ausgeht, das wir nicht akzeptieren konnten.* Somit kommt die Keramik nun doch nicht zum Einsatz.

Die Satellitenbauer haben alle drei SWARM-Satelliten rechtzeitig abgeliefert. Das war auch nötig, denn das Trio musste mit derselben Rakete

Germany. *We want to measure the Earth's natural magnetic field rather than the field artificially generated by humans—whenever electric current flows or metal with a magnetic signature is used, it creates a magnetic flux which gets in our way.* A satellite cannot do completely without electricity as it would then be unable to store or transmit data, but the team of engineers has a long track-record of problem-solving for earlier successful magnetism missions, such as the solar probe Ulysses.

Large sections of the satellite have been constructed of a carbon fiber material which has no effect on magnetic field measurements. *Constraints of size, volume, and performance were not a problem for us, as we know how to deal with these,* emphasizes Albert Zaglauer, *but there were several times when we might have wished for purer materi-als—for example, we planned one design using ceramics. It looked very good, and we were assuming that ceramics have no magnetic signature. We were all rather surprised when our results showed that the ceramic did emit magnetic interference, which was unacceptable.* Ceramics will thus not be used.

The satellite-builders delivered all three Swarm satellites on time. This requirement was necessary as the

sion des données seraient impossibles. Les ingénieurs possèdent néanmoins une vaste expérience des missions consacrées à l'étude des magnétosphères, acquise notamment avec la sonde solaire Ulysses.

Le satellite était constitué en grande partie de fibres de carbone, un matériau qui ne perturbe pas les mesures du champ magnétique. *Les contraintes de masse, volume et performance ne nous ont pas posé problème parce que nous maîtrisons ces aspects,* souligne Albert Zaglauer. *En revanche, nous aurions souhaité des matériaux plus purs. Nous avions conçu une structure en céramique. Cette solution nous semblait adéquate car nous pensions que la céramique était dépourvue de signature magnétique. Grande a été notre surprise quand nos mesures ont montré que la céramique émettait des signaux parasites que nous ne pouvions tolérer.* Il a donc fallu renoncer à la céramique.

Les bâtisseurs de satellites ont livré les trois satellites Swarm à temps. Ils n'avaient d'ailleurs guère le choix, puisque le trio devait rejoindre son orbite à bord d'un lanceur unique. Les scientifiques commencent les mesures du champ magnétique à 500 kilomètres d'altitude. Dans cette mince couche d'at-

magnéticos. Swarm es el nombre de un trío de satélites lanzados al espacio a finales de 2013.

El cliente es la ESA y los satélites exploradores del campo magnético pertenecen a la serie de los Earth Explorer, observadores orbitales tan pequeños como altamente especializados. El campo magnético terrestre es como una pantalla protectora invisible. Desvía las partículas cargadas de alta energía que llegan hasta nosotros desde las profundidades del universo y desde el Sol. Las tormentas de partículas solares presionan y dilatan el campo magnético de la Tierra. Pero las partículas peligrosas sólo suelen llegar a las capas más profundas de la atmósfera en la cercanía de los polos. Allí producen auroras boreales. Ese espectáculo de gran colorido, con cortinas de luz tornasoladas en tonos verdes y rojos, fascina a todos los espectadores. Para los vikingos eran señales de que los dioses partían a la guerra y, en cierto sentido, las auroras boreales son signos de la lucha entre las partículas infiltradas y el intrépido y luchador campo magnético que, en muchos aspectos, aún sigue inexplorado.

Swarm ha cambiado eso. Para Albert Zaglauer y sus colegas significa mucho trabajo: *Queremos medir el campo magnético natural de la Tierra*

starten. Die Wissenschaftler beginnen die Magnetmessungen in 500 Kilometern Höhe. Durch die Reibung an der extrem dünnen Lufthülle in jenen Höhen werden die Satelliten allmählich absinken. Das ist einkalkuliert, denn die Physiker wollen zum Ende der Mission in gut 300 Kilometern Höhe messen. Dort ist das begehrte Signal des Magnetfelds noch viel stärker. Vier Jahre lang sollen die drei Swarm-Satelliten im Einsatz sein. Bald nach Betriebsende werden sie in die Atmosphäre eintreten und verglühen.

Um die Widerstandsfläche so klein wie möglich zu halten, sind die Satelliten schlanke, fast stromlinienförmige Gebilde. Auf den ersten Blick sehen sie aus wie Gitarren – Luftgitarren der ganz besonderen Art. Zudem werden die Satelliten so massereich wie möglich gebaut. Auch das verlängert die Lebenszeit im All. Natürlich durften sie zusammen nicht das maximale Startgewicht der gebuchten Rockotrakete übersteigen. Aber es wurde sozusagen alles hinein gepackt, was irgendwie ging.

Der Projektleiter für Swarm hat eine bemerkenswerte Laufbahn hinter sich. Ursprünglich hat sich der Maschinenbauer mit Materialkunde beschäftigt. Um neuartige Aluminiumlegierungen zu untersuchen, war er an einem Schmelzexperiment im Spacelab beteiligt, einem in Deutschland gebauten Raumlabor, das mit den US-Raumfähren hin und wieder ins All geflogen ist. Albert Zaglauer blieb bei der Raumfahrt hängen. Einst entwickelte er eine Stereokamera für Europas Marssonde. Jetzt ist er für Magnetmessungen zuständig. Diese Instrumente liefern keine schönen Bilder, die die Titelseiten von Zeitungen und Magazinen zieren könnten. Dafür aber schicken sie äußerst bedeutende Messungen an die Bodenstation. So sehr Albert Zaglauer vom Magnetfeld der Erde fasziniert ist, so sehr stören ihn die Magneten auf der Erde. Bei der Arbeit an den Satelliten im Reinraum herrscht striktes Magnetverbot. Niemand darf dort mit magnetischen Materialien hinein. Der Projektleiter nimmt es ganz genau: Nicht einmal in seinem Büro hat er eine Magnettafel.

trio had to be launched with the same rocket. The scientists started measurements at a height of about 300 miles. Friction in the extremely thin atmosphere at such altitudes will gradually lower the satellite, but this effect has been included in the calculations—at the end of the mission, the physicists want to be measuring from a height of some 190 miles, where the sought-after magnetic field is much stronger. The three Swarm satellites are intended to function for four years.

To keep their aerodynamic resistance as low as possible, the satellites have narrow, almost streamlined profiles. At first glance they look like guitars—air guitars of a rather special kind. The satellites are also built with as much mass as possible, as all this lengthens their lifespan in space. They cannot exceed the maximum payload of the launch rocket, of course, but it would be true to say that as much as possible is packed in.

The Swarm project leader has had a remarkable career. A mechanical engineer with a background in material sciences, he worked on experiments examining the melting points of new aluminum alloys in Spacelab, a space laboratory built in Germany and periodically transported into space on American space shuttle missions. Albert Zaglauer has always stuck with space travel. He once developed a stereo camera for the European Mars probe and is now responsible for taking magnetic measurements. Such instruments don't produce the beautiful pictures which end up gracing the pages of magazines and newspapers, but the statistics they transmit to their base stations are of vital importance. Albert Zaglauer's fascination with the magnetic field around the Earth is matched only by his annoyance at the interference caused by the magnets on its surface, and magnets are completely banned from the sterile room where work is carried out on the satellites. No one is allowed to bring in magnetic materials of any kind and the project leader is extremely strict: there aren't even refrigerator magnets in his office.

mosphère, la résistance de l'air résiduel provoque une lente perte d'altitude des satellites. Ces données ont été intégrées aux calculs des chercheurs, qui souhaitent terminer la mission à une altitude d'environ 300 kilomètres, où l'intensité du signal magnétique recherché est nettement plus élevée. La mission des trois satellites Swarm durera quatre ans. Peu après leur mise hors service, ils pénétreront dans l'atmosphère où ils se désintégreront.

Pour réduire les frottements au maximum, les satellites sont dotés d'un profil aérodynamique. Ils ressemblent un peu à des guitares aériennes d'un genre particulier qui emportent un maximum de masse, autre garant de la durée de vie des satellites dans l'espace. Il n'était évidemment pas question qu'ils dépassent la charge utile maximale de leur lanceur Rockot. Mais ils ont emporté tous les bagages possibles.

Le parcours du chef de projet Swarm est assez remarquable. Ingénieur mécanique, il s'est tout d'abord intéressé à la science des matériaux. Dans le cadre d'études portant sur de nouveaux alliages d'aluminium, il a participé à une expérimentation sur la fusion des métaux dans le Spacelab, un laboratoire spatial construit en Allemagne qui a effectué plusieurs voyages dans l'espace à bord des navettes américaines. Albert Zaglauer est resté fidèle à la recherche spatiale. Il a développé une caméra stéréoscopique pour la sonde martienne de l'Europe. À présent, il est responsable des mesures magnétiques. Ses instruments ne fournissent pas d'images sensationnelles susceptibles de faire la une des magazines. Les données qu'ils nous transmettent sont toutefois d'une extrême importance. Le champ magnétique terrestre fascine autant Albert Zaglauer que le dérangent les ondes magnétiques sur la Terre. Ces dernières sont bannies de la salle blanche où sont construits les satellites. L'interdiction d'entrer avec des matériaux magnétiques est très stricte. Une règle que le chef de projet applique à la lettre : même son bureau est dépourvu de tableau magnétique.

y no el campo artificialmente generado por la humanidad. Cuando quiera que se produzca una tormenta eléctrica o se utilice un metal con firma magnética, se produce un campo artificial. Esto interfiere en nuestro trabajo. No obstante, ningún satélite puede renunciar a la corriente porque tampoco se podrían almacenar o transmitir los datos. Pero el equipo de ingenieros recurre a su gran experiencia en anteriores misiones de exploración del campo magnético, como por ejemplo la sonda solar Ulysses.

La mayor parte del satélite se fabricó con una estructura de fibras de carbono, material que no afecta a las mediciones del campo magnético. *No le dimos importancia a las limitaciones de masa, volumen y potencia, porque eso ya lo dominábamos*, subraya Albert Zaglauer. *A veces hubiésemos deseado un material puro. Por ejemplo, habíamos proyectado un diseño en cerámica. Tenía muy buen aspecto y además, partíamos de que la cerámica no presenta traza magnética. No sorprendimos mucho cuando nuestras mediciones mostraron que la cerámica también producía una interferencia magnética que no podíamos aceptar.* Por consiguiente, ahora ya no se utiliza la cerámica.

Los constructores de satélites entregaron al tiempo los tres satélites Swarm. Esto era necesario porque el trío se debía lanzar con el mismo cohete. Los científicos comienzan las mediciones magnéticas a 500 km de altura. Debido al rozamiento con la atmósfera extremadamente delgada que hay a aquella altura, los satélites descienden poco a poco, algo que se ha tenido en cuenta, pues los físicos quieren medir al final de la misión a 300 km de altura. Allí, la codiciada señal del campo magnético es aún más potente. Los tres satélites Swarm deben funcionar durante tres años. Cuando acaben la misión, entrarán en la atmósfera y se desintegrarán.

Para reducir la fricción, estos satélites son esbeltos, casi aerodinámicos. A primera vista parecen guitarras, pero con una masa abundante. Esto también prolonga su vida útil en el espacio. Naturalmente, todos juntos no pueden superar el peso máximo de lanzamiento del cohete Rockot reservado.

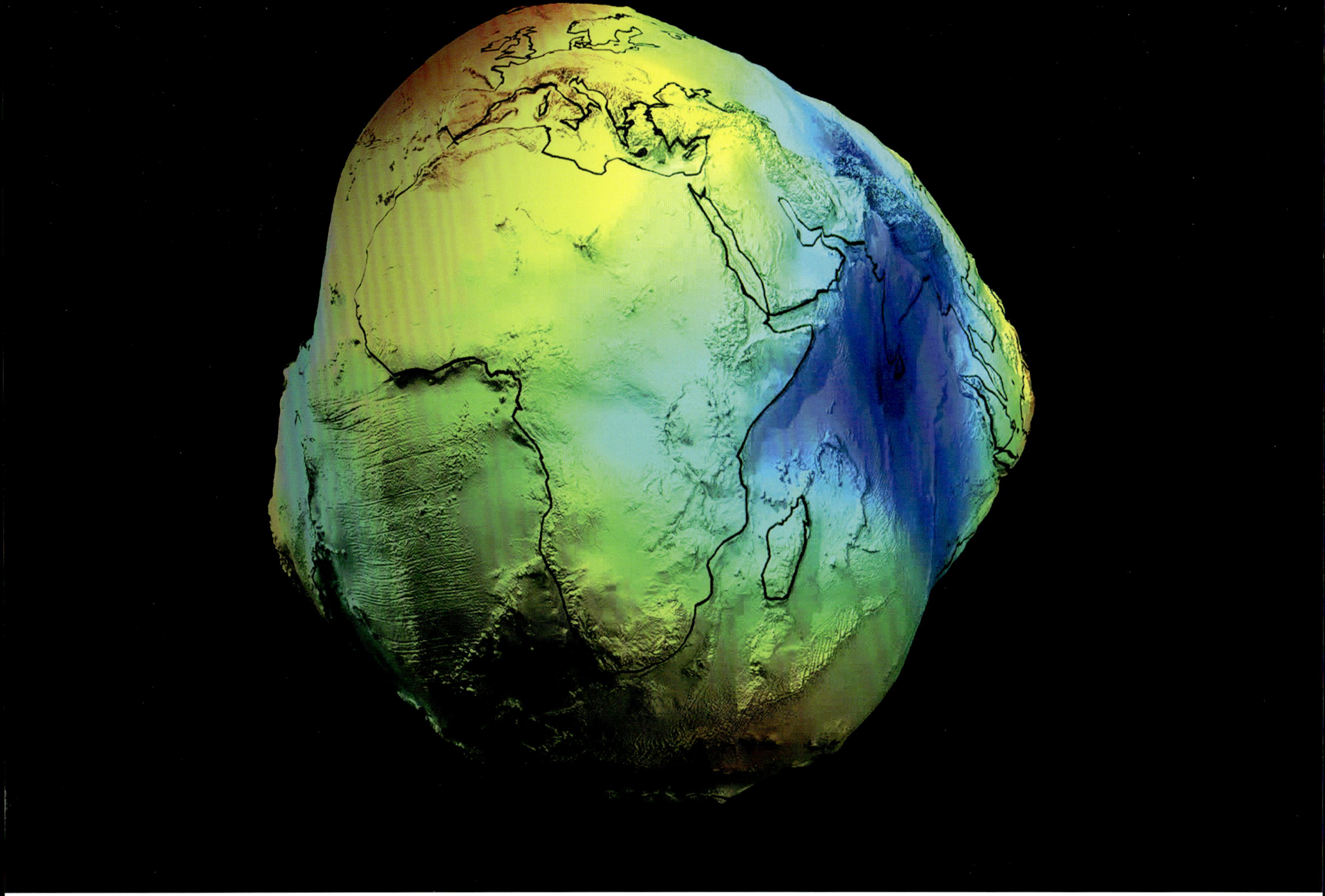

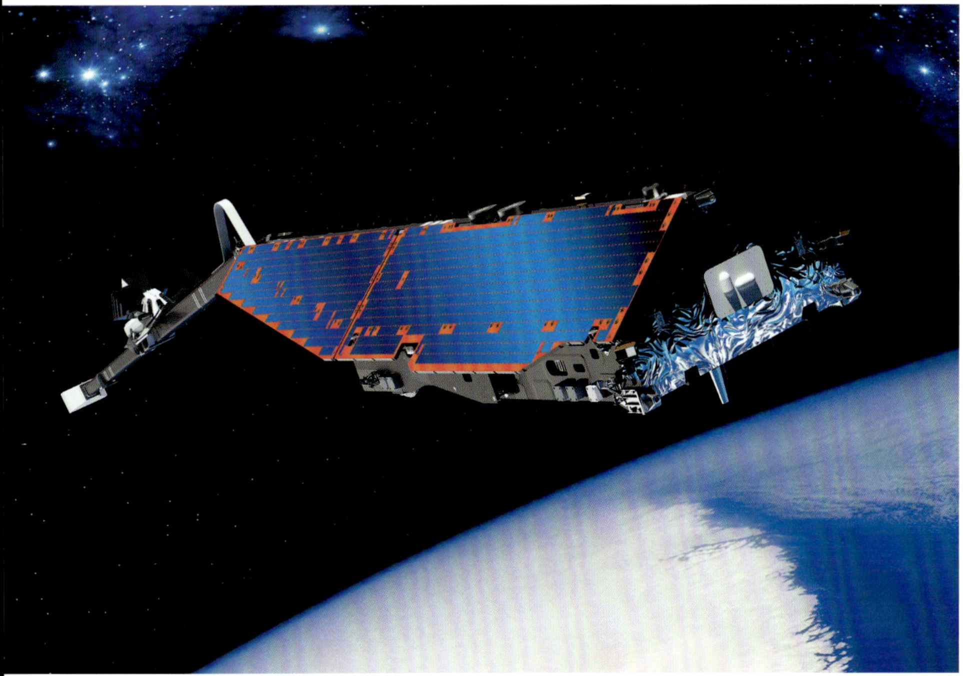

Links: »Luftgitarre« für das Magnetfeld – einer der drei Swarm-Satelliten, die 2013 gestartet sind (Zeichnung)

Left: Air guitar for the Earth's magnetic field — one of the three Swarm satellites launched in 2013 (illustration)

À gauche : « guitare aérienne » vouée au champ magnétique. L'un des trois satellites Swarm lancés en 2013 (vue d'artiste)

Izda.: «Guitarra aérea» para el campo magnético. Uno de los tres satélites Swarm lanzado en 2013 (dibujo)

Oben: 3D-Darstellung der Erdanziehungskraft – je tiefer die Senke, desto schwächer die Gravitation (GOCE)

Above: A 3-D representation of the Earth's gravitational force – the deeper the depression, the weaker the gravity (GOCE)

Ci-dessus : représentation en 3D du champ de pesanteur de la Terre. Plus la dépression est profonde plus la gravitation est faible (GOCE)

Arriba: Representación 3D de la fuerza de gravedad. Cuanto mayor es la hondonada, menor es la gravitación (GOCE)

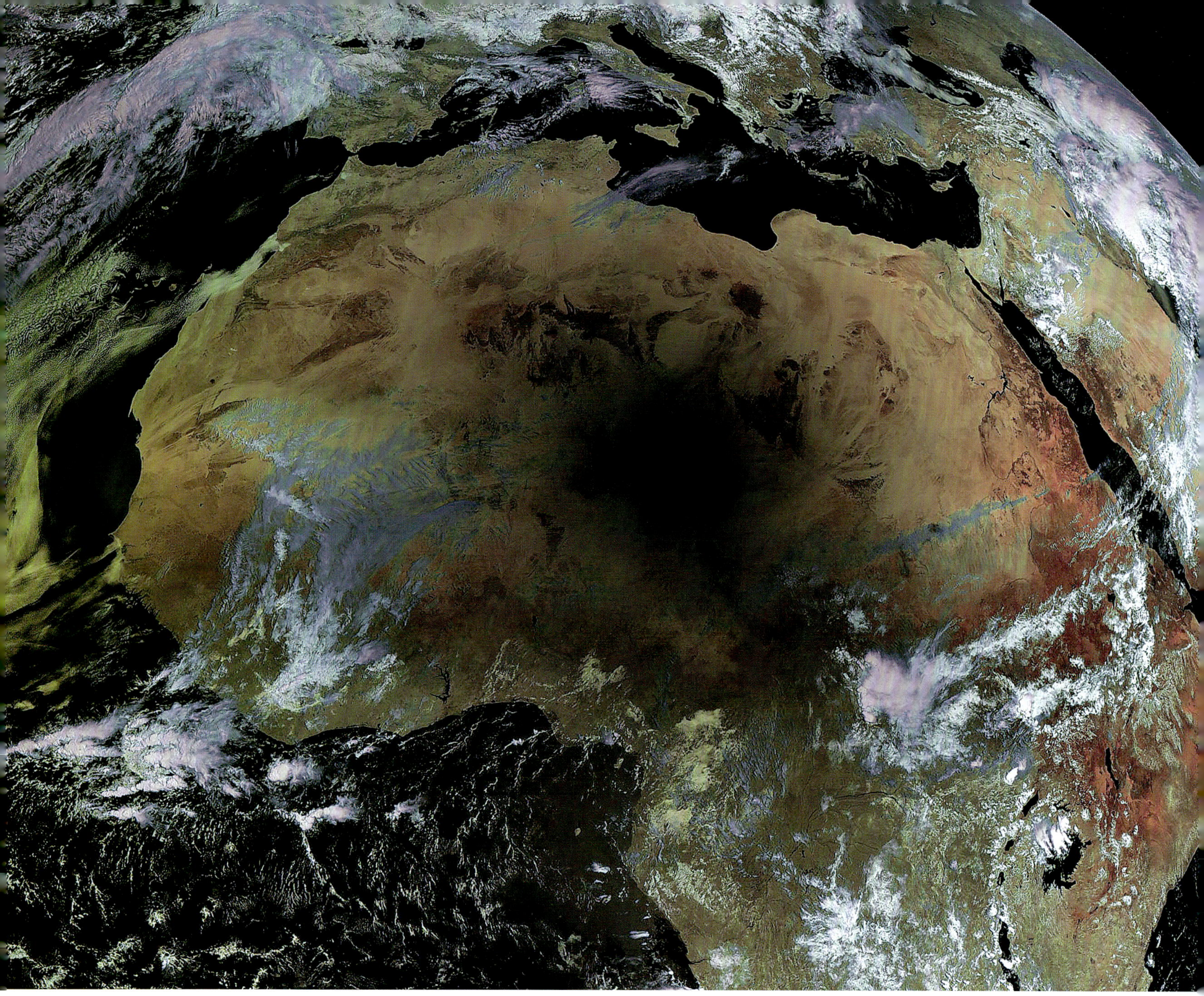

Oben: Sahara mit Mondschatten. Die Sonnenfinsternis vom 29. März 2006 (Meteosat)

Above: The Sahara with shadows of the Moon. Eclipse of the Sun from March 29, 2006 (Meteosat)

Ci-dessus : ombre de la lune au-dessus du Sahara. Éclipse totale de soleil du 29 mars 2006 (Météosat)

Arriba: Sombra de la Luna sobre el Sahara. Eclipse de Sol del 29 de marzo de 2006 (Meteosat)

Rechte Seite: Wassertemperatur im Mittelmeer am 9. Mai 2008, je stärker rot, desto heißer (Envisat)

Right: Mediterranean water temperatures on May 9, 2008; the deeper the red, the warmer the water (Envisat)

Page de droite : température de l'eau en Méditerranée le 9 mai 2008. Le rouge foncé indique les secteurs les plus chauds (Envisat)

Página dcha.: Temperatura del agua del Mediterráneo el 9.05.2008, cuanto más rojo, más caliente (Envisat)

Marco Polo, Humboldt, Envisat

Einst mussten sich die Entdecker auf jahrelange beschwerliche Reisen begeben, um unbekannte Gebiete unseres Planeten zu erkunden. Noch vor gut einem halben Jahrhundert gab es kaum eine Alternative zu Entdeckungsreisen. Wer die Wüste Gobi untersuchen wollte, der musste sich auf die Reise machen. Wer den Rand der Antarktis erforschen wollte, brauchte ein gutes Schiff und viel Geduld. Eine Portion Draufgängertum war bei Expeditionen in den Regenwald, durch die Wüste, durch das Eis oder zu fernen Gestaden sicher notwendig.

Auch heute noch brauchen die Forscher hin und wieder Messungen vor Ort. Aber unsere Erde lässt sich inzwischen von ganz speziellen Expeditionen mit rasenden Entdeckern erkunden. In einhundert Minuten geht es vom Pol über den Pazifik zum anderen Pol und dann über Afrika hinauf zum Ausgangspunkt. Das alles eben nicht am Boden, sondern in großer Höhe – aber die Erde immer fest im Blick. Haben sich einst Alexander von Humboldt, David Livingstone und Sven Hedin auf den Weg gemacht, um den Daheimgebliebenen von ihren Funden und Eindrücken zu berichten, so übernehmen das heute Entdeckungsreisende namens Envisat, TanDEM-X, GOCE & Co. Fast gleich geblieben ist die Reisedauer: Humboldts Expedition durch Lateinamerika währte fünf Jahre – die typische Dauer einer Satellitenmission.

Marco Polo, Humboldt, Envisat

Explorers used to have to undertake arduous voyages lasting years to research the uncharted areas of our planet, and even half a century ago there was little alternative to an expedition: anyone wanting to look at conditions in the Gobi Desert had to get packing. To explore the edge of the Antarctic, you needed a good boat and plenty of patience, and a fair bit of pluck must have been required for expeditions into the rainforest, the desert, across the ice, or into the unknown. Researchers still require the occasional local measurement even today, but now our Earth can be explored during rather special expeditions conducted by explorers flying at great speeds, traveling from one pole across the Pacific to the opposite pole before speeding over Africa to return to their starting point in less than 100 minutes. All this happens at a great height from the surface of Earth, without ever losing sight of the ground. Such trips are perilous, but fortunately no humans are now endangered, only machinery. Where once Alexander von Humboldt, David Livingstone, and Sven Hedin set off to report back to those at home of their findings and conclusions, now the expedition members have names like Envisat, TanDEM-X, and GOCE. The trips are of a similar length, however—the duration of a typical satellite mission is five years, the same as Humboldt's expedition to Latin America.

Marco Polo, Humboldt, Envisat

Jadis, les explorateurs s'embarquaient pour de longs voyages éprouvants à la découverte de territoires inexplorés. C'était il y a encore un demi-siècle la seule façon d'explorer le monde. Quiconque souhaitait étudier le désert de Gobi n'avait d'autre choix que de se rendre sur place. Quiconque souhaitait explorer les rives de l'Antarctique devait s'armer de patience et avoir un bon navire. Les expéditions dans la forêt vierge, à travers le désert, sur les glaces ou vers les rivages reculés exigeaient une bonne dose de témérité.

Les chercheurs d'aujourd'hui ont encore occasionnellement besoin de mesures *in situ*, mais notre Terre fait l'objet d'expéditions très particulières menées par des explorateurs véloces. Cent minutes leur suffisent pour faire l'aller-retour d'un pôle à l'autre en survolant le Pacifique puis l'Afrique. Un voyage qu'ils n'effectuent pas au sol, mais à haute altitude, sans jamais quitter la Terre des yeux. Ces expéditions sont certes toujours périlleuses, mais c'est désormais le matériel, et non les hommes, qui prend les risques. Si Humboldt, Livingstone et Hedin partaient jadis à l'aventure pour faire part de leurs découvertes à leurs concitoyens, les explorateurs d'aujourd'hui s'appellent Envisat, TanDEM-X ou GOCE. Seule la durée du voyage est restée la même : il a fallu cinq ans à Humboldt pour explorer l'Amérique latine, une durée comparable aux missions satellitaires.

El director del proyecto Swarm tiene una notable carrera tras de sí. Inicialmente estaba ocupado como ingeniero mecánico industrial especializado en ciencia de materiales. Tomó parte en un experimento de fundición en el Spacelab, un laboratorio espacial construido en Alemania, que se enviaba al espacio de vez en cuando con los transbordadores estadounidenses para desarrollar nuevas aleaciones de aluminio. Albert Zaglauer se quedó enganchado del espacio. Antiguamente había desarrollado una cámara estereoscópica para la sonda europea a Marte. Ahora es el responsable de las mediciones magnéticas. Estos instrumentos no proporcionan imágenes hermosas, como las que podrían adornar las primeras páginas de periódicos y revistas. Pero envían mediciones extremadamente significativas a la estación de tierra. En la misma medida en la que Albert Zaglauer está fascinado por el campo magnético de la Tierra, así le molestan los imanes en la misma. En el trabajo con satélites en la sala blanca, los imanes están estrictamente prohibidos. Nadie puede entrar allí con materiales magnéticos. El director del proyecto es firme: en su despacho ni siquiera tiene una pizarra magnética.

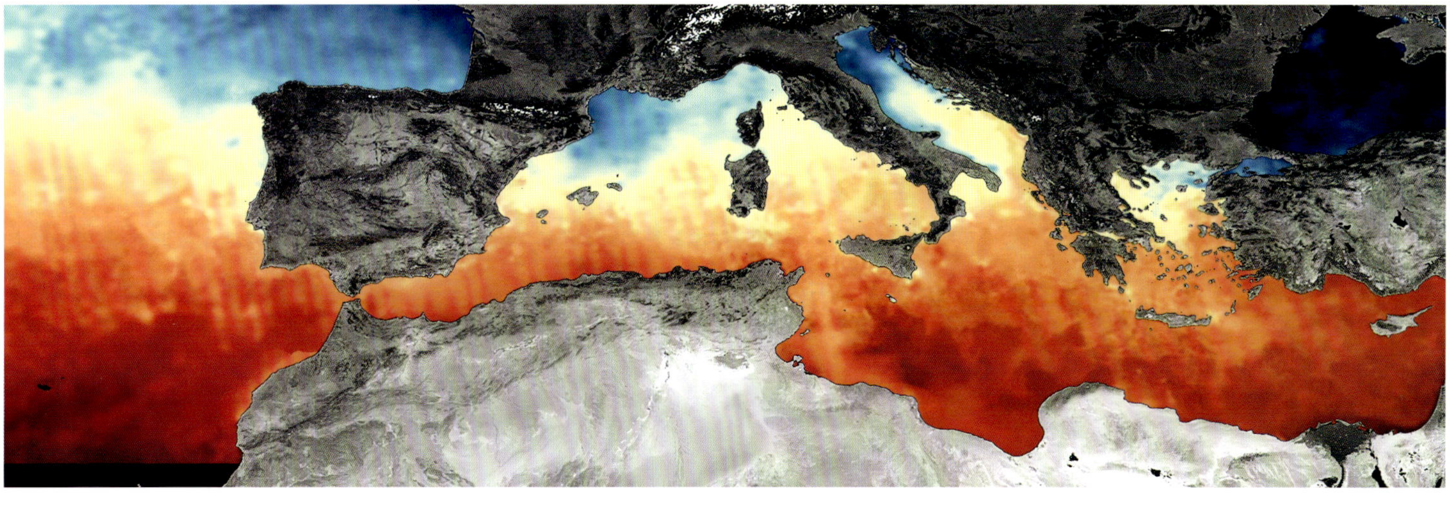

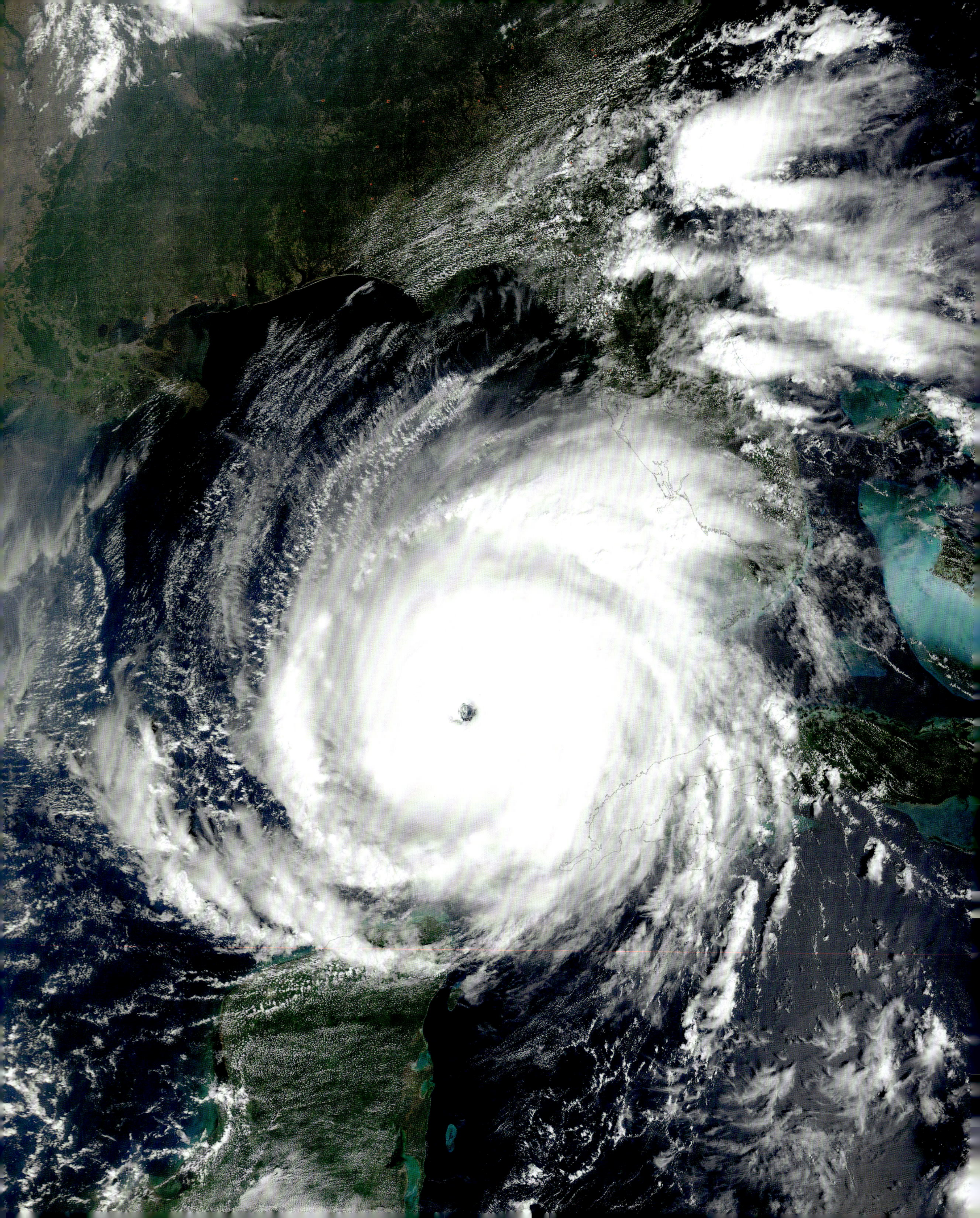

Klima, Karten, Katastrophen: Erdbeobachtung im Alltag

Climate, cartography, catastrophe: routine Earth observation

Climat, cartes, catastrophes : la Terre observée au quotidien

Clima, mapas, catástrofes: observación cotidiana de la Tierra

Ist die Erdbeobachtung nur ein schöner Luxus, den sich ein paar Wissenschaftler leisten? Oder eine nette Spielerei, die zwar wunderbare Fotos liefert, aber eigentlich nicht gebraucht wird? Ist es wirklich notwendig, ins All zu fliegen, um die Erde zu sehen? Daten aus dem All sind unverzichtbar – wegen der großen Übersicht und der Möglichkeit, in kurzer Zeit Daten von vielen Orten der Erde zu bekommen. Natürlich lassen sich an einer bestimmten Stelle auf der Erde Temperatur, Wind, Bodenfeuchte, Anziehungskraft, chemische Zusammensetzung der Luft, Beschaffenheit der Oberfläche, Vegetation etc. vor Ort viel genauer messen als von der Umlaufbahn aus. Das ist jedenfalls dann der Fall, wenn dieser Ort bequem zu erreichen ist, etwa der Hinterhof eines Forschungsinstituts. Schon beim Untersuchen von zehn oder hundert Orten gäbe es erheblichen logistischen Aufwand, um Instrumente und Personen zu transportieren.

Sobald mehr als nur lokale Messungen gebraucht werden, ist die Nutzung von Satelliten die einzige realistische Methode. Wollte man auf der gesamten Erdoberfläche Daten sammeln und sich dabei mit einem recht grobmaschigen Netz mit einhundert Kilometern Abstand zwischen den Stationen begnügen, hätte man es

Is observing the Earth a fine luxury only a few scientists can afford? Or perhaps an entertaining exercise which results in a few beautiful photographs of the Earth but is of little actual use? Do we really need to go into space to see the Earth? Data from the cosmos is essential, both because of its greater perspective and the opportunity it provides of retrieving information from many points on the Earth's surface within a short period of time. It is easier to take accurate measurements of temperature, wind speed, soil moisture, gravity, the chemical composition of the air, the nature of the Earth's surface, and the surrounding vegetation locally than from orbit, only when such locations are easy to reach—the backyard of the research institution, for example. Transporting instruments and personnel to examine even ten or a hundred sites would involve considerable logistical problems.

As soon as anything more than local measurements is required, the use of satellites is the only realistic approach. Anyone wishing to collect data from the entire Earth's surface, even if they settled for an extremely loose concatenation of stations 60 miles (100 kilometers) apart, would require no less than 50,000

L'observation de la Terre est-elle un luxe réservé à quelques chercheurs ? Ou un agréable passe-temps qui fournit de superbes photos, mais n'a au fond pas grande utilité ? Faut-il vraiment aller dans l'espace pour étudier la Terre ? La vision globale et la possibilité de recueillir en un temps record des informations sur divers points du globe rendent les données spatiales irremplaçables. On peut certes mesurer la température, la vitesse du vent, l'humidité du sol, la force de gravité, la composition de l'air, la nature de la surface, la végétation, etc. sur place avec bien plus de précision. À condition toutefois que le terrain observé soit facile d'accès. L'étude d'une dizaine voire d'une centaine de points différents se heurterait déjà à d'énormes difficultés logistiques liées au transport des personnes et instruments.

Quand les mesures locales ne suffisent plus, le recours aux satellites est la seule méthode réaliste. À supposer qu'on souhaite collecter des données sur l'ensemble de la surface terrestre avec un réseau de stations espacées tous les cent kilomètres, on compterait environ 50 000 points de mesure. Il faudrait réquisitionner des armées entières pour effectuer les relevés sur les continents et océans. Les satellites en revanche

¿Acaso la observación de la Tierra es un bonito lujo que sólo se pueden permitir un par de científicos? ¿O quizá un simpático pasatiempo que, aunque proporcione fotos maravillosas, en realidad no se utiliza? ¿Es realmente necesario volar al espacio para ver la Tierra? Los datos enviados desde el espacio resultan indispensables porque ofrecen una visión amplia y general, y por la posibilidad de obtener datos de diferentes lugares de la Tierra en un breve lapso de tiempo. Naturalmente, en cualquier región de la Tierra se puede medir in situ la temperatura, el viento, la humedad del suelo, la fuerza de gravedad, la composición química del aire, el estado de la superficie, la vegetación y otros factores, con mucha mayor precisión que desde la órbita. De todas formas, esto suele ser así cuando se trata de un lugar al que es fácil llegar, por ejemplo el patio trasero de un instituto de investigación. Cuando se trata de investigaciones en diez o en cien lugares distintos, se producen considerables gastos logísticos para transportar instrumental y personal.

En cuanto se necesitan más mediciones que las puramente locales, la utilización de satélites es el único método realista. Si se quisieran reunir datos de toda la superficie terrestre contentándose para ello con una red

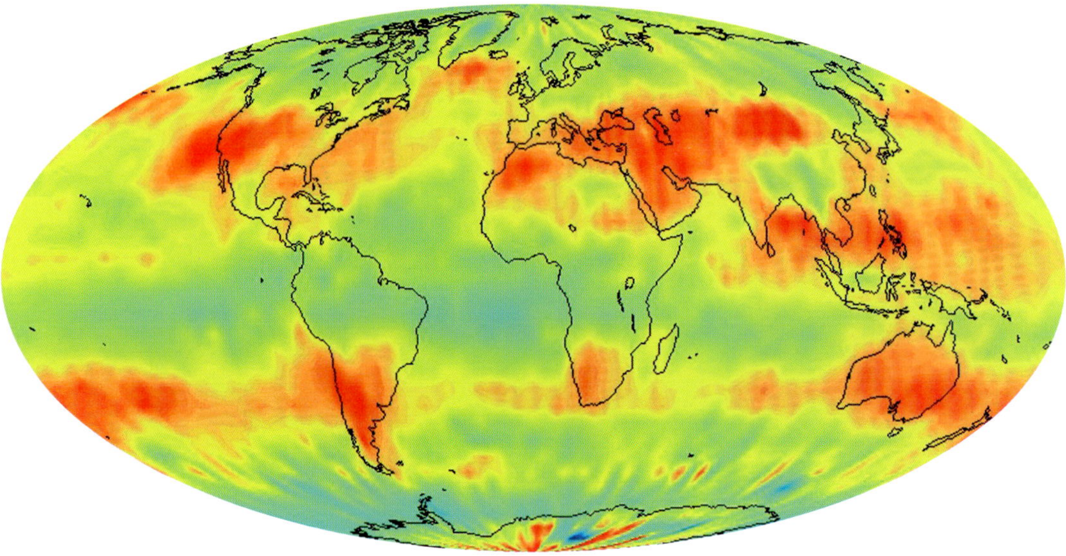

mit gut 50.000 Messpunkten zu tun. Da müssten schon ganze Armeen von Vermessern zu Land und zur See ausrücken. Dagegen liefern Satelliten die Daten recht schnell, buchstäblich im Vorbeifliegen. Zudem können sie die Messungen in regelmäßigen Abständen wiederholen und so zeitliche Veränderungen bestens dokumentieren.

Global messen, global verstehen

In unserem Alltag spielen die Daten der Erdbeobachtungssatelliten längst eine große Rolle, auch wenn das den meisten Menschen nicht bewusst ist. Ein Beispiel: Alle reden vom Klima. Heiße Sommer, Schneekatastrophen im Winter oder Starkregen im Frühjahr sorgen für viele Diskussionen. Gerät das Erdklima auf Abwege? Um das zu klären, bringt es gar nichts, die Atmosphäre nur am Wohnort zu untersuchen. Denn das Wetter und die chemischen Vorgänge in der Luft über Coventry, Göttingen, Lyon oder Toledo hängen eben auch von den Prozessen ab, die sich in der Sahara, den Anden, dem Nordpolarmeer oder in Indien abspielen.

Eine prominente Hauptrolle spielt das Gas Kohlendioxid. Es ist farblos, es riecht nicht, man schmeckt es

measuring points, and a whole army of surveyors would have to be dispatched by sea and land. By comparison, satellites provide information quickly and quite literally on the fly; in addition, they can repeat the observations at regular intervals and are best placed to track changes over time.

Measure globally, think globally

Data from Earth observation satellites have long played an important role in our everyday life, even if most people are unaware of it. For example: everybody is now talking about the climate, and hot summers, avalanches in winter, and heavy rain in spring are regular topics of discussion. Is the Earth's climate being disturbed? An answer will never be achieved by merely observing the weather where we live; the conditions and chemical processes occurring in the air over Coventry, Göttingen, Lyon, and Toledo are dependent on mechanisms taking place in the Sahara, the Andes, the Arctic Ocean, and India.

Carbon dioxide plays a crucial role. A colorless, odorless, and tasteless gas, it is produced anywhere where

livrent rapidement les données qu'ils recueillent sur leur route à intervalles réguliers, et enregistrent avec précision les évolutions temporelles.

Mesurer et comprendre à l'échelle planétaire

Les données d'observation de la Terre jouent depuis longtemps un rôle majeur dans notre vie quotidienne sans que nous le sachions toujours. Prenons la question du climat ô combien sujette à discussion : canicules estivales, tempêtes de neige, pluies torrentielles nourrissent les conversations. Sommes-nous témoins d'un dérèglement climatique ? Il ne suffit pas de mesurer l'atmosphère au-dessus de son toit pour répondre à cette question. Le temps et les processus chimiques observés à Coventry, Göttingen, Lyon ou Tolède sont en effet étroitement liés aux processus qui s'opèrent dans le Sahara, les Andes, l'océan Arctique ou en Inde.

Dans le rôle principal, un gaz célèbre : le dioxyde de carbone. Incolore, inodore, insipide, il est pourtant le compagnon omniprésent des feux et de toute activité humaine. Ce gaz est présent sur la Terre depuis des milliards d'années, bien avant que ne débute sa carrière peu glorieuse

de malla gruesa en la que hubiese cien kilómetros de distancia entre las estaciones, se tendría que hacer con más de 50.000 puntos de medición. Hacia allí deberían desfilar, por tierra y mar, ejércitos enteros de agrimensores. Por el contrario, los satélites suministran los datos muy rápidamente. Además, pueden repetir las mediciones de forma periódica, documentando así muy bien los cambios.

Medición global, comprensión global

Los datos de los satélites de observación terrestre juegan un importante papel en nuestro día a día, aunque la mayoría de las personas no sea consciente de ello. Un ejemplo: todo el mundo habla del clima. Los veranos calientes, las nevadas catastróficas en invierno o las intensas lluvias en primavera son siempre tema de conversación. ¿Va por mal camino el clima terrestre? Para aclarar esto, no aporta casi nada analizar la atmósfera solo en el lugar de residencia, porque la meteorología y los procesos químicos en el aire sobre poblaciones concretas dependen de procesos que ocurren en lugares muy alejados.

El dióxido de carbono es un gas que juega un papel destacado en la

Überflutete Gebiete (blau) im Flussdelta von Bangladesh. Der weiße Fleck links unten ist Kalkutta (Envisat, 2007)

Flooded areas (blue) in Bangladesh's Ganges Delta. The white area at the bottom left of the image is Kolkata (Envisat, 2007)

Régions inondées (en bleu) dans le delta fluvial du Bangladesh. La tache blanche en bas à gauche représente Calcutta (Envisat, 2007)

Comarcas inundadas (azul) en el delta fluvial de Bangladesh. La mancha blanca de la parte inferior izda. es Calcuta (Envisat, 2007)

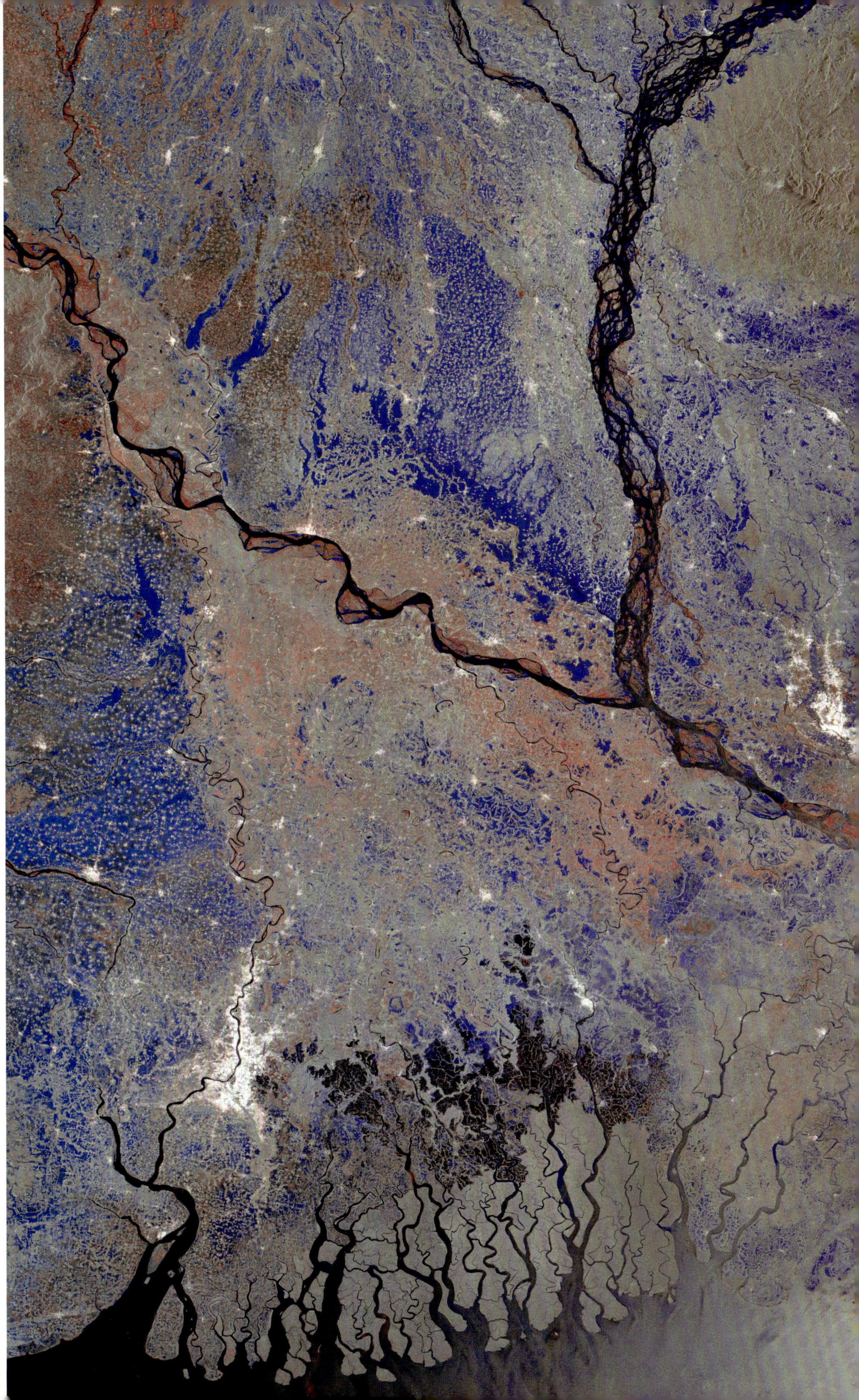

Linke Seite: Weltweite Konzentration von Kohlendioxid (je röter, desto stärker) in der Atmosphäre im Juli 2008 (Aqua)
Seite 136: Hurrikan Rita gehörte 2005 zur gefährlichsten Kategorie 5 und zog mitten durch den Golf von Mexiko (Terra)

Left: Global atmospheric carbon dioxide levels (high concentrations are shown in red) in July 2008 (Aqua)
Page 136: Hurricane Rita, which passed through the middle of the Gulf of Mexico in 2005, was one of the most dangerous category 5 storms (Terra)

Page de gauche : concentration mondiale de dioxyde de carbone (concentration la plus élevée en rouge) dans l'atmosphère en juillet 2008 (Aqua)
Page 136 : Rita, un cyclone de catégorie 5, la plus dangereuse, a traversé le golfe du Mexique en 2005 (Terra)

Página izda.: Concentración mundial de dióxido de carbono en la atmósfera (cuanto más roja, más intensa), en julio de 2008 (Aqua)
Página 136: El huracán Rita, en 2005, pertenecía a la peligrosísima categoría 5 y atravesó el golfo de México (Terra)

nicht – doch es entsteht überall, wo Menschen sind und Feuer gemacht wird. Dieses Gas gibt es auf der Erde seit Milliarden von Jahren, lange bevor es eine unrühmliche Karriere als »Klimakiller« gemacht hat. Atmen Tier und Mensch aus, wird Kohlendioxid in die Atmosphäre ausgestoßen. Pflanzen wiederum verbrauchen Kohlendioxid und liefern reinen Sauerstoff. Doch durch die massenhafte Freisetzung beim Verbrennen von Holz, Kohle, Erdgas und Öl hat im industriellen Zeitalter der Anteil des Kohlendioxids in der Atmosphäre stark zugenommen. Vermutlich ist dies eine der Hauptursachen für den Treibhauseffekt, der die Erde immer wärmer werden lässt. Ganz eifrige Klimaforscher verkünden Modelle, mit denen sie den weiteren Anstieg der mittleren Temperatur auf der Erde innerhalb dieses Jahrhunderts auf ein Zehntel Grad genau vorhersagen. Doch Prognosen sind schwierig, vor allem wenn sie die Zukunft betreffen – und wenn die Datenlage zumindest stark verbesserungsfähig ist, um es ganz vorsichtig zu formulieren.

Denn es gibt bis heute nur rudimentäre Daten darüber, wo es welche Mengen an Kohlendioxid, Methan oder anderen Spurengasen in der Atmosphäre gibt, die für das Klima von Bedeutung sind. Nur das SCIAMACHY-

there are people or where fires burn. The gas has existed on Earth for billions of years, long before it began its nefarious career as a "climate killer." Whenever humans or animals breathe out, carbon dioxide is expelled into the atmosphere. In contrast, plants absorb carbon dioxide and produce pure oxygen. However, the mass release of the gas through the burning of wood, coal, natural gas, and oil during the industrial age has greatly increased its occurrence in the atmosphere and it is thought that this is one of the principal causes of the greenhouse effect leading to global warming. Keen climate researchers produce models anticipating the further rise of average global temperatures during the century to an accuracy of a tenth of a degree, but predictions are difficult to make, especially about the future, and not least when the available data is patchy, to say the least.

At the moment, only rudimentary data is available on the presence and distribution of carbon dioxide, methane, and the other trace gases in the atmosphere which may be of importance to the climate. Only the SCIAMACHY instrument on board the European Envisat satellite was able to provide this global data, However, the satellite failed sud-

de perturbateur de climat. Les organismes humains et animaux qui respirent rejettent du dioxyde de carbone dans l'atmosphère. Les plantes consomment ce gaz et produisent de l'oxygène pur. Cependant, suite aux rejets massifs dus à la combustion de bois, charbon, pétrole et gaz naturel au cours de l'ère industrielle, la concentration de dioxyde de carbone dans l'atmosphère a fortement augmenté. Ce phénomène est considéré comme la cause principale de l'effet de serre responsable du réchauffement de la Terre. Certains spécialistes du climat publient des modèles pronostiquant une augmentation de la température moyenne de la Terre d'un dixième de degré au cours de ce siècle. Pourtant, rien de saurait être plus inexact que les prévisions d'avenir, surtout lorque les données disponibles laissent autant de place à l'amélioration.

À l'heure actuelle, les données relatives à la concentration de dioxyde de carbone, méthane et autres gaz traces dans l'atmosphère, essentielles à la compréhension du climat, restent rudimentaires. Seul le spectromètre SCIAMACHY embarqué sur Envisat fournissait des données globales. Cependant, le satellite est soudainement tombé en panne en avril 2012. La relève est prévue, mais nous ris-

atmósfera terrestre. Es incoloro, inodoro, insípido y se produce allí donde haya personas y se haga fuego. Este gas está presente en la Tierra desde hace miles de millones de años, bastante tiempo antes de que empezase su triste carrera como «asesino del clima». Cuando los animales y las personas espiran, expulsan dióxido de carbono a la atmósfera. Por otra parte, las plantas verdes, durante la fotosíntesis, consumen dióxido de carbono y producen oxígeno puro. Pero en la era industrial, la liberación masiva del gas por combustión de madera, carbón, gas natural y petróleo ha hecho aumentar la proporción de dióxido de carbono en la atmósfera. Esta es, presuntamente, una de las causas principales del efecto invernadero que calienta la Tierra. Afanosos investigadores climáticos publican modelos con los que pronostican el aumento adicional de la temperatura media de la Tierra a lo largo de este siglo con una exactitud de décimas de grado. Pero hacer pronósticos es complicado, sobre todo porque conciernen al futuro y porque el estado de los datos es, al menos, lo suficientemente mejorable como para formularlos con prudencia.

Hasta el momento actual, únicamente se dispone de datos rudimentarios sobre las cantidades de dióxido de carbono, metano y otros gases traza

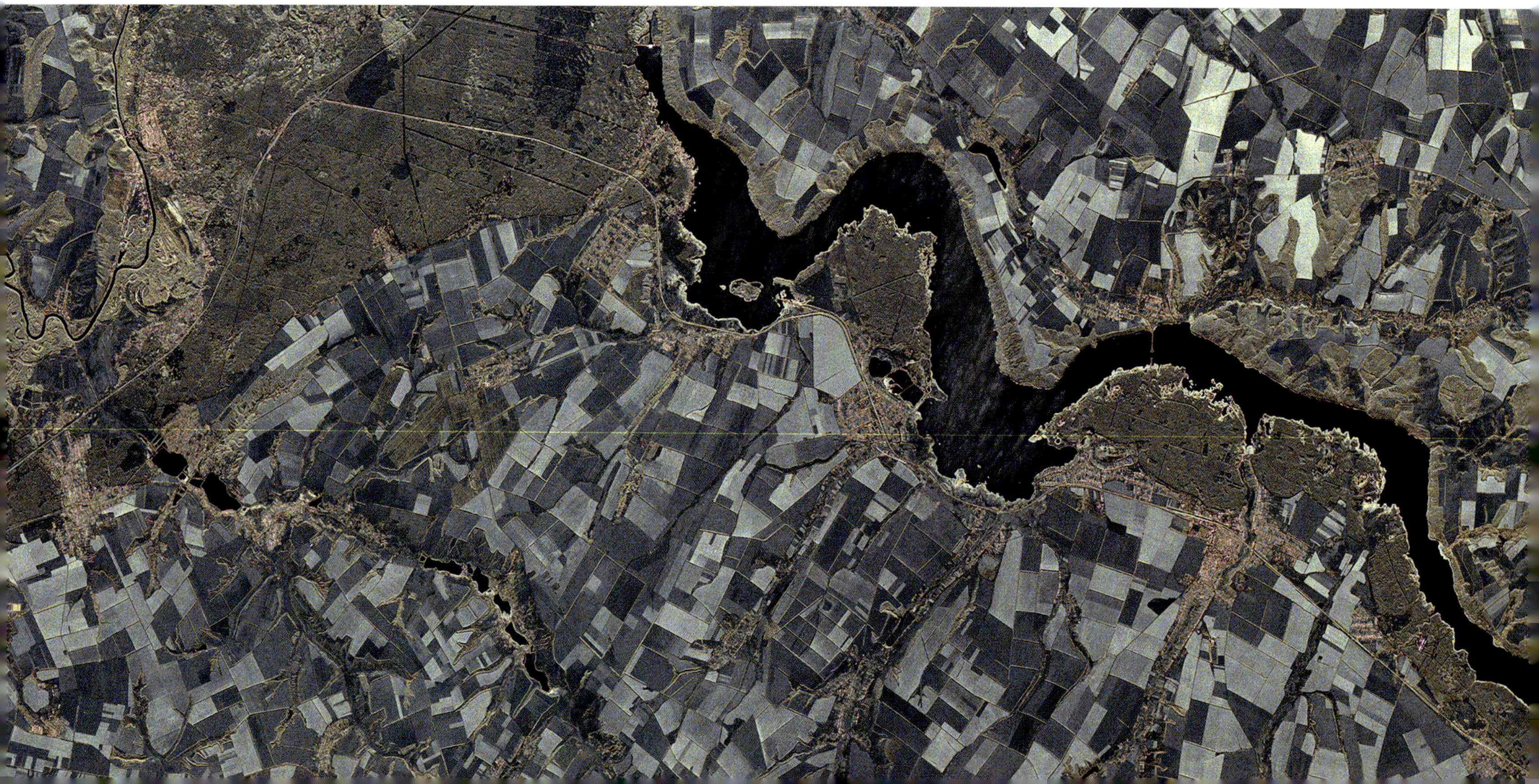

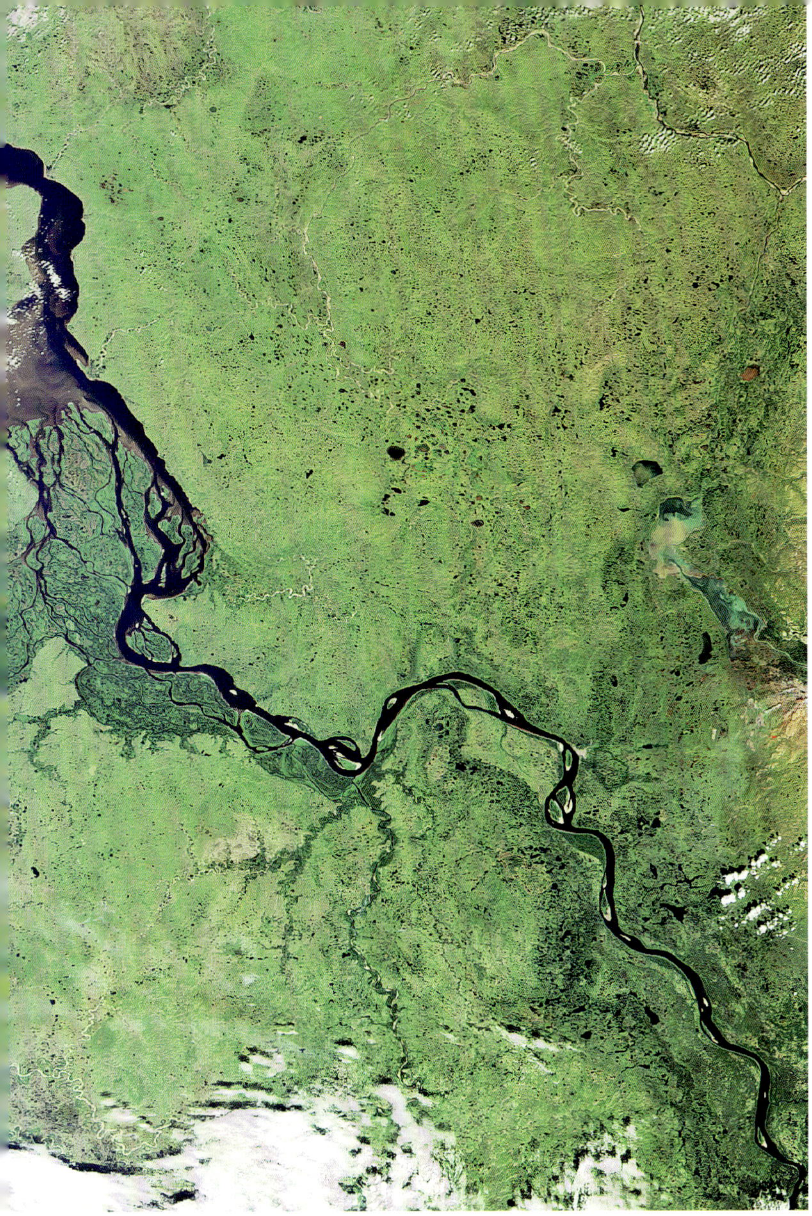

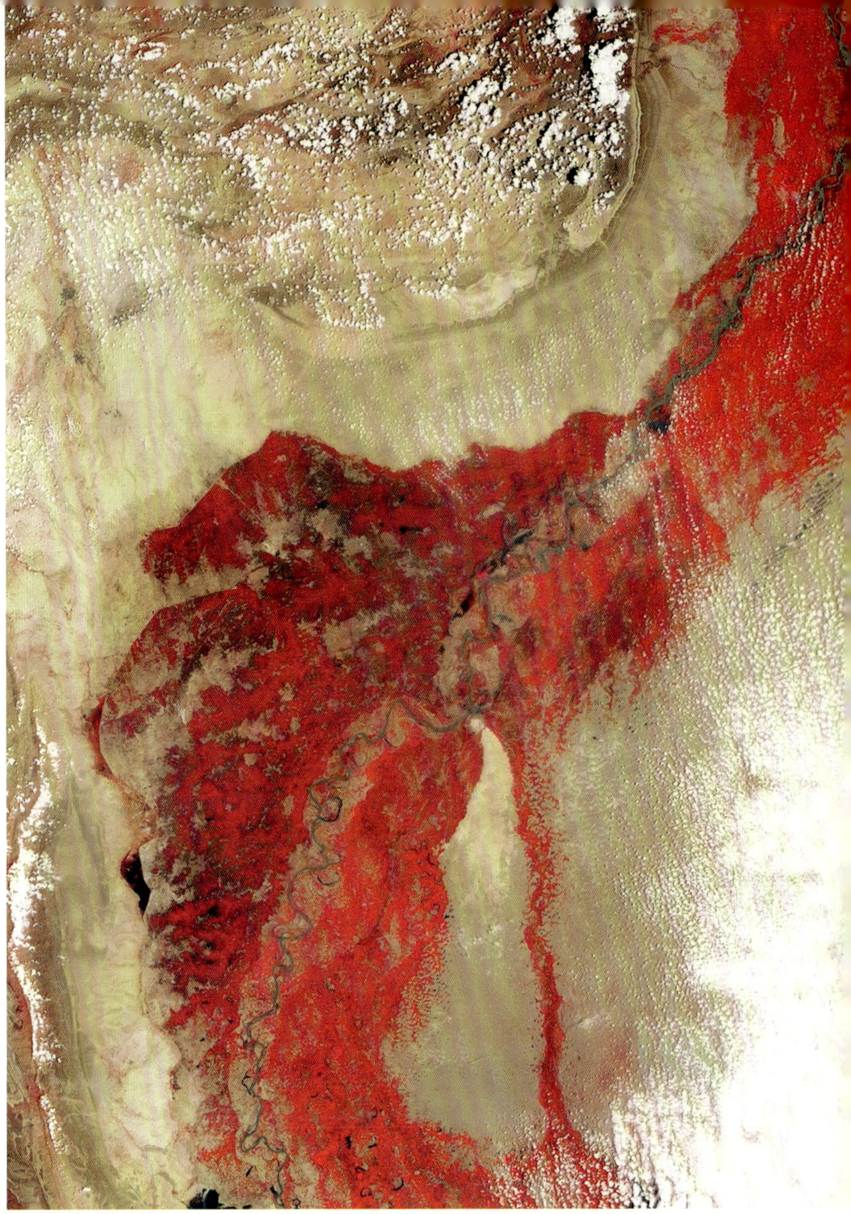

Linke Seite: Ein Stausee und Felder in der Ukraine (Radaraufnahme, TanDEM-X)

Left: Fields and a reservoir in Ukraine (radar image, TanDEM-X)

Page de gauche : champs autour du lac de retenue d'un barrage en Ukraine (image radar, TanDEM-X)

Página izda.: Embalse y tierras de labor en Ucrania (toma de radar, TanDEM-X)

Oben links: Der Dauerfrostboden in der Umgebung des Jenessei taut immer mehr auf (Envisat)

Above left: More and more of the permafrost surrounding the Yenisei is thawing (Envisat)

À gauche ci-dessus : le pergélisol a fond de plus en plus sur les rives du fleuve Ienisseï, en Sibérie (Envisat)

Arriba izda.: El permafrost de las cercanías del río Yeniséi se descongela cada vez más (Envisat)

Oben rechts: Unterlauf des Indus in normalen Jahren; rote Bereiche: viel Vegetation (Terra)

Above right: The course of the Indus in normal years; the red areas are densely vegetated (Terra)

À droite ci-dessus : cours inférieur de l'Indus en temps normal. En rouge : végétation abondante (Terra)

Arriba dcha.: Curso inferior del río Indo en años normales; zonas rojizas: mucha vegetación (Terra)

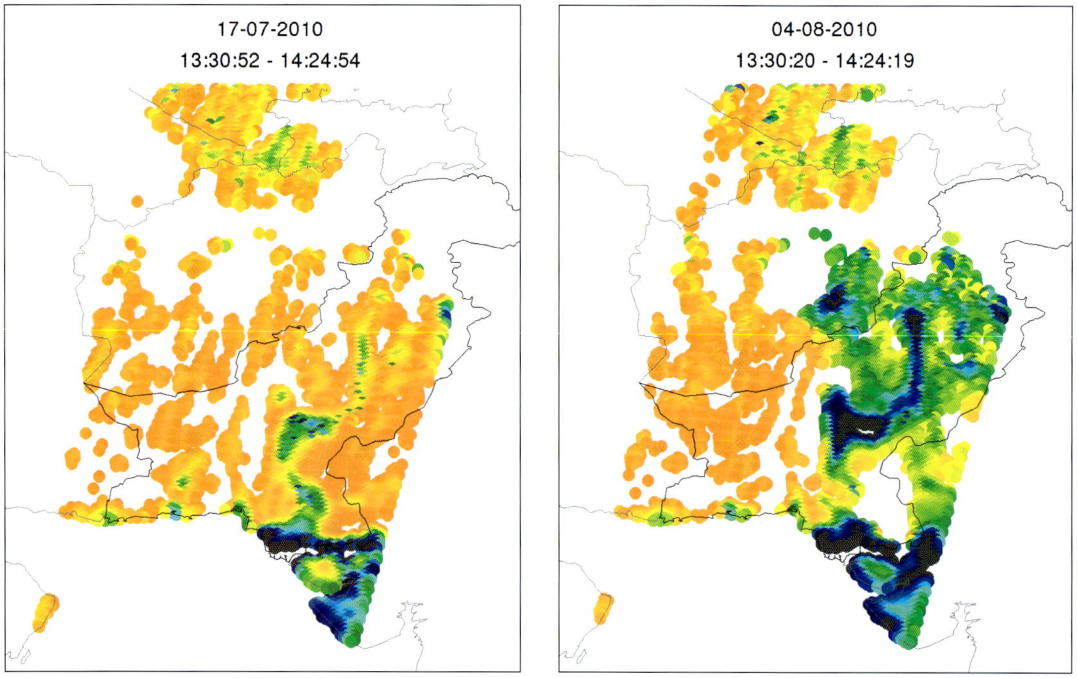

Flutkatastrophe 2010. *Links:* Die Bodenfeuchte in Pakistan, gemessen von SMOS, vor und während der Überflutungen. *Rechts:* Der Indus am 11. Juli (links) und 17. August (Terra)

The catastrophic flooding of 2010. *Left:* Soil moisture in Pakistan, measured by SMOS before (far left) and during (near left) the flood.
Right: The Indus on July 11 (near right) and again on August 17 (far right, Terra)

Inondations de 2010. Les images de gauche indiquent l'humidité du sol mesurée par SMOS au Pakistan avant et après la crue. *À droite :* l'Indus le 11 juillet (à gauche) et le 17 août (Terra)

Inundaciones de 2010. La página izquierda muestra la humedad del suelo en Pakistán, medida por SMOS, antes de y durante las inundaciones.
Dcha.: El Indo el 11 de julio (izda.) y el 17 de agosto (Terra)

Instrument auf dem europäischen Envisat-Satelliten lieferte hier globale Daten. Allerdings ist der Satellit im April 2012 plötzlich ausgefallen. Ersatz ist in Planung, aber es drohen noch etliche Jahre ohne ausreichende Informationen. Natürlich kann man von Stationen am Boden messen, wie die Atmosphäre dort zusammengesetzt ist. Aber gerade beim Methan wollen die Forscher genau wissen, wo es entsteht. Eine der natürlichen Quellen sind auftauende Böden der Tundra. Das Methan dort zu messen, belegt zum einen die Erwärmung und zeigt, dass sich das Problem verschärft, weil noch mehr Methan aus dem Boden in die Atmosphäre entweichen wird. So etwas lässt sich nur per Satellit erfassen. Ohne solche Daten bleiben Klimamodelle zwangsläufig recht grob.

Großes Mosaik, wenig Steine

Das gilt für etliche weitere Faktoren, die die Klimaforscher präzise berücksichtigen müssten. Bis heute ist unklar, was genau in den Polargebieten geschieht, den Kältekammern unserer Erde. Wo und wie sich Wolken bilden, ist noch immer nicht ganz verstanden. Ebenso ist den Wissenschaftlern nicht restlos klar, ob Wolken mehr

denly in 2012. A replacement is intended, but it seems there will be insufficient information for a number of years. Measurements of the composition of the atmosphere can of course be taken from ground stations, but in the case of methane, scientists are particularly interested in its origin. One natural source is to be found in the thawing soils of the tundra, and measuring the methane there is both proof of global warming and a sign of how the problem is getting worse, as more methane escapes from the soil into the atmosphere. Such results can only be gathered by satellite and without such data, climate models cannot help but be extremely inaccurate.

A large mosaic, few tiles

The same is true of a number of other factors which climate researchers would have to measure precisely. Even today, it is unclear exactly what is happening around the poles, the cold bunkers of our planet. Where and how clouds form is still not satisfactorily understood, nor have scientists fully established whether such vapors act as a blanket, warming the Earth's surface, or

quons de manquer de données pendant quelques années. Il est certes possible d'étudier la composition de l'atmosphère à l'aide de stations de mesure réparties sur le sol, mais les scientifiques souhaitent connaître les sources des rejets, notamment du méthane qui s'échappe par exemple de la toundra lors du dégel des sols. Les émissions de méthane mesurées à cet endroit confirment non seulement le réchauffement, mais révèlent également une aggravation de la situation due à une augmentation du méthane libéré par le sol. Et seuls les satellites sont aptes à fournir de telles observations, sans lesquelles les modèles climatiques sont condamnés à l'approximation.

Grande mosaïque, rares pierres

Il en va de même des autres facteurs qui mériteraient la rigoureuse attention des scientifiques du climat. Les processus qui régissent les régions polaires, les chambres froides de la Terre, restent énigmatiques. Le principe de formation des nuages n'est pas entièrement expliqué. Pas plus que leur rôle d'ailleurs : agissent-ils comme une couverture réchauffant la surface de la Terre, ou comme un miroir qui reflète la lumière solaire

que hay en la atmósfera y que son significativos para el clima. Tan solo el instrumento SCIAMACHY, del satélite europeo Envisat, proporcionaba datos globales. Sin embargo, el satélite falló repentinamente en abril de 2012. El relevo está previsto, pero se teme que durante un par de años no se disponga de suficientes datos. Naturalmente, es posible medir la composición de la atmósfera desde estaciones de medición en tierra. Pero en el caso del metano, los investigadores quieren saber exactamente dónde se produce. Una de sus fuentes naturales es la descongelación de los suelos de la tundra. Medir el metano allí prueba, por un lado, que el calentamiento es real y, por otro, que el problema se intensifica al escaparse más metano del suelo a la atmósfera. Algo así solo se puede captar mediante satélites. Sin estos datos, los modelos climáticos quedan incompletos.

Mosaico grande, pocas piedras

Esto sirve para algunos otros factores que los investigadores climáticos deberían tener en cuenta. Hasta hoy continúa sin esclarecerse por completo lo que ocurre en las regiones polares, verdaderas cámaras frigoríficas de la Tierra. Sigue sin comprenderse muy

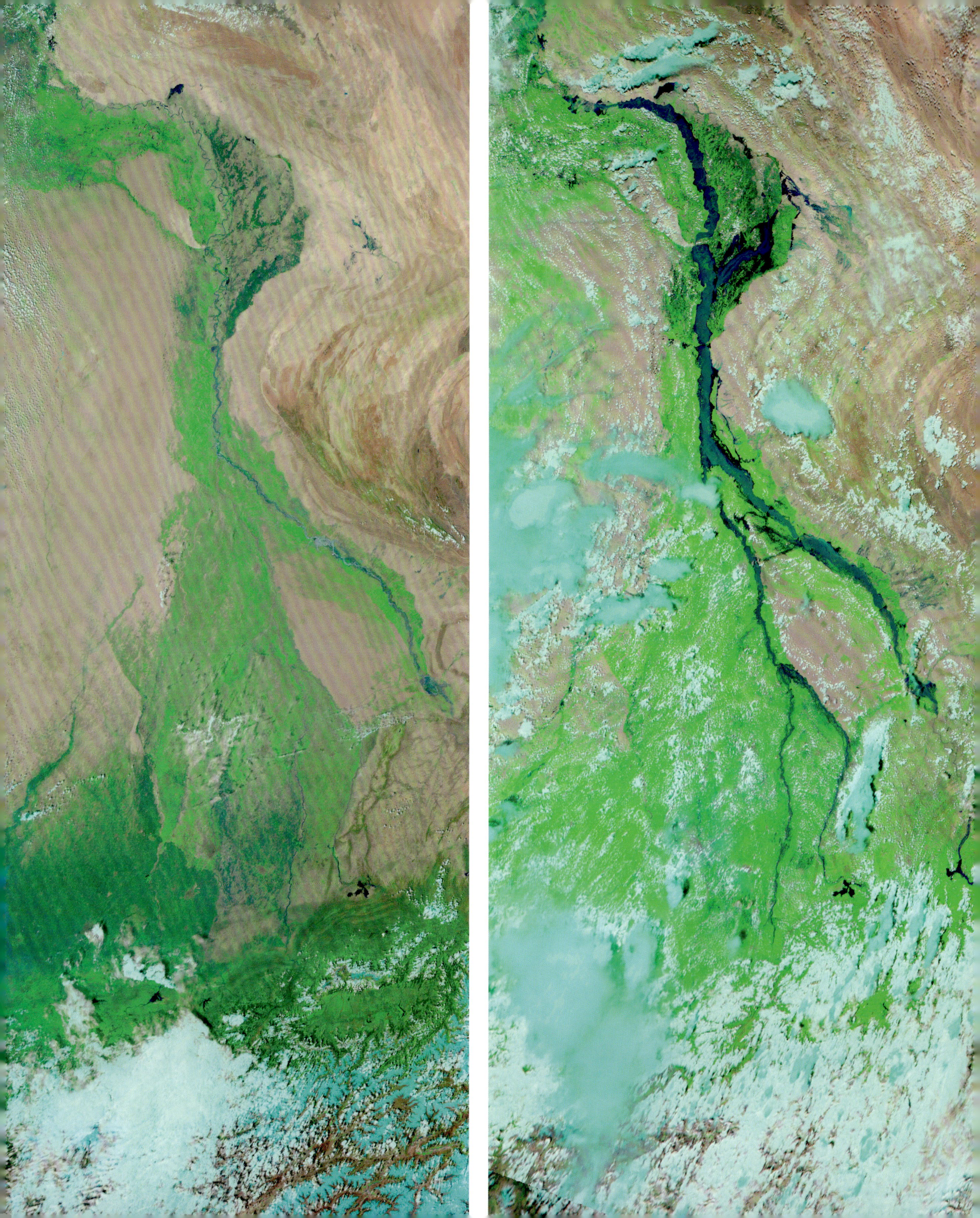

Unten: Mississippi-Flut 2008. Der Radar kann durch Wolken die aktuelle Lage erfassen (TerraSAR-X)

Below: Mississippi flooding in 2008. Radar can deliver up-to-the-minute images even through clouds (TerraSAR-X)

Ci-dessous : crue du Mississippi en 2008. Le radar établit l'état des lieux à travers les nuages (TerraSAR-X)

Abajo: Crecida del Misisipí en 2008. El radar puede registrar la situación actual a través de las nubes (TerraSAR-X)

Rechte Seite: Erdrutsch (lange braune Schneise) in Zhouqu, China, August 2010 (WorldView-2)

Right: Earthquake (the long brown rift) in Zhouqu, China, August 2010 (WorldView-2)

Page de droite : glissement de terrain (longue coulée brune) à Zhouqu, en Chine, en août 2010 (WorldView-2)

Página dcha.: Corrimiento de tierras (lengua marrón) en Zhouqu, China, agosto de 2010 (WorldView-2)

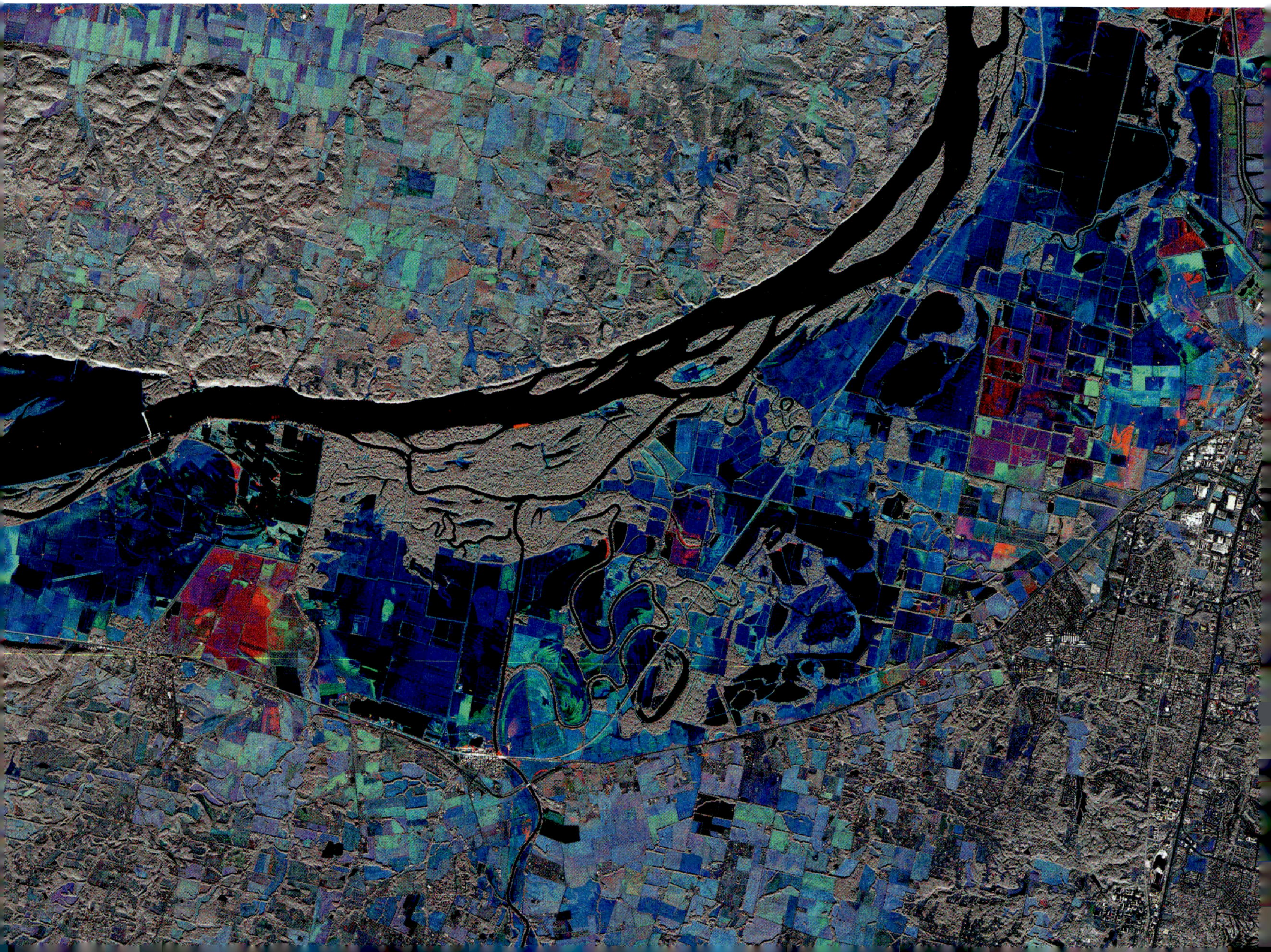

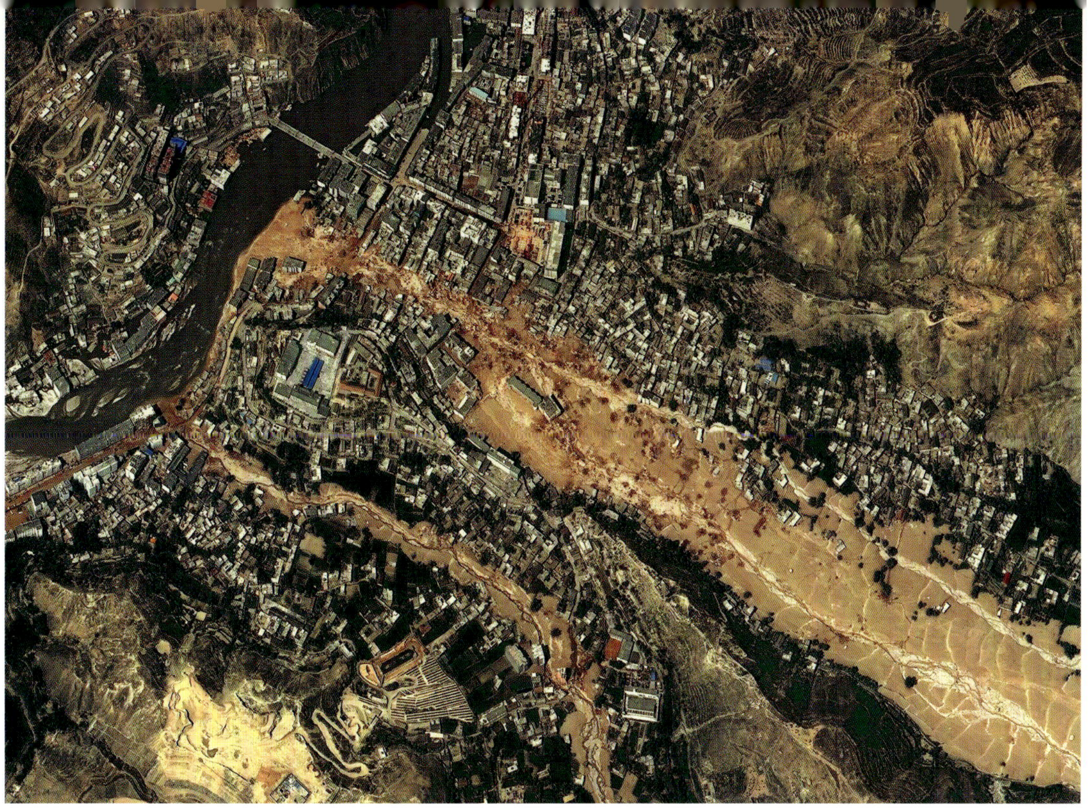

wie eine Decke wirken, die die Erdoberfläche wärmt, oder eher wie ein Spiegel, der das eintreffende Sonnenlicht ins All reflektiert und damit die Erdoberfläche kühlt. Verstärkt zusätzliche Bewölkung die Aufheizung der Erde, oder dämpft sie sie? Auf welche Weise sind die Vorgänge in den Ozeanen und in der Atmosphäre miteinander verknüpft? Auch liegt noch weitgehend im Dunkeln, wie die Sonne, unbestritten ein äußerst wichtiger Klimamotor, das Geschehen auf der Erde beeinflusst. Dafür sind auch die Astronomen verantwortlich, weil die Sonne sich als viel komplizierter herausstellt, als man lange gedacht hat.

Offene Fragen sind keine Schande für die Wissenschaft. Im Gegenteil: Sie inspirieren Forscher und Ingenieure, denn sie zeigen, wo noch Arbeitsbedarf besteht. Dass manche den Medien sehr zugewandte Klimaforscher gerne mal vergessen, auf offene Punkte hinzuweisen, und eine Genauigkeit ihrer Modelle vorgaukeln, die diese gar nicht haben können, mag menschlich verständlich sein. Doch auf dem Weg, das Klima und seine Phänomene zu verstehen, ist die Menschheit erst ein sehr kleines Stück vorangekommen. Das kann kaum verwundern. Unser Klima ist ein hoch komplexes System, das sich im Laufe

as a mirror, reflecting sunlight into space and cooling the ground below. Does excessive additional cloud cover increase global warming or reduce it? What is the connection between the mechanisms at work in the oceans and in the atmosphere? It is also unclear how the Sun, without doubt one of the principal drivers of our weather, influences events on Earth. Astronomers have partly contributed to this confusion, as the Sun has proved to be much more complex than was first thought.

Unanswered questions are no disgrace in science, indeed quite the opposite: they inspire researchers and engineers by showing them where more work is required. That many media-savvy climate researchers happily forget such unanswered questions and claim an accuracy for their models which is quite beyond them is perhaps understandable human nature, but mankind has come only a little way toward understanding the climate and its associated phenomena. This should come as no surprise—our climate is a highly complex system which has developed over eons and it is unlikely that we are going to discover all of its secrets in a few decades. *We need the data from the satellites already in*

vers l'espace pour la rafraîchir ? Le réchauffement de la Terre est-il amplifié ou atténué par une densité nuageuse accrue ? Quelle est la nature de l'interaction des processus océaniques avec les processus atmosphériques ? L'influence du soleil sur les événements climatiques terrestres est tout aussi méconnue, et la contribution des astronomes est indispensable pour comprendre un soleil beaucoup plus complexe qu'il n'y paraît.

La science n'a pas à rougir des questions sans réponse. Au contraire, car elles inspirent chercheurs et ingénieurs en leur indiquant des pistes de travail. La propension de certaines personnalités scientifiques médiatiques à omettre les questions en suspens et à attribuer à leurs modèles une précision irréaliste est peut-être humaine. Pourtant, dans l'exploration du climat et de ses phénomènes, les progrès de l'humanité restent minimes. Cela n'a rien de surprenant. Notre climat est un système extrêmement complexe, qui a évolué au cours des âges. Quelques décennies ne peuvent suffire à l'humanité pour expliquer ces phénomènes. *Nous aurons besoin des données fournies par les satellites en orbite et leurs successeurs pour combler nos lacunes dans l'étude du*

bien dónde y cómo se forman las nubes. Además, los científicos no tienen claro si las nubes actúan como una manta que no deja que escape el calor de la superficie terrestre o, al contrario, como un espejo que refleja al espacio la luz solar, enfriando así la superficie de la Tierra. ¿La nubosidad aumenta el calentamiento del planeta o lo amortigua? ¿De qué manera se interconectan los fenómenos oceánicos y los atmosféricos? Tampoco se conoce con precisión cómo el Sol, que sin duda es un importante motor del clima, influye en los acontecimientos de la Tierra.

Las cuestiones sin resolver no son, en absoluto, una deshonra para la ciencia. Al contrario. Inspiran a investigadores e ingenieros porque muestran en qué campos continúa siendo necesario trabajar. Es humanamente comprensible que algunos investigadores climáticos, demasiado mediáticos, olviden a veces indicar las cuestiones aún abiertas y simulen en sus modelos una exactitud que en realidad no tienen. Pero en el largo camino de comprensión del clima y sus fenómenos, la humanidad sólo ha avanzado un pequeño tramo. Esto no es sorprendente. Nuestro clima es un sistema de gran complejidad, que se ha desarrollado a lo largo de eones. El hombre apenas puede penetrar en algo tan complejo

der Äonen entwickelt hat. So etwas kann der Mensch kaum in wenigen Jahrzehnten durchdringen. *Wir brauchen die Daten von den bereits fliegenden und geplanten Satelliten, um unsere Wissenslücken beim System Erde zu schließen,* bekennt denn auch Johnny A. Johannessen. Er ist am Nansen Umwelt- und Fernerkundungszentrum in Norwegen tätig und gehört zu den Urgesteinen der Erdbeobachter. Bei den ERS- und Envisat-Missionen war er eine treibende Kraft.

Jeder neue Satellit, jedes neue Instrument, jede neue Messung fügt ein paar weitere Puzzle-Teile hinzu. Heute haben die Forscher meist nur ein paar Fragmente in der Hand und mühen sich, sie sinnvoll zusammenzufügen. Um im Puzzle-Bild zu bleiben: Erschwerend kommt hinzu, dass die Forscher weder das Motiv kennen noch wissen, wie viele Teile das Klima-Puzzle insgesamt hat. Um nicht missverstanden zu werden: Es spricht sehr viel dafür, dass große Mengen Kohlendioxids in der Atmosphäre unserer Erde nicht guttun. Eine ganze Reihe von Indizien spricht dafür, dass der beobachtete Klimawandel zu einem mehr oder weniger großen Teil von der Menschheit und ihrem Verhalten auf dieser Erde verursacht wurde. Harte Beweise gibt es in der Naturwissenschaft nicht. Die gibt es nur in der Mathematik. Naturwissenschaftler arbeiten immer mit Wahrscheinlichkeiten.

Andererseits sollten wir nicht warten, bis die letzten von Sachkenntnis unbelasteten Personen, gerade im politischen Leben, einräumen, dass der Wandel des Erdklimas eine Tatsache sei. Denn dann wäre es zu spät: nicht für die Erde und ihre Atmosphäre, aber für uns Menschen. Denn der Planet Erde ist so robust, dass er sich schnell regeneriert. Schnell auf der astronomischen Zeitskala: also innerhalb weniger hunderttausend Jahre. Das ist sehr kurz für einen Planeten, der seit viereinhalb Milliarden Jahren existiert. Denkt man sich die Erde als vierzig Jahre alten Menschen, dann wäre die vom Menschen verursachte Verschmutzung der Atmosphäre eine Art Unwohlsein von wenigen Stunden – eine kleine Magenverstim-

orbit and from the ones on the drawing board to fill the gaps in our knowledge about the Earth's system, admits Johnny A. Johannessen, who works at the Nansen Environmental and Remote Sensing Center in Norway. He was a devoted Earth observer and a driving force behind the ERS and Envisat missions.

Every new satellite, every new instrument, and every new measurement adds a few more pieces to the puzzle, but modern researchers tend to have only a few pieces in hand and are usually struggling to make sense of them. To stay with the puzzle image: further complicating factors include the fact that the researchers have no idea what the picture is or how many pieces the climate puzzle may have in total. Let's be perfectly clear: there is much to suggest that the presence of large amounts of carbon dioxide in the atmosphere of our Earth is not healthy. A large number of indicators point toward mankind and our activity on the Earth being to a greater or lesser degree the cause of this observed climatic change. However, there is no irrefutable proof in the study of environmental science; proof only exists in mathematics—environmental scientists are always working with probabilities.

On the other hand, we should not be waiting until even the least-qualified person, especially in political life, is obliged to admit that changes in the Earth's climate are a reality. That would be too late, not only for the Earth and its atmosphere, but also for us humans. The planet is robust and regenerates quickly, but this 'quickly' is on an astronomical scale—within a few hundred thousand years, a very brief period for a body which has existed for four and half billion years. If you think of the Earth as a 40-year-old, the atmospheric pollution caused by humans is an affliction of only a few hours—a slight stomach 'flu which will soon pass. For mankind, however, a few hundred thousand years is decidedly too long, and the global experiment

système terrestre, admet Johnny A. Johannessen du Centre environnemental et de télédétection de Nansen en Norvège. Vieux routier de l'observation de la Terre, il a joué un rôle moteur dans les missions ERS et Envisat.

Chaque nouveau satellite, chaque nouvel instrument, chaque nouvelle mesure apporte quelques pièces supplémentaires au puzzle. À l'heure actuelle, les chercheurs ne disposent souvent que de quelques fragments qu'ils s'efforcent d'assembler en un tout cohérent. Pour compliquer l'exercice, et poursuivre la métaphore précédente, les chercheurs ne connaissent ni le motif du puzzle climatique, ni le nombre de pièces. Dissipons immédiatement tout malentendu : tout porte à croire que de fortes concentrations de dioxyde de carbone dans l'atmosphère sont nocives à la santé de la Terre. De nombreux indices semblent indiquer que l'humanité et son comportement sur cette Terre sont en partie responsables du changement climatique observé. En sciences naturelles, les preuves irréfutables n'existent pas, elles sont réservées aux mathématiques. C'est pourquoi les scientifiques qui étudient la nature s'appuient toujours sur des probabilités.

D'un autre côté, mieux vaut ne pas attendre que tous les profanes en la matière, en particulier dans la sphère politique, admettent enfin la réalité du changement climatique. Car il sera trop tard : non pas pour la Terre et son atmosphère, mais pour nous autres humains. La planète est tellement robuste qu'elle se régénère rapidement. Rapidement sur l'échelle de temps astronomique, c'est-à-dire en quelques centaines de milliers d'années. Une courte durée pour une planète vieille de quatre milliards et demi d'années. Si la Terre était un homme de quarante ans, la pollution atmosphérique d'origine humaine correspondrait à une indisposition de quelques heures, une petite indigestion vite oubliée. Pour l'humanité en revanche, quelques centaines de milliers d'années sont beaucoup trop longues. C'est pourquoi il est urgent de mettre fin à l'expérimentation

en unas pocas décadas. *Necesitamos los datos de los satélites en vuelo actuales y de los proyectados para cerrar nuestras lagunas de conocimiento sobre el sistema Tierra,* confiesa Johnny A. Johannessen. Trabaja en el Centro Nansen de Medio Ambiente y Teledetección, en Noruega, y forma parte del selecto grupo de los observadores terrestres. Era uno de los impulsores de las misiones ERS y Envisat.

Cada nuevo satélite, cada nuevo instrumento, cada nueva medición, añade un par de piezas más al rompecabezas. La mayoría de las veces, los investigadores sólo tienen en sus manos un par de fragmentos y se afanan por relacionarlos. Siguiendo con la imagen del clima como un rompecabezas, lo que hace más difícil su resolución es que los investigadores no conocen el modelo, ni saben cuántas piezas tiene en total. Para que no se produzcan malentendidos: se habla mucho de que no es bueno que haya grandes cantidades de dióxido de carbono en la atmósfera de nuestra Tierra. Toda una serie de indicios hablan de que el cambio climático observado ha sido provocado, en mayor o menor medida, por la humanidad y su comportamiento en el planeta. Pero en las ciencias naturales no hay pruebas rigurosas. Eso solo ocurre en las matemáticas.

Por otra parte, no deberíamos esperar hasta que la última persona, sobre todo en el mundo político, admita que la transformación del clima terrestre sea un hecho. Entonces sería demasiado tarde. No para la Tierra y su atmósfera, sino para nosotros, los humanos. Porque el planeta Tierra es tan robusto que se regenera rápidamente. Rápido, entendido en la escala astronómica del tiempo, es decir, en un plazo de varios cientos de miles de años. Esto es muy poco para un planeta que existe desde hace cuatro mil quinientos millones de años. Si pensamos en la Tierra como en una persona de cuarenta años de edad, la contaminación atmosférica provocada por la humanidad sería una indisposición de unas pocas horas, una pequeña indigestión que desaparecería rápidamente. Pero para la humanidad, algunos cientos de miles de años son demasiados. Por eso, el experimento global

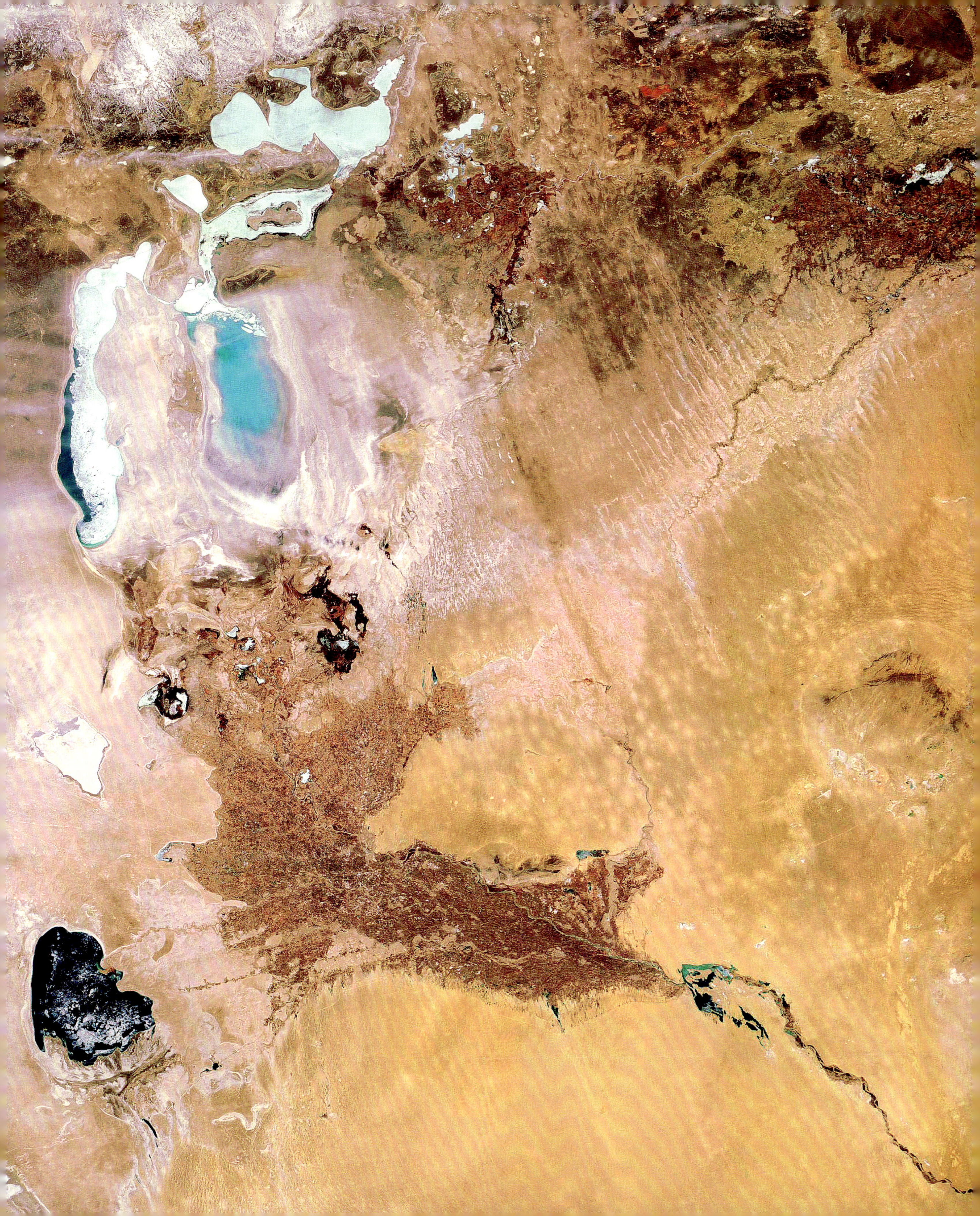

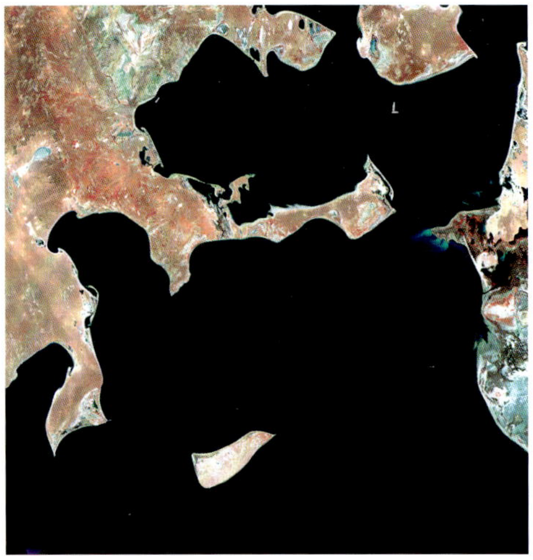 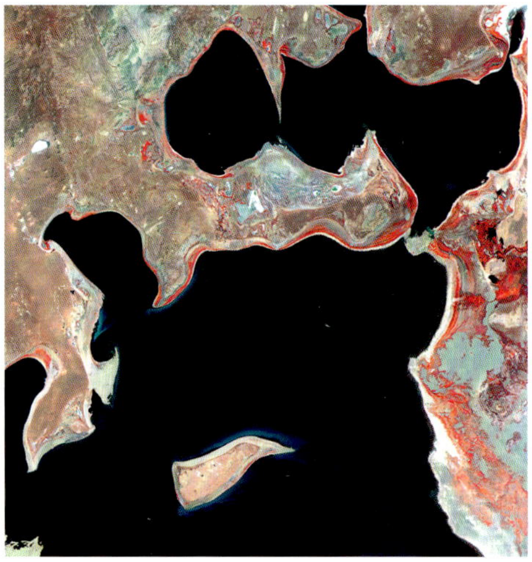 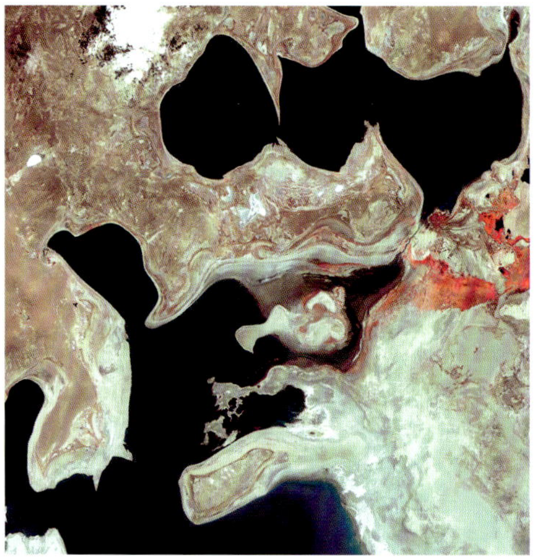

mung, die schnell wieder vergeht. Aber für die Menschheit sind einige hunderttausend Jahre entschieden zu lang. Deshalb sollte das globale Experiment, das wir gerade mit unserer Atmosphäre anstellen, schleunigst ein Ende haben. Geht im Labor ein Experiment daneben, schmeißt man das Reagenzglas einfach weg und fängt von vorne an. Wir haben aber nur genau eine Atmosphäre, die eher uns wegschmeißt als wir sie.

Global verstehen, lokal bewerten

Vielleicht helfen die Bilder und Daten der Erdbeobachtungssatelliten sogar, diese Erkenntnis in die breite Öffentlichkeit zu transportieren. Auf jeden Fall sind sie Forschern die wichtigste Grundlage, um das rätselhafte Klima zu verstehen. Ein Problem für die Forscher sind die oft langen Zyklen, die bei Prozessen in der Atmosphäre eine Rolle spielen. Manche Gaspartikel brauchen viele Jahre, um vom Erdboden in große Höhen zu gelangen, wo sie schweren Schaden anrichten. Eispanzer oder die großen Meeresströmungen haben ein solches Beharrungsvermögen, dass man nicht erwarten kann, schon nach wenigen Jahren eines starken Kohlendioxid-

we are currently carrying out in the atmosphere should be stopped as soon as possible. If an experiment goes wrong in the lab, we can just throw away the test tube and start again with a replacement; but we have only one atmosphere and if we spoil it, we are going to find out that we needed it much more than it needed us.

Think globally, act locally

Images and information from Earth observation satellites may even help to bring this conclusion to the attention of the general public; such information is certainly the most important resource used by scientists to understand our mysterious climate. One problem often presented to researchers is the sometimes lengthy cycle that atmospheric processes can follow: some gas particles may take many years to rise from the surface to the heights where they can cause great damage. Ice sheets and the great ocean currents are sufficiently tenacious that researchers cannot necessarily expect to see any direct consequences even after a large emission of carbon dioxide. It takes a while

générale à laquelle nous soumettons actuellement notre atmosphère. Quand une expérience échoue dans un laboratoire, on jette l'éprouvette et on reprend depuis le début. Mais nous n'avons qu'une atmosphère, et c'est elle qui risque littéralement de nous jeter.

Comprendre globalement, analyser localement

Les images fournies par les satellites contribueront peut-être à sensibiliser l'opinion publique. Les chercheurs, qui sont le pivot fondamental des études climatiques, sont souvent confrontés aux longs cycles qui influencent les processus atmosphériques. Il faut de nombreuses années à certaines particules de gaz pour s'échapper du sol et s'élever dans les airs où elles causent d'importants dégâts. Les calottes glaciaires et les grands courants marins ont une telle inertie que les effets directs de fortes émissions de dioxyde de carbone ne se font sentir que bien des années plus tard. Les océans se réchauffent lentement. Ce qui représente un risque, mais aussi une chance. On pourrait ainsi constater les effets alors qu'il est trop tard. Mais l'on pourrait aussi, sur la base de don-

al que estamos sometiendo a nuestra atmósfera debería finalizar de inmediato. Si un experimento fracasa en el laboratorio, simplemente se descarta el tubo de ensayo y se comienza de nuevo desde el principio. Pero nosotros únicamente tenemos una atmósfera y sería ella la que nos descartase a nosotros antes que nosotros a ella.

Comprensión global, evaluación local

Quizá las imágenes y los datos de los satélites de observación terrestre ayuden a trasladar estos conocimientos al público en general. En todo caso, son el fundamento más importante para ayudar a los investigadores a entender el misterio del clima. Otro problema añadido para los científicos son los largos ciclos que desempeñan un papel en los procesos atmosféricos. Algunas partículas de gas necesitan muchos años para llegar desde el suelo terrestre hasta las grandes alturas donde causan daños graves. Los casquetes de hielo o las grandes corrientes marinas tienen tal inercia que, las consecuencias directas de una intensa emisión de dióxido de carbono, solo se ven algunos años después. Los océanos tardan bastante en calen-

Seite 147: Der Aralsee und die kasachische Steppe 2010
Linke Seite: Der Nordrand des Sees 1973, 1987 und 2000
Rechts: Wie der Aralsee in den Jahren 2000 bis 2009 austrocknete. Die Küstenlinie von 1960 ist eingezeichnet (Envisat, Landsat, Terra)

Page 147: The Aral Sea and the Kazakh Steppe, 2010
Left: The northern shores of the sea in 1973, 1987 and 2000
Right: The extent to which the Aral Sea has dried out between 2000 and 2009, with an indication of the 1960 shoreline (Envisat, Landsat, Terra)

Page 147 : la mer d'Aral et les steppes kazakhes en 2010.
Page de gauche : partie nord de la mer en 1973, 1987 et 2000.
À droite : assèchement de la mer d'Aral entre 2000 et 2009. On distingue les côtes de 1960 (Envisat, Landsat, Terra)

Página 147: El mar de Aral y la estepa kazaja en 2010.
Página izda: La orilla norte del mar en 1973, 1987 y 2000.
Dcha.: Desecación del mar de Aral entre los años 2000 y 2009. Esta marcada la línea de la costa de 1960 (Envisat, Landsat, Terra)

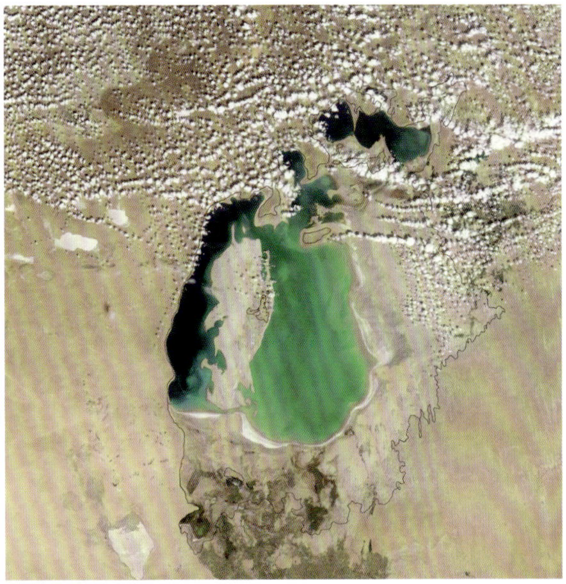

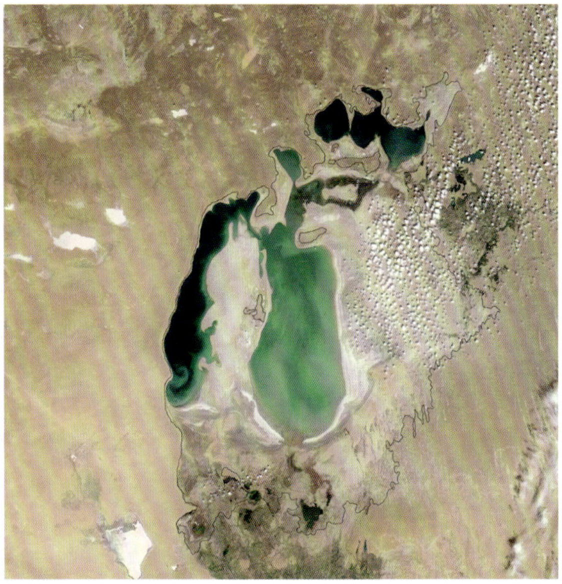

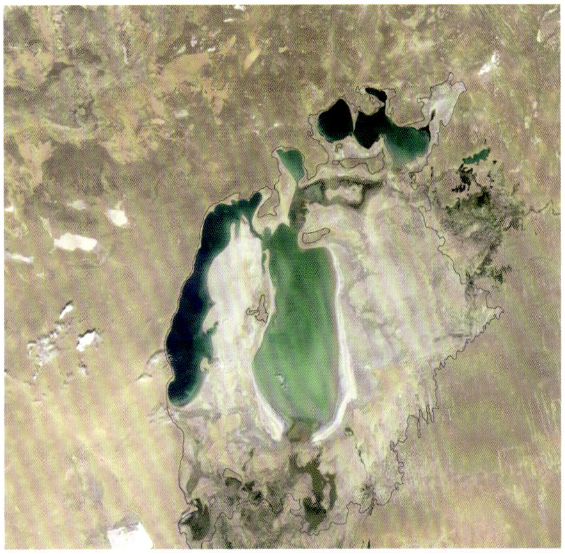

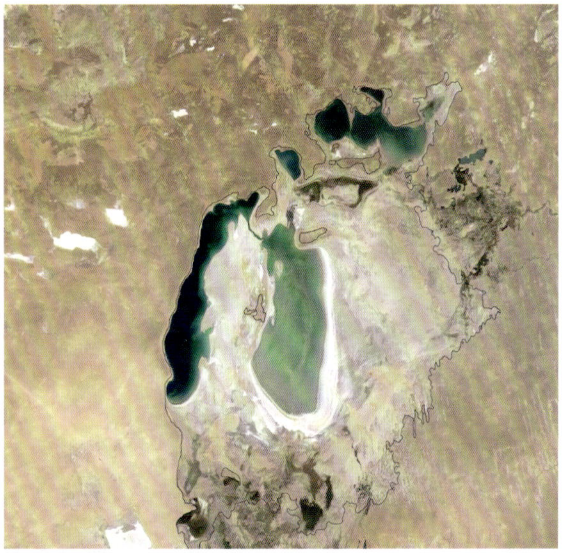

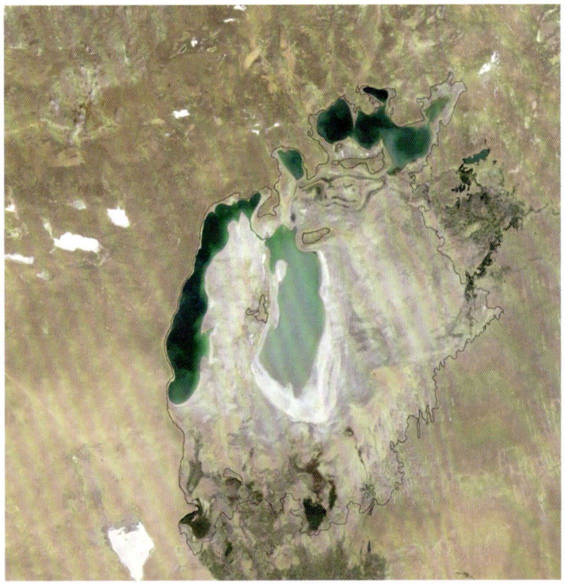

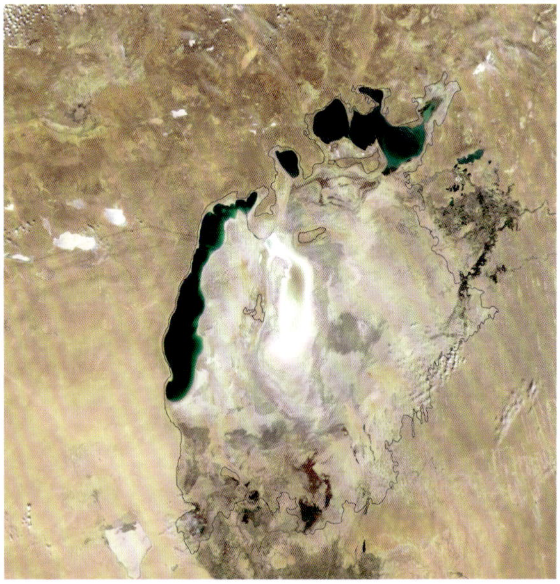

Seite 150/151: Die Neyriz-Salzseen im Süden Irans (Alos)

Pages 150–1: The Neyriz salt lakes in southern Iran (Alos)

Pages 150 et 151 : le lac salée de Neyriz dans le sud de l'Iran (Alos)

Páginas 150/151: Lagos salados de Neyriz, en el sur de Irán (Alos)

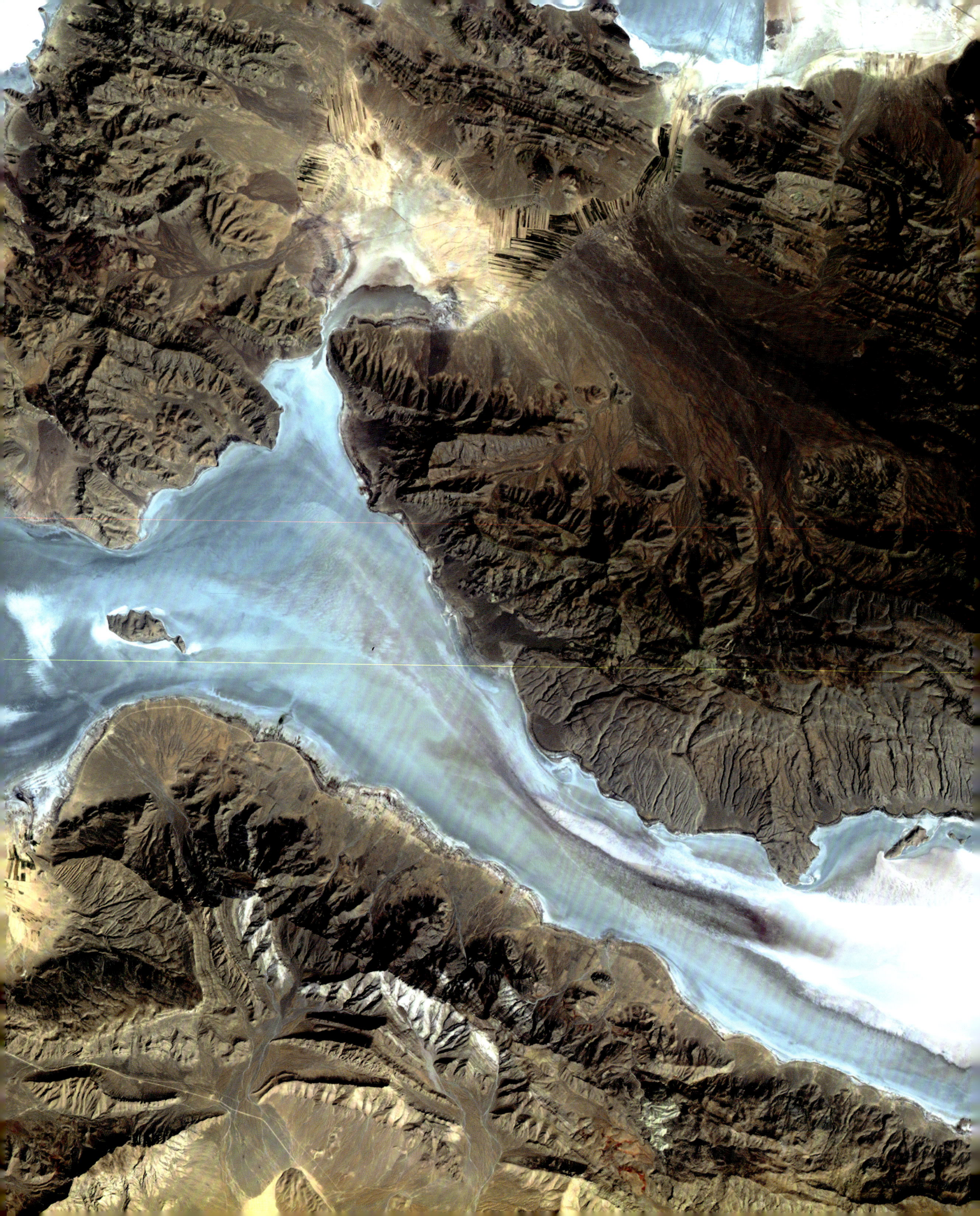

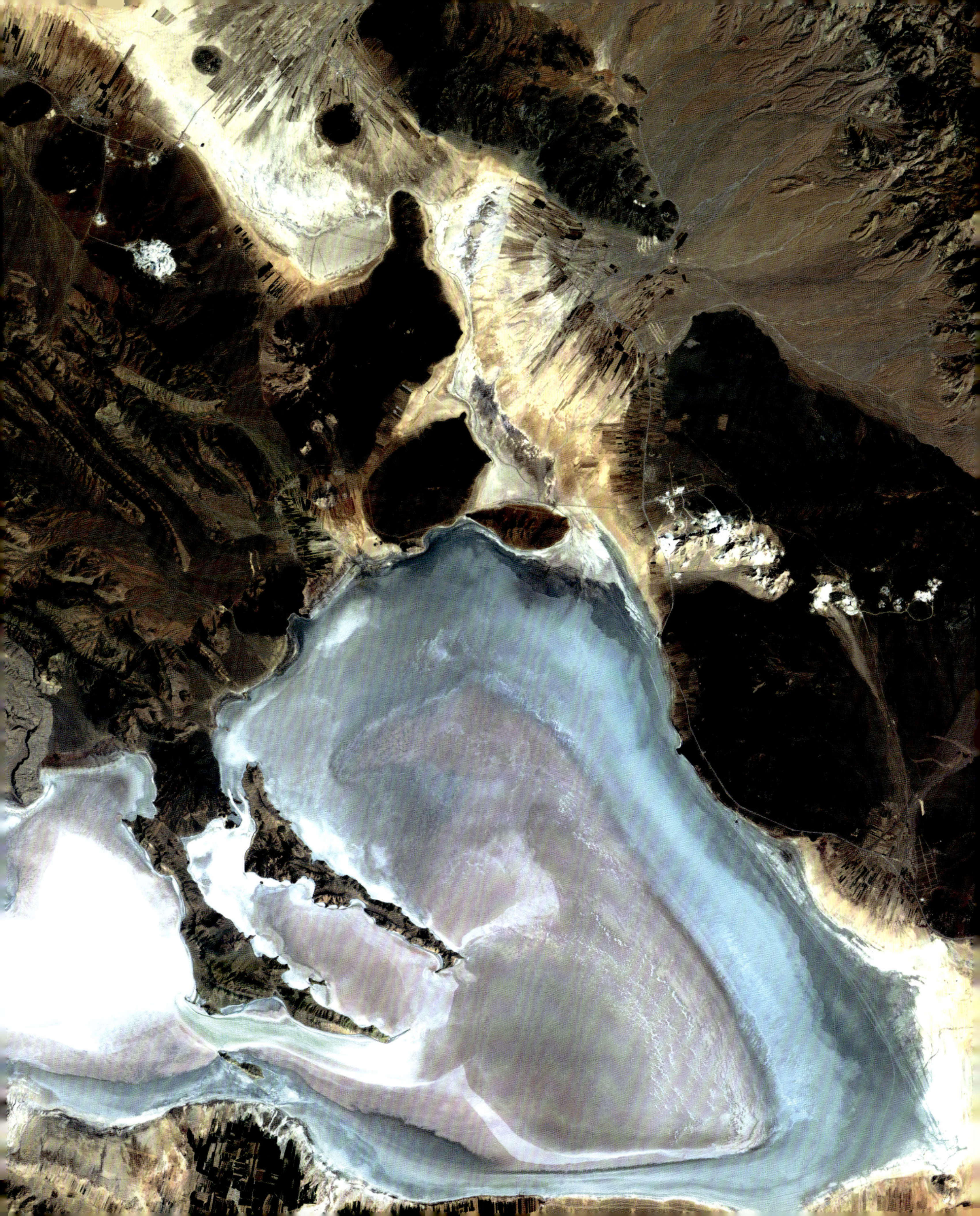

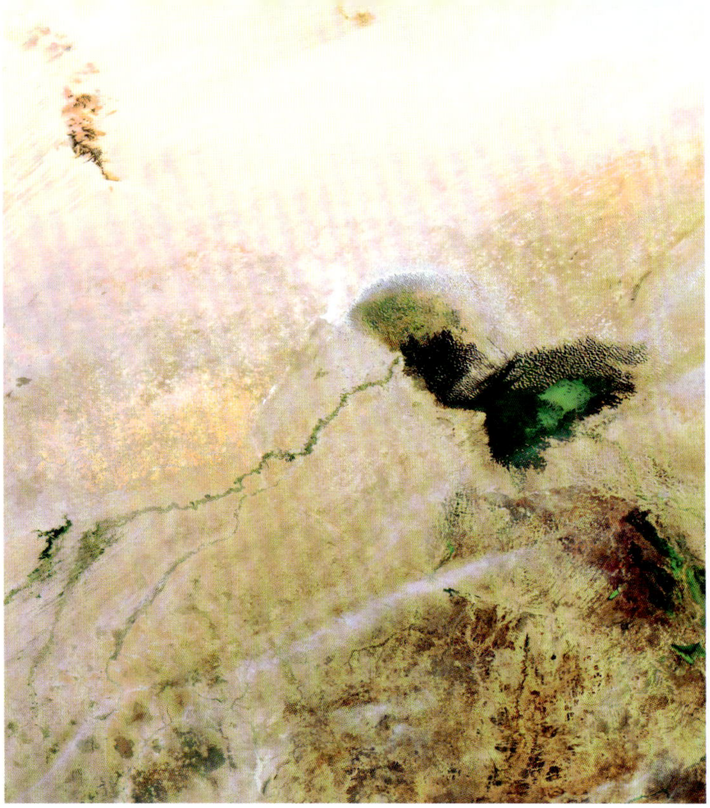

Auch der Tschadsee in Afrika trocknet durch Wasserentnahme aus den Zuflüssen immer weiter aus (Terra)

Lake Chad in Africa is also drying out as water is taken from its tributaries (Terra)

Le lac Tchad, en Afrique, s'assèche également en raison des ponctions d'eau sur ses affluents (Terra)

También el lago Chad de África se sigue desecando por la extracción de agua de sus afluentes (Terra)

Rechte Seite: »Land des Schreckens« im Süden Algeriens ohne Wasser und Vegetation. Bildausschnitt 50 km breit (Alos)

Right: The "Land of Fear" in southern Algeria, bereft of water or vegetation. The detail shown is 30 miles (50 km) wide (Alos)

Page de droite : « sinistre contrée » sans eau ni végétation dans le sud de l'Algérie. Image couvrant une largeur de 50 km (Alos)

Página dcha.: «Dominios del terror», en el sur de Argelia, sin agua ni vegetación. Encuadre de 50 km de ancho (Alos)

Ausstoßes direkte Folgen zu sehen. Es dauert, bis sich ganze Ozeane erwärmen. Das ist Risiko und Chance zugleich. Man könnte Effekte erst dann bemerken, wenn es zu spät ist. Aber man könnte auch mit den richtigen Erkenntnissen und konsequenten Entscheidungen gefährliche Klimafaktoren korrigieren, bevor globaler Schaden angerichtet ist, den zu reparieren die Menschheit kaum imstande wäre.

Die Eisflächen auf der Erde sind so eine Art Langzeitsensor. Gletscher verändern sich nicht binnen eines Jahres. Sie sind gute Indikatoren für langfristige Trends. Duncan Wingham, Physiker am University College in London, macht sich mit seinem CryoSat diesen Effekt zu Nutze: *Eis ist sehr wichtig für unser Klima – und umgekehrt. Wir Menschen sind dabei, mit der globalen Erwärmung viel Eis zu zerstören. Unser Satellit soll messen, wie schnell das vor sich geht.* Mit CryoSat erfahren die Forscher endlich, wo es Eis gibt, wie dick die auf dem Meer treibenden Eisschollen sind und ob die Eismassen an Land dicker oder dünner werden. Statt einfacher Fotos liefert CryoSat dreidimensionale Eisdaten, erklärt Duncan Wingham: *Auch wenn man die Polargebiete für weit entfernt hält: Verändert sich das Eis im arktischen*

for a whole ocean to warm up. This is both a danger and an opportunity: the effects might only be noticeable when it is too late, but with the right knowledge and determined intervention it might be possible to correct hazardous climatic factors before the sort of global damage ensues that mankind is hardly in a position to repair.

The Earth's ice caps are a kind of large-scale sensor. Glaciers change very little from one year to the next and are a good indicator of long-term trends. Duncan Wingham, a physicist at University College, London, makes use of this effect with his CryoSat: *Ice is very important for our climate—and vice versa. With global warming, we humans are destroying large amounts of ice and our satellite is intended to measure how fast this is happening.* CryoSat will finally allow scientists to find out the location of the ice, the thickness of the floes drifting in the sea, and whether the ice deposits on land are getting thicker or thinner. Instead of basic photos, CryoSat transmits 3-D ice data, as Duncan Wingham explains: *However far distant you might imagine the poles to be, if the ice changes in the Arctic Ocean it causes a chain reaction that alters the North Atlantic,*

nées exactes et de décisions cohérentes, corriger les effets néfastes des changements climatiques avant de causer des dégâts planétaires que nous ne saurions réparer.

Les surfaces glacées de la Terre agissent comme des senseurs longue durée. Les glaciers qui évoluent peu au cours d'une année sont de bons indicateurs de tendances à long terme. Duncan Wingham, un physicien de l'University College de Londres, tire parti de cet effet grâce à CryoSat : *La glace est essentielle pour notre climat, et réciproquement. Avec le réchauffement climatique, nous sommes en voie de détruire une grande quantité de glace. Notre satellite mesurera la vitesse à laquelle se produit cette évolution.* CryoSat révèle enfin aux scientifiques l'emplacement des glaces, l'épaisseur des couches de glace flottante et si les masses de glace terrestre augmentent ou diminuent. CryoSat ne fournit pas de simples photos, mais des données glaciaires tridimensionnelles, explique Duncan Wingham : *Même si les régions polaires semblent très éloignées, toute modification des glaces de l'océan Arctique entraîne une réaction en chaîne qui se répercute jusque dans l'Atlantique Nord et peut influencer les conditions météorologiques sur les latitudes tempérées*

tarse. Esto es, al mismo tiempo, un riesgo y una oportunidad. Las consecuencias únicamente se podrían observar cuando ya fuese demasiado tarde. Pero con conocimientos adecuados y decisiones coherentes, también se podrían corregir factores climáticos peligrosos antes de que causasen daños globales cuya reparación fuese casi imposible para la humanidad.

Las superficies heladas de la Tierra son una especie de sensores a largo plazo. Los glaciares no cambian en un año. Por eso son buenos indicadores de las tendencias a largo plazo. Duncan Wingham, físico en el University College de Londres, saca provecho de ese efecto con su CryoSat: *El hielo es muy importante para nuestro clima... y viceversa. Los seres humanos participamos con el calentamiento global en la destrucción de mucho hielo. Nuestro satélite debe medir la rapidez con la que esto sucede.* Con CryoSat, los investigadores averiguan dónde hay hielo, el grosor de los témpanos que flotan en el mar y si han aumentado o disminuido las masas terrestres de hielo. En lugar de simples fotografías, CryoSat proporciona datos tridimensionales de hielo, aclara Duncan Wingham: *Aunque parezca que las regiones polares están demasiado lejos, cuando el hielo cambia en el océano Ártico, al final termina por cambiar*

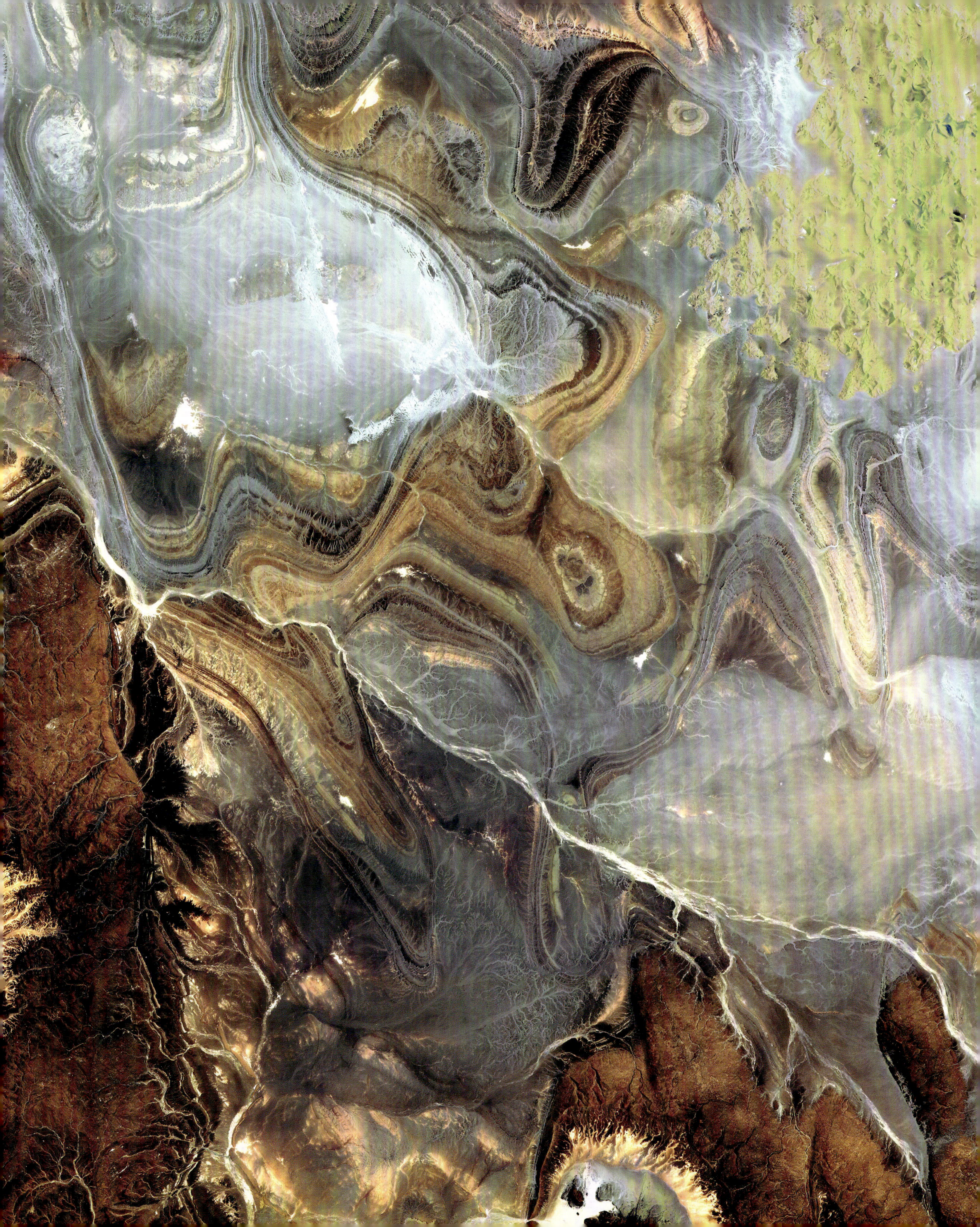

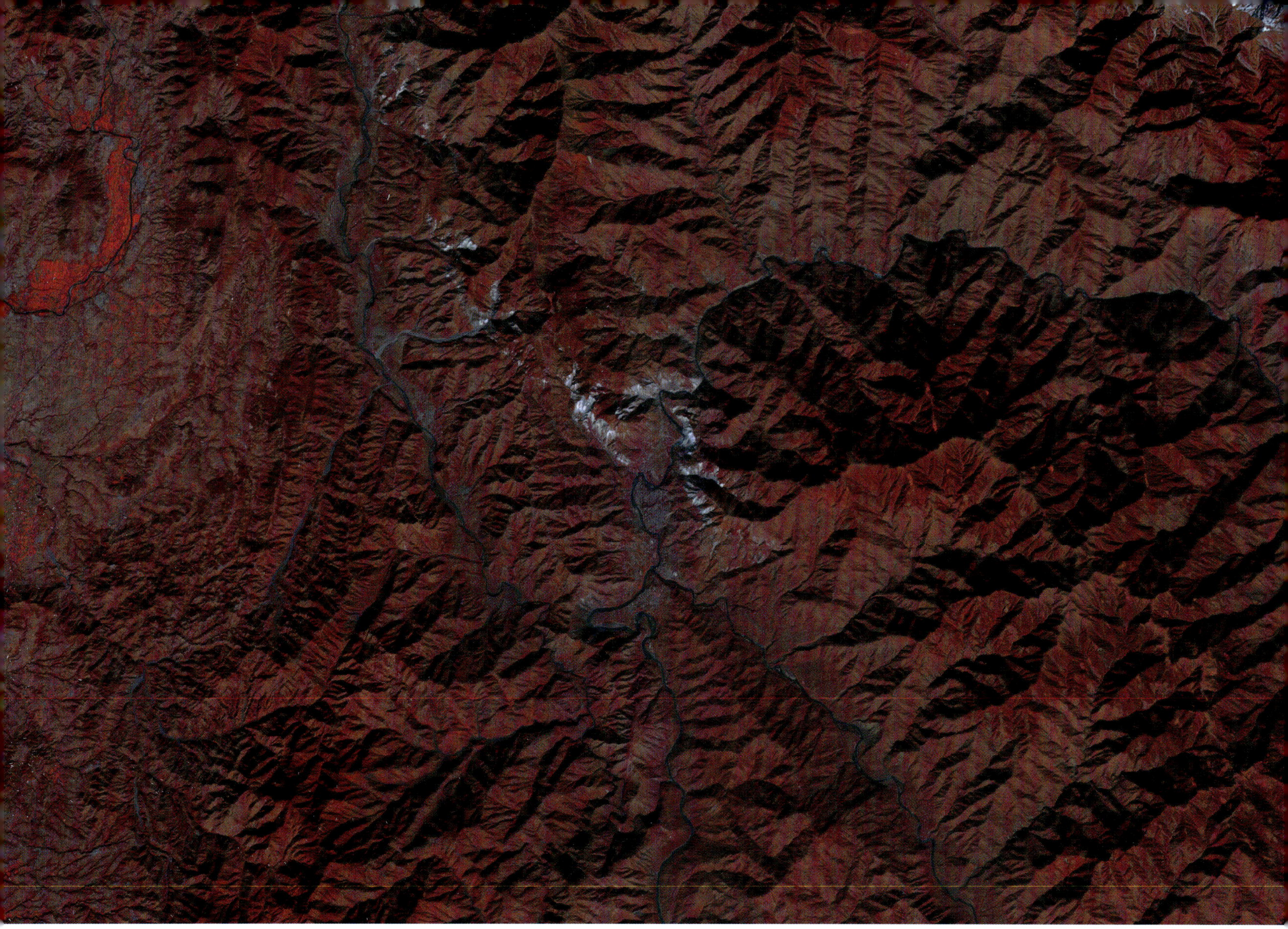

Oben, rechts und rechte Seite: Erdrutsch in Pakistan 2005. Graue Hänge sind abgerutscht (oben, Terra). Der Ort vor und nach dem Unglück (Ikonos)

All images: The Pakistani earthquake of 2005. The slopes in gray have slipped (above, Terra). The location before (near right) and after (far right) the disaster (Ikonos)

Ci-dessus, à droite et page de droite : glissement de terrain au Pakistan en 2005. Les versants en gris se sont affaissés (ci-dessus, Terra). Les lieux avant et après la catastrophe (Ikonos)

Arriba, dcha. y página dcha.: Corrimiento de tierras (laderas en gris) de Pakistán, en 2005 (arriba, Terra). El lugar, antes y después de la catástrofe (Ikonos)

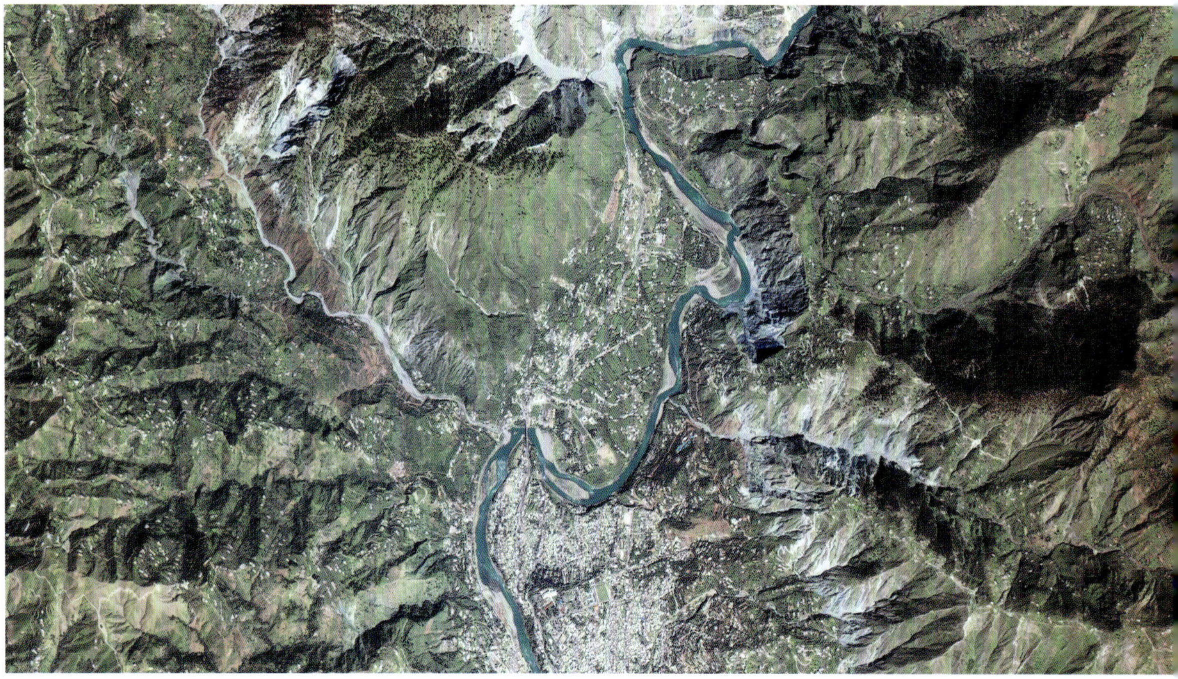

Ozean, dann ändert sich in einer Kettenreaktion schließlich auch der Nordatlantik – und das kann unser Wetter in gemäßigten Breiten wie London, Paris oder Berlin beeinflussen. Das Eis in den Polargebieten spielt eine zentrale Rolle für die Stabilität des Erdklimas. Es steuert den Wärmehaushalt der Erde und treibt die globalen Meeresströmungen an.

Klima ist alles – alles ist Klima

Verändert sich das Eis im Nordpolarmeer, besteht die Gefahr, dass die weit nach Norden reichenden Ausläufer des Golfstroms deutlich schwächer werden, was verheerende Folgen für das Wetter in Mitteleuropa hätte. Dann könnte es in Toulouse und Stuttgart bald so aussehen wie in Anchorage und Irkutsk. Zwar liefert CryoSat den Forschern erstmals konkreten Einblick in die Eisbildung und Eisschmelze in den Polargebieten. Doch in bestenfalls fünf Jahren Projektdauer lassen sich langfristige Klimatrends im Eis kaum beobachten. Da setzt Duncan Wingham auf die Kreativität der Kollegen: *Nachdem man erstmals den Ozongehalt in der Atmosphäre vom All aus gemessen und erkannt hatte, wie wichtig diese*

and that can affect our weather in temperate latitudes—in London, Paris, or Berlin. The ice around the poles is of central importance to the stability of the Earth's climate, controlling the Earth's energy budget and propelling ocean currents around the globe.

Climate is all—and everything is climate

If there is a change in the ice in the Arctic Ocean there is a danger that the furthest reaches of the Gulf Stream, which extend far into the north, will be weakened, and this would have devastating consequences for Central Europe—Toulouse and Stuttgart could end up looking like Anchorage or Irkutsk. Although CryoSat is the first to provide researchers with a concrete notion of ice formation and loss around the poles, even in a best-case scenario it would take a project lasting five years to detect long-term climate trends in the ice. Here, Duncan Wingham is relying on his colleagues' creativity: *After atmospheric ozone levels were first measured from space and it was recognized how important such data is, lots of satellites suddenly appeared with the ability to*

de Londres, Paris ou Berlin. Les glaces polaires jouent un rôle majeur dans la stabilité du climat terrestre. Elles assurent la régulation thermique de la Terre et actionnent les courants océaniques.

Tout est question de climat

L'altération des glaces dans l'océan Arctique risque d'affaiblir sensiblement la dérive Nord Atlantique qui prolonge le Gulf Stream vers le nord, ce qui aurait des conséquences dramatiques sur le climat de l'Europe centrale. Le temps à Toulouse et Stuttgart pourrait alors ressembler à celui d'Anchorage et d'Irkoutsk. CryoSat fournit certes aux scientifiques les premiers éléments concrets sur le processus de formation et de fonte des glaces, mais les cinq années de mission ne suffiront pas pour étudier les évolutions climatiques à long terme. C'est pourquoi Duncan Wingham mise sur la créativité de ses collègues : *Quand les premières estimations de la concentration de l'ozone dans l'atmosphère ont révélé l'importance de telles observations, les satellites capables de mesurer l'ozone ont soudain proliféré. Je suis persuadé qu'il se trouvera quelqu'un pour*

todo el Atlántico Norte en una reacción en cadena y eso puede influir en la meteorología de latitudes templadas, como Londres, París o Berlín. El hielo de las regiones polares juega un papel central en la estabilidad del clima terrestre. Controla el balance térmico de la Tierra y estimula la circulación global de las corrientes marinas.

El clima es todo. Todo es clima

Si se producen cambios en el hielo del océano Ártico, el peligro consiste en que los ramales de la corriente del Golfo que se extienden hasta las regiones del norte, se debiliten claramente, lo que podría traer consecuencias desastrosas para las condiciones meteorológicas centroeuropeas. Toulouse y Stuttgart podrían tener entonces un aspecto similar al de Anchorage e Irkutsk. Si bien CryoSat proporciona por primera vez a los investigadores conocimientos concretos sobre la formación y la fusión del hielo en las regiones polares, las tendencias climáticas a largo plazo apenas se podrían observar en el hielo con una duración de proyecto de, en el mejor de los casos, cinco años. Duncan Wingham apuesta aquí por la creatividad de los colegas: *Tras medir por primera vez desde el espacio la*

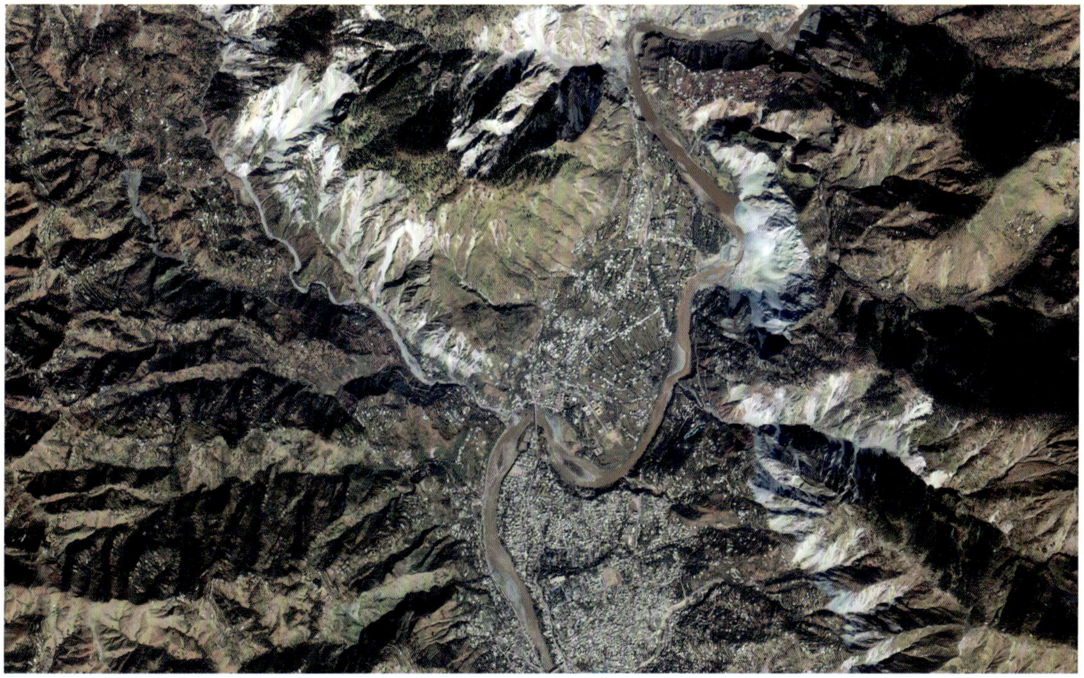

Seite 156/157: Port-au-Prince, Haiti, wenige Tage nach dem Erdbeben 2010. Hilfskräfte sind im Einsatz (GeoEye-1)

Pages 156–7: Port-au-Prince, Haiti, a few days after the 2010 earthquake. Aid workers have already arrived (GeoEye-1)

Pages 156 et 157 : Port-au-Prince, en Haïti, quelques jours après le séisme de 2010. Les équipes de secours sont sur les lieux (GeoEye-1)

Página siguiente 156/157: Puerto Príncipe, Haití, pocos días después del terremoto de 2010. Los equipos de socorro están operativos (GeoEye-1)

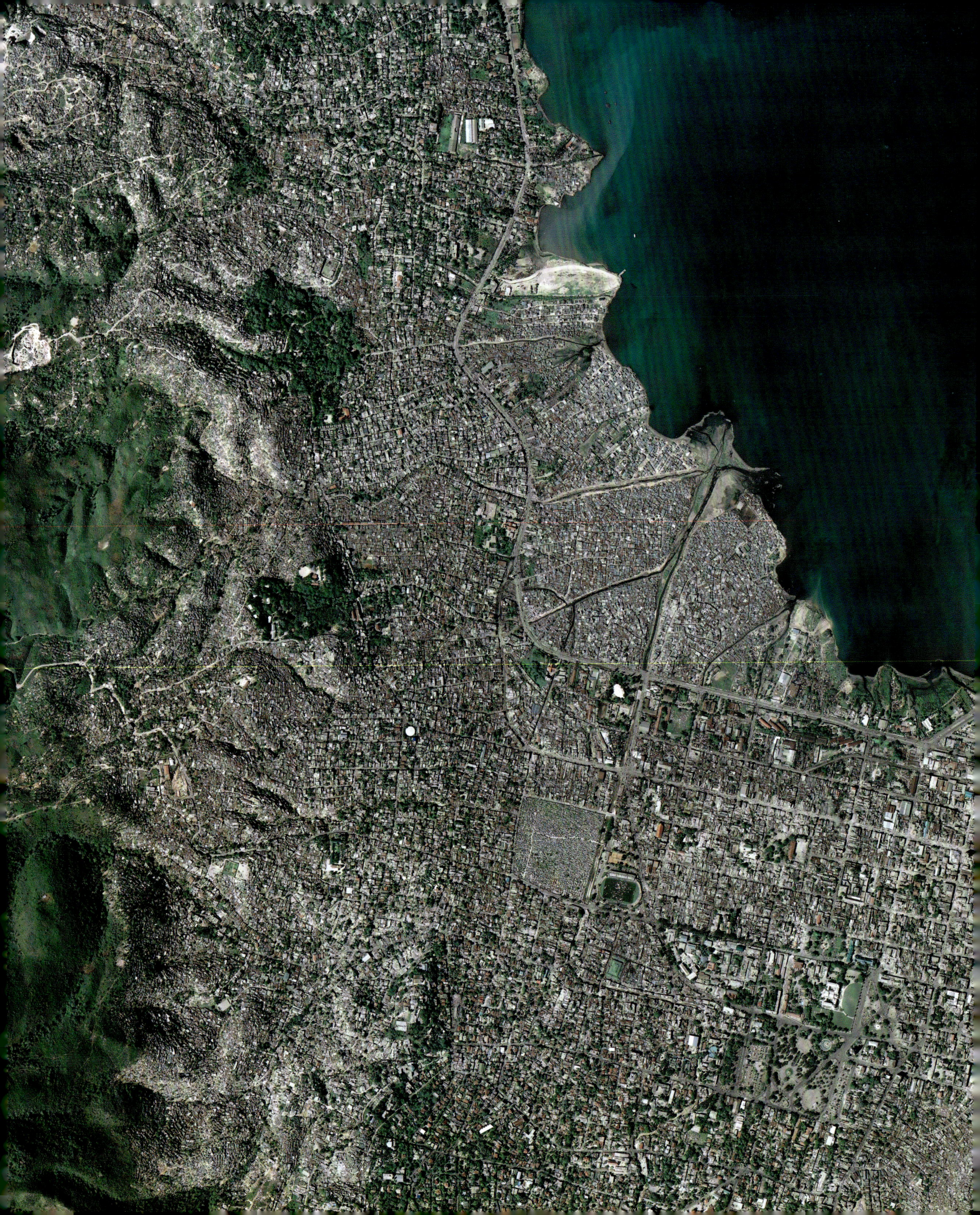

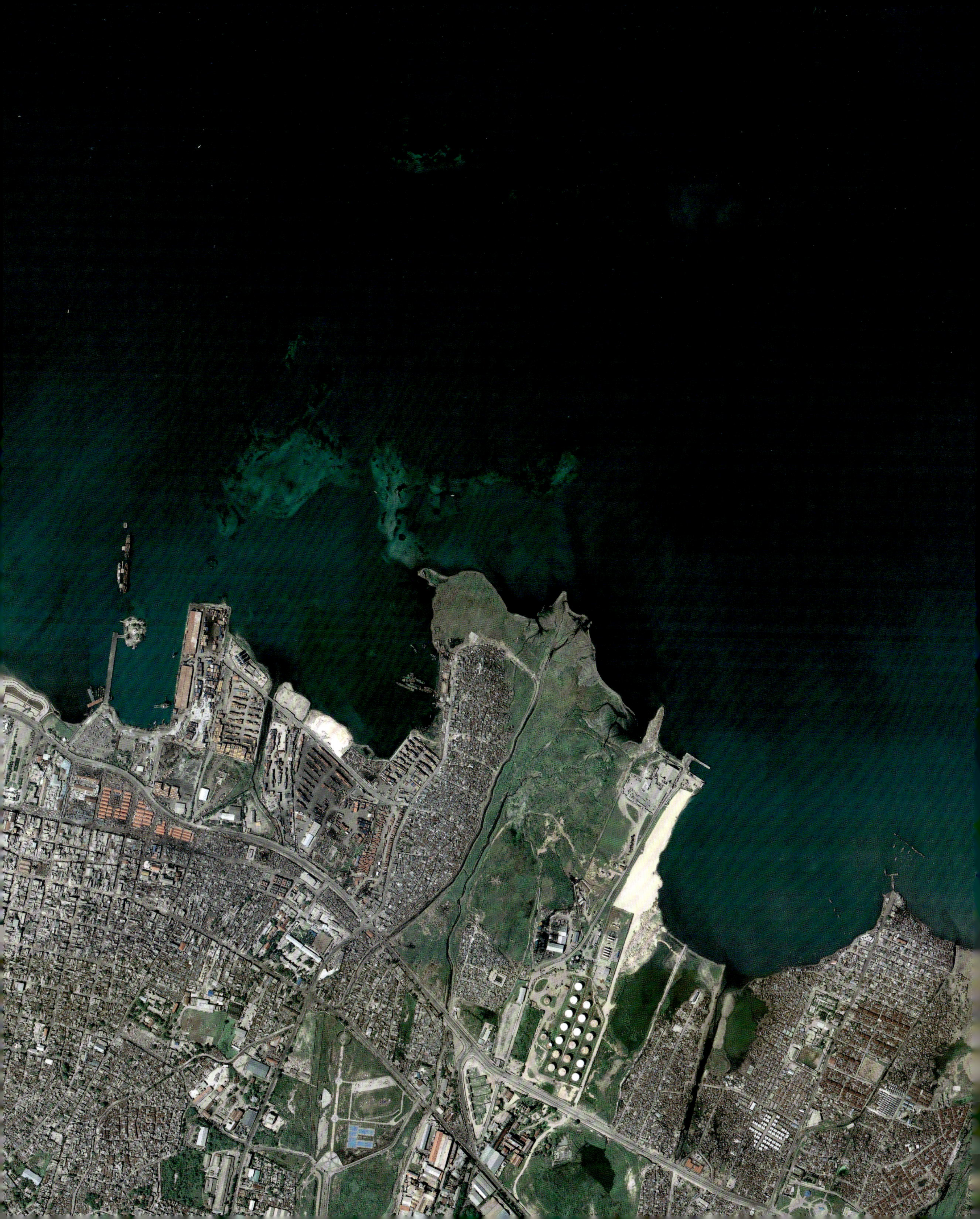

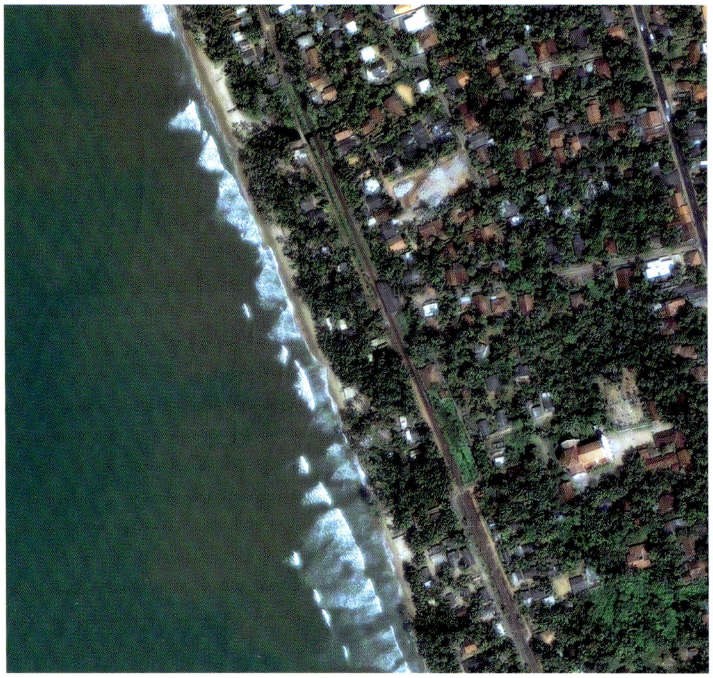

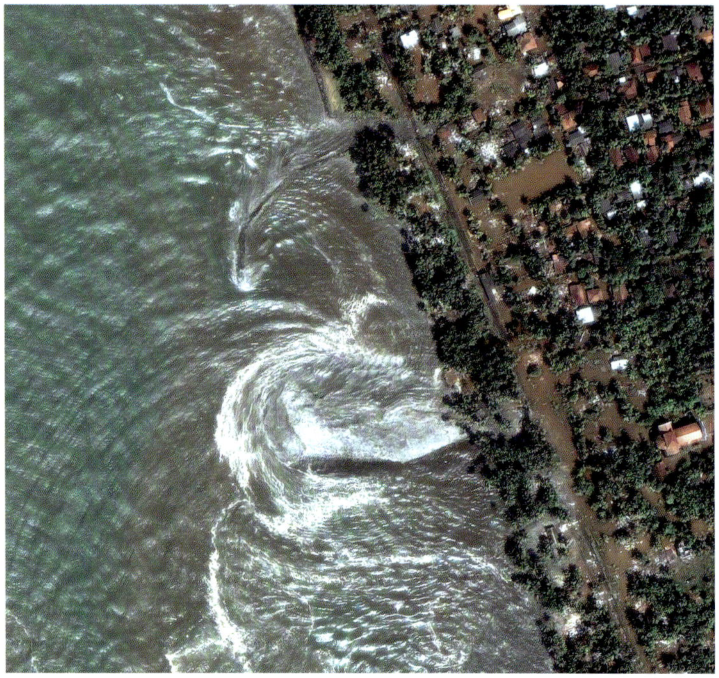

Daten sind, gab es plötzlich ganz viele Satelliten, die Ozon messen konnten. Ich bin sehr zuversichtlich, dass später jemand anderes eine weitere Eismission fliegt. Während phasenweise die Forscher fast manisch das Ozon gemessen haben, fand das Eis bisher vergleichsweise geringe Beachtung. Doch jetzt ist den Wissenschaftler klar, dass nur wer das Eis versteht, vielleicht eines Tages auch das Rätsel des Klimas lösen kann.

Auch die ERS-Satelliten und Envisat haben sich zum Teil dem Eis gewidmet. Es zeigt sich, dass die Gletscher des grönländischen Eispanzers immer schneller fließen. Dieses Phänomen bedarf dringend weiterer Beobachtung, denn im schlimmsten Falle heißt das, dass viel schneller viel mehr Eis vom Festland abrutscht und im Meer schmilzt. Dann wären die bisherigen Schätzungen zum Anstieg des Meeresspiegels deutlich zu niedrig. Messungen von Satelliten im Norden Grönlands sind keineswegs unwichtig oder allenfalls für die Inuit-Bevölkerung vor Ort relevant. Wenn Gletscher schneller fließen, bedroht das unmittelbar Städte wie Hamburg, Amsterdam oder Marseille. Egal welche noch so abgelegene Region die Satelliten im All ins Visier nehmen: Erdbeobachtung geht immer alle an. Denn wir leben auf *einer* Erde.

measure ozone. I am quite confident that someone else will launch another ice mission. At just the time when it seemed scientists were becoming almost manic in their attempts to measure ozone, comparatively little attention was being paid to ice; it is now clear to the research community that only those who understand ice may one day be in a position to solve the enigma of the climate.

Even the ERS satellites and Envisat have turned their attention to ice. They have discovered that the glaciers of the Greenland ice sheet are flowing ever faster, and it is imperative that further observations are made of this phenomenon. In the worst case, it would mean that more ice is slipping from the land and melting into the sea at a greater speed than before; this would mean that previous estimates of rises in sea level were entirely too conservative. Measurements taken from satellites over Northern Greenland may be of use here and will certainly benefit the local Inuit population. When glaciers flow faster, this presents a direct threat to cities like Hamburg, Amsterdam, and Marseilles. Earth observation is a matter for everyone, whichever remote region the orbiting satellites happen to be looking at; we all live on the same Earth.

lancer une autre mission glaciaire. Tandis que les scientifiques mesuraient l'ozone avec une frénésie par moment quasi obsessionnelle, l'attention portée à la glace était minime. Depuis, les scientifiques ont compris que seule l'étude des glaces permettra peut-être un jour de résoudre l'énigme climatique.

Les satellites ERS et Envisat se sont en partie intéressés à la glace. Il apparaît que les grands glaciers du Groenland glissent de plus en plus rapidement vers la mer. Ce phénomène requiert impérativement une étude approfondie, car il pourrait, dans le pire des cas, accélérer le glissement d'importantes masses de glace du continent vers la mer, où elles fondent. Lors d'un tel scénario, il faudrait revoir à la hausse les estimations de la montée du niveau des océans. Les mesures satellitaires réalisées au nord du Groenland ne sont en aucun cas négligeables, et ne sont pas seulement utiles pour la population inuit. Si le glissement d'un glacier s'accélère, il met en péril des villes comme Hambourg, Amsterdam ou Marseille. Quelle que soit la région sondée par les satellites en orbite, l'observation de la Terre est toujours l'affaire de tous. Car nous habitons tous une seule et même Terre.

concentración de ozono en la atmósfera y reconocer lo importantes que son estos datos, de repente se hicieron muchos satélites que podían medir el ozono. Confío en que no se tarde en lanzar otra misión dedicada a los hielos. Mientras los investigadores han medido periódicamente el ozono, el hielo ha suscitado hasta ahora mucha menos atención. Pero en la actualidad, los científicos tienen claro que sólo el que comprenda el hielo, quizá pueda resolver algún día el enigma del clima.

Los satélites ERS y Envisat también se han consagrado, en parte, al estudio del hielo. Hay pruebas de que los glaciares del casquete helado de Groenlandia discurren cada vez más rápido. Este fenómeno requiere urgentemente de otras observaciones porque, en el peor de los casos, esto significa que cada vez se desliza con más rapidez una mayor cantidad de hielo desde tierra firme y se derrite en el mar. Entonces, las estimaciones anteriores sobre la subida del nivel del mar serían insuficientes. Las mediciones de los satélites en el norte de Groenlandia no carecen de importancia o relevancia para la población esquimal local. Si los glaciares se deslizaran con más rapidez, se verían directamente amenazadas ciudades como Hamburgo, Ámsterdam o Marsella. No importa lo aisladas que estén las regiones.

Tsunami-Katastrophe,
26. Dezember 2004
Linke Seite: Kalutara in Sri Lanka vor und während des Tsunamis (QuickBird)
Rechts: Küste von Phang-Nga in Thailand vor und nach dem Tsunami. Intakte Vegetation erscheint rot. Graue Gebiete sind verwüstet (Terra)

The tsunami of December 26, 2004.
Left: Kalutara in Sri Lanka before (far left) and during (near left) the tsunami (QuickBird)
Right: The coast of Phang Nga in Thailand before (near right) and after (far right) the tsunami. Intact vegetation is shown in red. The gray areas have been devastated (Terra)

Tsunami du 26 décembre 2004.
Page de gauche : Kalutara au Sri Lanka, avant et pendant le tsunami (QuickBird)
À droite : plage de Phang Nga en Thaïlande avant et après le tsunami. En rouge, la végétation intacte. Les zones grises sont dévastées (Terra)

Tsunami: 26 de diciembre de 2004.
Página izda.: Kalutara en Sri Lanka, antes y después del tsunami (QuickBird).
Dcha.: Costa de Phang Nga, en Tailandia, antes y después del tsunami. En rojo, la vegetación intacta. En gris, las zonas devastadas (Terra)

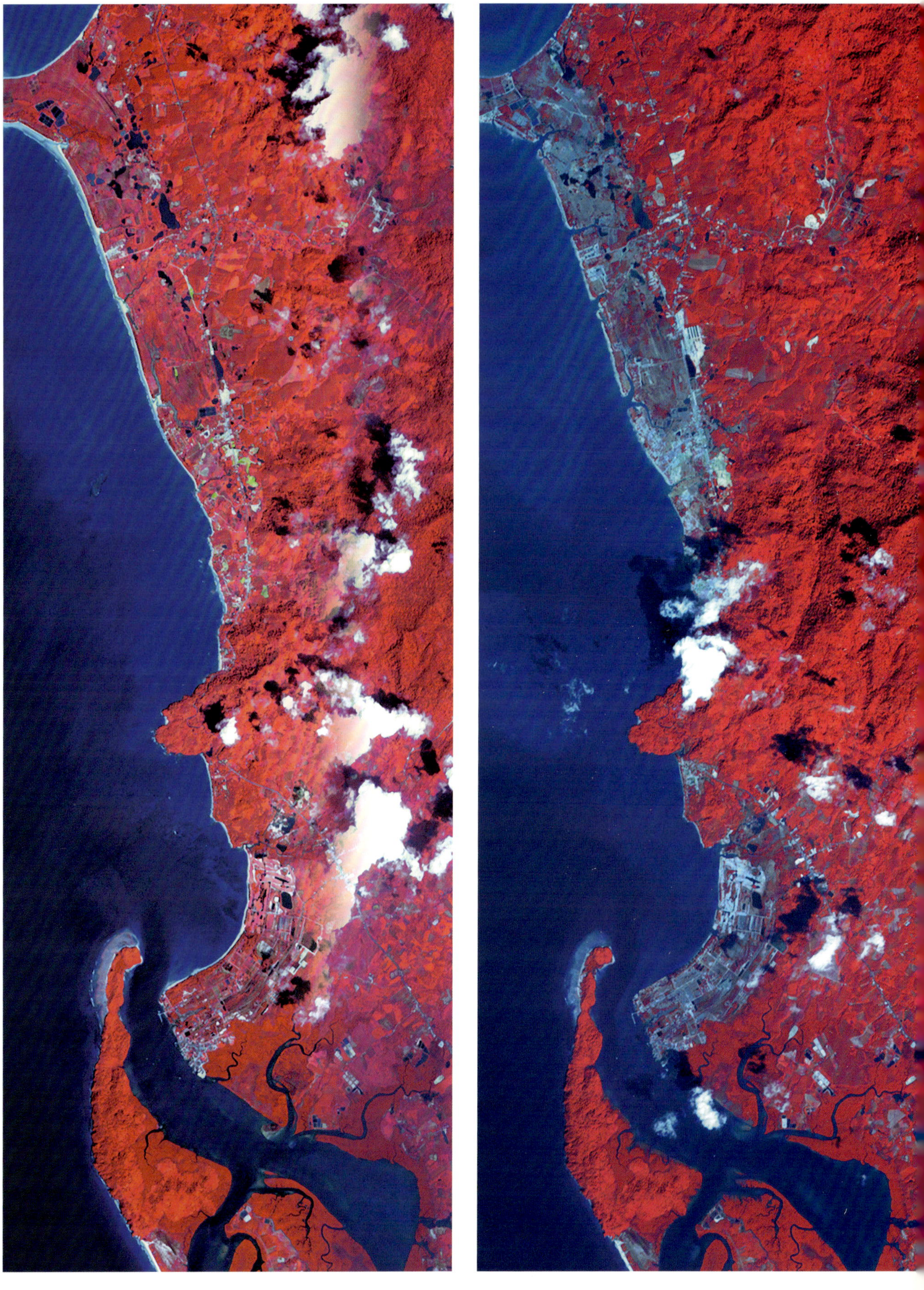

Beifang für die Wissenschaft

Selbst der GOCE-Satellit, der gemessen hat, wie stark die Anziehungskraft der Erde über einzelnen Regionen ist, ist ein Klimasatellit – auch wenn man die Gravitation sicher nicht auf den ersten Blick mit Wolken, Regen oder Dürren in Verbindung bringt, räumt Mark Drinkwater, der Projektwissenschafter dieser Mission bei der ESA, ein: *Das Schwerefeld liefert uns viele Informationen über die großräumigen Meeresströmungen und die Prozesse in der Erdkruste. Wir vermessen zudem die Form der Erde auf etwa zwei Zentimeter genau. Diese Daten sind extrem wichtig, um Veränderungen im Klima besser zu erkennen.* GOCE lieferte eine Art Eichung, um Daten anderer Satelliten – etwa der Meereshöhe – in den richtigen Bezug zu setzen. Bisher gab es keine hochgenaue Eichung dieser Daten für alle Regionen der Erde. Kurz gesagt: Man wusste nie genau, worauf sich Angaben des Meeresspiegels bezogen. Wegen unterschiedlicher Bezugssysteme waren solche Daten etwa aus Südostasien und Nordamerika nie gut zu vergleichen. Das ist fatal, wenn man die Folgen des Klimawandels auf die Meere genauer untersuchen will. Denn schmelzende Eismassen lassen das Wasser nicht nur in einem Ozean höher schwappen.

A scientific bonus

The GOCE satellite, which measured the gravitational pull of the Earth in specific regions, was a climate satellite, too, even if people at first glance wouldn't necessarily associate gravitation with rainfall, drought, or clouds. Mark Drinkwater, the head scientist on this ESA project, explains: *The gravity field can tell us a lot about large-scale ocean currents and processes in the Earth's crust. We also measure the shape of the Earth to an accuracy of about 1 inch [2.5 centimeters]. Such data is extremely important in helping us to understand climate change.* GOCE provided a kind of base line against which data from other satellites—sea level for example—can be compared. Previously, such data was incompatible across all the regions of the Earth—in other words, people couldn't tell what reference point sea level measurements were measured against, and differing systems meant that it was difficult to reconcile data from Southeast Asia and North America, for example. This is fatal if you are trying to examine the consequences of climate change for the oceans—when the ice caps melt, the water level doesn't rise in just one sea.

Un plus pour la science

Même le satellite GOCE, qui étudiait le champ gravitationnel de la Terre, s'intéressait au climat. On ne perçoit peut-être pas immédiatement le lien entre la pesanteur et les nuages, la pluie et la sécheresse, concède Mark Drinkwater, un scientifique travaillant pour la mission de l'ESA, *mais le champ gravitationnel nous fournit des informations abondantes sur la circulation océanique à l'échelle planétaire et les processus observés dans la croûte terrestre. Nous mesurons par ailleurs la forme de la Terre, ou géoïde, à deux centimètres près. Ces données sont indispensables pour mieux identifier les changements climatiques.* GOCE a fourni une sorte d'étalon qui servira de référence aux données transmises par les autres satellites, notamment les mesures du niveau de la mer, pour lesquelles il n'existait jusqu'ici aucune surface de référence précise à l'échelle du globe. Autrement dit, on ne savait jamais exactement sur quoi reposaient les mesures du niveau de la mer. De plus, l'existence de différents systèmes de référence rendait la comparaison des données prises par exemple en Asie du Sud-Est et en Amérique du Nord presque impossible. Ce qui est plutôt gênant pour étudier avec précision

Datos adicionales para la ciencia

Incluso el satélite GOCE, que medía la intensidad de la fuerza de gravedad de la Tierra en determinadas regiones, era un satélite climático. Aunque seguramente en un primer análisis no se relacione la gravitación con las nubes, la lluvia o la sequía, el científico responsable del proyecto de esta misión en la ESA, Mark Drinkwater, admite: *El campo gravitatorio proporciona mucha información sobre las grandes corrientes marinas y los procesos de la corteza terrestre. Además, mide la forma de la Tierra con una precisión de casi dos centímetros. Estos datos son extremadamente importantes para detectar mejor las modificaciones del clima. GOCE* proporcionaba un tipo de calibración para fijar la referencia correcta de los datos de otros satélites. Hasta ahora no existía una calibración de precisión de estos datos para todas las regiones de la Tierra. Resumiendo: nunca se sabía exactamente a qué indicaciones de nivel del mar se referían. Estos datos eran difíciles de cotejar, por ejemplo entre el sudeste de Asia y Norteamérica, debido a los diferentes sistemas de referencia. Esto es fatal cuando se quieren investigar con precisión las consecuencias del cambio climático en los mares, porque la fu-

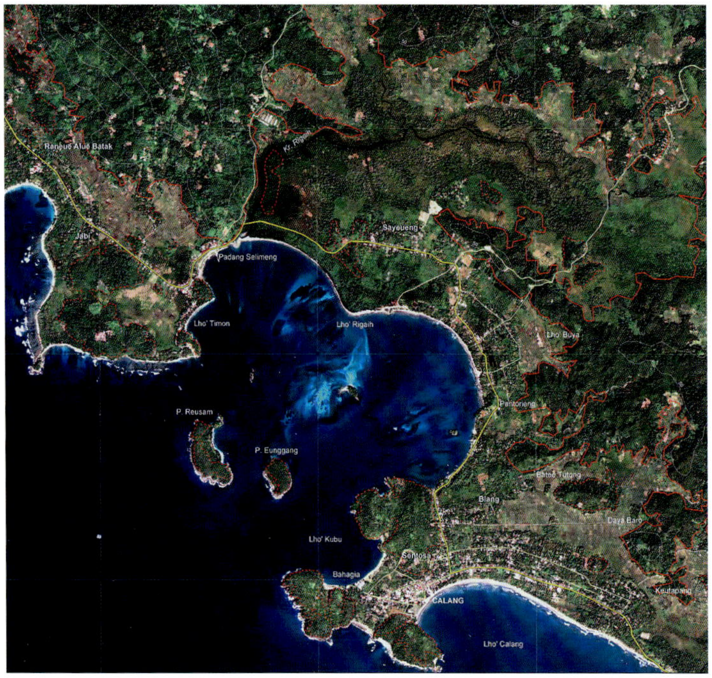

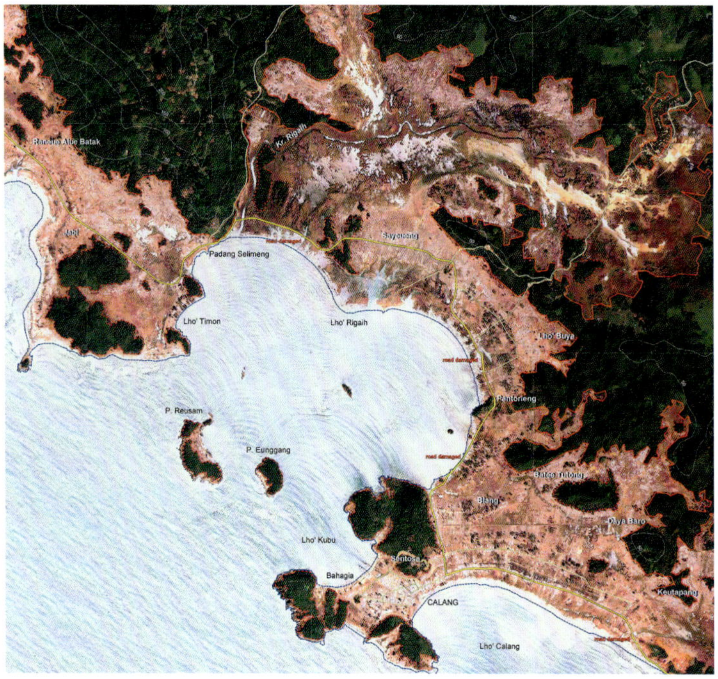

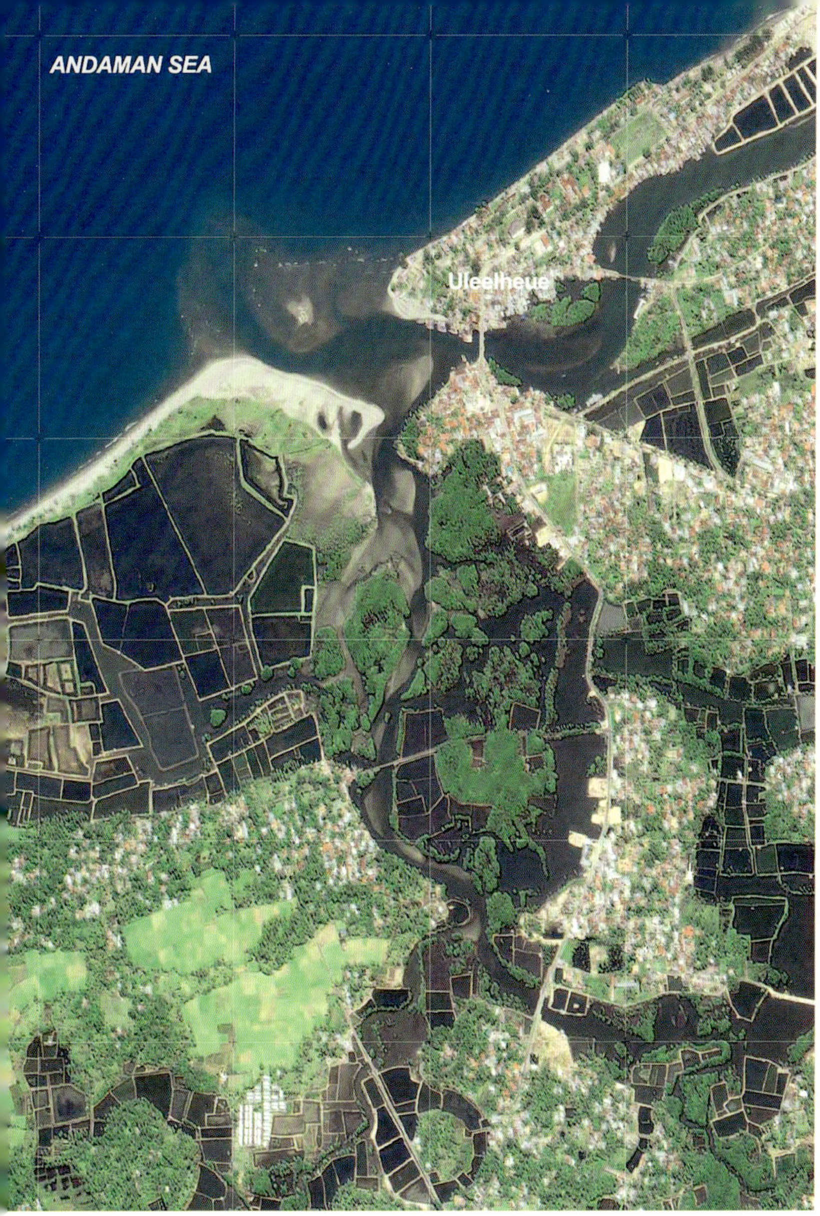

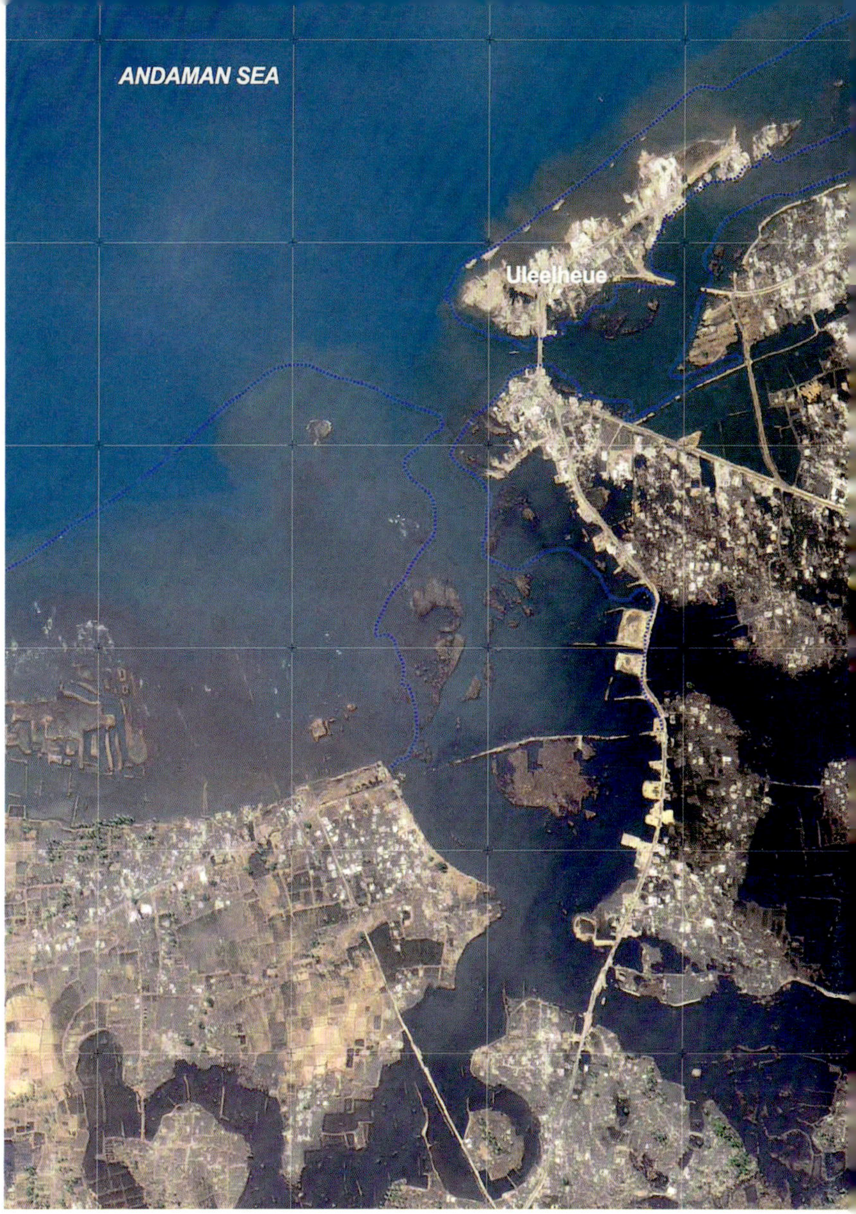

Linke Seite: Calang auf Sumatra vor und nach dem Tsunami. Satellitenbilder liefern Karten für Hilfskräfte (QuickBird)

Left: Calang on Sumatra before (far left) and after (near left) the tsunami. Satellite images provide useful maps for aid workers (QuickBird)

Page de gauche : la ville de Calang sur l'île de Sumatra avant et après le tsunami. Les satellites fournissent des cartes pour les secours (QuickBird)

Página izda.: Calang, Sumatra, antes y después del tsunami. Los satélites proporcionan mapas para los equipos de socorro (QuickBird)

Oben links: Teile von Banda Aceh in Indonesien am 10. Januar 2003. Die Stadt liegt an der Andamanen-See (Ikonos)

Above left: Parts of Banda Aceh in Indonesia on January 10, 2003. The city lies on the Andaman Sea (Ikonos)

À gauche ci-dessus : districts de Banda Aceh, en Indonésie le 10 janvier 2003, en bordure de la mer d'Andaman (Ikonos)

Arriba izda.: Parte de Banda Aceh, en Indonesia, el 10.01.2003. La ciudad está a orillas del mar de Andamán (Ikonos)

Oben rechts: Banda Aceh, drei Tage nach dem Tsunami: Große Teile der Stadt sind von der Flutwelle zerstört worden (Ikonos)

Above right: Banda Aceh, three days after the tsunami; large portions of the city have been destroyed by flooding (Ikonos)

À droite ci-dessus : Banda Aceh trois jours après le tsunami : une large partie de la ville a été détruite par le raz-de-marée (Ikonos)

Arriba dcha.: Banda Aceh, tres días después del tsunami: gran parte de la ciudad ha sido destruida por las enormes olas (Ikonos)

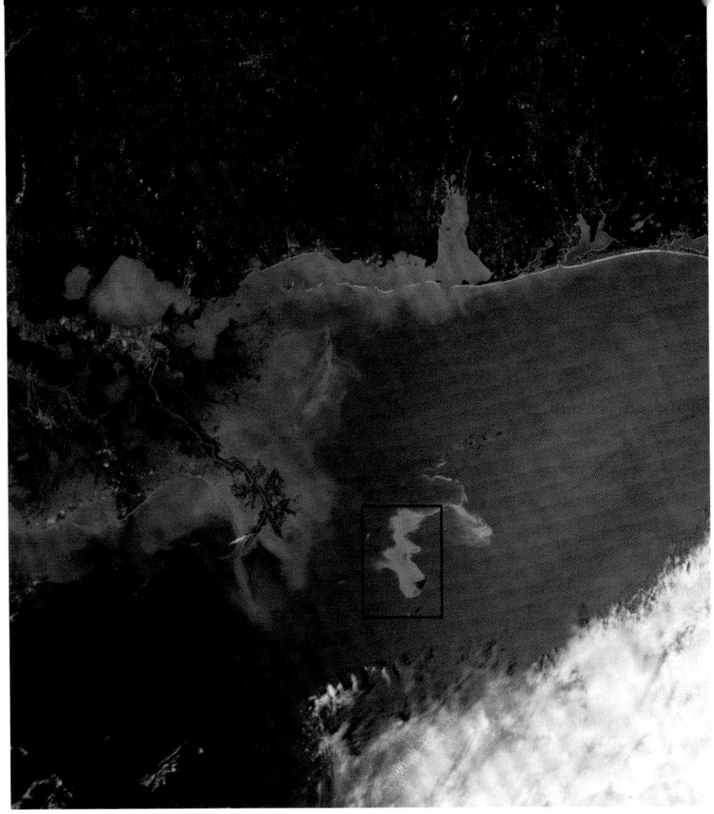

Ölteppich im Golf von Mexiko nach der Explosion der Bohrinsel Deep Water Horizon im April 2010 (Aqua)

An oil slick in the Gulf of Mexico after the Deepwater Horizon drilling rig explosion in April 2010 (Aqua)

Marée noire dans le golfe du Mexique après l'explosion de la plate-forme pétrolière Deep Water Horizon en avril 2010 (Aqua)

Mancha de petróleo en el golfo de México tras la explosión de la plataforma Deep Water Horizon, en abril de 2010 (Aqua)

Die Messungen der Schwerkraft bringen die Forscher auch weiter, um das Innere der Erde genauer zu verstehen. Da sind vor allem die geologisch aktiven Regionen von Interesse, in denen Schollen der Erdkruste aufeinanderprallen oder sich verhaken, erklärt Mark Drinkwater: *Mit unseren Daten werden wir besser die Prozesse verstehen, die zu Vulkanausbrüchen und Erdbeben führen.* Ist GOCE jetzt ein Gravitationssatellit, der nebenher Klimaforschung macht? Oder ein Klimasatellit, der en passant das Schwerkraftfeld erfasst? Dieser Satellit macht deutlich, dass die Erforschung unserer Erde längst interdisziplinär ist. Zwar sind die Missionen meist auf eine konkrete Fragestellung zugeschnitten und verfügen oft nur über ein Instrument. Aber die Daten, die gesammelt werden, lassen sich vielfältig nutzen. Es ist fast so wie zu Zeiten des Universalgelehrten und Forschungsreisenden Alexander von Humboldt. Auch die Entdecker von heute, die einige hundert Kilometer höher als ihr berühmter Vorgänger zu Beginn des 19. Jahrhunderts im Einsatz sind, arbeiten in sehr vielen Wissenschaftsbereichen.

Die großen Erdbeobachtungssatelliten ERS und Envisat lieferten Daten an Wissenschaftler aus allen Disziplinen, die irgendetwas mit der Erde zu

Measuring the force of gravity also helps scientists toward a better understanding of the Earth's interior. Here, geologically active areas, where sections of the Earth's crust are colliding or getting caught on one another, are of the most interest, as Mark Drinkwater elucidates: *Our data will better explain the processes causing volcanic eruptions and earthquakes.* So is GOCE a gravity satellite that conducts climate research? Or a climate satellite that happens to measure gravitational fields? In fact, such a satellite clearly demonstrates how exploring the Earth is a multi-disciplinary undertaking. Although the missions are usually planned to address a specific question, often only using one measuring instrument, the data collected has a variety of uses. Things are perhaps not so different from the days of the polymath expedition leader Alexander von Humboldt. The explorers of today are traveling the world a few hundred miles (a few kilometers) higher than their predecessors in the 19th century, yet they still work in a great number of different fields. ERS and Envisat, the largest Earth observation satellites, has provided data for scientists in every area of expertise associated with the Earth.

les effets du changement climatique sur la mer, car la fonte des glaces ne fait évidemment pas monter le niveau d'un seul océan.

Les mesures de gravité permettent par ailleurs aux chercheurs de mieux comprendre les processus affectant l'intérieur de la Terre. Dans ce domaine, ce sont les régions soumises à une grande activité géologique, où les plaques tectoniques se heurtent ou se chevauchent, qui nous intéressent, explique Mark Drinkwater : *Nos données nous aiderons à mieux comprendre les processus responsables des éruptions volcaniques et des séismes.* GOCE est-il un satellite voué à l'étude de la gravité qui accessoirement s'adonne à la recherche climatique, ou un spécialiste du climat qui mesure le champ de pesanteur ? Ce satellite est la parfaite illustration de l'interdisciplinarité qui caractérise depuis longtemps l'exploration de la Terre. Les missions sont certes généralement conçues pour répondre à une question concrète et disposent souvent d'un instrument unique. Mais l'exploitation des données collectées est multiple. À l'exemple d'Alexander von Humboldt, savant universel et explorateur scientifique, les voyageurs d'aujourd'hui, qui naviguent quelques centaines de kilomètres plus haut que leur fameux prédécesseur du

sión de las masas de hielo no afecta tan sólo al nivel de un único océano.

Las mediciones de la fuerza de gravedad hacen avanzar a los investigadores en la mejor comprensión del interior de la Tierra. En este caso, son de interés sobre todo las regiones geológicamente activas, en las que las placas de la corteza terrestre chocan o entran en contacto. Explica Mark Drinkwater: *Con nuestros datos vamos a entender mejor los procesos que conducen a las erupciones volcánicas y a los terremotos.* ¿Es entonces GOCE un satélite para medir la gravitación que, además, hace investigación climática? ¿O un satélite climático que, de paso, registra el campo gravitatorio? Este satélite revela que la investigación de la Tierra es, desde hace mucho tiempo, interdisciplinar. Aunque la mayoría de las veces las misiones estén hechas a la medida de un planteamiento concreto y con frecuencia dispongan de un único instrumento, los datos que se recogen pueden utilizarse de muchas formas. Esto es casi lo mismo que sucedía en tiempos del erudito y expedicionario universal Alexander von Humboldt. También los descubridores de la actualidad, que están operativos unos cientos de kilómetros más arriba que su conocido predecesor de comienzos del siglo XIX,

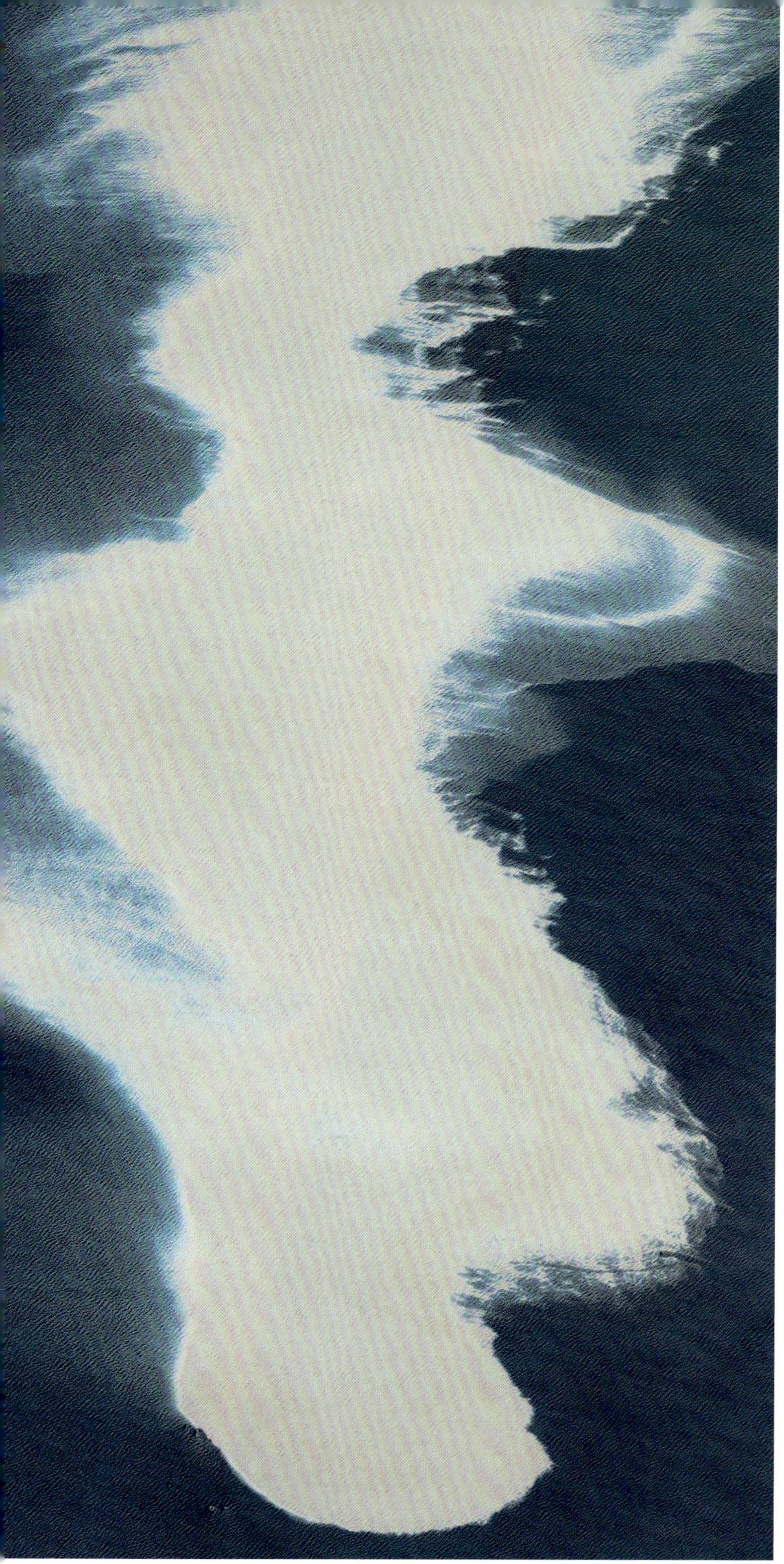

Links: Radaraufnahme des auf Seite 162 rot markierten 60 km langen Ölteppichs (EO-1)

Left: A radar image of the 37-mile-long (60-km) oil slick, marked by a red box on page 162 (EO-1)

À gauche : image radar de la marée noire longue de 60 km encadrée de rouge sur la page 162 (EO-1)

Izda.: Toma de radar de la mancha de 60 km marcada en rojo en la página 162 (EO-1)

Unten: Der Ölteppich driftet monatelang über das Meer (Raumstation, Envisat)

Below: The oil slick drifted for months at sea (space station, Envisat)

Ci-dessous : la marée noire dérive durant des mois sur la mer (station spatiale, Envisat)

Abajo: La mancha de aceite deriva por el mar durante meses (estación espacial, Envisat)

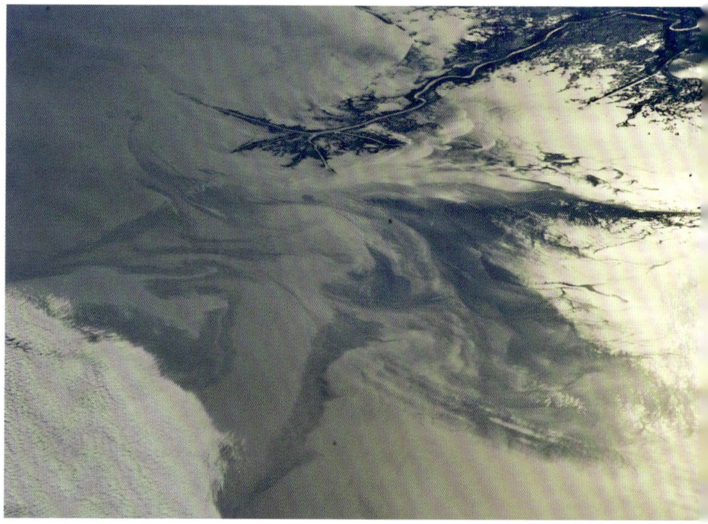

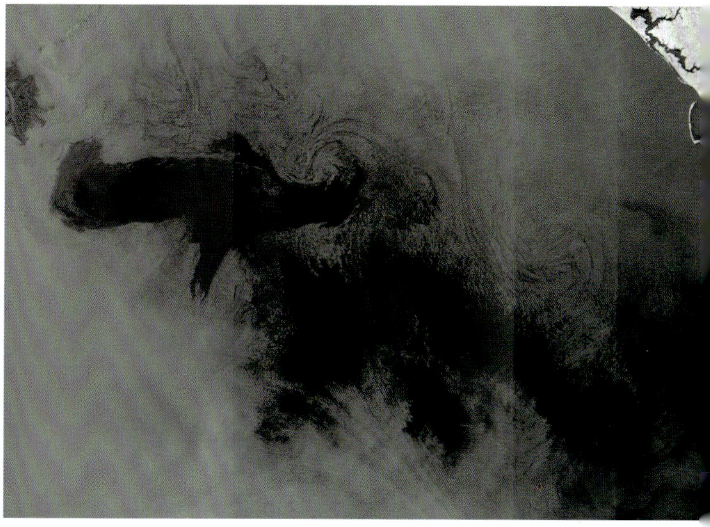

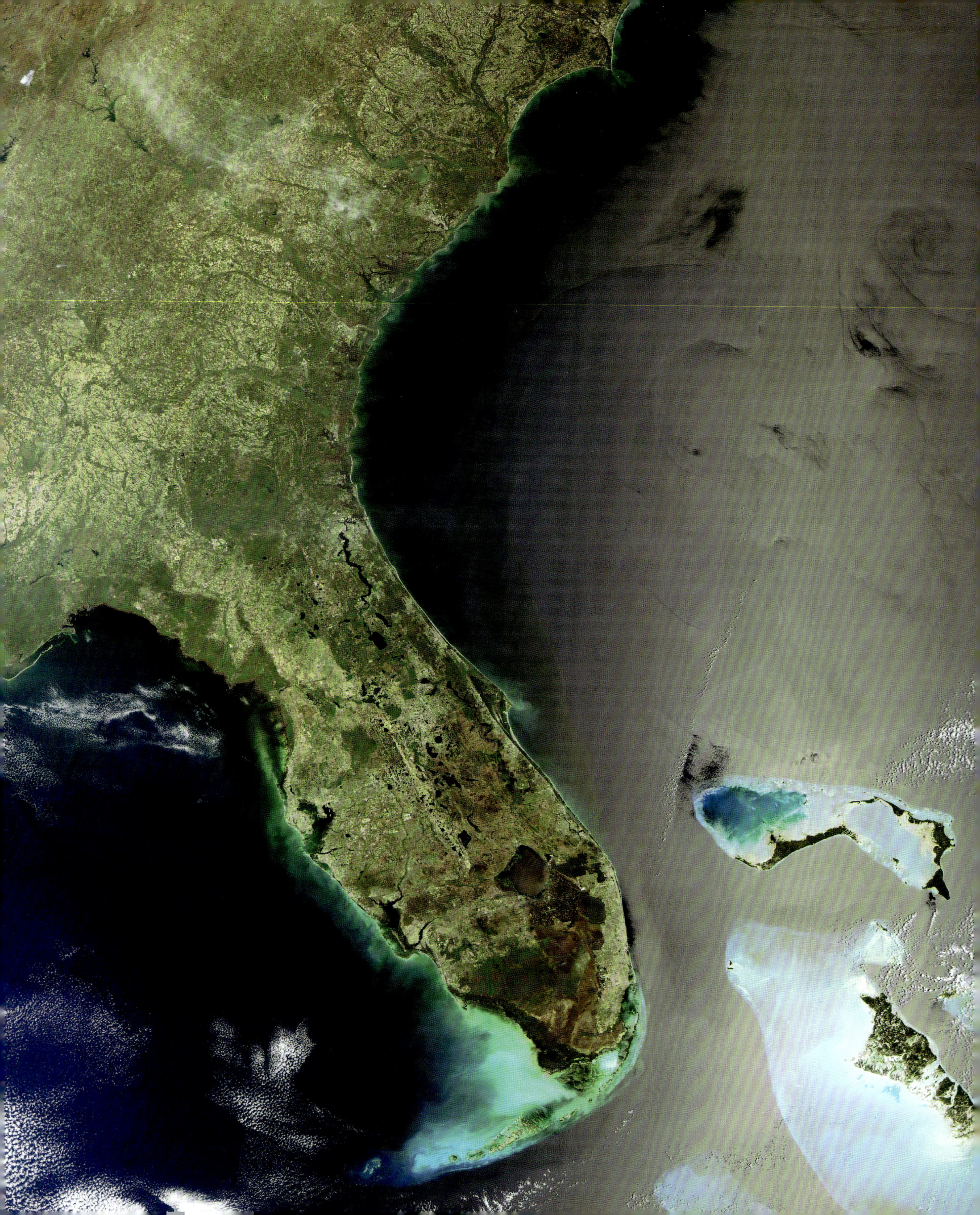

tun haben. Spektakulär sind die hochgenauen Radaraufnahmen, die diese Satelliten, aber auch der japanische Kollege Alos oder der deutsche TerraSAR liefern. Mit Hilfe einer aufwändigen Technik lässt sich per Radar eine ganze Geländeformation zentimetergenau erfassen. Nach verheerenden Erdbeben wie 1999 in der Türkei oder 2010 in Haiti und Chile zeigen die Daten, wie sich die Erdoberfläche minimal gesenkt oder gehoben hat. Solche Informationen sind enorm wichtig, um zu verstehen, was genau beim Beben passiert ist. Nur so lassen sich die Vorgänge in der Erdkruste rekonstruieren. Die Forscher hoffen, ein genaueres Bild davon zu bekommen, was ein Erdbeben auslöst. Auch Vulkanologen nutzen die Radardaten begeistert, weil sie präzise zeigen, ob sich ein Berg etwas aufbläht oder was im Innern eines Kraters passiert, der aus Sicherheitsgründen nicht mehr vor Ort zu erforschen ist.

Für Radardaten gibt es so viele kommerzielle Anwendungen, dass die beiden deutschen Satelliten TerraSAR-X und TanDEM-X gemeinsam von Airbus und dem Deutschen Zentrum für Luft- und Raumfahrt (DLR) konzipiert, gebaut und bezahlt wurden. Staatliche Stellen und privatwirtschaftliche Unternehmen arbeiten

These satellites (along with the Japanese Alos and Germany's TerraSAR-X) provide spectacular and fantastically accurate radar images which, with the aid of sophisticated technology, allow us to build up a picture of a landscape that is accurate to the inch (a few centimeters). After devastating earthquakes, such as that in Turkey in 1999 or those in Haiti and Chile in 2010, such data reveals the tiny degree to which the Earth's surface has risen or sunk—this information is of huge importance when trying to understand what happened during the quake and is the only way of reconstructing the processes taking place in the Earth's crust. Scientists are hoping to build up a better picture of what causes an earthquake. Vulcanologists are also enthusiastic users of the data, as it is a precise indication of whether a mountain is swelling, or what might be happening in a crater which can no longer be visited for safety reasons.

There are so many commercial applications for radar data that state institutions and private enterprise are able to cooperate—the German satellites TerraSAR-X and TanDEM-X were conceived, built, and paid for jointly by Airbus and the German Aerospace Center (*Deutsches Zentrum*

début du XIXe siècle, touchent à de nombreuses disciplines scientifiques.

Les grands satellites ERS et Envisat ont fourni des données à tous les chercheurs des sciences de la Terre. Les images radar haute résolution transmises par ces satellites et leurs confrères japonais Alos et allemand TerraSAR sont impressionnantes. Grâce à des techniques sophistiquées, les radars sont capables de cartographier des régions entières au centimètre près. Après de terribles tremblements de Terre, tels qu'en Turquie en 1999 ou en Haïti et au Chili en 2010, les données relevées indiquent les zones où la surface terrestre s'est légèrement soulevée ou affaissée, des informations essentielles pour comprendre le mécanisme sismique, reconstituer les processus qui régissent la croûte terrestre, et expliquer les causes du déclenchement des séismes. Les vulcanologues sont également friands de données radar qui indiquent le gonflement des volcans et décèlent l'activité à l'intérieur des cratères inapprochables pour des raisons de sécurité.

Les applications commerciales des données radar sont tellement variées qu'Airbus et le Centre de recherche aérospatial allemand (DLR) se sont associés pour concevoir, construire et financer les deux satellites allemands

trabajan en disciplinas científicas muy variadas.

Los grandes satélites de observación terrestre ERS y Envisat han proporcionado a los científicos datos de todas las disciplinas relacionadas con la Tierra. Las tomas de radar de alta precisión de estos satélites son espectaculares, pero también las que suministra su colega japonés Alos o el alemán TerraSAR. Con ayuda de una costosísima tecnología, se puede registrar por radar una completa información del terreno con una exactitud de centímetros. Tras los desastrosos terremotos de 1999 en Turquía o de 2010 en Haití y Chile, los datos muestran cómo la superficie terrestre se ha hundido o elevado ligeramente. Esta información es enormemente importante para entender lo que pasa durante los seísmos y, más tarde, reconstruir los procesos de la corteza terrestre. Los investigadores esperan conseguir así una imagen más exacta de aquello que desencadena los terremotos. También los vulcanólogos utilizan entusiasmados los datos de radar, porque muestran con precisión si una montaña expulsa gases o lo que ocurre en el interior de un cráter, sucesos que por razones de seguridad no se puede investigar in situ.

Existen tantas aplicaciones comerciales de los datos de radar, que los

Linke Seite: Das Wasser strömt um Florida herum in den Atlantik. Das Öl nahm diesen Weg zum Glück nicht (Envisat)

Left: Water flows around Florida and out into the Atlantic; fortunately the oil did not follow this path (Envisat)

Page de gauche : le courant marin contourne la Floride puis pénètre dans l'Atlantique. Le pétrole n'a heureusement pas emprunté la même route (Envisat)

Página izda.: El agua corre, rodeando Florida, hacia el Atlántico. Por suerte, el petróleo no sigue ese camino (Envisat)

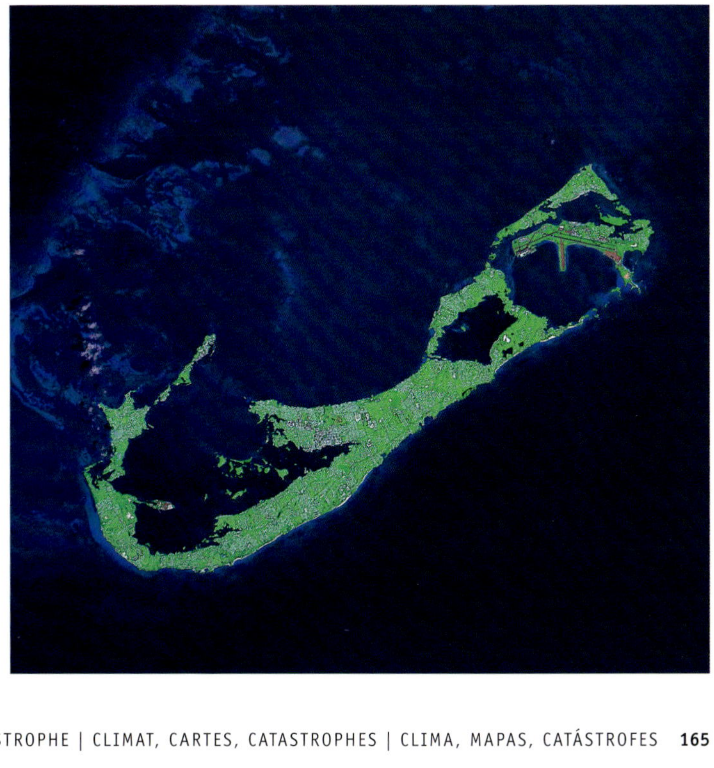

Eine der Inseln der Bahamas vor der Küste Floridas. Ganz rechts ist der Flughafen zu erkennen (Landsat)

One of the Bahama Islands off the Florida coast. The airport is visible on the far right (Landsat)

Une île des Bahamas au large de la Floride. À l'extrême droite, on aperçoit l'aéroport (Landsat)

Una de las islas Bahamas delante de la costa de Florida. A la derecha se puede ver el aeropuerto (Landsat)

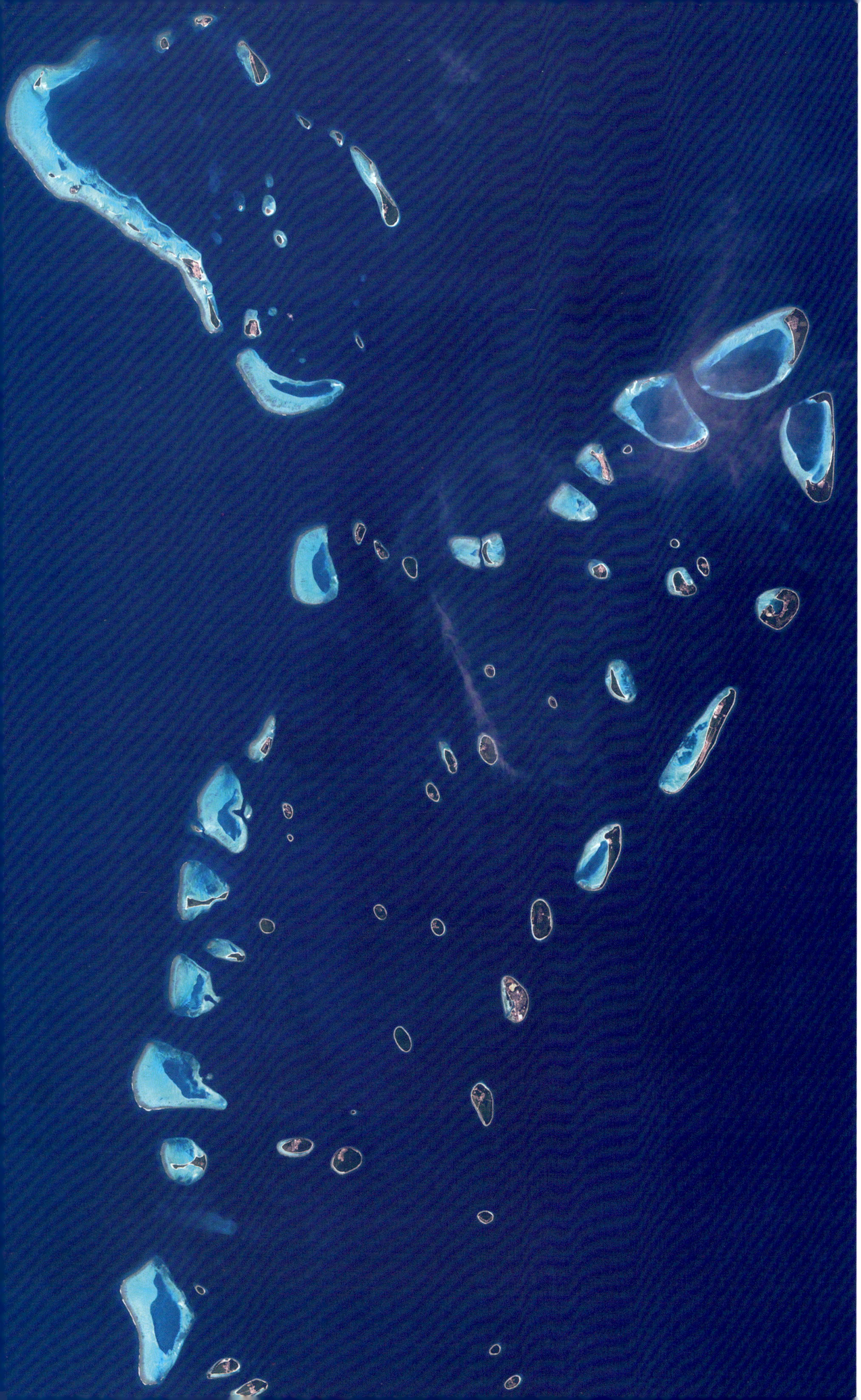

Bedrohte Paradiese: Die Malediven (links), Tureia, Tikihau, Atafu, Mataiva (unten von links) und alle anderen Atolle könnten bei ansteigendem Meeresspiegel bald versinken (Landsat, EO-1, Raumstation)

Paradise under threat: The Maldives (left), Tureia, Tikehau, Atafu, Mataiva (below, from left), and every other atoll could soon disappear beneath rising sea levels (Landsat, EO-1, space station)

Paradis menacé : si le niveau des océans monte, les Maldives (à gauche), Tureia, Tikihau, Atafu, Mataiva (ci-dessous de g. à d.) et tous les autres atolls pourraient bientôt être engloutis (Landsat, EO-1, station spatiale)

Paraísos amenazados: Las Maldivas (izda.), Tureia, Tikehau, Atafu, Mataiva (abajo, de izda. a dcha.) y el resto de los atolones podrían hundirse pronto por la subida del nivel del mar (Landsat, EO-1, estación espacial)

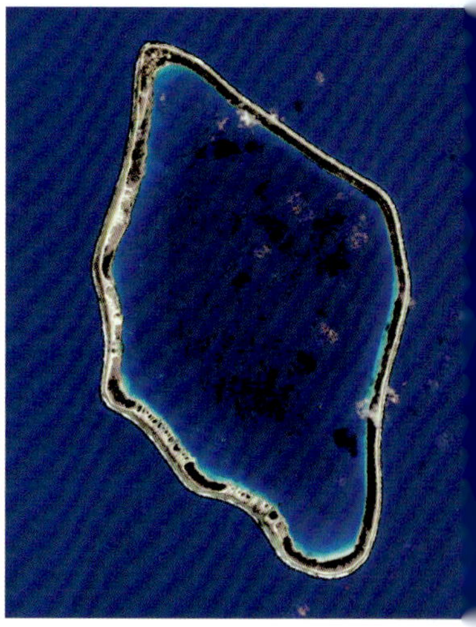

zusammen. Solche Daten, die Auflösung liegt oft bei etwa einem Meter, sind zum Beispiel die Grundlage für topografische Karten. Die Radarsensoren zeigen dreidimensional alle Oberflächendetails. Betreiber von Mobilfunknetzen, aber auch Geologen und Hydrologen, die sich mit der Bewegung von Wasser auf und unter dem Erd-boden befassen, nutzen gern solche Daten. Mit speziellen Mess- und Analysemethoden liefern die Radardaten zudem Informationen darüber, ob sich die Oberfläche hebt oder senkt. Entsprechende Daten sind vor allem für bebaute oder zu bebauende Flächen wichtig. Radarbilder aus dem All zeigen auch, wo es welche Vegetation gibt. Wälder in der Arktis lassen sich nur per Radar genau erfassen. Die Satelliten ermöglichen eine Art globaler Waldinventur. Das hat wieder einmal Bezug zum Klima – denn die Radarbilder helfen den Forschern, den Kohlenstoffkreislauf auf der Erde nachzuvollziehen: Wo wird Kohlenstoff frei, wo wird er in Pflanzen gebunden?

Natürlich greift auch das Militär gern auf Radardaten zurück, zeigen sie doch selbst bei Nacht und Nebel, wo der Gegner seine Stellungen hat oder ob ein Panzer im Wald verborgen ist. Die Bundeswehr betreibt sogar ein eigenes Netz von fünf Radarsatel-

für Luft- und Raumfahrt). Such data, whose resolution is often accurate to within a few feet (a meter or so), provides the basis for topographic maps, for example—the radar sensors reveal every surface detail in 3-D. Cell phone providers are also keen to use such information, not to mention the geologists and hydrologists dealing with the movement of water on and beneath the Earth's surface. Special techniques in measuring and analyzing the radar data provide information about the rise and fall of the Earth's surface, and such data is especially important in built-up areas or for development land. Radar images from space also show where vegetation is growing, and only this method can keep accurate track of tree growth in the Arctic. Satellites provide a kind of global stocktaking for forestry, and this too is of importance for our climate; the radar images allow scientists to track the Earth's carbon cycle, showing where the element is being emitted and where it is being stored in plants.

The military is of course also not averse to using radar information, as in all weathers and conditions it reveals where the enemy may have taken up a position or if there is a tank concealed in a wood. The German army even maintains its own

TerraSAR-X et TanDEM-X. Leurs données, qui atteignent en général une résolution d'un mètre, servent notamment à la réalisation de cartes topographiques. Les capteurs radar fournissent des images tridimensionnelles du relief terrestre qu'utilisent les opérateurs de réseaux radio mobiles, tout comme les géologues et les hydrologues qui étudient le cycle de l'eau sur la Terre. Les informations radar permettent en outre, grâce à des méthodes de mesure et d'analyse spécifiques, de déceler les soulèvements et affaissements de la surface terrestre. Des données particulièrement importantes pour les surfaces bâties et les terres cultivables. Les images radar relèvent par ailleurs la position des différents types de végétation. Elles sont seules aptes à déterminer l'étendue exacte des forêts arctiques, dressant ainsi une sorte d'inventaire forestier mondial. Ces données, toujours liées au climat, aident les scientifiques à comprendre le cycle du carbone sur la Terre.

Dans le recours aux données radar, l'armée n'est pas en reste. Essayez donc vous-même de repérer la nuit et par temps de brouillard la position de l'ennemi ou un char au milieu de la forêt. L'armée allemande exploite ainsi une constellation de cinq satel-

dos satélites alemanes TerraSAR-X y TanDEM-X fueron diseñados, construidos y pagados conjuntamente por Airbus y el Centro Aeroespacial Alemán (DLR). Colaboraron organismos oficiales y la empresa privada. Los datos obtenidos son, por ejemplo, la base para los mapas topográficos. Los sensores de radar muestran tridimensionalmente todos los detalles de la superficie. Estos datos son utilizados tanto por los operadores de redes de telefonía móvil, como por geólogos e hidrólogos que se ocupan del movimiento de las aguas sobre y bajo el suelo terrestre. Además, gracias a especiales métodos de medición y análisis, los datos de radar informan si la superficie se eleva o se hunde. Los datos correspondientes son importantes, sobre todo, para superficies urbanizadas o por urbanizar. Las imágenes de radar desde el espacio también muestran la distribución de la vegetación. Los satélites hacen posible una especie de inventario forestal global. Esto hace referencia nuevamente al clima, ya que las imágenes de radar ayudan a los investigadores a comprender el ciclo del dióxido de carbono en la Tierra. En definitiva, dónde se libera carbono y dónde está ligado a las plantas.

Naturalmente, el ejército también recurre a los datos de radar, que

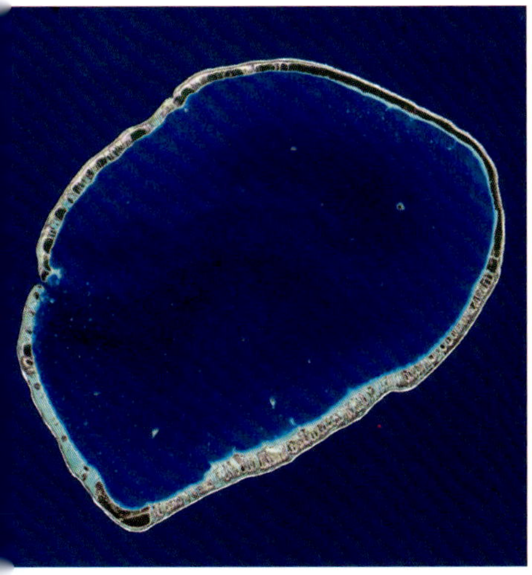

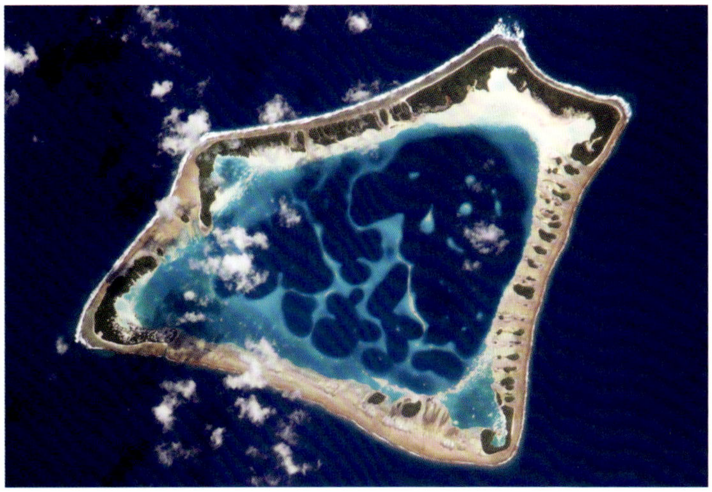

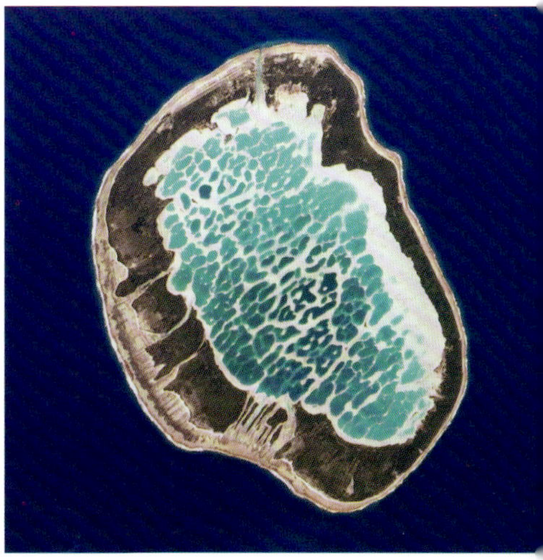

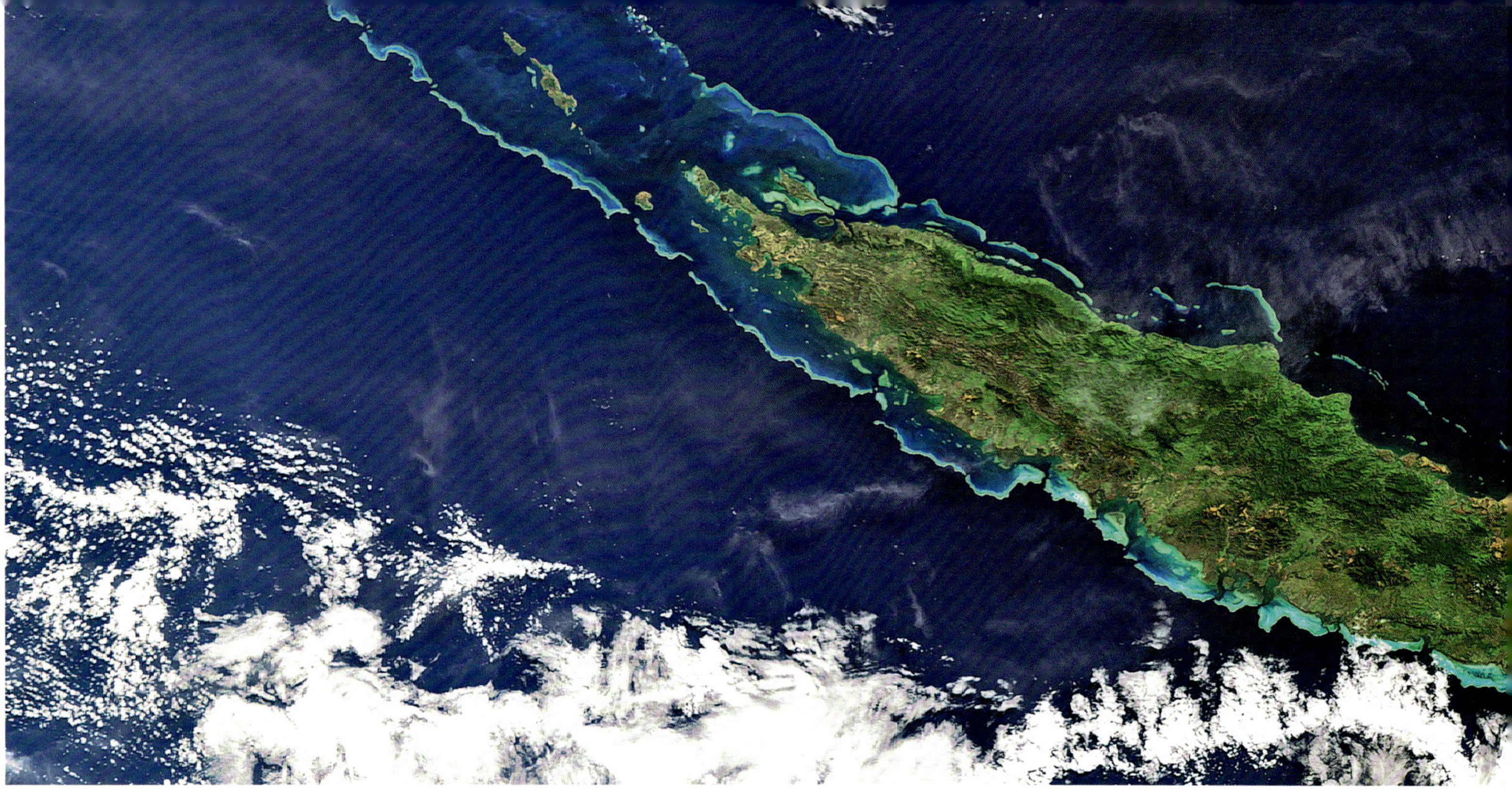

liten namens SAR-Lupe. Die Satelliten wurden von der Bremer Firma OHB gebaut. Der Begriff Lupe ist gut gewählt. Zwar teilen die Generäle nicht mit, wie scharfsichtig sie mit ihren fliegenden Vergrößerungsgläsern auf den Boden blicken. Man darf aber davon ausgehen, dass Details deutlich kleiner als ein Meter zu erkennen sind. Mit fünf Satelliten lässt sich binnen zehn Stunden jeder beliebige Ort auf der Erde präzise erfassen. Auch die französischen Streitkräfte nutzen diese Daten. Dafür hat die Bundeswehr Zugriff auf die Bilder der optischen Aufklärungssatelliten der Helios-II-Serie, die Frankreich betreibt. Sie zeigen noch Einzelheiten bis zu etwa dreißig Zentimetern Größe. Militärische Einsätze wären ohne Erdbeobachtungssatelliten kaum mehr möglich beziehungsweise für die Soldaten viel zu gefährlich.

Wer darf die Daten nutzen?

Sicher sehen es einige mit etwas Sorge, wenn Militärs solche Daten nutzen. Dual Use ist das Stichwort. Damit ist gemeint, dass sich manche Daten sowohl zivil als auch militärisch nutzen lassen. Doch das gilt inzwischen für fast alle Bereiche. Bei Einsätzen im Irak oder in Afghanistan

network of five satellites, known as SAR-Lupe. The satellites were built by the Bremen firm OHB and the name, which means "magnifying glass," is well chosen; the top brass are not revealing just how precisely their flying eyeglass can observe details on the ground, but it can be assumed that objects smaller than a few feet (a meter or so) can be pinpointed, and with five satellites, any point on Earth can be captured in detail within ten hours. The data is shared with the French armed forces and in return the Germans have access to images from the Helios II series of photographic spy satellites maintained by France. These can show objects as small as a foot (30 centimeters) in diameter. Military operations would be impossible or far too dangerous for the troops involved without satellite imaging.

Who should be allowed to use the data?

There are of course some who would have reservations about the use of such information by the military, but the usual description is "dual-use technology." This means that much of the data is available to

lites nommés SAR-Lupe, construits par la société allemande OHB, et qui portent bien leur nom : « Loupe ». Les généraux ne veulent pas dire avec quelle résolution ces instruments observent le sol à travers leurs verres grossisseurs, mais on peut supposer qu'ils atteignent une précision nettement inférieure à un mètre. Les données de ces cinq satellites, capables d'inspecter n'importe quel point du globe en l'espace de dix heures, sont également à la disposition des forces armées françaises. En contrepartie, la Bundeswehr dispose des images fournies par les systèmes spatiaux français de reconnaissance optique Helios II, qui repèrent des détails d'environ trente centimètres. Sans les satellites d'observation de la Terre, les opérations militaires sont devenues quasi impossibles, et beaucoup trop dangereuses pour les troupes.

Qui a accès aux données ?

L'utilisation par l'armée de ce type de données en inquiétera sans doute plus d'un. On parle ici de double usage car les données sont exploitées à des fins civiles et militaires. Rares sont les domaines qui échappent désormais à cette réalité. Les forces

muestran, incluso de noche y con niebla, las posiciones del enemigo o si un carro de combate se esconde en el bosque. El ejército de la República Federal de Alemania explota su propia red de cinco satélites de radar, denominada SAR-Lupe. Los satélites fueron construidos por la empresa OHB de Bremen. El término lupe (lupa) está muy bien escogido. Aunque los generales no informen sobre la agudeza visual de estos satélites, hay que suponer que pueden detectar detalles con precisiones inferiores a un metro. Cinco satélites pueden registrar con precisión un lugar cualquiera de la Tierra en un plazo de diez horas. Las Fuerzas Armadas francesas también utilizan estos datos. Por ello, el ejército de la República Federal tiene acceso a las imágenes de los satélites-espía ópticos de la serie Helios-II que Francia gestiona. Estos muestran objetos con dimensiones de hasta treinta centímetros. Las operaciones militares ya no serían posibles sin los satélites de observación terrestre o, en todo caso, serían demasiado peligrosas para los soldados.

¿Quién puede utilizar los datos?

Seguro que algunos ven con preocupación que los militares utilicen estos

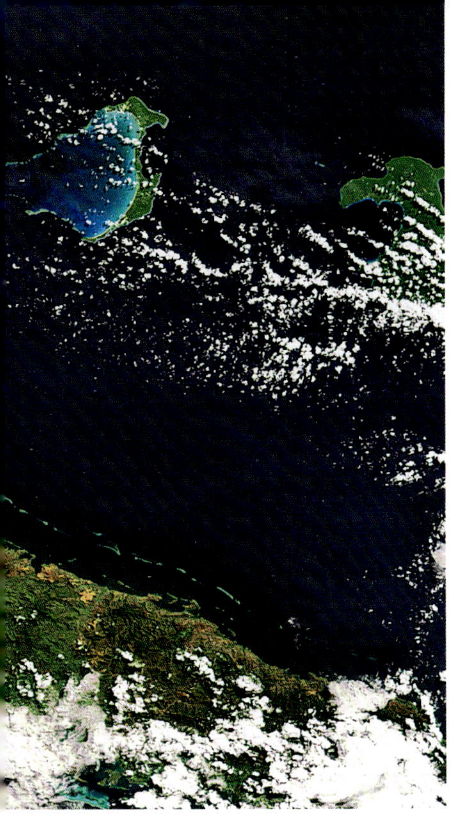

Neukaledonien mit seinen Korallenriffen. Die Insel ist immerhin 50 km breit (Envisat)

The coral reefs of New Caledonia. The island is 30 miles (50 km) wide (Envisat)

La Nouvelle-Calédonie et ses récifs coralliens. L'île s'étend sur une largeur de 50 km (Envisat)

Nueva Caledonia y sus arrecifes de coral. La isla tiene, al menos, 50 km de ancho (Envisat)

Seite 171: Sand weht von Westafrika über die Kapverden manchmal bis in die Karibik (Envisat)

Page 171: Sand blown from West Africa across the Cape Verde Islands can sometimes reach the Caribbean (Envisat)

Page 171 : les vents de sable d'Afrique de l'Ouest soufflant sur le Cap-Vert atteignent parfois les Caraïbes (Envisat)

Página 171: Arena arrastrada desde África occidental sobre Cabo Verde, a veces hasta el Caribe (Envisat)

werden natürlich auch Wetter- und Klimadaten genutzt. Sind die Meteosats deswegen Militärsatelliten? Schnelle Aufklärung über die Lage vor Ort mögen Truppenverbände beim Einsatz brauchen – genauso wie Katastrophenhelfer, die nach Überflutungen, Erdbeben, Gerölllawinen, Erdrutschen und Vulkanausbrüchen ein Lagebild benötigen, um schnell zu wissen, welche Gebiete zu erreichen und welche Straßen unpassierbar sind. Europas Weltraumorganisation ESA ist laut ihrer Gründungscharta allein friedlichen Zwecken verpflichtet. Sie betreibt Grundlagenforschung und entwickelt Technologien. Sofern militärische Einrichtungen direkten Zugriff auf Satellitendaten brauchen, starten sie in der Regel eigene Missionen, etwa zu Aufklärungszwecken oder um die Kommunikationen zu den Einsatzgebieten weltweit zu verbessern.

Um Erdbeobachtungsdaten möglichst schnell nutzen zu können, müssten sie umgehend zur Verfügung stehen. Doch bei wissenschaftlichen Satelliten dauert es oft Stunden oder Tage, bis die Aufnahmen zur Erde gefunkt und ausgewertet sind. Europa will jetzt mit seinem Copernicus-Programm in eine neue Ära der Erdbeobachtung eintreten. Die Dienste sollen operationell werden, wie es

both civilian and military users and is now true of many applications: weather and climate data was used as a matter of course during operations in Iraq and Afghanistan. Does this mean that weather satellites are also military satellites? Obtaining rapid intelligence about the lie of the land can be of as much use to troops in action as a local picture is to aid workers who need to find out quickly which areas to visit and which streets are impassable after floods, earthquakes, mudslides, and volcanic eruptions. According to its charter, the European Space Agency (ESA) is committed exclusively to peaceful ends, conducting fundamental research and developing technology. When military institutions require direct access to satellite data, they generally launch their own missions, either for intelligence-gathering purposes or to improve communications between theaters of operation.

To make the quickest use of Earth observation data, it should be available straight away. Research satellites can often take hours or even days to process and transmit their images to Earth. The Copernicus program represents a new era in Earth observation. Its services bring Earth observation up to the

armées utilisent de toute évidence les données météorologiques et climatiques lors de leurs opérations en Irak ou Afghanistan. Météosat est-il pour autant un satellite militaire ? La reconnaissance rapide de la situation sur le terrain est indispensable à l'intervention des unités militaires, tout comme les services de secours d'urgence requièrent une image des zones accessibles et des voies praticables pour intervenir rapidement suite aux inondations, séismes, glissements de terrain et éruptions volcaniques. D'après sa convention, l'ESA a pour mission d'encourager la recherche et le développement de technologies spatiales à des fins exclusivement pacifiques. Les organisations militaires qui requièrent un accès direct aux données spatiales, lancent en général des missions spécifiques visant, entre autres, à améliorer les capacités de reconnaissance ou les communications sur les théâtres d'opération dans le monde.

Pour être rapidement exploitables, les données d'observation de la Terre doivent être immédiatement disponibles. Or la transmission et l'analyse des images prises par les satellites scientifiques durent souvent plusieurs heures, voire plusieurs jours. Avec son programme Copernicus, l'Europe

datos. La palabra clave es dual use. Esto quiere decir que tales datos se pueden utilizar tanto civil como militarmente. En la actualidad, es válido para casi todos los ámbitos. Por ejemplo, para las operaciones en Irak o en Afganistán también se utilizaron los datos meteorológicos y climáticos. ¿Consideraríamos entonces a los Meteosat como satélites militares? Los cuerpos armados pueden necesitar un reconocimiento rápido de la situación en el lugar de operaciones, al igual que el personal de emergencia que tras inundaciones, terremotos, aludes, corrimientos de tierras y erupciones volcánicas, necesita un cuadro de situación para saber rápidamente a qué regiones hay que acceder y qué vías son intransitables. La Agencia Espacial Europea ESA está obligada, por su carta fundacional, a perseguir sólo fines pacíficos. Gestiona investigación básica y desarrolla tecnologías. Si las instituciones militares necesitan tener acceso a los datos de los satélites, por regla general lanzan sus propias misiones.

Para utilizar los datos de observación terrestre lo más rápido posible, éstos deberían estar disponibles de inmediato. Pero en los satélites científicos suelen pasar horas, o incluso días, hasta que las grabaciones se envían a la Tierra y se analizan.

heißt. Damit will man das nachvollziehen, was in der Meteorologie oder in der Telekommunikation schon seit Jahrzehnten Standard ist.

Die ersten Satelliten zur Aufnahme von Wolkenmustern oder Infrarotbildern des Wasserdampfgehalts in der Atmosphäre waren aufwändige Experimente. Gleiches gilt für die Missionen, die die ersten Fernsehübertragungen über Kontinente hinweg ermöglicht haben. Die technischen Hürden waren hoch und die Analyse der Daten meist nicht einfach. Im Laufe der Zeit führten die Forscher und Ingenieure solche Beobachtungen routinemäßig durch. Heute werden die Daten am Fließband hergestellt. Es fällt kein Wetterbericht aus, weil der Satellitenfilm fehlt. Wolken- und Wetterdaten zu sammeln, ist keine wissenschaftliche Herausforderung mehr. Telekommunikation via Satellit ist längst ein einträgliches Geschäft, das privatwirtschaftlich betrieben wird. Agenturen und Firmen bieten diese Dienstleistungen an und vermarkten sie. Ein Instrument zur Messung von Windfeldern auf dem Meer war einst eine große Neuerung auf Envisat. Heute gehen solche Daten laufend bei den Meteorologen ein – denn ein ganz ähnliches Instrument kreist an Bord von MetOp-A um die Erde. Das ist ein Wettersatellit auf einer niedrigen polaren Umlaufbahn, der die Arbeit der Meteosats ergänzt. MetOp-A hat nicht das Wetter besser gemacht, aber den Wetterbericht. Die Prognosen sind viel besser als ihr Ruf – vor allem wegen der Satellitendaten.

Erst Experiment, dann Service

Das Copernicus-Programm, gemeinsam aufgelegt von ESA und Europäischer Union, soll in einigen Jahren viele Daten der Erdbeobachtung praktisch rund um die Uhr und fast in Echtzeit liefern. Messungen der Bodenfeuchte oder des Schwerefelds werden beim Routinebetrieb noch nicht dabei sein, wohl aber Missionen, die die Erdoberfläche in vielen Spektralkanälen fotografieren oder

standards that have been the norm in meteorology and telecommunications for decades.

The first satellites to record cloud patterns or infrared images of the atmosphere's moisture levels were sophisticated and expensive experiments, rather like the launches which made the first intercontinental television broadcasts possible—the technical hurdles were immense and the data was in no way easy to analyze. As time progressed, such observations became routine for scientists and engineers, and the data is now rolling off a production line, as it were—these days, a weather forecast is never missed because the satellite film is missing, and collecting cloud and climate data is no longer a great scientific challenge. Developed and marketed by agencies and private companies, satellite telecommunications have always been a profitable enterprise run by commercial interests. An Envisat instrument that could measure wind field at sea was at one time a great innovation, but now meteorologists collect such data as a matter of course, since a very similar instrument is orbiting the Earth aboard MetOp-A, a low Earth satellite in a polar orbit whose data is used in conjunction with that from Meteosat. MetOp-A can't improve the weather, but it has improved weather reporting, and forecasts are now much better than their reputation would suggest, mainly due to satellite data.

Once an experiment, now a service

The Copernicus program, a joint venture between the ESA and the European Union, is intended to provide large amounts of real-time Earth observation data in the near future, more or less 24 hours a day. Its everyday duties will not include measurement of soil moisture or gravity fields, but it will photograph the Earth's surface in a number of bandwidths and make radar images. In addition, there are instruments

entend entrer dans une nouvelle ère de l'observation de la Terre. L'objectif consiste à rendre les services opérationnels, autrement dit suivre la voie tracée par la météorologie et les télécommunications depuis plusieurs décennies.

Les photos de formations nuageuses et les images infrarouges de la concentration de vapeur d'eau atmosphérique fournies par les premiers satellites résultaient de longues expérimentations. Il en va de même des premières transmissions télévisées intercontinentales. Les défis techniques étaient nombreux et l'analyse de données complexe. Avec le temps, ces observations sont devenues le pain quotidien des scientifiques et ingénieurs. Les animations satellites ne font défaut à aucun bulletin météo, et collecter des données météo n'a plus rien d'une gageure scientifique. Les télécommunications par satellite sont depuis longtemps une activité rentable pour les opérateurs privés qui commercialisent ces prestations. L'instrument mesurant la vitesse du vent sur la mer à bord d'Envisat était alors une grande innovation. Aujourd'hui, les météorologues reçoivent de telles données en continu, grâce à un instrument similaire qui gravite autour de la Terre à bord de MetOp-A, un satellite météorologique qui, sur une orbite polaire basse, complète le travail de Météosat. MetOp-A n'a pas amélioré la météo, mais les prévisions météorologiques. Elles sont en effet bien meilleures que leur réputation, ce qu'elles doivent avant tout aux satellites.

D'abord expérience, puis service

Le programme Copernicus est censé fournir dans quelques années de nombreuses données d'observation de la Terre en temps réel et sans interruption. Les mesures de l'humidité des sols et de la gravité ne figureront pas encore parmi elles, contrairement aux missions fournissant des images radar ou multispectrales de la surface terrestre, et à celles qui surveillent

Europa quiere ingresar ahora en una nueva era de la observación terrestre con su programa Copérnico. Con ello se pretende seguir lo que ya es norma en la meteorología o las telecomunicaciones desde hace décadas.

Los primeros satélites para grabación de patrones de nubes o de imágenes infrarrojas de contenido de vapor de agua en la atmósfera eran experimentos costosos. Lo mismo se puede decir de las misiones que hicieron posibles las transmisiones de televisión entre continentes. Los obstáculos técnicos eran importantes y, la mayoría de las veces, el análisis de los datos no resultaba sencillo. Con el paso del tiempo, los investigadores e ingenieros convirtieron en rutinarias estas observaciones. En la actualidad, los datos se producen en cadena. Reunir datos meteorológicos y de nubes ya no resulta un desafío científico. La telecomunicación vía satélite es, desde hace tiempo, un lucrativo negocio explotado por manos privadas. Las agencias y las empresas ofrecen estas prestaciones de servicio y las comercializan. Antiguamente, fue una gran novedad un instrumento para la medición de vientos en el mar instalado en el Envisat. En la actualidad, los meteorólogos reciben continuamente esos datos, ya que un instrumento muy parecido gira alrededor de la Tierra a bordo del MetOp-A. Este es un satélite meteorológico que, desde una órbita polar baja, complementa el trabajo de los Meteosat. MetOp-A no ha mejorado el tiempo, pero sí el pronóstico meteorológico.

Primero experimentación, después servicio

El programa Copérnico, desarrollado conjuntamente por la ESA y la Unión Europea, debe proporcionar en los próximos años muchos datos de observación terrestre, prácticamente 24 horas al día y casi en tiempo real. Las mediciones de la humedad del suelo o del campo gravitatorio todavía no se incluyen en un funcionamiento rutinario, pero sí las misiones que fotografían la superficie terrestre

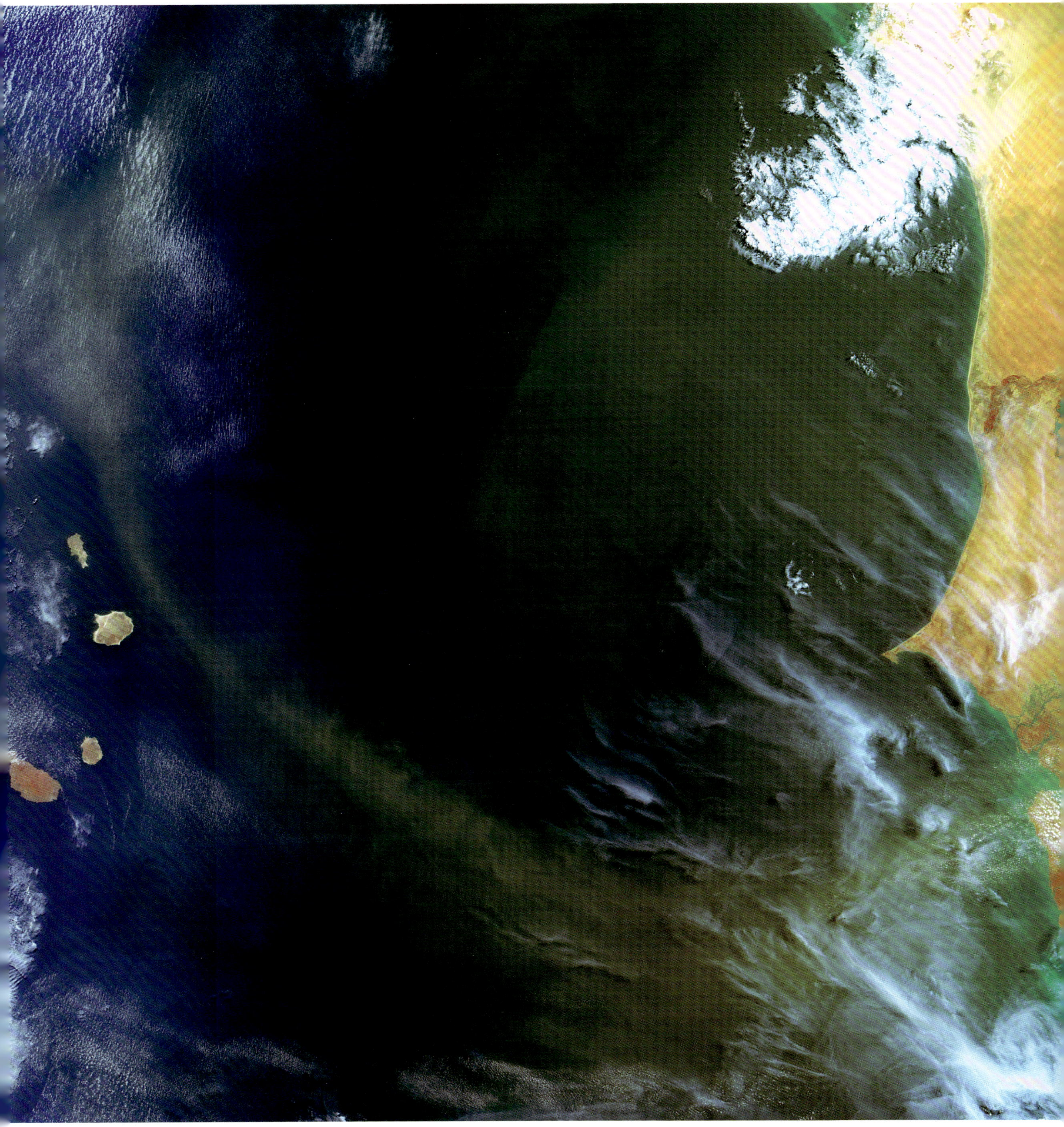

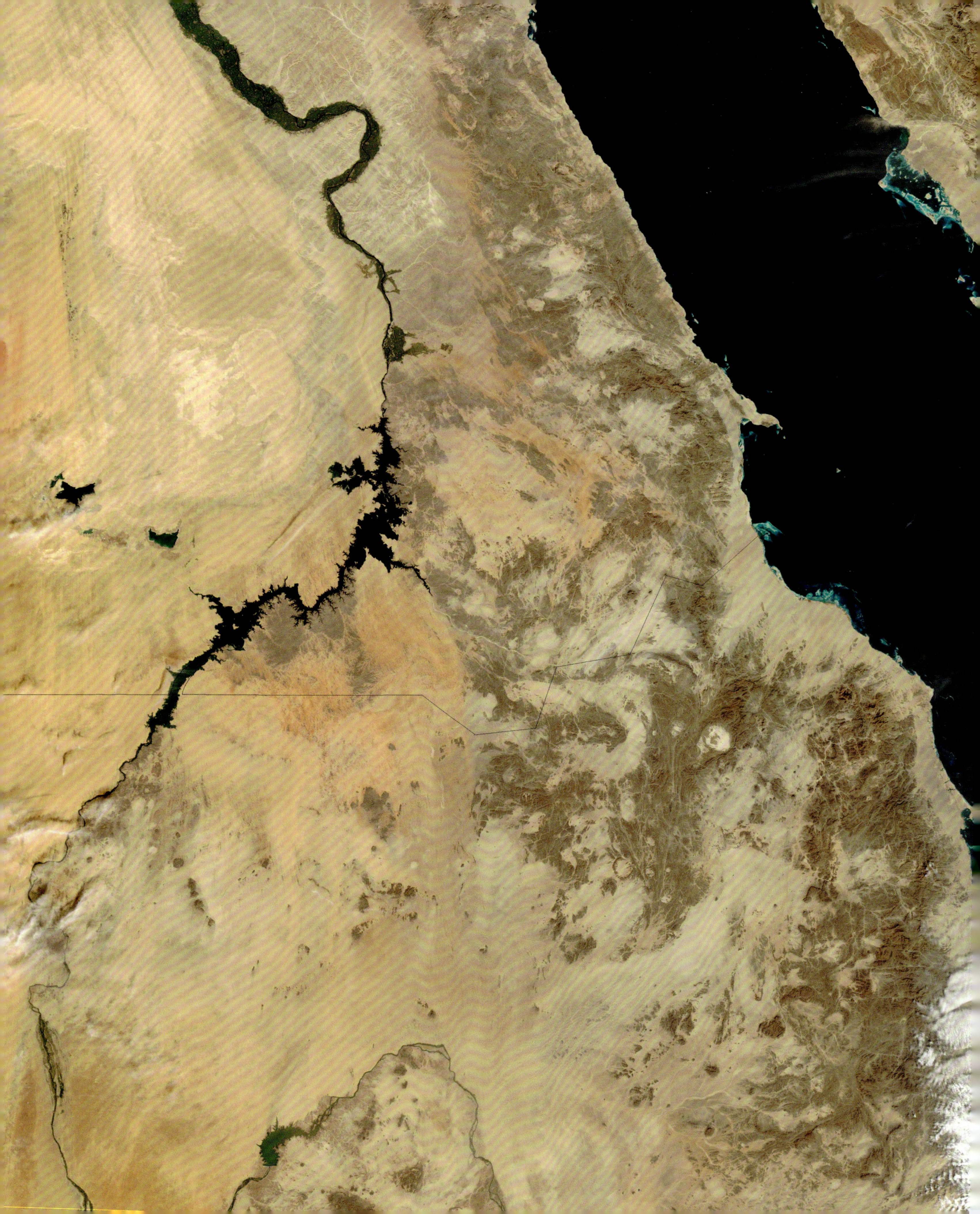

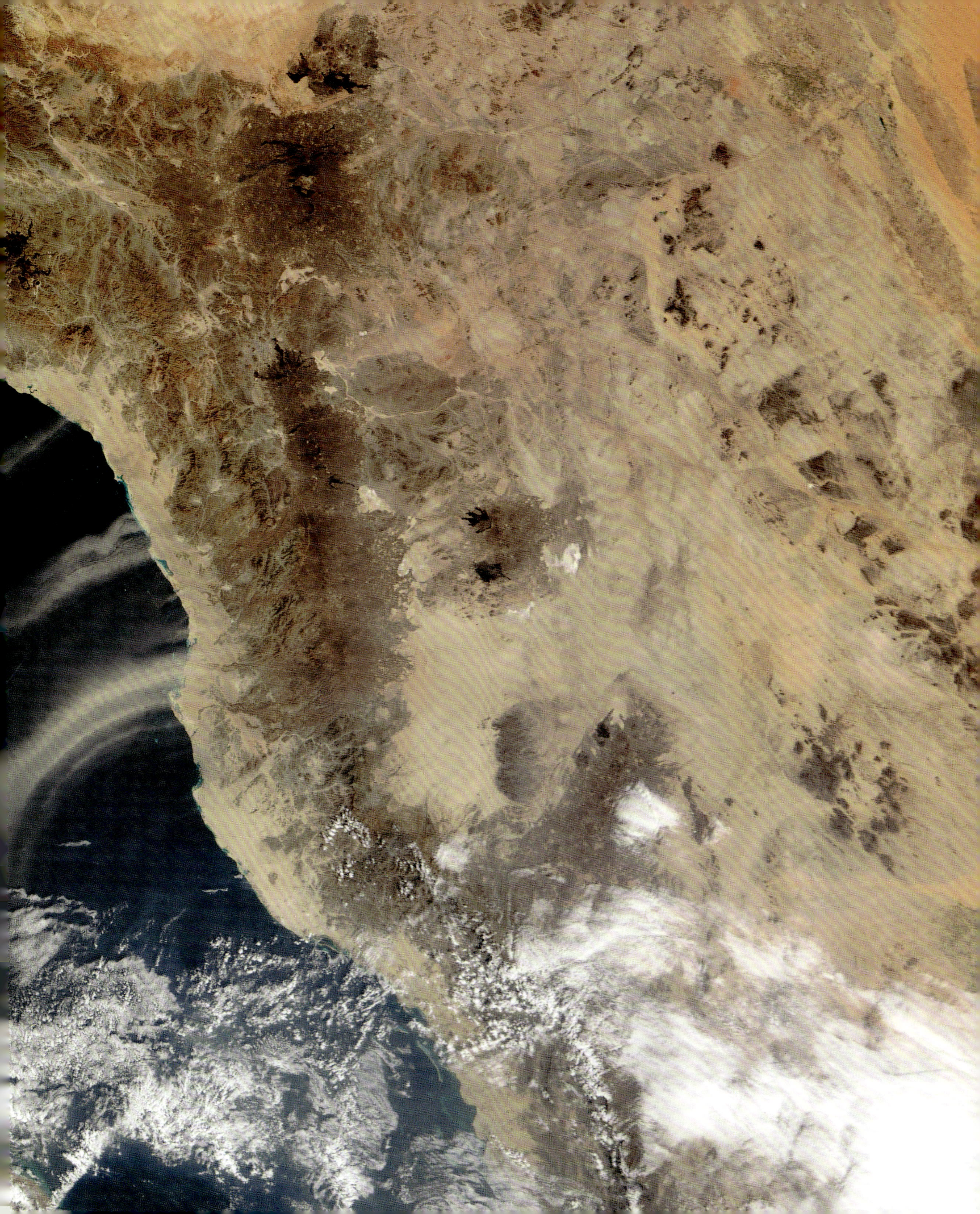

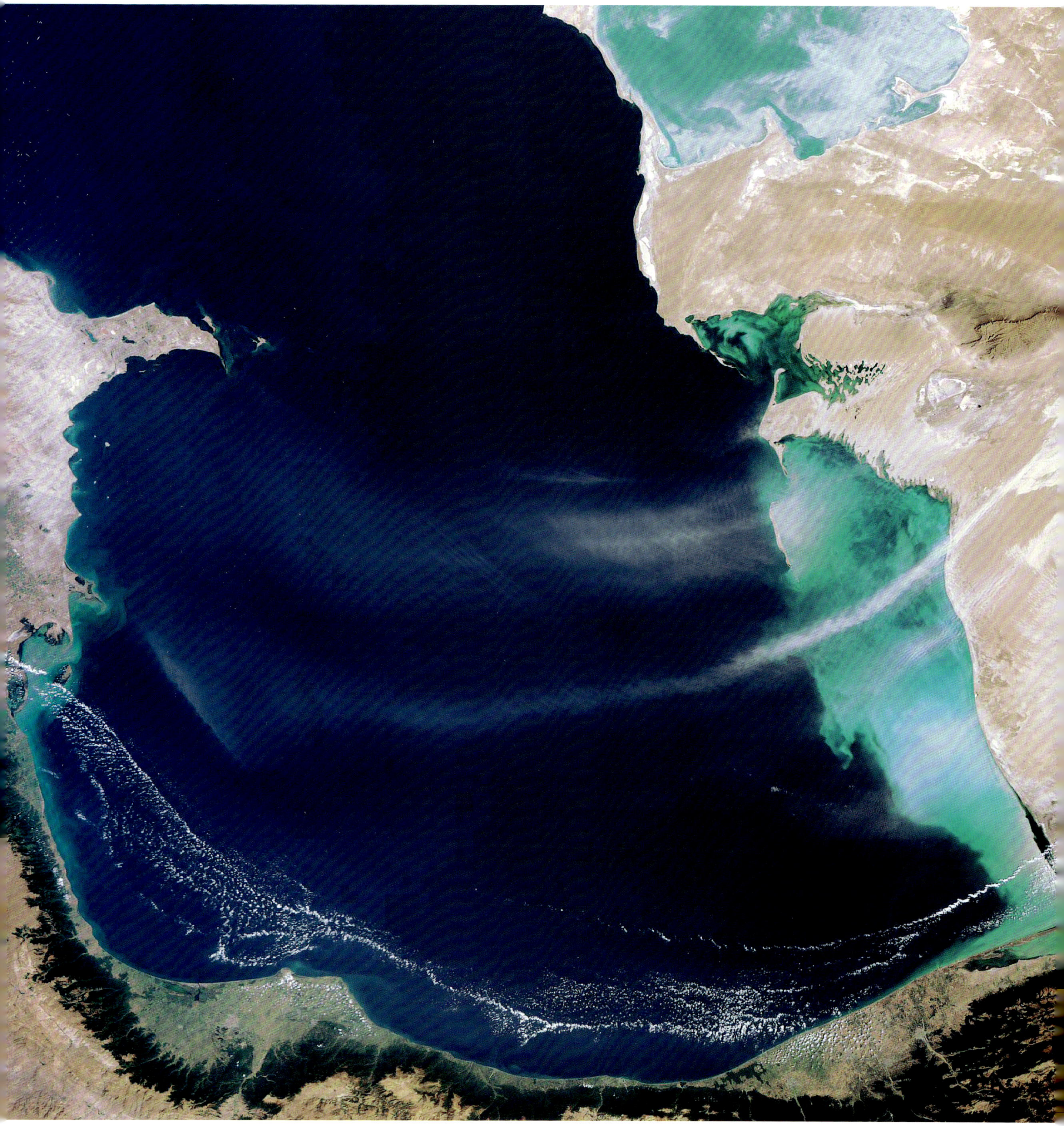

per Radar Bilder erstellen. Zudem gibt es Instrumente, die großflächig die chemischen Vorgänge in der Atmosphäre, die Entwicklung der Vegetation an Land oder das Auftreten von Chlorophyll in den Meeren überwachen. Heinz Sontag, bei Airbus in Friedrichshafen für den Bau des Sentinel-2 genannten Copernicus-Satelliten zuständig, ist überzeugt, dass die Bilder seiner Hochleistungskamera vielseitig genutzt werden: *Die Daten werden frei verfügbar sein. Das ist eine ganz tolle Sache. Viele kleinere und größere Firmen bereiten bereits solche Daten für spezielle Anwendungen auf. Die werden noch viele gute Ideen haben.*

In den USA waren die Bilder der Landsat-Satelliten zunächst unter Verschluss. Anfang der neunziger Jahre hat der US-Kongress entschieden, die Daten freizugeben. *Dann ist die Anzahl der Anwendungen regelrecht explodiert. Das wird bei Sentinel-2 so ähnlich sein,* freut sich Heinz Sontag. EU und ESA haben aus dem Fehler der Kollegen in Übersee gelernt. Die Copernicus-Daten sind für alle da. Ausgiebige Nutzung ist ausdrücklich erwünscht. Meereskundler, Regionalplaner, Stadtentwickler, Land- und Forstwirte, Geografen, Geologen, Firmen auf der Suche nach Ressourcen im Boden, Aufsichtsbehörden, Fischereibetriebe

to track che-mical processes in the atmosphere, vegetation growth on land, and the presence of chlorophyll at sea on a large scale. Heinz Sontag, who is directing the construction of the Copernicus satellites known as Sentinel-2, in Friedrichshafen, Germany, has no doubt that the images from his high-performance cameras will have a multitude of uses: *The data will be freely available, which is entirely a good thing. There are plenty of small firms who process such data for special applications, and they won't be short of good ideas.*

Images from the Landsat orbiters were initially restricted in the USA, but at the beginning of the 1990s, Congress decided to release the data. *The number of uses just exploded, and we expect it to be the same with Sentinel-2,* grins Heinz Sontag. The EU and ESA have learnt from the mistakes of their colleagues overseas: the Copernicus data will be available to all and as much use as possible should be made of it. Oceanographers, regional and urban planners, agriculturists and forestry researchers, geographers, geologists, companies looking for resources in the ground, the security services, fishing fleets, and a host of others will fall upon this information from

à l'échelle du globe les processus chimiques dans l'atmosphère, le développement végétal et la concentration en chlorophylle de l'eau de mer. Heinz Sontag, responsable chez Airbus de la construction du satellite Sentinelle-2 de Copernicus, est convaincu que les images de ses caméras ultraperformantes trouveront des applications variées : *Les données seront à la disposition de tous. Ce qui est formidable. Les nombreuses entreprises qui traitent ces données pour des applications spécifiques ne manqueront pas de bonnes idées.*

Aux États-Unis, les images transmises par les satellites Landsat étaient protégées, jusqu'à ce que le congrès américain décide au début des années 1990 d'ouvrir ces données au public. *Le nombre d'applications a ensuite littéralement explosé. Ce sera la même chose pour Sentinelle-2,* se réjouit Heinz Sontag. L'Union européenne et l'ESA ont tiré les leçons des erreurs de leurs confrères d'outre-Atlantique. Les données Copernicus seront en libre accès, et leur large utilisation explicitement souhaitée. Océanographes, planificateurs régionaux, développeurs urbains, agriculteurs, sylviculteurs, géographes, géologues, entreprises en quête de ressources souterraines, autorités de contrôle, exploitations halieu-

en muchos canales espectrales o que elaboran imágenes por radar. Además, existen instrumentos que controlan a gran escala los procesos químicos atmosféricos, el desarrollo de la vegetación en tierra o la aparición de clorofila en los océanos. Heinz Sontag, responsable de la construcción en Airbus, Friedrichshafen, del satélite Sentinel-2 de Copérnico, está convencido de que las imágenes de su cámara de alto rendimiento tendrán múltiples utilizaciones: *Los datos serán de dominio público. Esto es algo estupendo. Muchas firmas pequeñas preparan estos datos para aplicaciones especiales. Todavía tendrán muy buenas ideas.*

En Estados Unidos, las imágenes de los satélites Landsat estaban en principio clasificadas. A comienzos de la década de 1990, el congreso estadounidense decidió autorizar su difusión. *Después estalló realmente el número de aplicaciones. Algo parecido ocurrirá con el* Sentinel-2, dice ilusionado Heinz Sontag. UE y ESA han aprendido de los errores de sus colegas de ultramar. Los datos Copérnico estarán disponibles para todos. Se desea, expresamente, que sean muy utilizados. Oceanógrafos, planificadores regionales, urbanistas, agricultores y forestales, geógrafos, geólogos, empresas en busca de recursos en el

Linke Seite und rechts: Staubstürme über dem südlichen Kaspischen Meer und über dem Persischen Golf (Envisat)

Left and right: Dust storms over the southern Caspian Sea and the Persian Gulf (Envisat)

Page de gauche et à droite : tempêtes de sable dans le sud de la mer Caspienne et dans le golfe Arabo-Persique (Envisat)

Página izda. y dcha.: Tormenta de arena sobre el sur del mar Caspio y sobre el golfo Pérsico (Envisat)

Seite 172/173: »Staubbrücke« von Saudi-Arabien über das Rote Meer in den Sudan. Ganz links: Niltal mit Stausee (Terra)

Pages 172–3: A "dust bridge" across the Red Sea from Saudi Arabia on the right to the Sudan on the left. In the far left, a reservoir in the Nile valley (Terra)

Pages 172 et 173 : « pont de poussière » sur la mer Rouge entre l'Arabie saoudite et le Soudan. À gauche : barrage dans la vallée du Nil (Terra)

Páginas 172/173: «Puentes de arena» sobre el mar Rojo, desde Arabia Saudí a Sudán. A la izda.: Valle del Nilo con embalse (Terra)

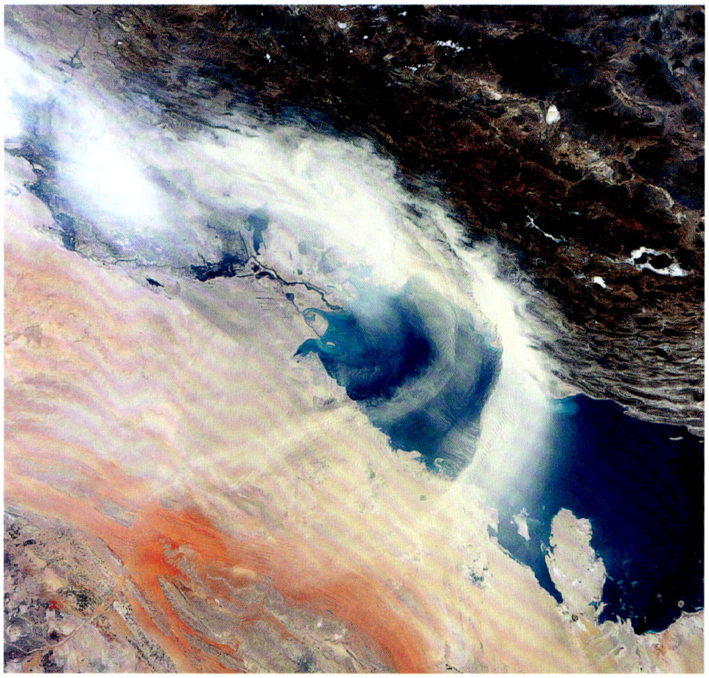

etc. werden sich auf die himmlischen Daten stürzen und sie für ihre Zwecke nutzen.

Im Bereich der Ozeane gibt es eine Vielzahl geplanter Anwendungen. Die Satelliten ermöglichen die weltweite Überwachung des Schiffsverkehrs, was bisher nur in küstennahen Gewässern möglich ist. Damit lassen sich zum Beispiel Fischereiflotten verfolgen und so die Einhaltung von Fangverboten kontrollieren. Dann gibt es keine blinden Flecken auf den Weltmeeren mehr, in denen Kriminelle unbeobachtet illegal fischen und so die Bestände dramatisch reduzieren können. Daten aus dem All zeigen zudem, wo Algenblüten drohen und den Fischen die Luft zum Atmen nehmen könnten. Treibeisfelder werden erfasst und ermöglichen gezielte Warnungen und Routenempfehlungen für die Schifffahrt. Umweltsünder, die auf hoher See ihre Tanks reinigen, werden aus dem All leicht zu erwischen sein. Radarwellen sind empfindlich genug, um Öl auf der Meeresoberfläche zu entdecken. Werden diese Daten umgehend an die Behörden am Boden weitergeleitet, lassen sich die Täter zügig fassen. Das hatte man sich ursprünglich schon von Envisat erhofft, aber dort waren die Abläufe der Datenübermittlung und Auswertung in der Regel zu lang.

the sky to use for their own purposes.

A number of applications are already planned for the oceans. Satellites allow shipping to be tracked globally, a feat once only possible in coastal waters. This means that fishing fleets can be tracked and adherence to fishing quotas monitored; there will be no more blind spots on the high seas where criminals can fish illegally, thereby dramatically reducing stocks of fish. Data from space also reveals where algae growth is threatening to choke the oxygen that fish need to breathe. Fields of drift ice can be tracked and specific warnings and course recommendations can be passed on to shipping traffic. Polluters who flush their tanks at sea are easy to catch from space as radar waves are very sensitive to oil on the surface of the sea. If such data is passed on to the authorities on the ground in a timely manner, the malefactors are soon tracked down. It had been hoped initially that Envisat would be able to provide this service, but the process of transmitting and analyzing the data proved to be just too laborious.

tiques, etc. se rueront sur les données célestes qu'ils exploiteront à leur profit.

Dans le domaine maritime, les applications attendues sont nombreuses. Les satellites permettront de surveiller le trafic maritime mondial, ce qui n'est actuellement possible que dans les zones côtières. Il sera par exemple aisé de suivre les flottes de pêche et de contrôler le respect des quotas. Il n'y aura plus d'espaces non surveillés, dans lesquels les criminels pratiquent la pêche illégale, décimant les réserves mondiales. Les données spatiales révèlent en outre les zones où la prolifération des algues risque d'étouffer les poissons. Le repérage des glaces flottantes permet d'émettre des alertes ciblées et des recommandations pour le trafic maritime. Il sera également plus facile de prendre sur le fait les navires pollueurs qui dégazent au large. La transmission immédiate des données recueillies par les ondes radar, assez sensibles pour déceler le pétrole flottant sur la mer, permettra aux autorités compétentes d'arrêter rapidement les coupables. Envisat avait déjà fait naître de tels espoirs, mais la durée de transmission et d'analyse de ses données était généralement trop longue.

suelo, organismos de inspección y empresas pesqueras, entre otros muchos, se abalanzarán sobre los datos celestes y los utilizarán para sus propios fines.

En lo que se refiere a los océanos, hay un gran número de aplicaciones proyectadas. Los satélites permiten la supervisión mundial del tráfico marítimo, algo que hasta ahora sólo era posible en las aguas cercanas a la costa. Así se podrá seguir, por ejemplo, a flotas pesqueras y controlar el cumplimiento de las prohibiciones de pesca. Por tanto, en los océanos ya no existirán puntos ciegos. Los datos desde el espacio también muestran dónde la amenaza de proliferación de algas podría quitarles a los peces el oxígeno para respirar. Se registran los campos de hielo permitiendo advertencias selectivas y recomendaciones de rutas de navegación. Desde el espacio, se podrá sorprender fácilmente a los criminales medioambientales que limpian sus depósitos en alta mar. Las ondas de radar son lo suficientemente sensibles como para descubrir el petróleo en la superficie marina. Si estos datos se transmiten sin demora, los culpables pueden ser capturados rápido. Así se quiso para el Envisat, pero sus plazos de transmisión y análisis de datos eran demasiado largos.

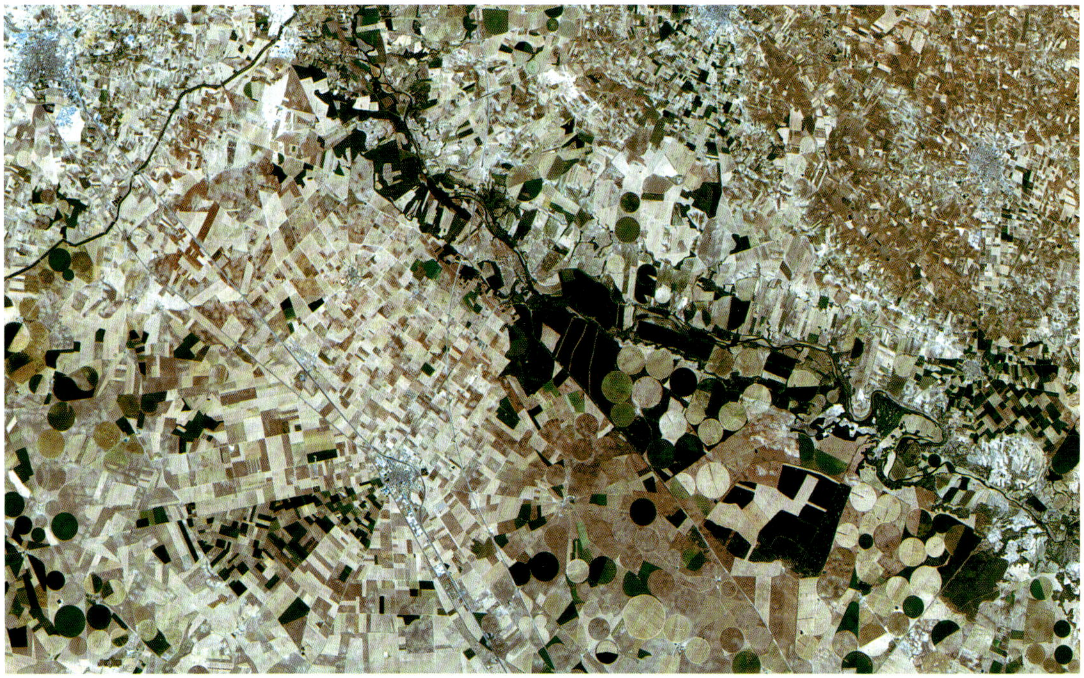

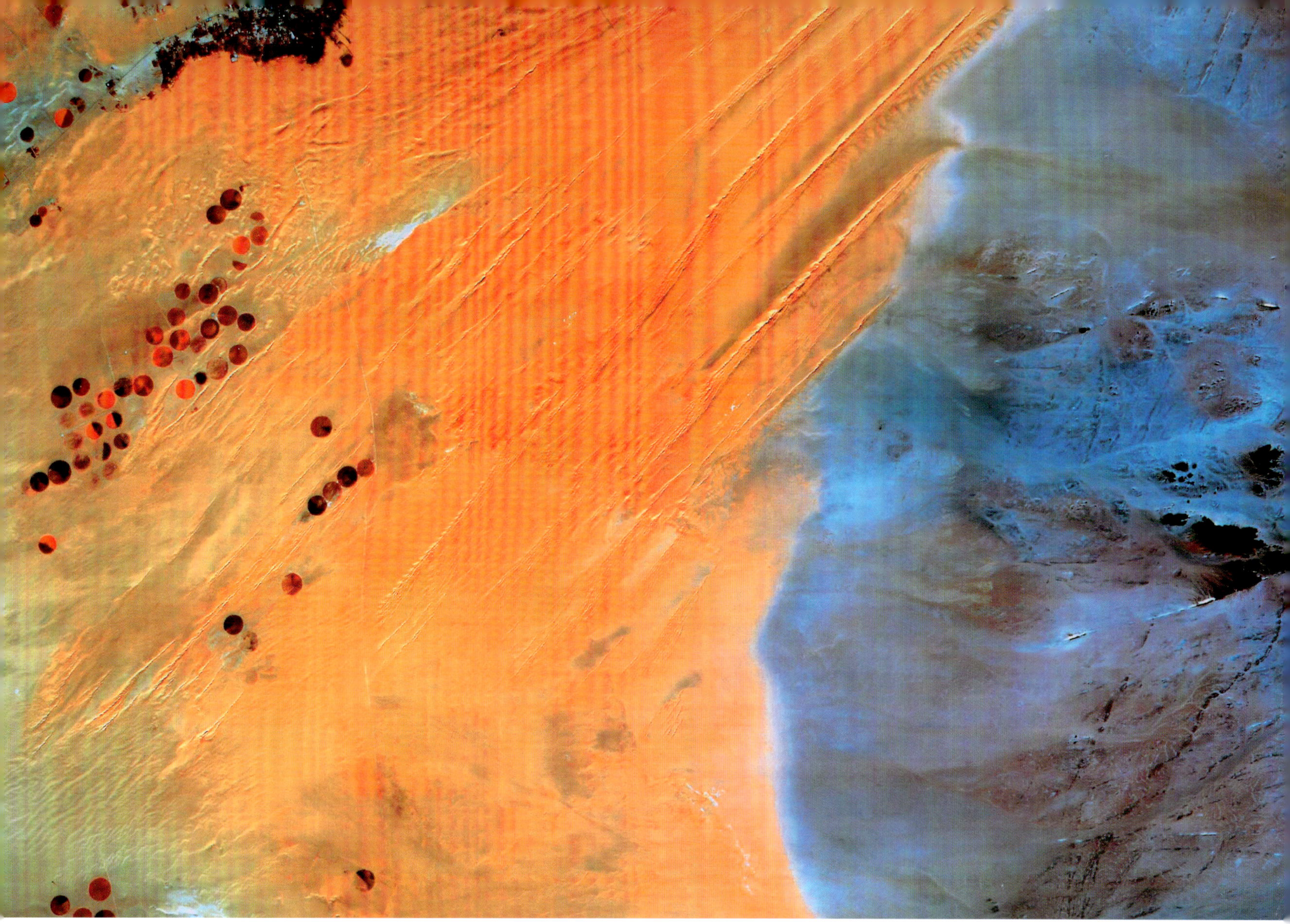

Linke Seite: Bewässerung in Albacete in Spanien. Die runden Felder fallen auf (Envisat)

Left: Irrigation at Albacete in Spain. The circular fields are clearly visible (Envisat)

Page de gauche : irrigation à Albacete, en Espagne. On remarque les champs circulaires (Envisat)

Página izda.: Regadíos en Albacete, España. Destacan los campos circulares (Envisat)

Oben: Durch die Bewässerung kreisrunde, oft mehr als 500 m große Felder in Saudi-Arabien (DLR)

Above: Irrigation has created circular fields in Saudi Arabia which can be more than 1600 ft (500 m) in diameter (DLR)

Ci-dessus : irrigation à pivot central sur des grands champs de plus de 500 m en Arabie saoudite (DLR)

Arriba: Campos en Arabia Saudí, de más de 500 m, debidos a los regadíos circulares (DLR)

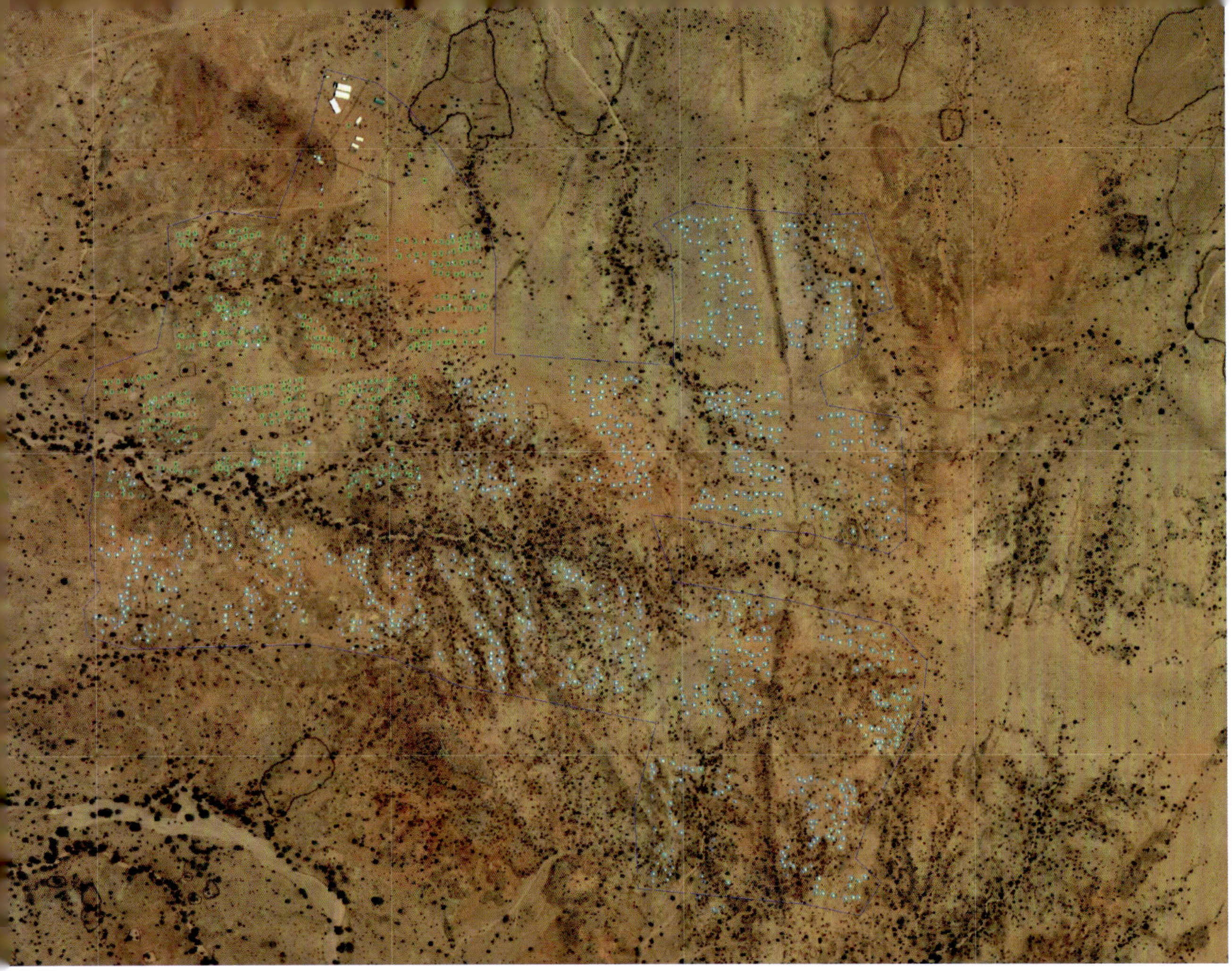

1800 Zelte stehen in diesem Camp für Darfur-Flüchtlinge im Tschad in Afrika. Experten ermitteln anhand von Satellitendaten die benötigte Menge an Hilfsgütern (QuickBird)

1800 tents have been pitched in this camp in Chad, Africa for refugees from Darfur. Satellite data allows experts to allot the right amount of supplies (QuickBird)

1 800 tentes forment ce camp de réfugiés du Darfour, au Tchad, en Afrique. Les experts calculent à l'aide des données satellite les quantités d'aide humanitaire nécessaires (QuickBird)

En este campo para refugiados de Darfur en el Chad, África, se levantan 1.800 tiendas. Los expertos determinan las necesidades de ayuda humanitaria con datos de los satélites (QuickBird)

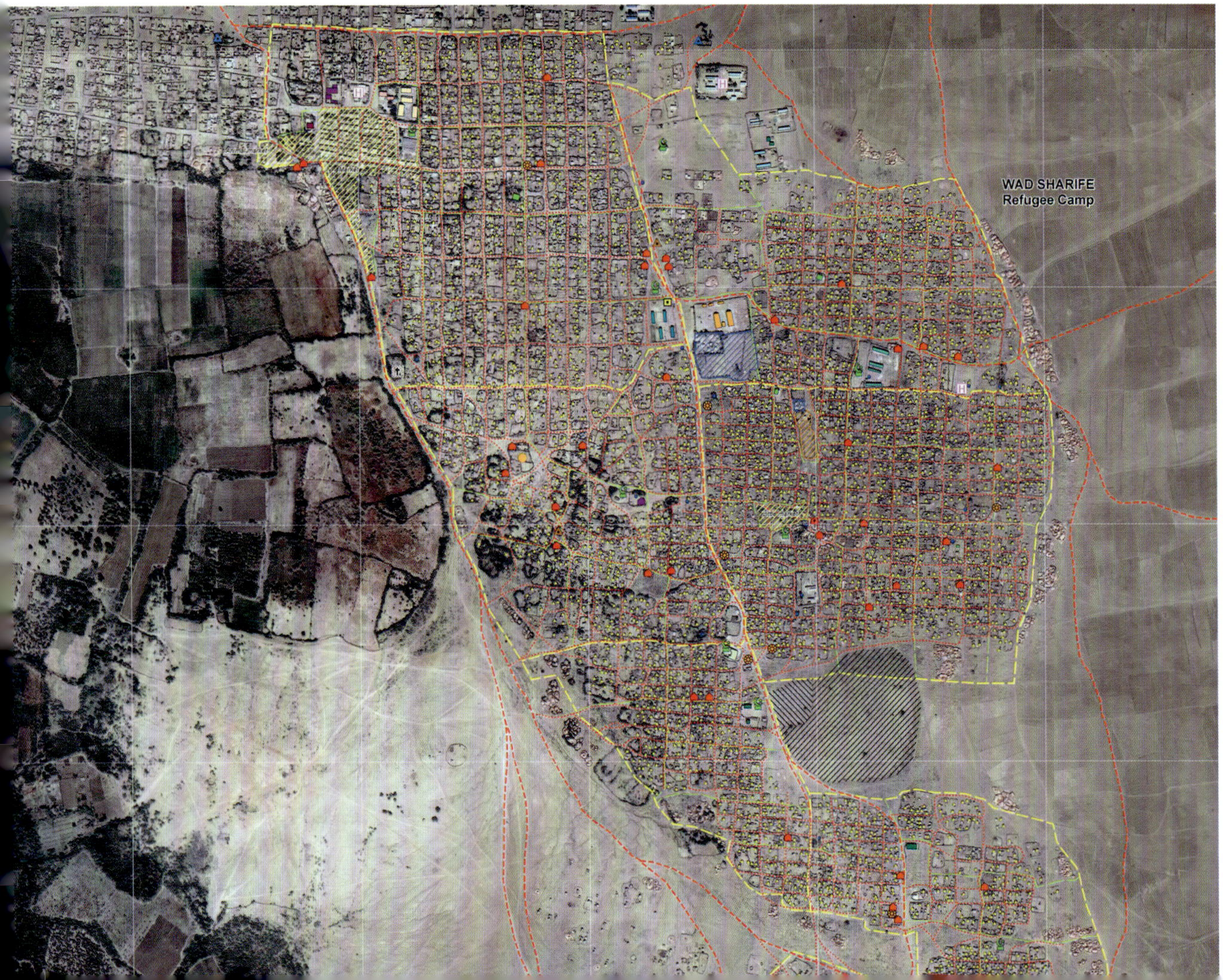

Flüchtlingscamp im Sudan: Das Satellitenbild mit 60 cm Auflösung zeigt Wege, Schulen, Moscheen, Mühlen, Märkte etc. Es dient Hilfskräften als Karte vor Ort (QuickBird)

A refugee camp in the Sudan. The satellite's resolution of 2 ft (60 cm) reveals paths, schools, mosques, mills, markets, and the like, providing images which can be used as maps by local aid workers (QuickBird)

Camp de réfugiés au Soudan : l'image satellite d'une résolution de 60 cm montre les routes, écoles, mosquées, moulins, marchés, etc. Elle sert de carte aux équipes de secours sur place (QuickBird)

Campo de refugiados en Sudán: la imagen de satélite, de 60 cm de resolución, muestra caminos, escuelas, mezquitas, molinos, mercados, etc. Sirve de mapa a los equipos de socorro (QuickBird)

Satelliten retten Leben

Existenziell wichtig sind die Anwendungen im Bereich der humanitären Hilfe. Mehr als vierzig Millionen Menschen sind weltweit auf der Flucht – oft vor Krieg oder einem brutalen Regime, und immer häufiger zwingen auch Naturkatastrophen wie Dürren, Überflutungen oder Erdbeben Menschen, ihre Heimat zu verlassen. Mit den Copernicus-Satelliten lässt sich gezielt vor Stürmen und Überflutungen warnen. Dürren zeichnen sich langfristig ab, und Satellitendaten ermöglichen, frühzeitig Hilfe zu organisieren. Leider werden die Daten den betroffenen Menschen oft nicht so direkt helfen, dass die drohenden Unglücke noch abgewendet werden. Einen Sturm kann man nicht aufhalten. Bei einer drohenden Dürre aber ließe sich etwas tun. Immerhin ermöglichen die aus dem All gesammelten und von Copernicus zur Verfügung gestellten Informationen Staaten und Hilfsorganisationen einen schnellen und gezielten Einsatz. Die ESA stellt schon heute vielen afrikanischen Staaten die Daten, die sie mit ihren Satelliten über diese Länder gewinnt, kostenlos zur Verfügung. Leider ist das keine Garantie, dass die Staaten die Daten auch nutzen und in vernünftiges politisches Handeln umsetzen.

Naturkatastrophen gibt es nicht nur in weniger entwickelten Ländern weit außerhalb Europas. So etwas passiert überall. Jahr für Jahr kommt es zu rund 50.000 Waldbränden in Europa. Überflutungen und Schlammlawinen haben bei uns im vergangenen Jahrzehnt fast eintausend Menschen getötet und gut 25 Milliarden Euro Schaden angerichtet. Wenn ein Satellit die Schäden nur einer Überschwemmung minimieren könnte, hätte er sich schon bezahlt gemacht. Mit Copernicus setzen die Europäische Union und ESA auf ein gutes Vorwarnsystem für Flutkatastrophen. Satellitendaten werden deutlich zeigen, wo sich Regenmassen zusammenbrauen, wo schon viel Wasser steht, wie es abfließen könnte und welchen Einfluss ungünstige Windverhältnisse haben. Das alles ist schon lange keine bloße Theorie mehr. Bei der vom Hurrikan

Satellites save lives

Satellites are of crucial importance to humanitarian aid. There are more than 40 million asylum-seekers across the world, fleeing their homes because of war, brutal regimes, and—more and more frequently—natural disasters such as droughts, floods, or earthquakes. The Copernicus satellites provide accurate warnings of storms and floods; droughts take longer to set in, and the satellite data allows aid to be organized in time. Unfor-tunately, such data is often unable to help the people concerned directly and is of course unable to avert threatening disaster—you can't stop a hurricane—but as drought begins to take hold, something can always be done. The data gathered in space and made available by Copernicus is able to help states and aid organizations provide a speedy and concentrated response. The ESA is already providing a number of African states with the data it has gathered about their countries free of charge, although the provision of data isn't a guarantee of political action, of course.

Natural disasters don't just occur in developing countries far beyond Europe, they can happen anywhere. There are some 50,000 forest fires in Europe every year, and over the last decade, floods and mudslides have cost about 1000 people their lives and caused damage amounting to some 25 billion euros. If a satellite could minimize the damage caused by only one flood, it would already have paid for itself. The European Union and the ESA are establishing a good early warning system for floods with Copernicus—satellite data will reveal early indications of where heavy rain is about to fall, where there are already large amounts of water, where it might run off, and what influence unfavorable winds might have. All this is no longer the stuff of theory: during the catastrophic flooding caused by Hurricane Katrina in 2005, data from European satellites alerted people to the second wave of flooding and permitted evacuation in time.

Les satellites sauvent des vies

Les applications en matière d'aide humanitaire sont quant à elles vitales. Plus de quarante millions de réfugiés à travers le monde fuient les guerres ou les régimes tyranniques, et sont fréquemment contraints de quitter leur terre natale suite à des catastrophes naturelles. Les satellites Copernicus facilitent l'émission d'alertes ciblées en cas de tempête ou d'inondation. Capables de déceler les premiers signes de sécheresse, ils permettent d'organiser l'aide humanitaire à un stade précoce. Malheureusement les données satellitaires ne pareront pas les dangers qui menacent les populations. Nul ne peut arrêter la tempête. En revanche, il est possible d'agir en cas de sécheresse imminente. Les informations satellitaires de Copernicus permettent en effet aux États et organisations humanitaires d'intervenir rapidement et de manière ciblée. L'ESA accorde d'ores et déjà à de nombreux États africains un libre accès aux données satellitaires qui les concernent. Hélas, rien ne garantit que les données sont utilisées dans le cadre d'une action politique appropriée.

Les catastrophes naturelles ne sont pas l'apanage des pays en développement situés hors d'Europe. Au fil des ans, l'Europe a connu environ 50 000 feux de forêt. Inondations et coulées de boue ont coûté la vie à un millier de personnes au cours de la dernière décennie, causant des dégâts à hauteur de 25 milliards d'euros. Un satellite qui contribue à minimiser les dommages provoqués par la crue d'un seul fleuve est rentabilisé. Avec Copernicus, L'Union européenne et l'ESA misent sur un système d'alerte efficace contre les inondations. Les données satellitaires indiqueront précisément les zones de concentration des eaux de pluie, celles où le niveau des eaux est élevé, les zones d'écoulement possible, et l'influence potentielle des vents. Nous ne sommes plus dans la théorie depuis longtemps. Lors des inondations causées en 2005 par le passage de l'ouragan Katrina à la Nouvelle-Orléans, les satellites européens ont détecté la

Los satélites salvan vidas

Desde un punto de vista vital, las aplicaciones en el sector de la ayuda humanitaria son muy importantes. En todo el mundo hay más de cuarenta millones de refugiados que huyen de la guerra o de regímenes brutales y, cada vez con más frecuencia, también personas que se ven obligadas a abandonar su patria por catástrofes naturales, como sequías, inundaciones o terremotos. Con los satélites Copérnico se puede advertir selectivamente sobre tormentas e inundaciones. Las sequías se conocen con antelación y los datos de los satélites permiten organizar la ayuda temprana. Por desgracia, en muchas ocasiones los datos no pueden ayudar directamente a las personas afectadas evitando las desgracias inminentes. Nadie puede detener una tormenta. Sin embargo, se puede hacer algo ante una amenaza de sequía. Así y todo, la información recogida por Copérnico en el espacio y puesta a disposición de estados y organizaciones humanitarias permite una respuesta rápida y selectiva. Actualmente, la ESA pone ya a disposición de muchos estados africanos, de forma gratuita, los datos que obtiene con sus satélites al pasar por esos países. Por desgracia, esto no garantiza que los estados también utilicen los datos para transformar razonablemente la acción política.

Las catástrofes naturales no sólo ocurren en los países menos desarrollados y muy lejos de Europa. Suceden en todas partes. Año tras año se producen casi 50.000 incendios forestales en el continente europeo, y las inundaciones y las avalanchas de lodo mataron en la pasada década a casi mil personas y ocasionaron daños por valor de 25.000 millones de euros. Si un satélite pudiese minimizar los daños de una sola inundación, ya habría valido la pena. Con Copérnico, la Unión Europea y la ESA apuestan por un sistema de alerta temprana para inundaciones. Los datos de los saté-lites mostrarán claramente dónde se ciernen las masas de lluvia, dónde hay ya mucha agua, cómo podría desplazarse y la influencia que

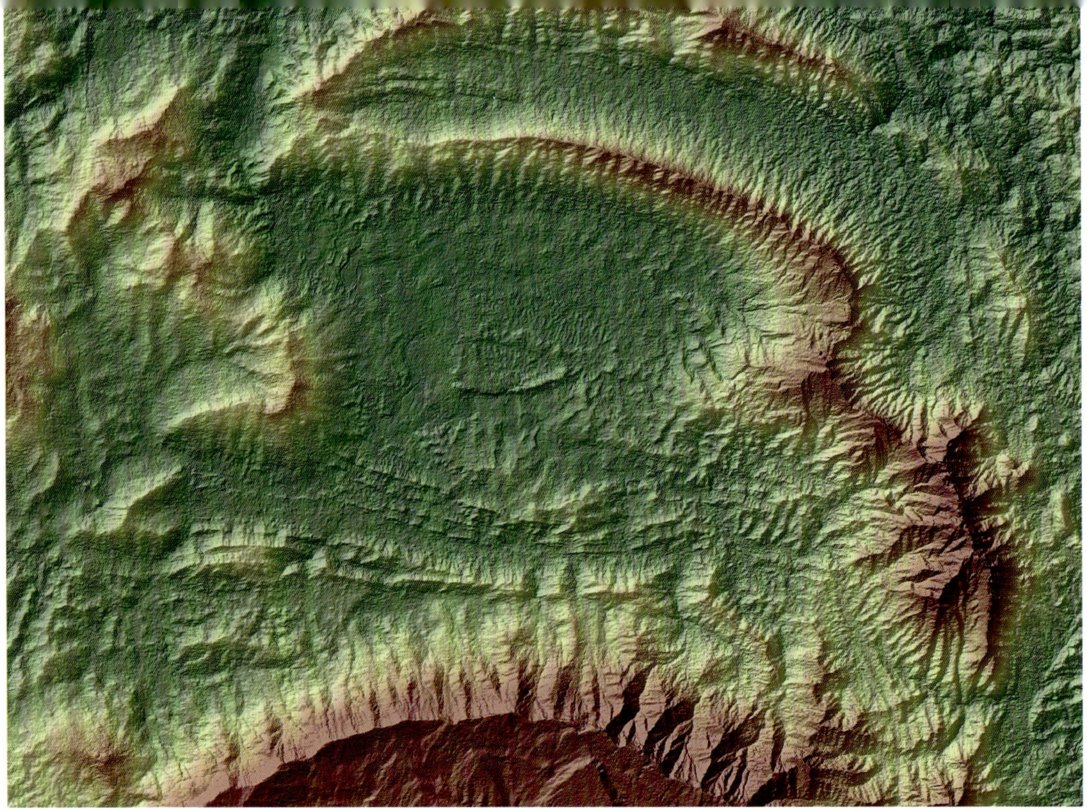

Katrina verursachten Flutkatastrophe in New Orleans im Jahr 2005 haben Daten europäischer Satelliten vor der zweiten Überflutung gewarnt und eine rechtzeitige Evakuierung ermöglicht. So etwas muss in wenigen Jahren Standard sein. Die Daten sind da. Man muss nur sicherstellen, dass die Satelliten sie kontinuierlich aufnehmen und die Daten am Boden umgehend zur Verfügung stehen und ausgewertet werden.

Eine neue Ära

Aus dem All zeigt sich auch, welche Folgen der Klimawandel für Europa hat. Wo treten Regenfälle verstärkt auf? Wo kommt es zu Dürren? Wo drohen Hitzewellen, die gerade in Großstädten für viele Menschen eine tödliche Gefahr darstellen? Welche Regionen sind für künftigen Ackerbau geeignet? Welche nicht oder nicht mehr? Daten dieser Art versetzen Entscheidungsträger in die Lage, gefährliche Entwicklungen zu erkennen und Gegenmaßnahmen zu planen. Der Bereich der Energiegewinnung gehört ebenfalls zu den Kernthemen, bei denen Satellitendaten eine immer größere Bedeutung zufallen werden. Wo gibt es geeignete Flächen für Solarfarmen, um im großen Stil Son-

This sort of thing should be routine in a few years time—the data is there for the taking. The only requirement is that the satellites should continually record the data and make it immediately available.

A new era

Observation from space also shows the consequences of climate change. Where is the heavy rainfall? Where are the droughts? Or the heatwaves, which present a mortal danger for big cities with large populations? Which regions can be developed for agriculture? Which are no longer suitable? Information of this sort allows decision-makers to recognize hazardous developments and to plan countermeasures. Energy provision is another key area where the role played by satellite data is taking on ever greater importance. Where can solar panels be deployed to harvest solar energy to the greatest advantage? Where should wind farms be placed out to sea? How can shipping traffic be optimized to help to protect deepwater installations?

With Copernicus data, targeted alerts of air pollution or increased ultraviolet radiation from the Sun are now a possibility. These and all the

deuxième vague d'inondations, et permis d'évacuer à temps la population. Ce genre d'intervention devrait devenir monnaie courante dans quelques années. Les données sont disponibles. Il suffit que les satellites les enregistrent en continu et qu'elles soient immédiatement accessibles et analysées sur Terre.

Une nouvelle ère

Les données spatiales dévoilent par ailleurs les effets du changement climatique en Europe. Quelles régions sont soumises à un renforcement des pluies, à la sécheresse, à la canicule ? Quelles régions seront à l'avenir propices à l'agriculture ? Quelles régions ne le seront pas ou plus ? Grâce à de telles données, les décideurs sont en mesure d'identifier les évolutions dangereuses et de prendre des contre-mesures. La production énergétique dépend, elle aussi, de plus en plus des données satellitaires pour identifier les surfaces adaptées à l'installation de fermes solaires ou de parcs éoliens, et optimiser le trafic maritime pour protéger les plates-formes pétrolières en haute mer.

Les alertes ciblées émises en cas de pollution atmosphérique ou de rayonnement ultraviolet intense

tienen las circunstancias de viento desfavorables. Hace tiempo que esto es algo más que una mera teoría. En las inundaciones de Nueva Orleans ocasionadas por el huracán Katrina en el año 2005, los datos de los satélites europeos advirtieron antes de la segunda inundación e hicieron posible la oportuna evacuación. Dentro de pocos años, esa será la norma.

Una nueva era

Desde el espacio también se muestran las consecuencias del cambio climático en Europa ¿Dónde se presenta un aumento de las precipitaciones? ¿Dónde hay sequía? ¿Qué lugares están amenazados por olas de calor, que representan un peligro mortal para muchas personas en las grandes urbes? ¿Qué regiones son aptas para una futura explotación agrícola? ¿Cuáles no en el momento actual y en el futuro? Los datos de este tipo ponen a los responsables en condiciones de reconocer desarrollos peligrosos y planificar contramedidas. El sector de la obtención de energía es igualmente uno de los asuntos nucleares en el que los datos de los satélites cada vez tendrán mayor significado. Con ellos se podrán conocer los lugares más adecuados para insta-

Linke Seite: Radaraufnahme des Geländes im Regenwald bei Sabah in Malaysia. Ausschnitt 40 km breit (TerraSAR-X)

Left: A radar image of the rainforest around Sabah in Malaysia. The area illustrated is 25 miles (40 km) across (TerraSAR-X)

Page de gauche : image radar du relief dans la forêt tropicale de l'État de Sabah en Malaisie, couvrant une largeur de 40 km (TerraSAR-X)

Página izda.: Toma de radar del suelo de la selva tropical de Sabah, en Malasia. Detalle de 40 km de ancho (TerraSAR-X)

Unten: Lake Disappointment ist ein Salzsee in der Wüste Westaustraliens (Terra)

Below : Lake Disappointment is a salt lake in the deserts of Western Australia (Terra)

Ci-dessus: le lac Disappointment est un lac salé du désert de l'Australie-Occidentale (Terra)

Abajo: El lago Disappointment es un lago salado del desierto de Australia occidental (Terra)

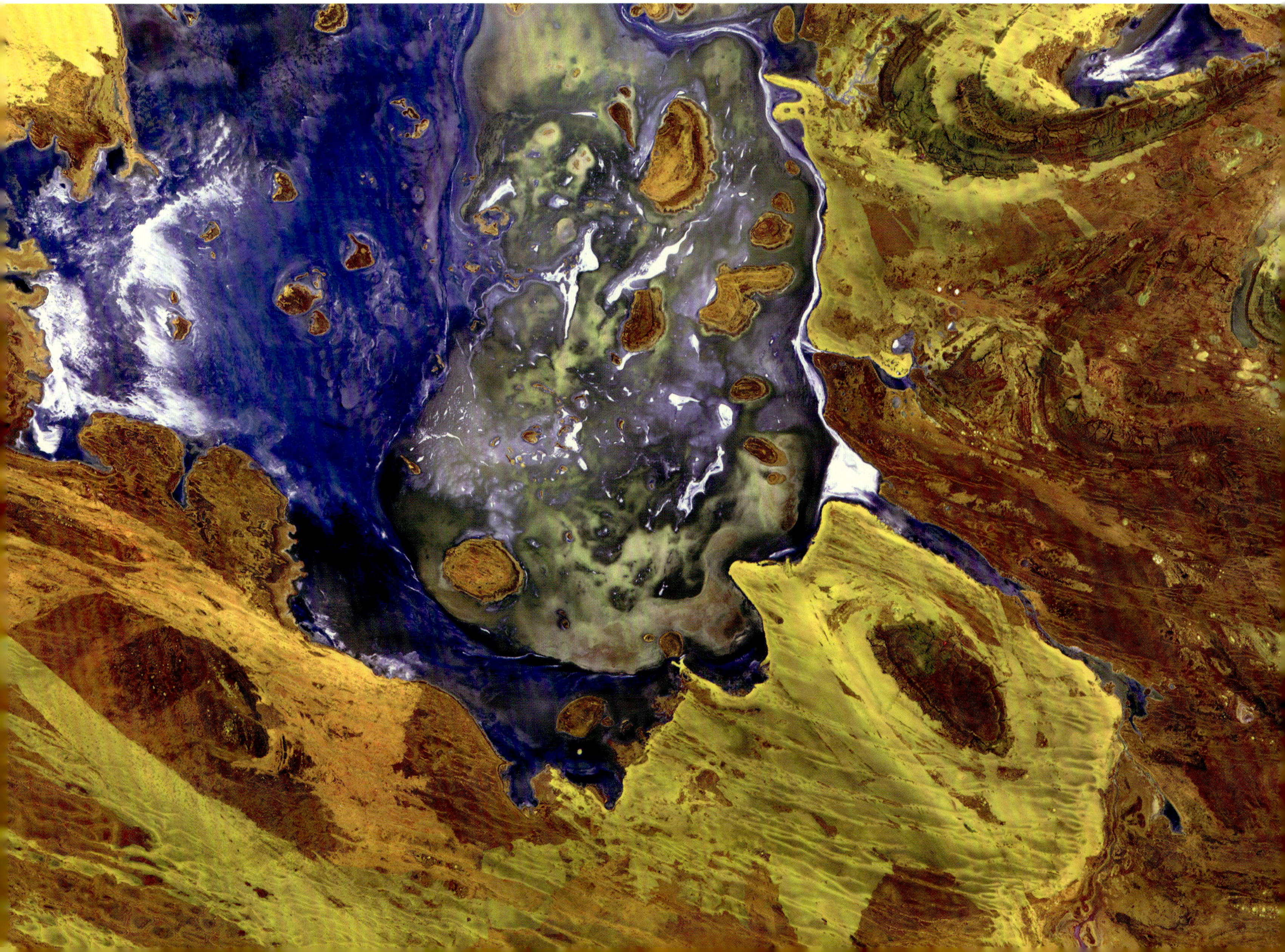

Unten: Ein weiterer MetOp-Satellit wartet gut verpackt in der Werkshalle in Friedrichshafen auf seinen Start

Below: Another well-protected MetOp satellite awaiting its launch in a Friedrichshafen workshop

Ci-dessous : un nouveau satellite MetOp bien empaqueté attend son lancement sur le site de production de Friedrichshafen

Abajo: Otro satélite MetOp espera su lanzamiento, bien empaquetado, en la nave de Friedrichshafen

Rechte Seite: Sentinel-2 (geplanter Start 2015) soll die Erde in verschiedenen Spektralbereichen extrem genau untersuchen

Right: Sentinel-2 (to be launched in 2015) will closely examine the Earth using a number of different spectra

Page de droite : Sentinelle-2 (lancement prévu en 2015) étudiera la Terre au moyen d'un instrument à plusieurs canaux spectraux

Página dcha.: El Sentinel-2 (lanzamiento previsto en 2015) debe analizar la Tierra con extrema precisión en diferentes bandas espectrales

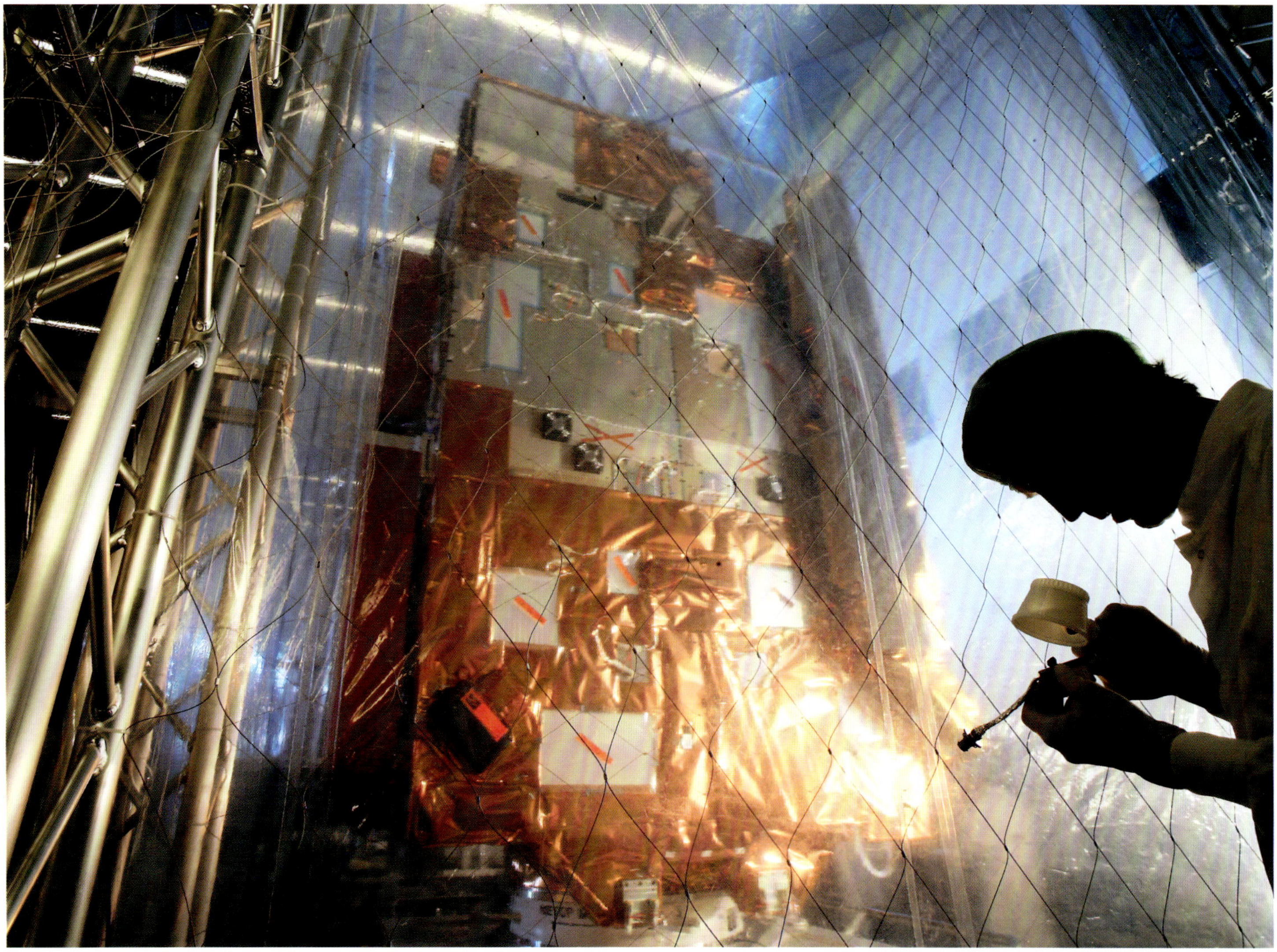

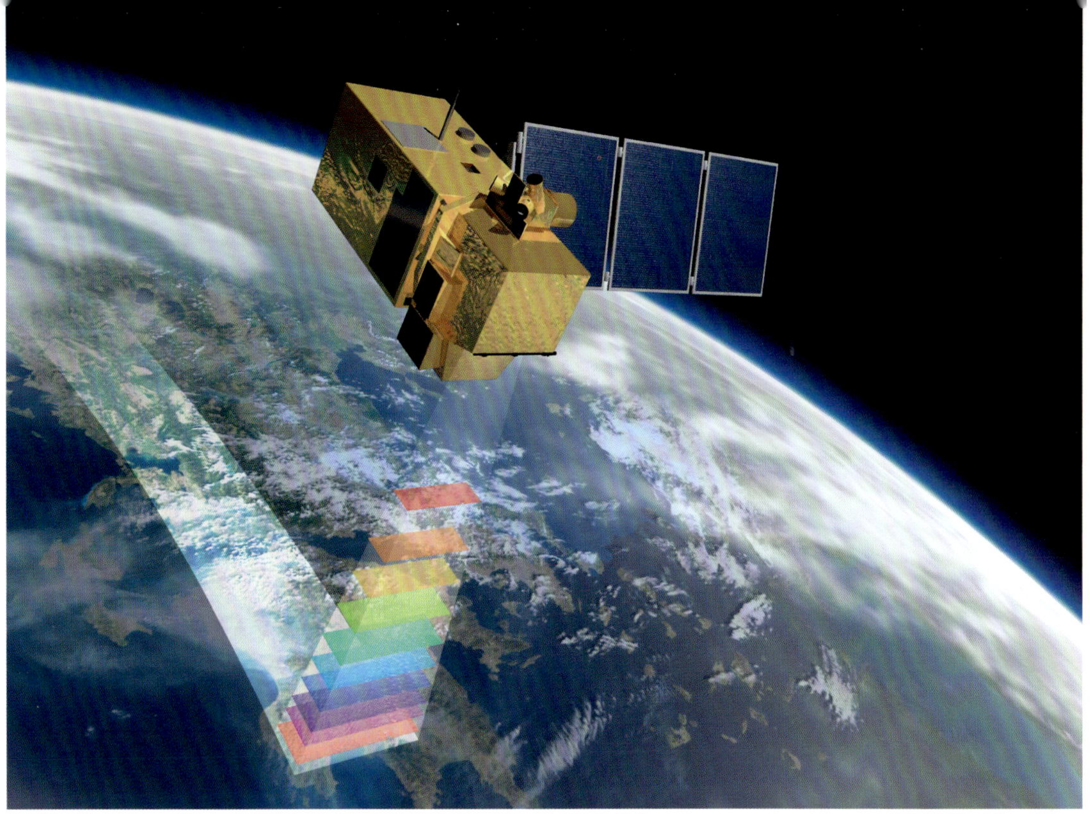

nenenergie zu nutzen? In welchen Gebieten lohnen sich Windanlagen auf dem Meer? Wie lässt sich der Schiffsverkehr optimieren, um Plattformen auf hoher See zu schützen?

Die Copernicus-Daten ermöglichen schon jetzt gezielte Warnungen vor Luftverschmutzung oder einer erhöhten Ultraviolett-Einstrahlung der Sonne. Diese und alle anderen Warndienste kann sich jeder je nach persönlicher Empfindlichkeit und Gefährdung zuschneiden. Ein Almbauer braucht sicher keine Sturmflutwarnung. Bewohner am Meer werden eher selten vor Lawinen gewarnt werden wollen. Auch Bereiche wie Stadtentwicklung, Landflucht oder Bevölkerungswanderung werden von Satellitendaten profitieren. Aus dem All ist unmittelbar zu sehen, wo Häuser gebaut oder Dörfer verlassen werden. Selbst die Tourismusindustrie nutzt die Daten von oben: Sie zeigen, ob besonders attraktive Landschaften von Umweltverschmutzung, Klimawandel oder großem Flächenverbrauch gefährdet werden.

Wenn die Satelliten erst fliegen und die Daten erhältlich sind, werden sich unzählige weitere Anwendungen ergeben. Der Kreativität der Nutzer sind keine Grenzen gesetzt. Einschränkungen bei der Verfügbarkeit der Daten gibt es nur, wenn die Informationen hohe Sicherheitsrelevanz haben.

other warning services can be tailored to individual exposure and danger levels—a mountain farmer will probably not need a flood warning, and coastal dwellers are equally unlikely to require an avalanche warning. Areas of concern, including urban development, rural depopulation, and migration, will all benefit from satellite data—from space, it is easy to tell where houses are being built or villages deserted. Even the tourist industry can make use of this data from above, as it reveals where beautiful landscapes are threatened by pollution, climate change, or greedy land use.

Once the satellites are launched and the data is available, countless further applications will become apparent—the creativity of the user of such data knows almost no bounds. The availability of data is to be restricted only where such information is of "strategic importance," whatever that may mean.

The first satellites in the Copernicus program launched in April 2014. If all goes to plan, the first five missions will be in space by 2019 and in an ideal scenario, some two dozen Earth observation satellites from all over the world will have joined them by 2025. These will take their place among the many other satellites

reposent déjà sur les données Copernicus. À partir des services d'alerte existants, chacun peut se créer un système d'alarme personnalisé. Un agriculteur de montagne n'aura que faire d'une alerte aux raz-de-marée, tout comme les alertes avalanche intéresseront peu les habitants des littoraux. Les satellites seront les témoins du développement urbain, de la désertification des campagnes et des migrations de population. Nouvelles maisons et villages désertés sont en effet aisément repérables sur les images satellites. L'industrie touristique utilise également les images spatiales pour surveiller les paysages menacés par la pollution, le changement climatique ou l'urbanisation intensive.

Quand les satellites en orbite commenceront à fournir des données, de multiples applications nouvelles verront le jour. La créativité des utilisateurs est pour ainsi dire débridée. L'accès aux données sera uniquement restreint dans le cas d'informations relevant d'un niveau élévé de sécurité stratégique.

Le premier satellite du programme Copernicus a été lancé en avril 2014. Si tout se passe comme prévu, les cinq premières missions seront en orbite en 2019, et d'ici 2025 environ deux douzaines de nouveaux satel-

lar granjas en las que aprovechar al máximo la energía solar, o en qué regiones pueden ser rentables las instalaciones eólicas marinas y también, entre otras muchas cuestiones, cómo optimizar el tráfico marítimo para proteger las plataformas en alta mar.

En la actualidad, los datos Copérnico ya permiten hacer advertencias selectivas de contaminación atmosférica o de aumento de la radiación ultravioleta del Sol. Cada uno puede diseñar a medida estos y otros servicios de advertencia. Los datos del satélite también se usan en ámbitos como el urbanismo, el éxodo rural o los flujos migratorios. Incluso la industria turística se aprovecha de ellos: muestran hasta qué punto están en peligro paisajes especialmente atractivos debido a la contaminación medioambiental, el cambio climático o la excesiva ocupación de terrenos.

Cuando los satélites vuelan y los datos están disponibles, se revelan innumerables aplicaciones adicionales. No hay prácticamente límites a la creatividad de los usuarios. Las restricciones en la disponibilidad de los datos sólo se dan si la información que contienen lleva aparejada una relevancia de seguridad lo suficientemente alta como para llamarse así.

El primer satélite del programa Copérnico se lanzó en abril de 2014.

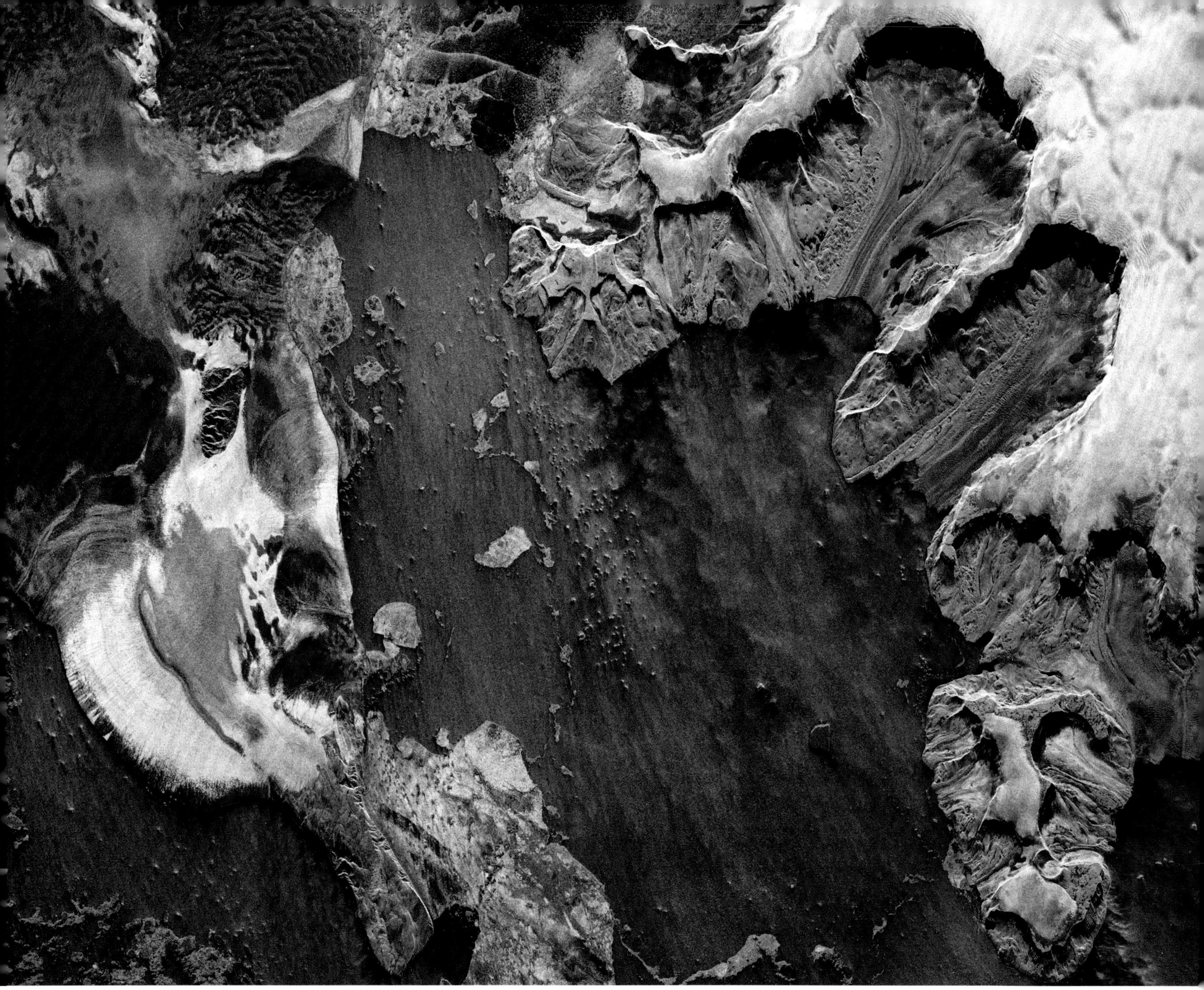

Oben: Radarbild einer skurril strukturierten Eislandschaft in der Antarktis

Above: A radar image revealing fascinating patterns in the Antarctic ice

Ci-dessus : image radar de la curieuse structure d'un paysage de glace en Antarctique

Arriba: Imagen de radar de un paisaje helado, de estructura grotesca, en la Antártida

Rechte Seite: Radaraufnahme von Gewitterwolken vor Panama (TerraSAR-X)

Right: A radar image of storm clouds off Panama (TerraSAR-X)

Page de droite : image radar de nuages d'orage au-dessus de Panama (TerraSAR-X)

Página dcha.: Toma de radar de nubarrones delante de Panamá (TerraSAR-X)

Der erste Satellit im Rahmen von Copernicus ist im April 2014 gestartet. Wenn alles gut geht, sind im Jahr 2019 die ersten fünf Missionen im All. Im Idealfall beziehen bis 2025 etwa zwei Dutzend neue Erdbeobachtungssatelliten aus allen Teilen der Welt Posten im All. Hinzu kommen viele weitere Satelliten, die bereits im Einsatz sind und deren Daten ebenfalls zur Verfügung stehen. Man kann nur ahnen, wie bedeutend die Erdbeobachtung im kommenden Jahrzehnt werden wird. Schon jetzt nutzen mehr als 3000 Forschungsprojekte die Daten der ESA-Satelliten. Über einhundert Terabyte an Daten kommen jedes Jahr hinzu, das entspricht der Datenmenge von etwa 200.000 Compact Discs. Allein im Bereich Klima gibt es ein Archiv, in dem sich für alle Regionen der Welt die Daten der vergangenen dreißig Jahre befinden: Daten zur Wolkenbedeckung, zu Treibhausgasen in der Atmosphäre, zum Ozongehalt, zur Temperatur des Oberflächenwassers der Ozeane, der Höhe des Meeresspiegels, des Vorkommens von Meereis, zur Farbe der Ozeane, zum Verhalten von Gletschern und Polkappen, zum Aussehen der Erdoberfläche etc.

Es soll später keiner sagen, er habe um den Zustand unserer Erde nicht gewusst. Die Daten sind vorhanden, und es kommen immer mehr und immer bessere Informationen hinzu. Die Erdbeobachtung aus dem Weltraum liefert die Grundlage. Wir Menschen müssen entscheiden, ob und wie wir die Daten auch nutzen wollen.

already in service and providing data. We can only imagine how important Earth observation will become in the next few decades—there are already more than 3000 research projects using the data from the ESA's satellites. More than 100 terabytes of information are added every year, about the same amount as can be stored on 200,000 CDs. Climate research alone has an archive in which 30 years of data for every region of the world is stored: cloud cover, greenhouse gas levels in the atmosphere, ozone distribution, ocean surface temperatures, sea levels, sea ice patterns, ocean coloration, records of glacier and polar ice cap behavior, the appearance of the Earth's surface, and very much more besides.

Let no one say at a later date that we did not know what state the Earth was in. The data is already there, and more and more accurate data is being added constantly. Earth observation from space has provided a start, but it is up to us humans to decide what we are going to do with the information.

lites d'observation de la Terre seront à leur poste. Ils rejoindront les nombreux satellites déjà opérationnels. On peut tout juste entrevoir l'importance que prendra l'observation de la Terre dans les dix années à venir. Plus de 3 000 projets de recherche utilisent déjà les données fournies par les satellites de l'ESA. Plus d'une centaine de téraoctets de données sont transmises chaque année, soit environ le volume de 200 000 disques compacts. Dans le domaine climatique, un répertoire rassemble et archive les données collectées sur l'ensemble du globe au cours des trente dernières années : couverture nuageuse, présence de gaz à effet de serre, concentration d'ozone, température de surface des océans, niveau de la mer, présence de glaces flottantes, couleurs des océans, comportement des glaciers et calottes polaires, aspect de la surface terrestre, etc.

À l'avenir, personne ne pourra prétendre ne pas connaître l'état de la Terre. Les données sont disponibles et de nouvelles informations de qualité toujours meilleure ne cessent de les compléter. L'observation spatiale apporte la base. Il nous incombe, à nous autres habitants de la Terre, de déterminer l'utilisation que nous souhaitons faire de ces données.

Si todo va como está previsto, en el año 2019 las cinco misiones estarán en el espacio. En el mejor de los casos, hasta 2025 ocuparán su puesto en el espacio casi dos docenas de nuevos satélites de observación terrestre de todas las partes del mundo. A ellos hay que añadir muchos otros satélites que ya están operativos y cuyos datos se encuentran igualmente disponibles. Ya podemos vislumbrar lo significativa que será la observación terrestre en la próxima década. Ahora mismo, más de 3.000 proyectos de investigación utilizan los datos de los satélites de la ESA. Cada año se añade más de un terabyte de datos, lo que corresponde a un volumen de datos aproximado de 200.000 discos compactos. Únicamente en el campo del clima, existe un archivo en el que se encuentran datos de los últimos treinta años de todas las regiones del mundo: datos sobre nubosidad, gases de efecto invernadero en la atmósfera, concentración de ozono, temperatura de las aguas superficiales de los océanos, altura del nivel del mar, color de los océanos, comportamiento de glaciares y casquetes polares, aspecto de la superficie terrestre, etc.

En el futuro nadie podrá decir que no sabía nada sobre el estado de nuestra Tierra. Los datos están disponibles y cada vez se añade más y mejor información. La observación de la Tierra desde el espacio nos proporciona la base. Nosotros, los seres humanos, debemos decidir si queremos utilizar los datos y cómo queremos hacerlo.

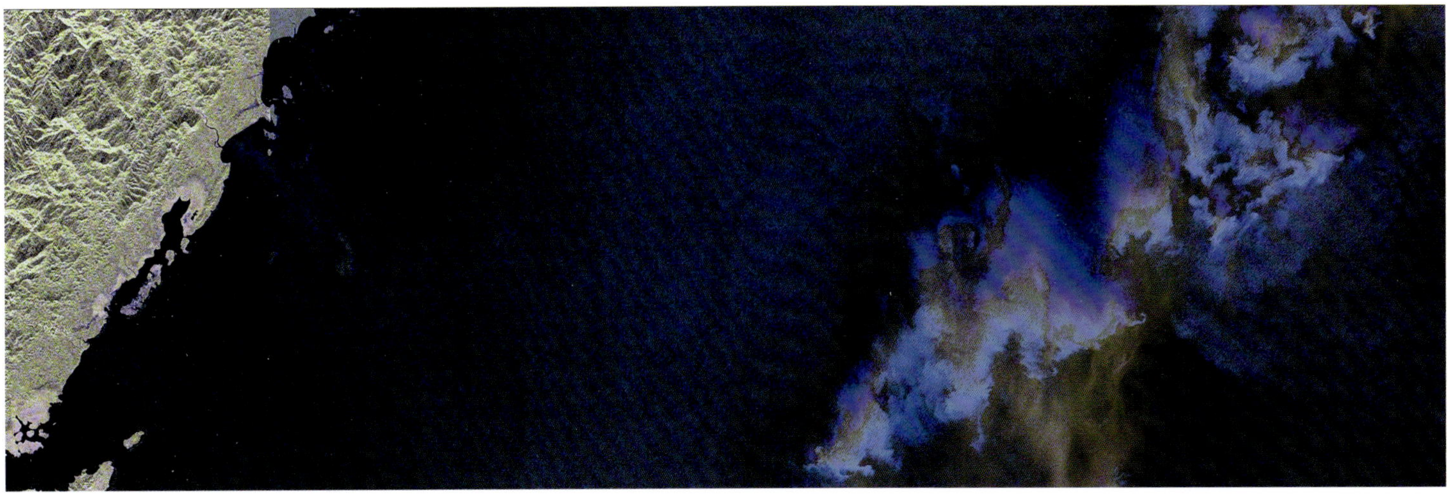

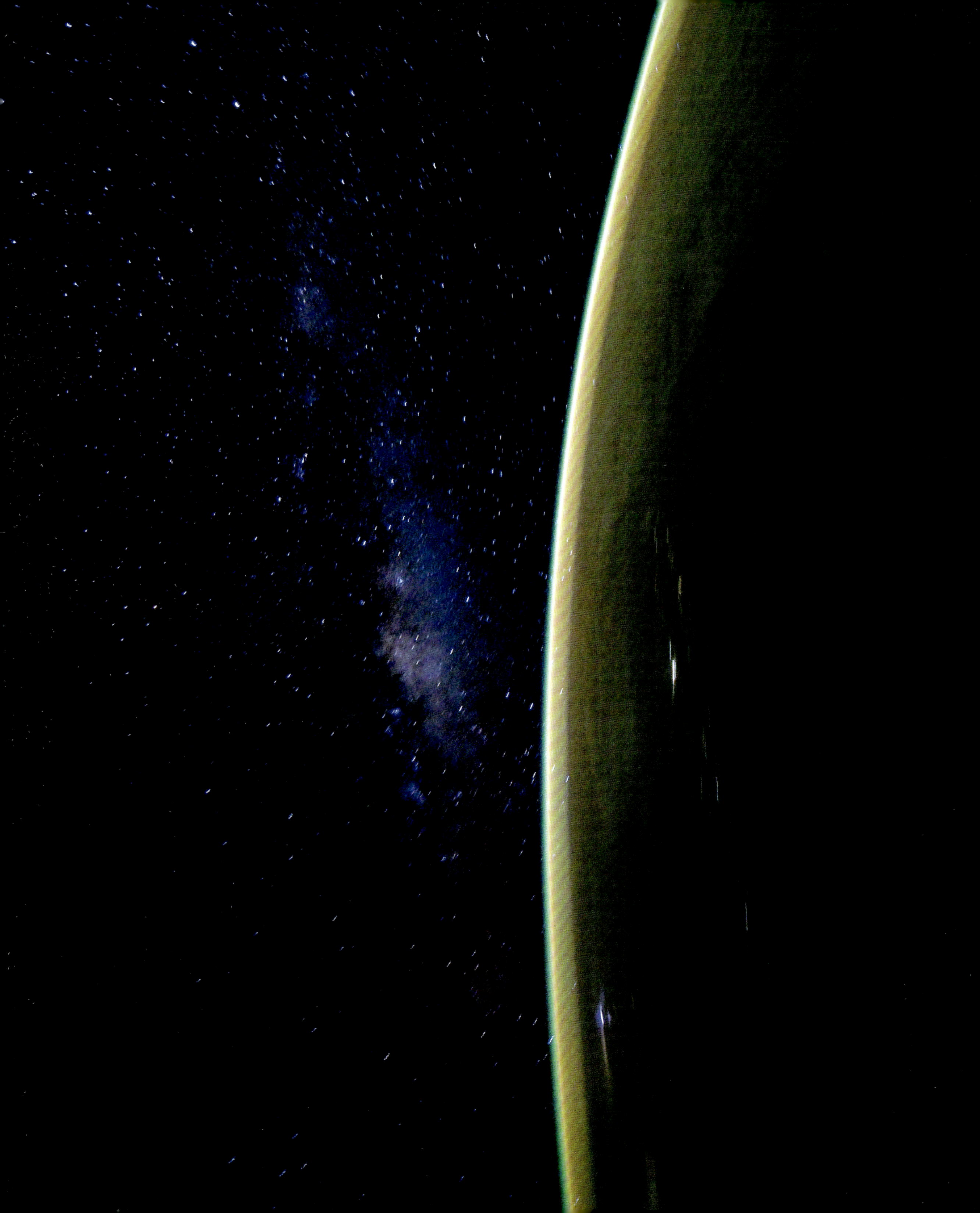

Mensch, Erde, Kosmos: Wie alles zusammenhängt

Humans, the Earth, the cosmos: how are they all connected?

Être humain, Terre et cosmos : un tout indissociable

El ser humano, la Tierra, el cosmos: interrelaciones

In den Außenbezirken der Milchstraße, am inneren Rand des großen Orion-Spiralarms, leuchtet ein ganz gewöhnlicher gelber Stern, einer von etwa 200 Milliarden Sternen in unserer Galaxis. Um ihn kreisen acht Planeten und jede Menge Eis- und Gesteinsbrocken. Der dritte Planet, eine blaue Murmel, ist etwas ganz Besonderes: Auf ihm gibt es Leben. Es ist unsere Erde. Sie reist im Schlepptau der Sonne durch das Universum. Etwa 230 Millionen Jahre dauert ein Lauf um das galaktische Zentrum. Fast zwanzig Runden hat die Erde seit ihrer Entstehung gedreht. Trotz aller Gefahren, die im All lauern, ist unser Planet in dieser Zeit zu einem wunderbaren, lebens- und liebenswerten Platz im Kosmos geworden. Auf den ersten Blick scheint die Erde ihr Dasein allein auf weiter Flur zu fristen: Zum Mond sind es fast 400.000 Kilometer. Die Venus, unser engster Nachbar, kommt uns bestenfalls auf 38 Millionen Kilometer nahe. Die Sonne ist 150 Millionen Kilometer entfernt und der nach der Sonne nächste Stern hat bereits den astronomischen Abstand von mehr als vier Lichtjahren, was knapp 40 Billionen Kilometern entspricht. Eine Billion ist eine Eins mit zwölf Nullen.

Die Entfernungen mögen groß sein. Doch wir sind keineswegs isoliert im

On the inner edge of the great Orion spiral arm in the outer reaches of the Milky Way there shines an unremarkable yellow star, one of about 200 billion in our galaxy. It is orbited by eight planets and any amount of rubble, ice, and rock, but the third planet, a blue dot, is something rather special: it bears life. It is our Earth, chained to our Sun as it races through the universe. One circumnavigation of the galactic central point takes about 230 million years and the Earth has made almost 20 revolutions since its creation. Despite all the dangers lurking in space, our planet has, over time, become the most astonishing and beautiful place for life in the cosmos. At first glance, the Earth seems to be living out its existence in some isolation: it is almost 250,000 miles (400,000 kilometers) to the Moon, while Venus, our nearest planet neighbor, is 24 million miles (38 million kilometers) away at its closest. There are 93 million miles (150 million kilometers) between us and the Sun, while the nearest stars after that (the Alpha Centauri stars) lie at an astronomical distance of four light years from the Sun, about 25 trillion miles (40 trillion kilometers), where a trillion is a 1 with 12 zeros.

Dans les faubourgs de la Voie lactée, sur le bord intérieur du bras spiral d'Orion, brille une étoile jaune assez commune parmi les quelque 200 milliards d'étoiles de notre galaxie. Huit planètes et quantité de blocs de glace et de pierre gravitent autour d'elle. La troisième planète, un calot bleu, a quelque chose de particulier : elle abrite la vie. C'est notre Terre. Elle voyage à travers l'univers dans le sillage du soleil. Cette course autour du centre galactique dure près de 230 millions d'années. Depuis sa formation, la Terre a déjà effectué près de vingt révolutions. Malgré tous les dangers qui guettent dans le cosmos, notre planète est devenue au cours du temps un lieu splendide où il fait bon vivre. À première vue, la Terre semble gérer seule son quotidien. 400 000 kilomètres la séparent de la Lune, et Vénus, notre plus proche voisine, séjourne à quelque 38 millions de kilomètres. Le soleil est à 150 millions de kilomètres et la prochaine étoile se trouve à une distance astronomique de plus de quatre années lumière, soit près de 40 billions de kilomètres, sachant qu'il faut douze zéros pour faire un billion.

Les distances sont peut-être immenses, mais nous ne sommes pas isolés dans l'univers. Tout ce qui se

En la periferia de la Vía Láctea, en el borde interior del gran brazo espiral de Orión, brilla una estrella amarilla completamente normal, una más entre los 200 millardos de estrellas de nuestra galaxia. Alrededor de ella giran ocho planetas y un montón de fragmentos de hielo y roca. El tercer planeta, una canica azul, tiene una particularidad: en él existe vida. Es nuestra Tierra, que viaja por el universo a remolque del Sol. Su marcha alrededor del centro de la galaxia dura aproximadamente 230 millones de años. Desde su nacimiento, la Tierra ya le ha dado casi veinte vueltas. En ese tiempo, y a pesar de todos los peligros que acechan en el espacio, nuestro planeta se ha convertido en un maravilloso, habitable y encantador lugar en el cosmos. A primera vista, la Tierra parece llevar una existencia solitaria: hasta la Luna hay casi 400.000 km de distancia. Venus, nuestro vecino más cercano, está a 38 millones de kilómetros, y eso en las circunstancias más favorables. El Sol está a 150 millones de kilómetros de distancia y la siguiente estrella más allá del Sol se encuentra ya a más de cuatro años luz, lo que corresponde a 40 billones de kilómetros escasos.

Las distancias pueden ser grandes, pero en modo alguno estamos aisla-

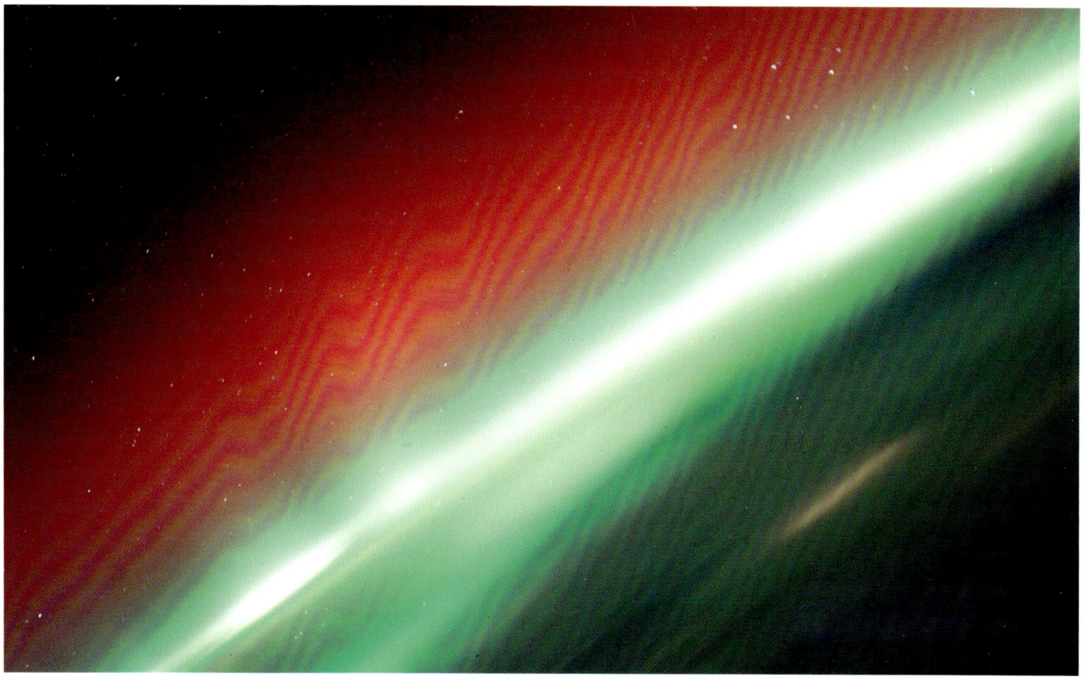

All. Denn was immer auf dieser Erde passiert, geht zurück auf kosmische Ereignisse. Wir alle sind im wahrsten Wortsinne Kinder des Weltalls: ob Kohlenstoff in den Molekülen des Erbgutes, Sauerstoff in der Luft, Eisen im Blut, Kalzium in den Knochen, Titan in der Hüfte oder Gold in den Zähnen. Alle chemischen Elemente – außer Wasserstoff und Helium, die bereits beim Urknall entstanden sind – sind Überreste längst erloschener Sterne. Diese Asche früherer Sterngenerationen war Rohstoff für jene Urwolke, aus der sich vor gut viereinhalb Milliarden Jahren die Sonne und mit ihr die Erde und die übrigen Planeten gebildet haben.

Drei Schwestern

Seit etwa drei Milliarden Jahren hat sich Leben auf der Erde ausgebreitet und immer weiterentwickelt. Nach einem Kalauer der Astronomen ist allerdings noch offen, ob es auf unserem Planeten wirklich intelligentes Leben gibt. Man mag manchmal daran zweifeln, wenn man sieht, wie die Menschheit mit ihrer himmlischen Heimat umgeht. Man wird es vorbehaltlos bejahen, wenn man sich die raffiniert konstruierten Satelliten ansieht, die

The distances might be great, but we are in no way deserted in space—everything that happens on Earth goes back to events in the cosmos. In the truest sense, we are all children of the universe: the carbon in our DNA, the oxygen in the air, the iron in our blood, the calcium in our bones, even the titanium in our hips, and the gold in our teeth—all the chemical elements, apart from hydrogen and helium, which were present at the Big Bang, are remnants of long-dead stars. The ash remaining from a generation of heavenly bodies provided the raw material for an ancient cloud from which was formed the Sun, and with it the Earth and the other planets, some four and a half billion years ago.

Three sisters

Life first appeared and began to develop on Earth about three billion years ago, although an old astronomer's joke wonders if we will ever discover intelligent life on our planet. You might well wonder, when you consider how mankind treats its home in space, but you would have no doubt on seeing the sophisticated construction of the satellites

produit sur Terre tire son origine d'événements cosmiques. Nous sommes tous *stricto sensu* des enfants de l'espace : carbone dans les molécules de notre génome, oxygène dans l'air, fer dans le sang, calcium dans les os, titane dans les hanches et dents en or. Tous les éléments chimiques, sauf l'hydrogène et l'hélium nés lors du big-bang, sont les vestiges d'étoiles éteintes. Le nuage originel était composé de poussières d'étoiles de première génération qui ont donné naissance
à notre soleil ainsi qu'à la Terre et aux autres planètes il y a quatre milliards et demi d'années.

La triade des sœurs

Apparue sur la Terre il y a trois milliards d'années, la vie n'a cessé de se développer. Les astronomes aiment plaisanter en se demandant si l'on trouvera un jour une preuve de la présence de vie intelligente sur la Terre. On peut effectivement en douter au vu du traitement qu'inflige l'humanité à son berceau céleste. Mais les doutes se dissipent complètement à la vue des satellites ultrasophistiqués qui gravitent autour de la Terre pour surveiller son état.

Au cours des années passées, les

dos en el espacio. Porque todo lo que pasa en esta Tierra se atribuye a los acontecimientos cósmicos. Todos nosotros somos hijos del universo en el sentido más genuino de la palabra: ya sea carbono en las moléculas del patrimonio genético, oxígeno en el aire, hierro en la sangre, calcio en los huesos, titanio en las caderas y oro en los dientes, todos los elementos químicos, excepto el hidrógeno y el helio, que ya se produjeron en la Gran Explosión, son restos de estrellas extinguidas desde hace mucho tiempo.

Tres hermanos

Estas cenizas de antiguas generaciones de estrellas fueron la materia prima de la protonebulosa de la que hace más de cuatro mil quinientos millones de años se formó el Sol y, con él, la Tierra y los demás planetas. La vida surgió en la Tierra hace 3.000 millones de años y continúa evolucionando.

Durante los últimos años, las sondas en órbita han trasmitido muchos datos desconcertantes a las estaciones de tierra. De repente, el planeta aparece ante los ojos de los investigadores como una estructura mucho más compleja de lo que se había pensado. La observación desde el exte-

Linke Seite und unten: Spektakuläre Bilder von Polarlichtbögen, aufgenommen von Astronauten auf der Raumstation

Left and below: Spectacular images of polar lights captured by astronauts on the International Space Station

Ci-dessous et page de droite : photographie des arcs formés par les aurores polaires prise par des astronautes depuis la station spatiale

Página izda. y abajo: Espectaculares imágenes de auroras boreales tomadas por astronautas desde la estación espacial

Seite 186: Die Erde und die Milchstraße, fotografiert von der Raumstation. Daher sind die Sterne etwas verzerrt

Page 186: The Earth and the Milky Way, photographed from the International Space Station (which is why the stars are slightly out of focus)

Page 186 : photographie de la Terre et de la Voie lactée effectuée depuis la station spatiale, ce qui explique la déformation des étoiles

Página 186: La Tierra y la Vía Láctea fotografiadas desde la estación espacial. Por eso las estrellas están algo deformadas

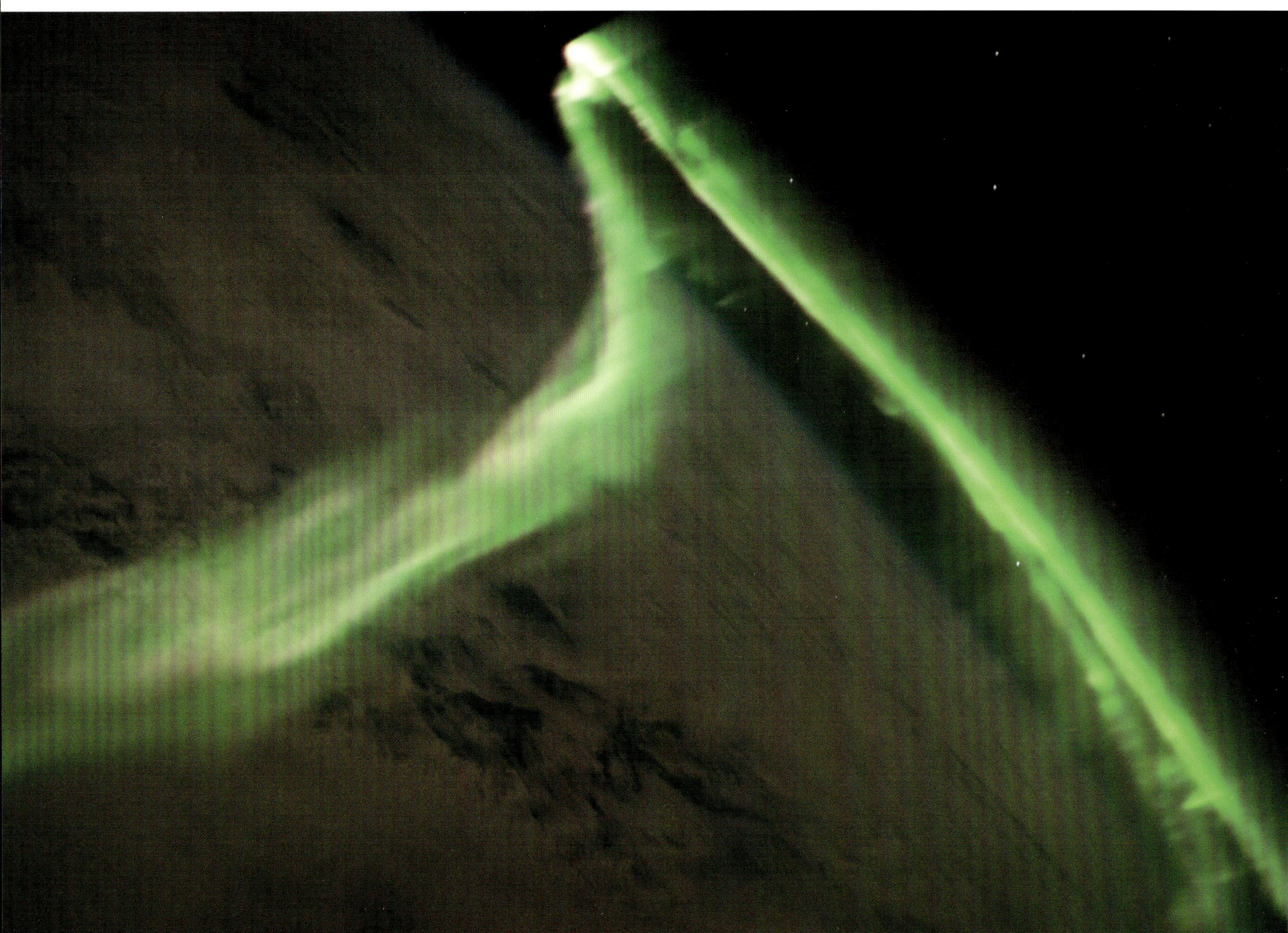

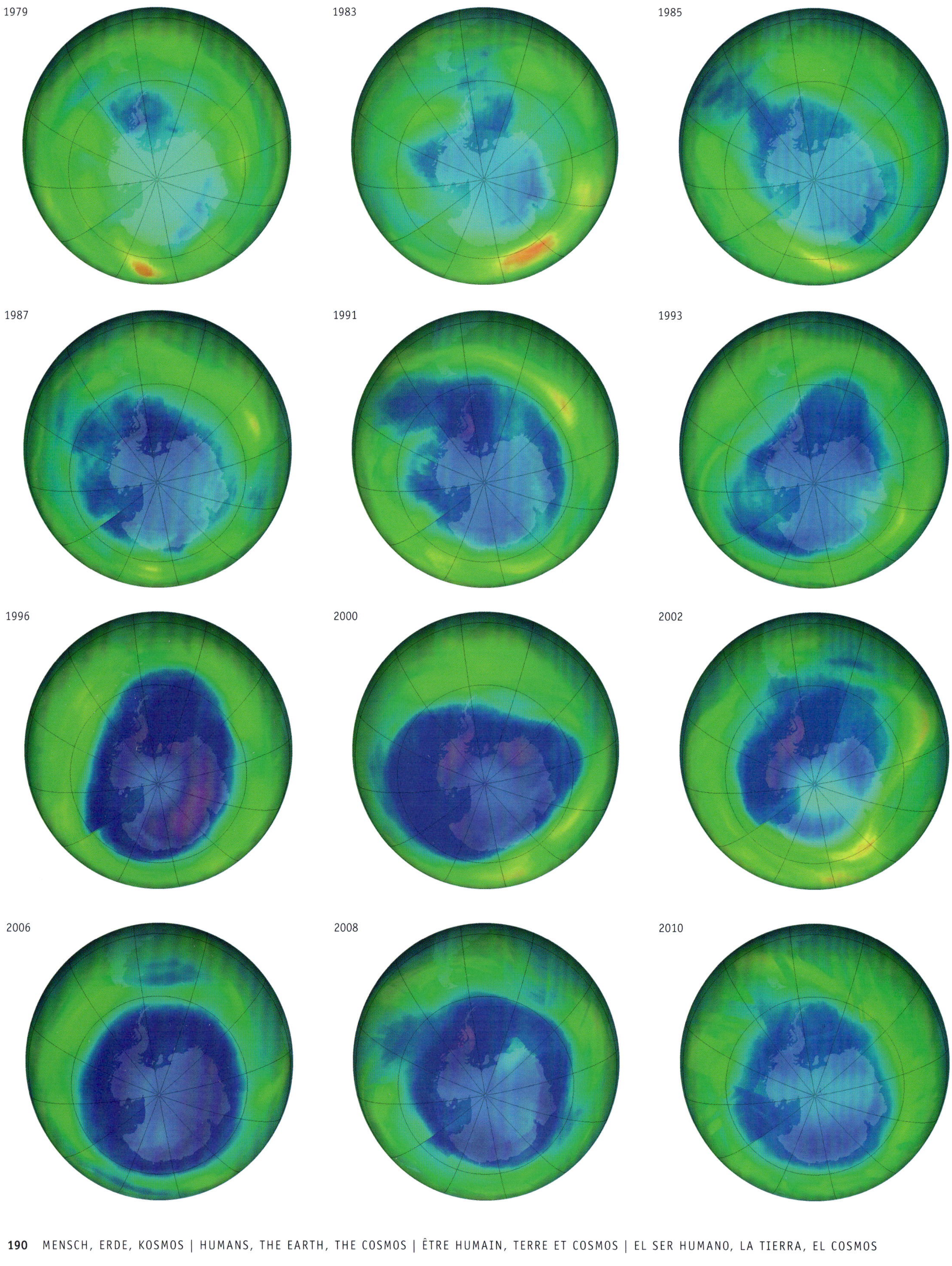
1979
1983
1985
1987
1991
1993
1996
2000
2002
2006
2008
2010

die Erde umkreisen und ihren Zustand überwachen.

In den vergangenen Jahren haben die Sonden in der Umlaufbahn viele verblüffende Daten an die Bodenstationen gefunkt. Unsere Erde erscheint den Forschern mit einem Mal noch viel komplexer, als zuvor gedacht. Fast scheint es, als habe man oft den Wald vor lauter Bäumen nicht gesehen. Der Blick von außen hilft, die Übersicht zu bekommen und die großen Zusammenhänge zu erkennen. Daher wird die Bedeutung der Erdbeobachtung in den nächsten Jahren noch deutlich zunehmen. Viele Missionen sind bereits in Bau oder in Planung. Schon bald werden die Phänomene auf unserer Erde praktisch rund um die Uhr überwacht, und die Daten werden live zur Verfügung stehen. Es wird so sein, wie es immer in der Forschung ist: Genaues Hinsehen hilft, offene Fragen zu klären. Aber jedes gelöste Problem ergibt etliche neue Fragestellungen. Denn die Neugier hört nie auf, Wissenschaft ist nie fertig.

In den vergangenen Jahrzehnten hat ein großes Umdenken stattgefunden. Früher war die Erforschung der Erde vor allem in der Hand von Geografen und Biologen. Man reiste in ferne Länder, untersuchte die Landschaft, zeichnete die Witterungsverhältnisse auf und beobachtete Flora

which circle the globe and monitor its condition.

These orbiting probes have transmitted amazing data to their base stations over the years, and our Earth is now regarded by scientists as far more complex than we first suspected. It almost seems as if we could not see the wood for the trees. A viewpoint from above helps us to gain perspective and see the bigger picture, and Earth observation will only increase in importance in the next few years. A number of missions are already under construction or in the planning phase and the Earth's phenomena will soon be under more or less 24-hour supervision, with the resultant data available in real time. Things won't change for research, however: close observation can answer open questions, but every solved problem opens up numerous new avenues of enquiry. Curiosity knows no bounds, and science is never "finished."

A sea change has occurred in recent decades: exploring the Earth used to be the preserve of geographers and biologists who traveled to distant lands to examine the landscape, record weather conditions, and observe flora and fauna. The planets in space were the

sondes spatiales nous ont transmis des images stupéfiantes. La Terre est soudain apparue aux scientifiques dans toute sa complexité. Les arbres apparemment leur cachaient la forêt. Le regard extérieur fournit une vision globale qui permet de discerner les forces qui gouvernent l'univers. C'est pourquoi l'observation de la Terre gagnera en importance dans les années à venir. De nombreuses missions sont en cours ou en projet. Bientôt, les phénomènes terrestres seront surveillés presque jour et nuit, et les données disponibles en temps réel. Et comme toujours dans le domaine de la recherche : les études approfondies permettent d'éclaircir certains points, mais chaque problème résolu engendre une nouvelle série de questions. Car la curiosité n'a pas de limites et la science n'est jamais définitive.

Nous avons assisté à un revirement de pensée au cours des dernières années. Auparavant, l'exploration de la Terre était avant tout l'apanage des géographes et biologistes qui se rendaient dans des contrées reculées, étudiaient paysages et climats, et observaient la faune et la flore, tandis que les astronomes étudiaient l'univers. La distinction entre « ici-bas » et « là-haut » était claire, elle a désormais quasi disparu. Les

rior ayuda a hacerse una idea general y a reconocer las grandes interconexiones. Por eso, la importancia de la observación terrestre se intensificará todavía más en los próximos años. Muchas misiones están ya en construcción o en proyecto. Ya pronto se podrán observar los fenómenos que ocurren en la Tierra prácticamente las 24 horas del día y los datos estarán disponibles en directo. Ocurrirá como siempre pasa con la investigación: fijarse bien, ayuda a aclarar preguntas abiertas. Pero cada problema resuelto suscita algunos nuevos planteamientos. Como la curiosidad es insaciable, la ciencia nunca termina.

En las pasadas décadas se produjo un gran cambio en la manera de pensar. Antes, la exploración de la Tierra estaba, sobre todo, en manos de geógrafos y biólogos. Se viajaba a países lejanos, se examinaba el paisaje, se apuntaban las circunstancias meteorológicas y se observaba la flora y la fauna. Los astrónomos se encargaban de los planetas en el espacio. Se distinguía entre «aquí abajo» y «allí arriba». De esto ya no queda casi nada y las disciplinas se combinan, como la astrobiología y la bioastronomía, en las que:investigadores de diferentes especialidades se apoyan para descubrir los fundamentos de la vida en nuestra Tierra y utilizar

Linke Seite: Das Ozonloch über der Antarktis jeweils an einem 30. August. *Rechts:* Der UV-Index am 8. November 2005 (Envisat)

Left: The hole in the ozone layer over the Antarctic pictured at yearly intervals on August 30. *Right:* The UV index on November 8, 2005 (Envisat)

Page de gauche : le trou dans la couche d'ozone au-dessus de l'Antarctique le 30 août de différentes années. *À droite :* l'indice UV le 8 août 2005 (Envisat)

Página izda: Agujero de ozono sobre la Antártida cada 30 de agosto. *Dcha.:* Índice UV el 8.11.2005 (Envisat)

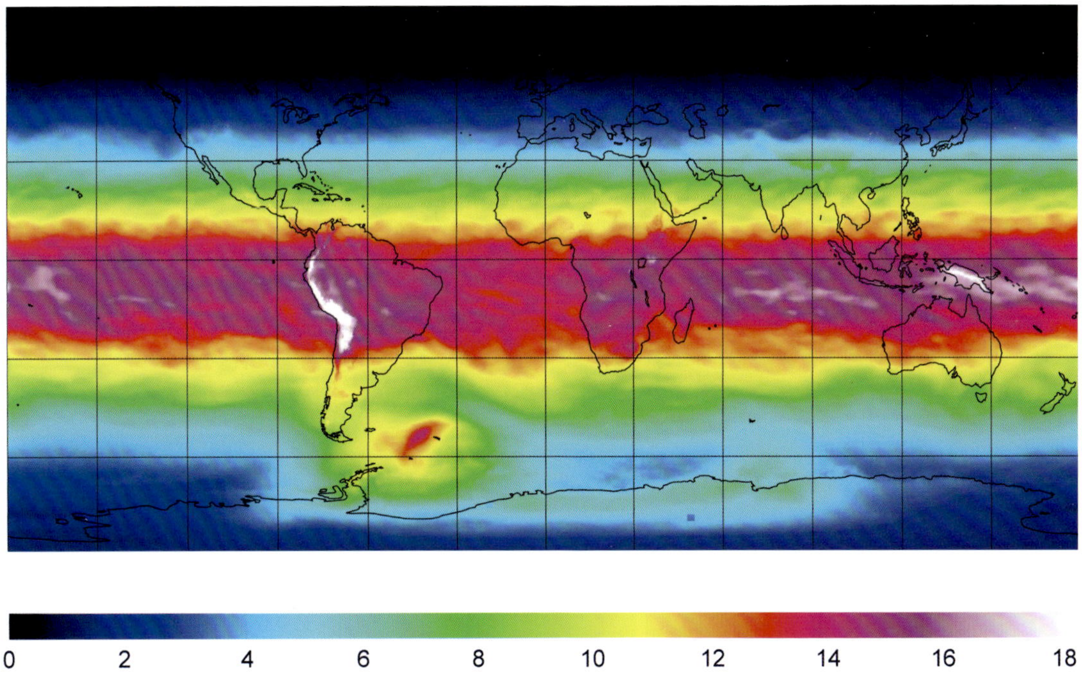

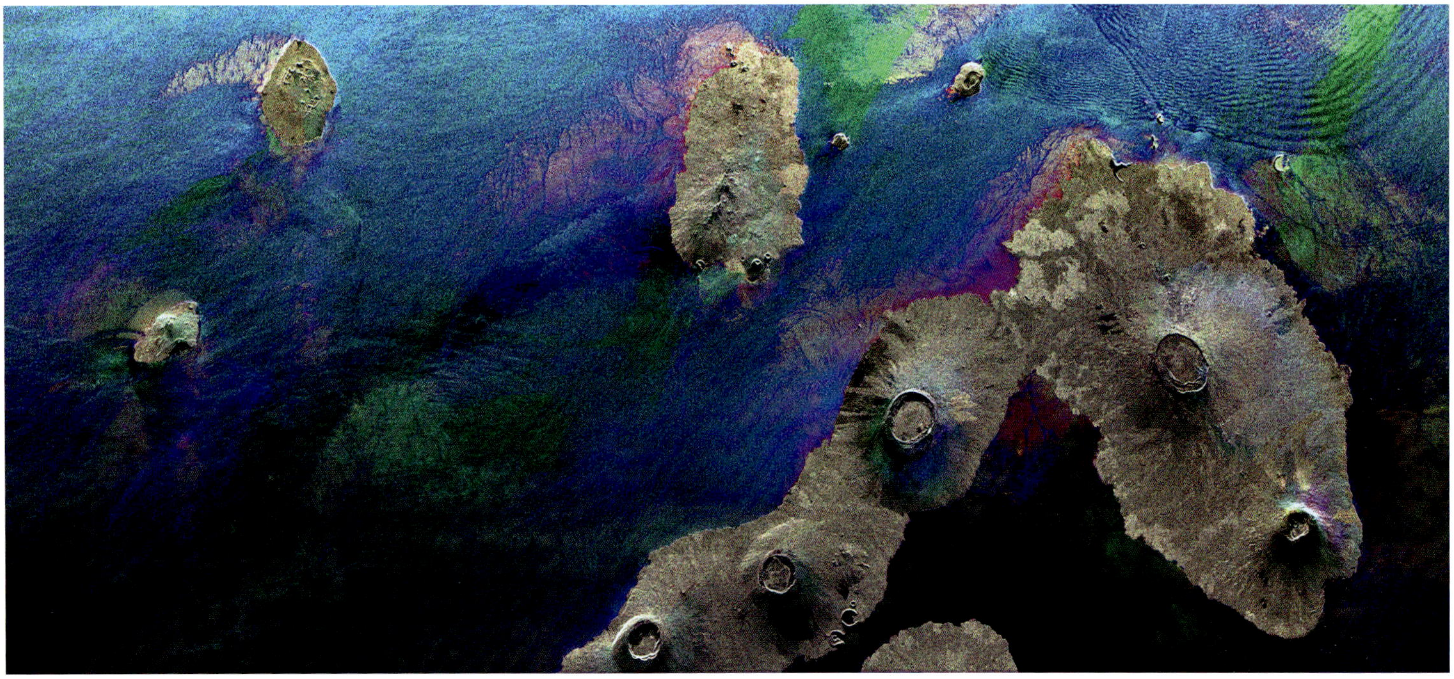

und Fauna. Für die Planeten im All waren die Astronomen zuständig. Man unterschied ein »hier unten« und »da oben«. Davon ist fast nichts mehr übrig. Die Disziplinen verschmelzen. Heute gibt es so etwas wie Astrobiologie oder Bioastronomie: Forscher verschiedener Fachrichtungen versuchen, die Grundlagen des Lebens auf unserer Erde zu erkunden und mit diesen Erkenntnissen nach Leben im Universum zu suchen.

Das fängt gleich vor unserer Haustür an: Die Erde und ihre kosmischen Nachbarn darf man nicht getrennt sehen, man muss sie gemeinsam untersuchen, betont Augustin Chicarro, aus Spanien stammender leitender Wissenschaftler der ESA-Mission Mars Express: *Mars ist der erdähnlichste Planet, was die Geologie an der Oberfläche und das Klima betrifft. Venus ähnelt der Erde am meisten, wenn es um die innere Struktur und die Geodynamik geht.* Venus, Erde und Mars sind Schwesterplaneten – mit ähnlicher Größe und vergleichbarem Abstand von der Sonne. Aber wie menschliche Schwestern können sie dennoch sehr verschieden sein. *Mich fasziniert zutiefst, welche unterschiedliche Entwicklungswege Erde, Mars und Venus eingeschlagen haben – und wie Mars zur bitterkalten Welt geworden ist, die wir heute beobachten,* so Chicarro.

business of astronomers and a line was drawn between “down here” and “up there.” Little of this division now remains and the disciplines are merging; these days, there is even such a thing as astrobiology or bioastronomy. Scientists in a number of fields are attempting to explore the basis of life on Earth and to use these findings to search for life in the universe.

All this is starting on our doorstep. The Earth and its cosmic neighbors should be seen as of a piece and explored together, stresses Augustin Chicarro, the Spaniard in charge of the ESA’s Mars Express mission. *Mars is the planet most similar to the Earth as far as its geology, surface, and climate are concerned. Venus has the closest resemblance in respect of its inner structure and geodynamics.* Venus, Mars, and the Earth are sister planets, with similar dimensions and entirely comparable distances from the Sun. Just like human sisters, however, they can also be very different. *I am absolutely fascinated by the different developments that the Earth, Mars, and Venus have undergone, and how Mars has become the bitterly cold planet that we see today,* explains Augustin Chicarro.

frontières entre les disciplines s’estompent pour former l’astrobiologie : une science interdisciplinaire qui étudie les fondements de la vie sur Terre pour mieux explorer l’univers en quête de vie extraterrestre.

L’observation repose d’abord sur l’analogie : la Terre et ses voisines cosmiques ne peuvent être étudiées séparément, souligne Augustín Chicarro, le responsable scientifique espagnol de la mission Mars Express de l’ESA : *Mars est la planète qui ressemble le plus à la Terre en termes de géologie de surface et de climat. Vénus est comparable à la Terre dans sa structure interne et sa géodynamique.* Vénus, Mars et la Terre sont comme trois sœurs. Elles ont la même taille et se trouvent à une distance similaire du soleil. Mais elles peuvent, tout comme les sœurs humaines, être très dissemblables. *L’évolution extrêmement différente de la Terre, Mars et Vénus me fascine au plus haut point, de même que la métamorphose de Mars en ce monde glacial que nous observons aujourd’hui,* déclare Augustín Chicarro.

Comparée à Mars et Vénus, la Terre est un paradis. Composée essentiellement de dioxyde de carbone, l’atmosphère martienne est extrêmement fine. La vie y est impossible, tout

esos conocimientos para buscar vida en el universo.

La Tierra y sus vecinos cósmicos deben ser considerados en conjunto y no por separado, subraya Agustín Chicarro, el director científico español de la misión Mars Express de la ESA: *Marte es el planeta más parecido a la Tierra en lo que respecta a la geología superficial y al clima. Venus se asemeja más a la Tierra cuando se trata de estructura interior y geodinámica.* Venus, Tierra y Marte son planetas hermanos. Tienen un tamaño parecido y una distancia muy similar con respecto al Sol. Pero, al igual que los hermanos humanos, también pueden ser muy distintos. *Me llama profundamente la atención los diferentes caminos de evolución que han seguido la Tierra, Marte y Venus y cómo Marte se ha convertido en el mundo extremadamente frío que hoy podemos observar*, explica Agustín Chicarro.

En comparación con Marte y Venus, la situación en la Tierra es paradisíaca. La atmósfera marciana es extremadamente delgada y está compuesta, casi en exclusiva, por dióxido de carbono. Allí no es posible la vida, al menos, no en la superficie. Venus es el verdadero infierno del cielo. Su atmósfera es 90 veces más densa que la de la Tierra. Las temperaturas

Die Vulkane Gelai (oben), Kitumbeine (unten rechts) und Ol Doinyo Lengai in Tansania, Afrika (Envisat)

The volcanoes of Gelai (top), Kitumbeine (bottom right), and Ol Doinyo Lengai in Tanzania, Africa (Envisat)

les volcans tanzaniens Gelai (ci-dessus), Kitumbeine (en bas à droite) et Ol Doinyo Lengaï, en Afrique (Envisat)

Los volcanes Gelai (arriba), Kitumbeine (abajo dcha.) y Ol Doinyo Lengai, en Tanzania, África (Envisat)

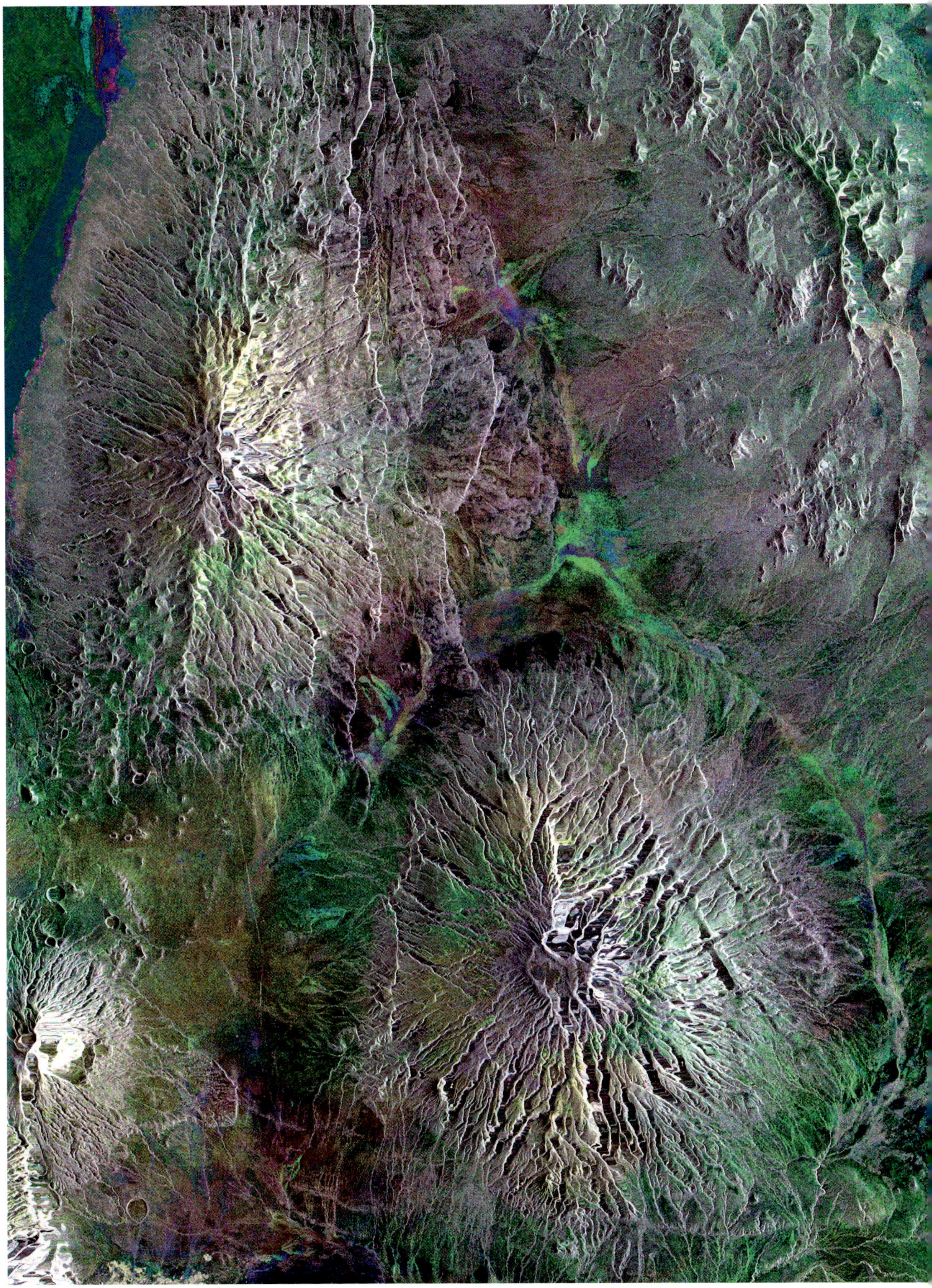

Linke Seite: Die Galapagos-Inseln in einer Radaraufnahme. Auf der Hauptinsel gibt es fünf Vulkane (Envisat)

Left: A radar image of the Galapagos Islands. There are five volcanoes on the main island (Envisat)

Page de gauche : image radar des îles Galápagos. L'île principale compte cinq volcans (Envisat)

Página izda: Toma de radar de las islas Galápagos. En la isla principal hay cinco volcanes (Envisat)

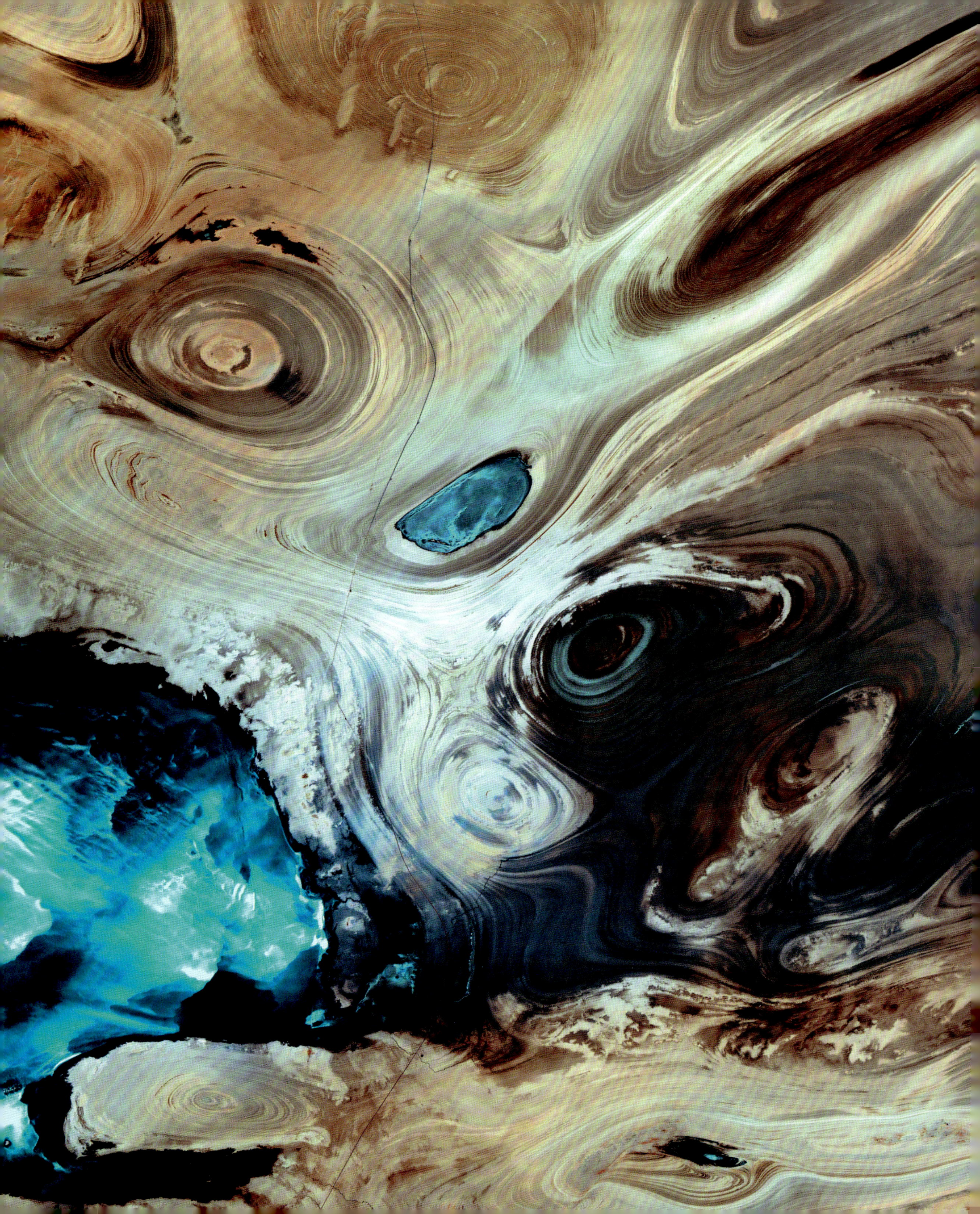

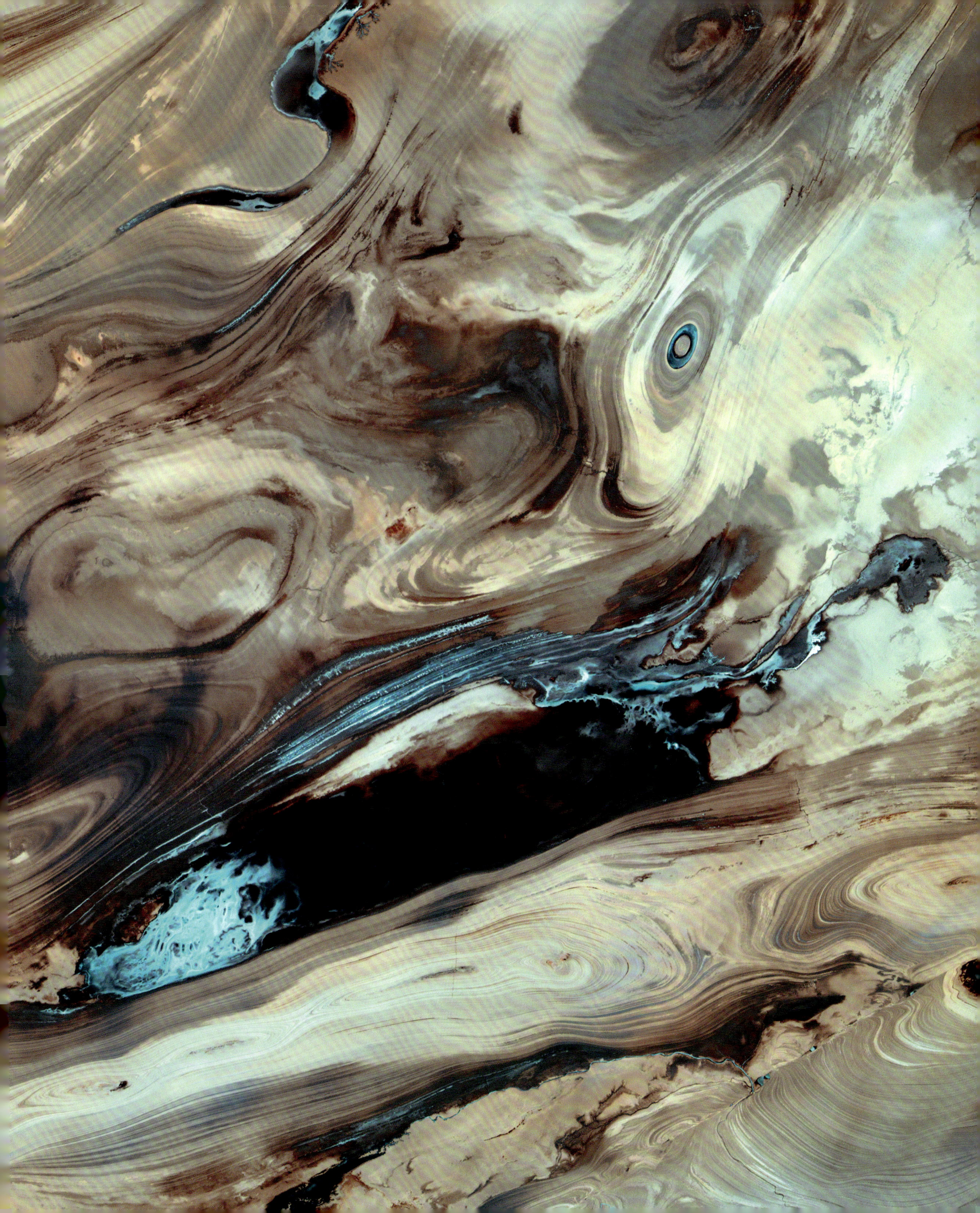

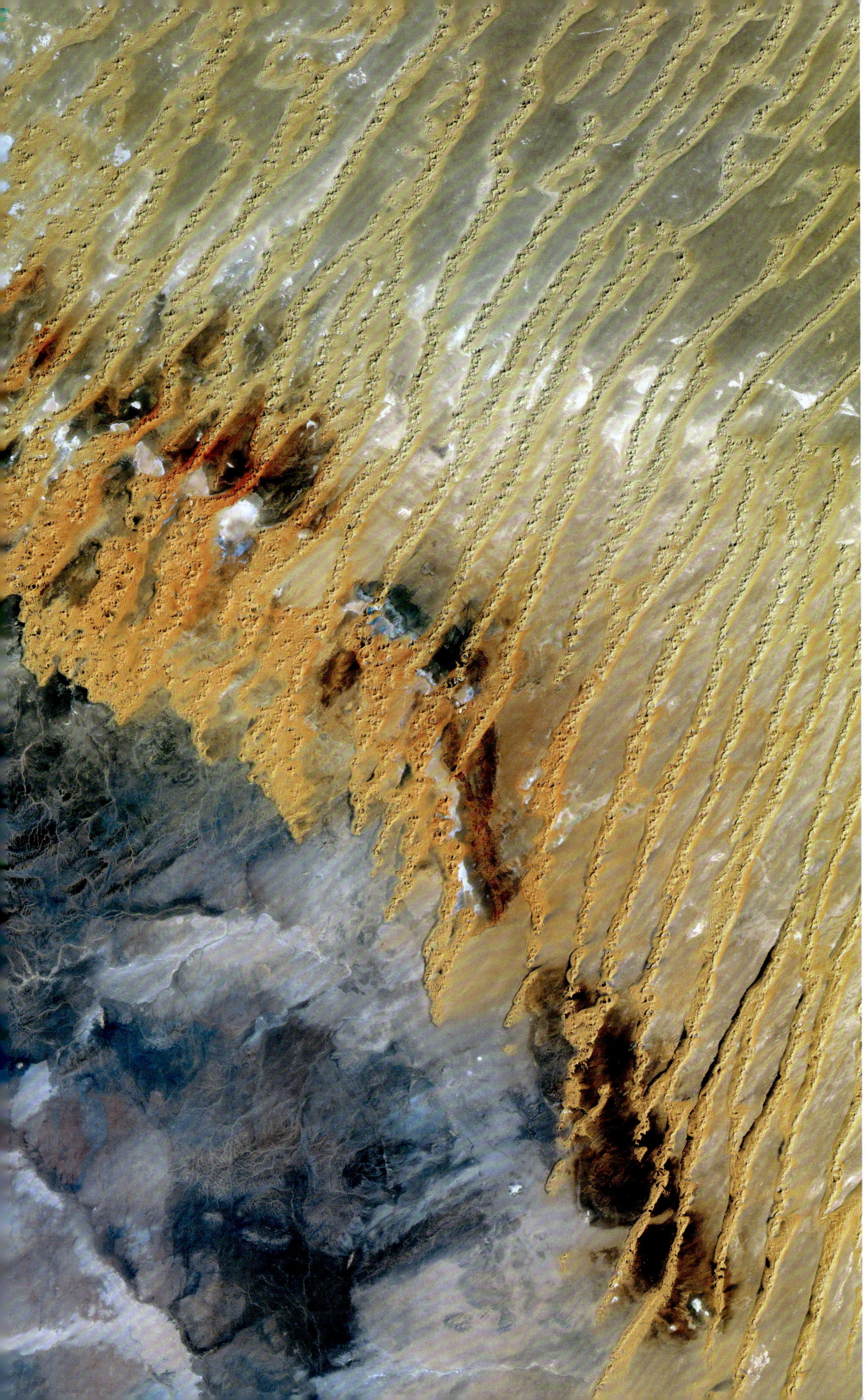

Mars? Nein, es ist die Erde. Sanddünen in Algerien wirken außerirdisch (Landsat)

Is it Mars? No, it's Earth. Sand dunes in Algeria appear extraterrestial (Landsat)

Mars ? Non, c'est la Terre. Les dunes de sable algériennes ont un aspect extraterrestre (Landsat)

¿Marte? No, es la Tierra. Las dunas de arena de Argelia parecen extraterrestres (Landsat)

Seite 194/195: Bizarre Strukturen in Dasht-e Kevir, der Großen Salzwüste im Iran. Im ganz linken Bildteil verläuft eine Straße (dunkle Linie) von unten nach oben zwischen den Salzpfannen entlang (Landsat)

Pages 194–5: Strange structures in Dasht-e Kevir, Iran's Great Salt Desert. The section of the photo far left shows a road (the dark line) running from top to bottom between the salt pans (Landsat)

Pages 194 et 195 : Curieuses structures du Dasht-e-Kavir, le grand désert salé d'Iran. À l'extrême gauche de l'image une route (ligne foncée) serpente de haut en bas entre les cuvettes de sel (Landsat)

Páginas 194/195: Extravagantes estructuras en el desierto de Kavir, el mayor desierto salado de Irán. Por la izquierda de la imagen transcurre una carretera (línea oscura) de abajo a arriba serpenteando entre las salinas (Landsat)

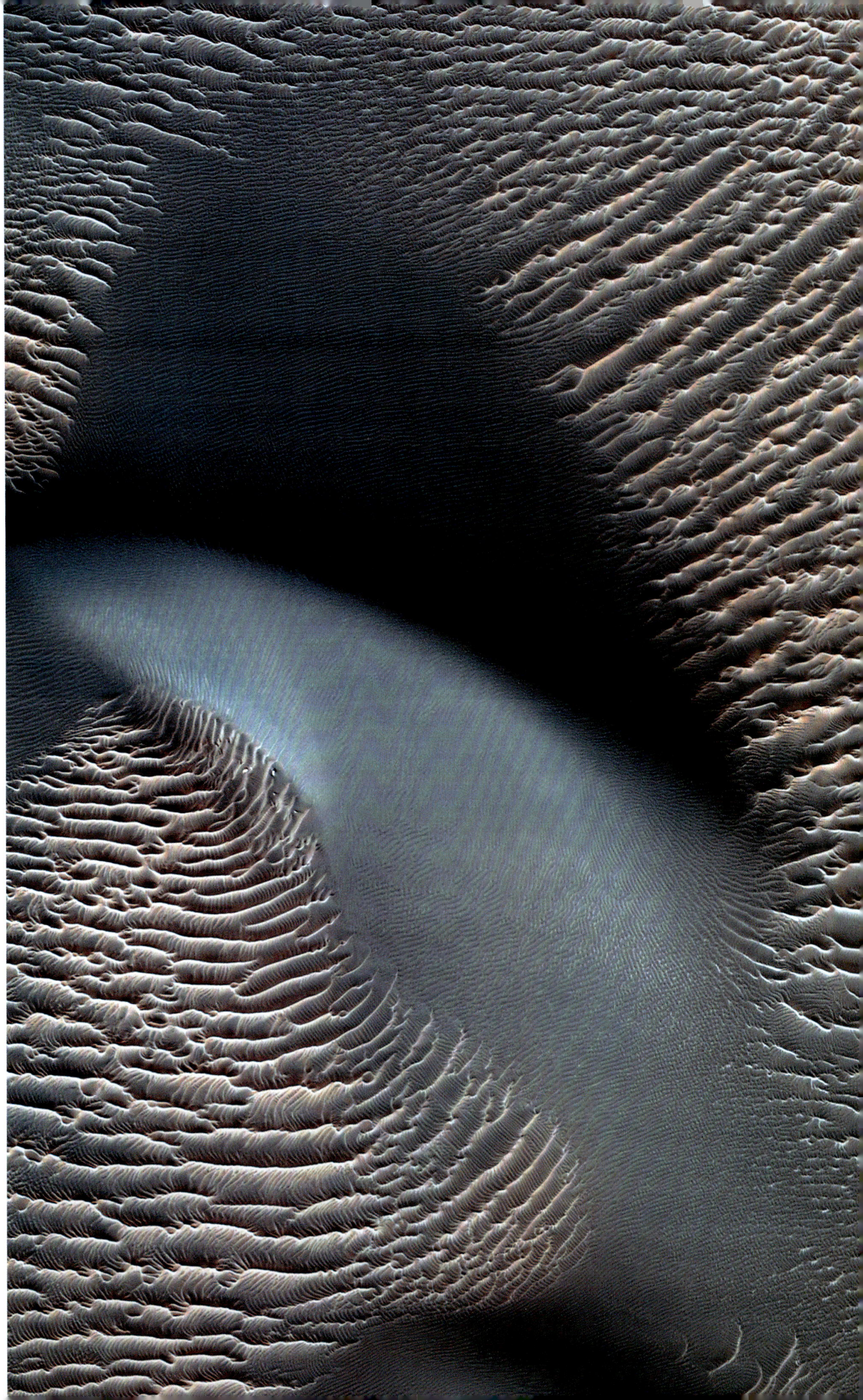

Erde? Nein, es ist der Mars. Sanddünen auf unserem Nachbarplaneten sehen ganz irdisch aus (MRO)

The Earth? No—Mars; the sand dunes on our planetary neighbor look strangely familiar (MRO)

La Terre ? Non, c'est Mars : les dunes de sable de la planète rouge ont un aspect tout à fait terrestre (MRO)

¿La Tierra? No, es Marte. Las dunas de arena en nuestro planeta vecino parecen casi terrestres (MRO)

Seite 198: Great Sandy Desert in Australien. Die Linien sind lange Dünen, der einzige Sand in dieser Wüste aus Felsen und Gestrüpp. Die hellen Gebiete sind Spuren von Buschbränden (Landsat)

Page 198: The Great Sandy Desert in Australia. The lines are long dunes, the only sand in this desert of rock and spinifex. The light areas are scars from bush fires (Landsat)

Page 198 : le Grand désert de sable en Australie. Les traits horizontaux sont des dunes, seuls espaces de sable dans ce désert de roches et de broussailles. Les zones claires sont des traces de feux de brousse (Landsat)

Página 198: Gran Desierto Arenoso de Australia. Las líneas son dunas largas, la única arena que hay en este desierto de rocas y matorrales. Las regiones más claras son indicios de incendios (Landsat)

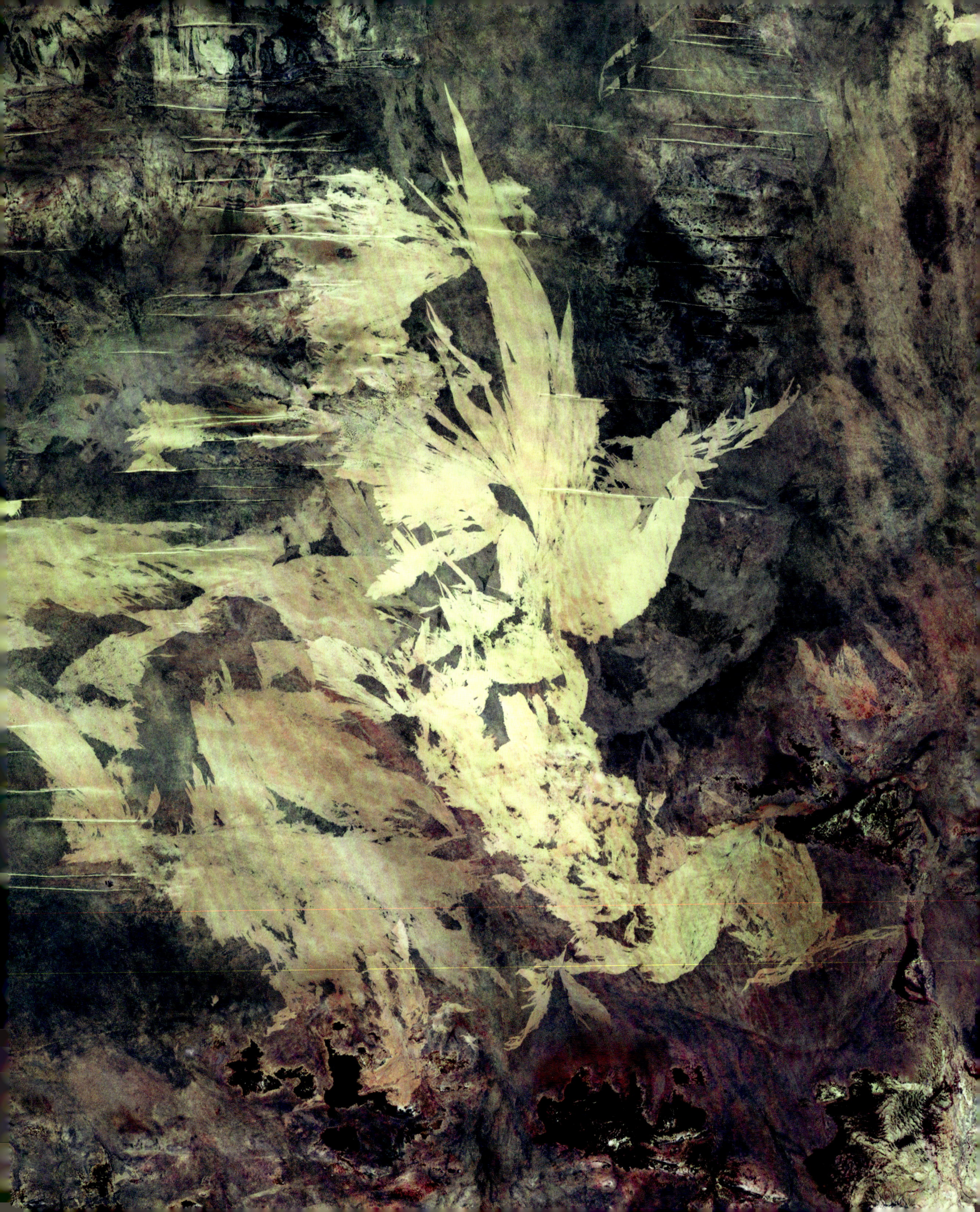

Verglichen mit Mars und Venus herrschen auf der Erde paradiesische Zustände. Die Marsatmosphäre ist extrem dünn und besteht fast nur aus Kohlendioxid. Leben ist dort nicht möglich, jedenfalls nicht an der Oberfläche. Venus ist die wahre Hölle am Himmel. Die Atmosphäre ist 90-mal dichter als die unserer Erde. Die Temperaturen liegen bei fast 500 Grad Celsius. Daran gemessen ist ein Backofen ein Kühlschrank. In der ebenfalls fast nur aus Kohlendioxid bestehenden Atmosphäre tobt dauerhaft ein Orkan, gegen den die Stürme auf der Erde laue Lüftchen sind. Als Krönung fegen dicke Wolken aus Schwefelsäure über den Venus-Himmel.

Die große Frage ist, wie sich diese Planeten, die mit ganz ähnlichen Bedingungen begonnen haben, so unterschiedlich entwickeln konnten. Die Forscher sorgen sich vor allem um eine Frage: Wird die Erde stets ein lebensfreundlicher Planet bleiben? Oder droht ihr langfristig ein Schicksal ähnlich dem der Venus oder des Mars? Um das zu klären, müssen sie verstehen, was auf der Erde passiert – und auf den anderen Planeten, betont Jean-Pierre Bibring, Physiker an der Universität von Paris in Orsay. Er hat ein Instrument für die Mars-Express-Sonde gebaut, das die chemische Zusammensetzung der Oberfläche und der Atmosphäre untersucht: *Die Missionen zu den anderen Planeten haben uns wie nie zuvor einen Blick auf die Geschichte des Sonnensystems und ganz besonders unserer Erde ermöglicht. Es ist höchste Zeit, dass wir endlich die vielen fast im Verborgenen ablaufenden Prozesse auf unserer Erde verstehen.*

Lieber Venus oder Mars?

Der Blick nach oben schärft den Blick nach unten. Wer Venus und Mars beobachtet, erforscht dabei automatisch auch unsere Erde. Die ESA hat fast identische Sonden zu beiden Planeten geschickt. Das geschah nicht nur aus Gründen der Sparsamkeit. Es war auch wissenschaftlich geboten. Denn nun lassen sich die gewonnenen Daten bestens in Zusammenhang

Conditions on Earth are like paradise compared with Mars and Venus. Consisting predominantly of carbon dioxide, Mars's atmosphere is extremely thin and life is unsustainable there, at least on the surface. Venus is a true vision of hell: the atmosphere is 90 times denser than on Earth and temperatures rise to about 900° F (500° C). Compared to this, an oven is like a refrigerator. The atmosphere, which also consists mostly of carbon dioxide, is continually rent by hurricanes in comparison with which storms on Earth are gentle breezes. To top all this off, thick clouds of sulfuric acid lurk on every horizon.

The great question is how these planets, which all started under similar conditions, could develop so differently. Researchers are much exercised by one question in particular: will the Earth always remain a planet that can support life? Or will its fate eventually be similar to that of Mars or Venus? To answer this, Jean-Pierre Bibring, a physicist at the University of Paris at Orsay, reminds us that we have to understand what is happening on Earth and on other planets. He has developed an instrument for the Mars Express probe which will investigate the chemical composition of the planet's surface and atmosphere. *Missions to the other planets have afforded us a unique insight into the history of the solar system and especially that of the Earth. We are rapidly approaching the time when we will finally understand the many processes which are happening almost in secret on our Earth.*

Venus or Mars?

Looking down sharpens our focus for when we look up, and observing Venus and Mars automatically involves observing our own planet. The ESA has sent virtually identical probes to each planet, which was both cost-effective and of scientific interest. The results obtained could then be most easily compared, declares a contented Gerhard

au moins en surface. Vénus est un véritable enfer. Son atmosphère est 90 fois plus dense que celle de la Terre. Elle atteint des températures de près de 500 degrés, à côté desquels nos fours passeraient pour des frigidaires. Un ouragan qui ferait pâlir la plus furieuse de nos tempêtes souffle sans répit dans son atmosphère composée presque exclusivement de dioxyde de carbone. Pour couronner le tout, d'épais nuages de soufre balaient le ciel vénusien.

La question primordiale est de savoir pourquoi ces planètes, à l'origine si semblables, ont connu une évolution si contrastée. Une énigme préoccupe les chercheurs par-dessus tout : la Terre restera-t-elle une planète accueillante, ou connaîtra-t-elle un destin similaire à celui de Mars et Vénus ? Pour répondre à cette question, il faut comprendre ce qui se passe sur la Terre, et sur les autres planètes, explique Jean-Pierre Bibring, professeur de physique à l'Université Paris-Sud. Il a construit pour la sonde Mars Express, un instrument capable d'analyser la composition chimique de la surface et de l'atmosphère : *Les missions vers d'autres planètes nous ont ouvert la porte de l'histoire du système solaire, et plus particulièrement de notre Terre. Il est grand temps que nous percions les secrets des processus qui régissent notre planète.*

Vénus ou Mars ?

En regardant le ciel nous affinons notre vision de la Terre. L'observation de Vénus et Mars conduit naturellement à explorer la Terre. L'ESA a envoyé des sondes quasi identiques vers ces deux planètes. Cette décision n'était pas simplement économique, elle répondait à une exigence scientifique. Nous sommes ainsi en mesure de mettre les données recueillies en relation, se réjouit Gerhard Schwehm, responsable des missions planétaires à l'ESA : *Nous faisons de la planétologie comparée. Nous étudions l'atmosphère de chaque planète pour élargir nos connaissances sur la Terre.*

Toutes les planètes sont nées du

alcanzan casi los 500 °C. En comparación, uno de nuestros hornos sería allí como un frigorífico. En la atmósfera de Venus, compuesta como la terrestre casi exclusivamente por dióxido de carbono, gira de forma permanente un huracán para el que las tormentas de la Tierra serían como vientecillos tibios. Para colmo, por el cielo de Venus corren densas nubes de ácido sulfúrico.

La gran pregunta es cómo estos planetas, que comenzaron con unas condiciones bastante similares a las terrestres, han podido evolucionar de un modo tan distinto. Y los investigadores se preguntan, ¿continuará siendo la Tierra un planeta propicio para la vida? ¿O está amenazada a largo plazo por un destino similar al de Venus o Marte? Para aclararlo, hay que entender primero qué ocurre en la Tierra y también en otros planetas, subraya Jean-Pierre Bibring, físico de la Universidad de París en Orsay. Él ha construido un instrumento para la sonda Mars Express, que analiza la composición química de la superficie y de la atmósfera: *Las misiones a los demás planetas han hecho posible, como nunca antes, una mirada sobre la historia del Sistema Solar y, especialmente, de nuestra Tierra. Finalmente, ha llegado la hora de entender los muchos procesos, casi siempre ocultos, que se desarrollan en la Tierra.*

¿Venus o Marte?

La mirada hacia arriba agudiza la mirada hacia abajo. Observando a Venus y Marte, automáticamente se investiga también nuestra Tierra. La ESA ha enviado a ambos planetas sondas casi idénticas. No sólo lo hizo por razones de ahorro, también era científicamente necesario, porque ahora se pueden relacionar mejor los datos obtenidos, como celebra Gerhard Schwehm, que dirige en la ESA el funcionamiento de las misiones planetarias: *Hacemos investigación comparada de planetas. Contemplamos la atmósfera de cada uno de ellos y pretendemos saber más sobre la Tierra estableciendo paralelismos.*

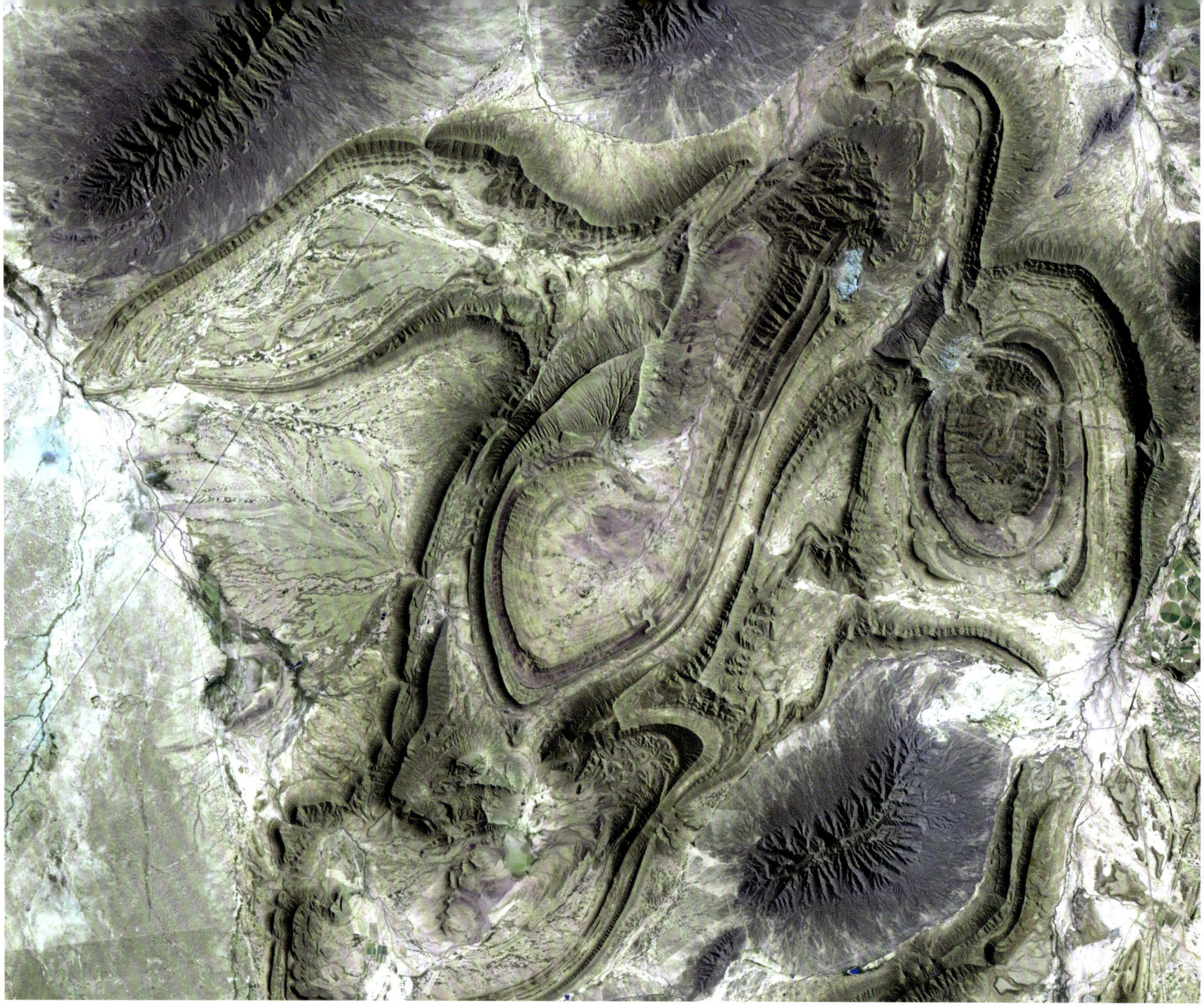

Dieser 30 km große steinerne Vogel fliegt durch Coahuila in Mexiko (Landsat)

This great stone bird flying through Coahuila in Mexico is 19 miles (30 km) long (Landsat)

Cet oiseau de pierre de 30 km vole dans l'État de Coahuila au Mexique (Landsat)

Este pájaro petrificado, de 30 km, vuela sobre Coahuila, en México (Landsat)

Seite 202/203: Gletscher San Quintin und San Rafael (rechts) in Patagonien (TerraSAR-X)

Pages 202–3: The San Quintín (left) and San Rafael (right) Glaciers in Patagonia (TerraSAR-X)

Pages 202 et 203 : les glaciers de San Quintin et San Rafael (à droite) en Patagonie (TerraSAR-X)

Páginas 202/203: Glaciares de San Quintín y San Rafael (dcha.), en Patagonia (TerraSAR-X)

Rechte Seite: Der Drachensee nahe der Stadt Bratsk im Süden Sibiriens (Landsat)

Right: Dragon Lake near the city of Bratsk in southern Siberia (Landsat)

Page de droite : le lac du Dragon près de la ville de Bratsk, au sud de la Sibérie (Landsat)

Página dcha.: El lago del Dragón, junto a la ciudad de Bratsk, en el sur de Siberia (Landsat)

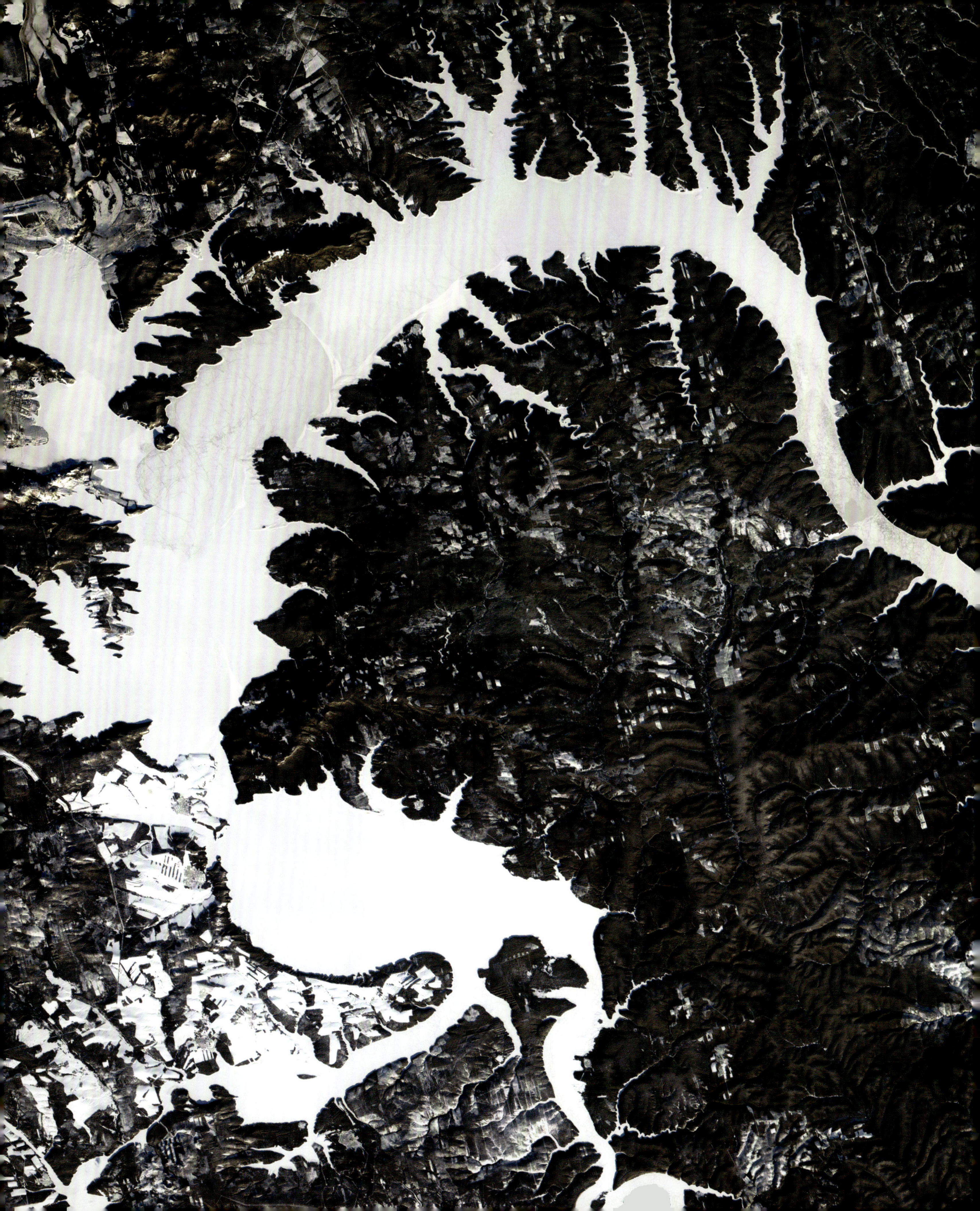

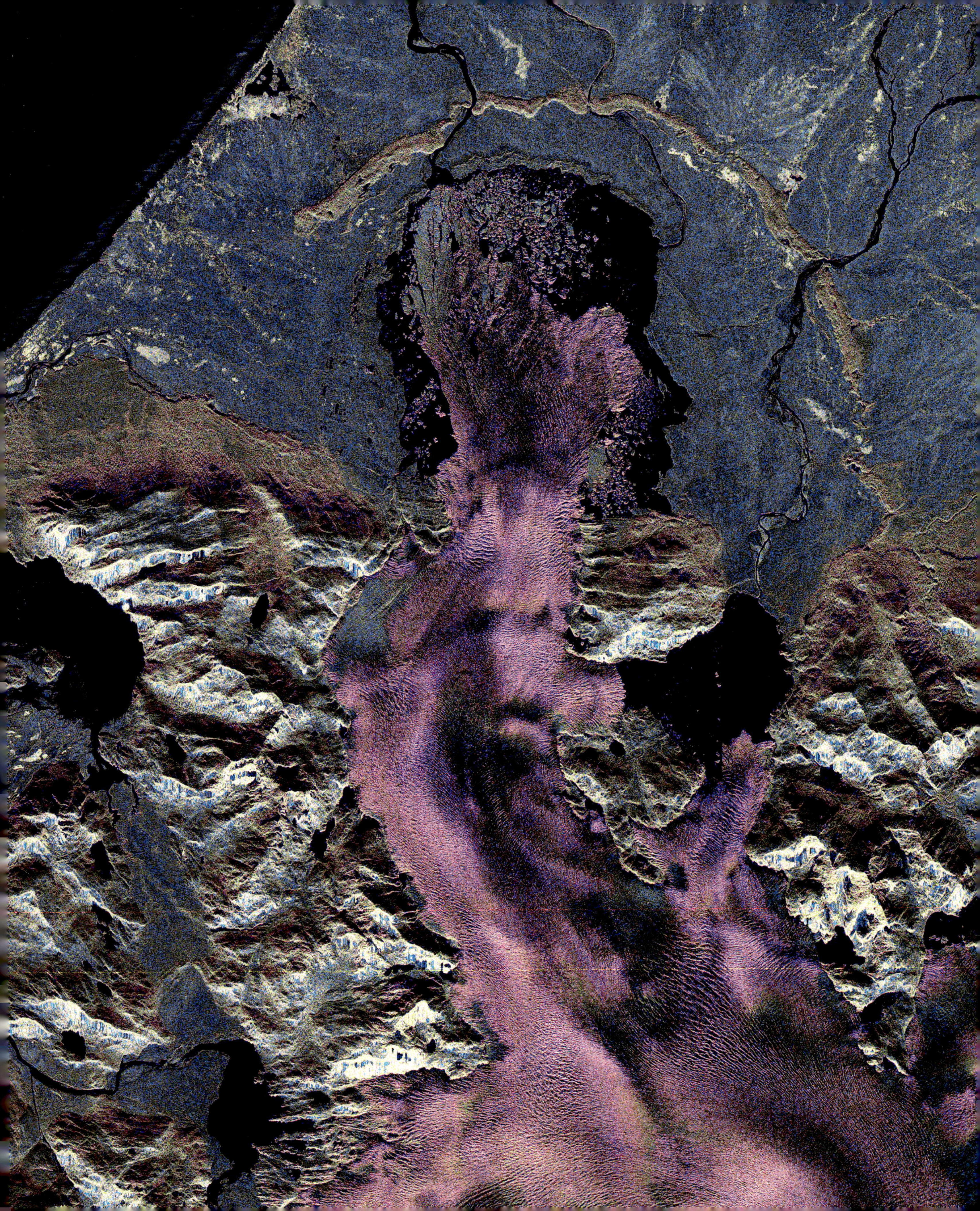

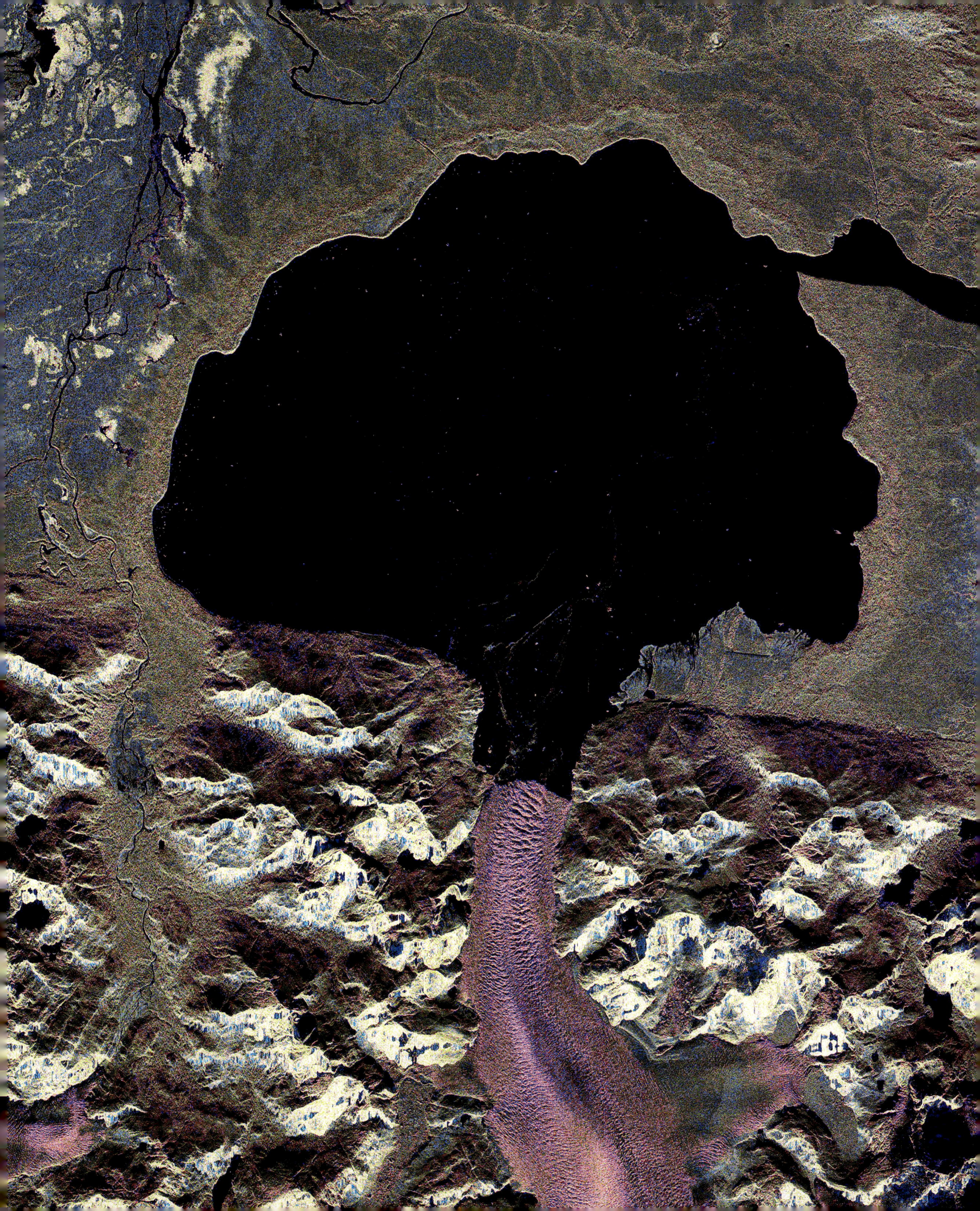

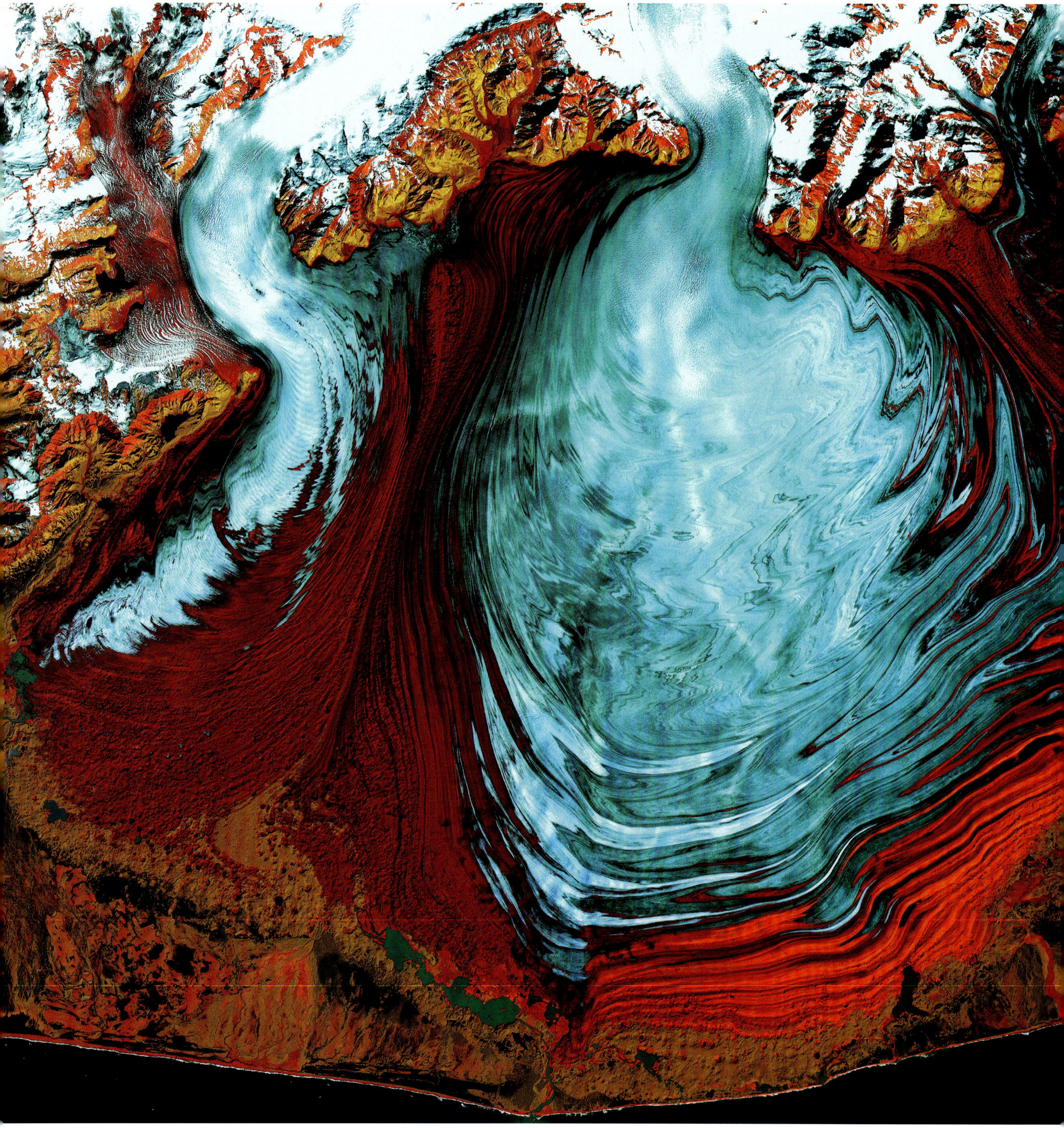

Linke Seite: Die 30 km breite Zunge des Malaspina-Gletschers im Süden Alaskas (Landsat)

Left: The tongue of the Malaspina Glacier in southern Alaska is 19 miles (30 km) wide (Landsat)

Page de gauche : le lobe glaciaire d'une largeur de 30 km du glacier Malaspina, dans le sud de l'Alaska (Landsat)

Página izda: Lengua de 30 km de ancho del glaciar Malaspina, en el sur de Alaska (Landsat)

Im antarktischen Lambert-Gletscher, dem größten der Erde, fließt Eis wie Wasser – nur viel langsamer (Landsat)

The ice of the world's largest glacier, the Lambert in the Antarctic, flows like water—only much more slowly (Landsat)

Dans le glacier antarctique Lambert, le plus grand du monde, la glace coule comme de l'eau, mais beaucoup plus lentement (Lansat)

En el glaciar antártico Lambert, el mayor de la Tierra, el hielo corre como el agua, aunque más lento (Landsat)

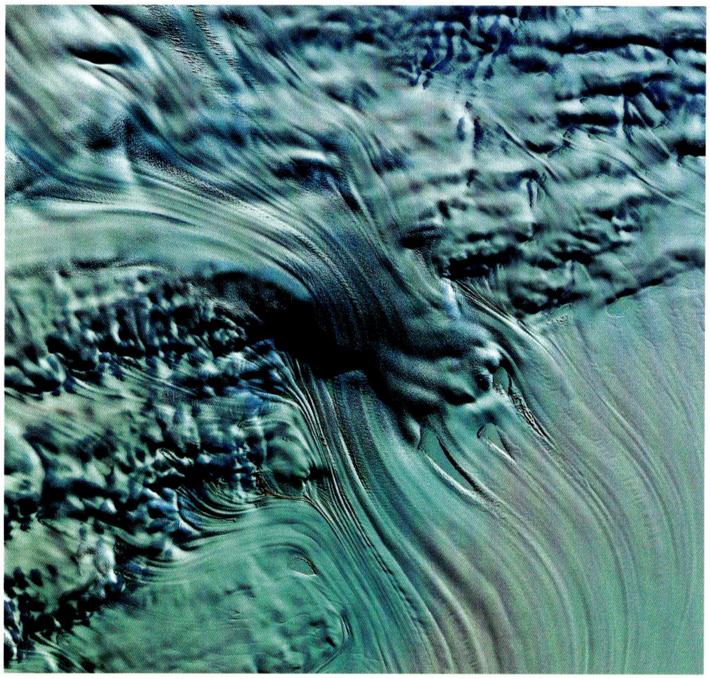

setzen, freut sich Gerhard Schwehm, der bei der ESA den Betrieb der Planetenmissionen leitet: *Wir machen vergleichende Planetenforschung. Wir sehen uns die Atmosphären der einzelnen Planeten an und wollen durch den Vergleich mehr über die Erde lernen.*

Alle Planeten sind vor viereinhalb Milliarden Jahren aus derselben Urwolke entstanden. Die chemische Ausgangslage war für die felsigen Planeten Merkur, Venus, Erde und Mars recht ähnlich. Gleiches gilt für die vier Gasriesen Jupiter, Saturn, Uranus und Neptun. Alle Planeten wurden von der Sonne bestrahlt. Allerdings ist die Intensität der Strahlung je nach Entfernung von der Sonne recht unterschiedlich. Die Forscher haben acht Modellfälle im Sonnensystem, wobei man Merkur als nackten Felskörper nicht mitzählen darf. Dafür springt der Saturnmond Titan mit seiner dichten Atmosphäre ein.

Achtmal können die Forscher versuchen zu verstehen, wie sich die Gasmassen bei unterschiedlichem Druck, anderen Temperaturen und verschiedenen Vorgängen im Innern der Planeten verhalten, welche Rolle die Anwesenheit eines Magnetfelds spielt und welche chemischen Prozesse bei welchen Bedingungen ablaufen. Vielleicht lässt sich so irgendwann das komplexe Wechselspiel in den Atmo-

Schwehm, who is in charge of planetary missions at the ESA. *We are undertaking comparative exploration of the planets, examining the atmospheres of individual heavenly bodies and using the contrast to help us to learn more about the Earth.*

All the planets grew from the same cloud of dust 4.5 billion years ago and the chemical basis of the rocky planets (Mercury, Venus, Earth, and Mars) was very similar; the same is true of the four gas giants Jupiter, Saturn, Uranus, and Neptune. All the planets were illuminated by the Sun, although the intensity of the Sun's rays varies considerably according to the planet's distance from its home star. Researchers thus have eight models to choose from in the solar system; Mercury, a rocky ball with no atmosphere, doesn't count, but Titan, a moon of Saturn with a dense atmosphere, is included.

Therefore, scientists have eight opportunities to attempt to understand how masses of gas behave when exposed to varying pressures, temperatures, and seismic activity within the planets, what role a magnetic field may play, and which chemical processes take place under particular conditions. This may help

même nuage originel il y a quatre milliards et demi d'années. À l'origine, la composition chimique des planètes telluriques Mercure, Vénus, Mars et la Terre était assez semblable. Il en va de même des quatre géantes joviennes Jupiter, Saturne, Uranus et Neptune. Toutes les planètes ont été soumises au rayonnement solaire. Toutefois, l'intensité du rayonnement varie sensiblement en fonction de l'éloignement au soleil. Les scientifiques disposent dans le système solaire de huit exemples de cas, encore qu'il faille exclure Mercure, simple corps rocheux. Mais Titan, la lune de Saturne dotée d'une dense atmosphère, est une bonne alternative.

Les chercheurs ont donc huit opportunités de comprendre le comportement des masses de gaz soumises à des pressions, températures et processus internes variés, le rôle du champ magnétique et les conditions régissant l'apparition des processus chimiques. Peut-être parviendront-ils un jour à expliquer dans le détail la complexité des échanges atmosphériques. Les processus ter-restres sont à l'évidence les mieux connus. Mais l'étude des autres planètes met souvent en lumière d'intéressants phénomènes, jusqu'alors inobservés sur la Terre ou tenus pour insignifiants.

Todos los planetas surgieron hace más de cuatro mil quinientos millones de años de la misma protonebulosa. La situación química de partida era muy similar para Mercurio, Venus, Tierra y Marte, los planetas rocosos. Lo mismo se puede decir de los cuatro gigantes gaseosos, Júpiter, Saturno, Urano y Neptuno. Todos los planetas fueron iluminados por el Sol. Pero la intensidad de la radiación es muy diferente según la distancia con respecto al astro rey. Los investigadores tienen ocho casos modelo en el Sistema Solar, teniendo en cuenta que no se debe contar con Mercurio como cuerpo rocoso desnudo. Le sustituye con su densa atmósfera Titán, una luna de Saturno.

Los investigadores pueden intentar entender, en ocho escenarios diferentes, cómo se comportan las masas gaseosas a presiones diferentes, otras temperaturas y distintos procesos internos de los planetas, qué papel juega la presencia de un campo magnético y bajo qué circunstancias se desarrollan los diferentes procesos químicos. Quizá algún día se pueda comprender en detalle el complejo juego de intercambio en las diferentes atmósferas. Naturalmente, los investigadores saben mucho más de los procesos en la Tierra. Pero a menudo es en otros planetas donde

sphären im Detail nachvollziehen. Über die Vorgänge auf der Erde wissen die Forscher natürlich am meisten. Aber oftmals weisen erst Befunde bei anderen Planeten auf interessante Phänomene hin, die auch auf der Erde ablaufen, aber davor übersehen oder als nicht so bedeutend bewertet wurden. So suchen die Wissenschaftler bei Mars oder Venus nach Indizien, die zeigen, welche Entwicklungsphasen die Planeten durchlaufen haben. Der große Traum ist, die Geschichte unserer Schwesterplaneten möglichst lückenlos zu rekonstruieren.

Gefahren von außen

Angesichts der Verknappung der Ressourcen auf unserer Erde, der nicht zu leugnenden Veränderung des Klimas und einer zunehmenden Verschmutzung der Atmosphäre haben einige ganz Eifrige schon den Mars als künftigen Zufluchtsort der Menschheit auserkoren. Allerdings ist die Luft dort viel zu dünn und voller Kohlendioxid. Das soll sich ändern – dank »Terraforming«. Durch gezieltes Einbringen bestimmter chemischer Stoffe oder durch spezielle Organismen will man aus dem eher unwirtlichen Mars eine Art Garten Eden machen. Das klingt nach Science-Fiction, und das

us one day to explain in detail the complex interplay of processes in our atmosphere. Scientists naturally know more about the mechanisms on Earth, but finds on other planets often turn up interesting phenomena which also occur on our planet but had been overlooked or thought of as less important. Scientists are now looking for indicators on Mars and Venus of the developmental phases undergone by these planets, and their great dream is to reconstruct the history of our sister planets as completely as possible.

Dangers from outside

As resources become ever more scarce on our planet, as atmospheric pollution worsens, and as it becomes increasingly impossible to deny that the climate is undergoing changes, some dreamers have already chosen Mars as the future refuge of the human race. The air there is admittedly much too thin and full of carbon dioxide, but this will one day change, thanks to "terraforming." Judicious introduction of certain chemical compounds or special organisms will make a Garden of Eden of the hostile environment on Mars. This may sound

C'est ainsi que les scientifiques cherchent sur Mars et Vénus des indices témoins des grandes étapes de l'évolution des planètes. Leur plus grand rêve serait de reconstituer parfaitement l'histoire de nos planètes sœurs.

Les dangers extérieurs

Au vu de la raréfaction des ressources terrestres, du changement climatique, et de la pollution croissante de l'atmosphère, quelques enthousiastes voient déjà en Mars le futur refuge de l'humanité. Cependant, l'air sur Mars est rare et regorge de dioxyde de carbone. La « terraformation » se propose d'y remédier. Grâce à l'apport ciblé de substances chimiques et d'organismes spécifiques, elle se propose de transformer l'environnement inhospitalier de Mars en une sorte de jardin d'Éden. Cela relève bien entendu de la science fiction. D'une part, une telle transformation serait beaucoup trop longue pour faire de Mars une alternative réaliste pour accueillir l'humanité. D'autre part, pour modifier à dessein les processus de Mars et de son atmosphère, encore faudrait-il les connaître avec exactitude. Gerhard Schwehm se contente de hocher

descubren primero interesantes fenómenos que también se dan en la Tierra, aunque habían sido pasados por alto o considerados poco significativos. La mayor aspiración es reconstruir, del modo más completo posible, la historia de nuestros planetas hermanos.

Peligros externos

En vista de la escasez de recursos de la Tierra, del innegable cambio climático y de una contaminación creciente en la atmósfera, algunas personas especialmente entusiastas ya han elegido Marte como futuro refugio de la humanidad. Sin embargo, allí el aire está mucho más enrarecido que en la Tierra y lleno de dióxido de carbono. Esto debería cambiar mediante la «terraformación», que consiste en la aportación selectiva de determinadas sustancias químicas o la acción de organismos especiales, para hacer del inhospitalario Marte un nuevo jardín del edén. Esto suena a ciencia ficción y todavía lo es. Por un lado, una transformación así duraría demasiado tiempo, al menos demasiado para que Marte fuese una alternativa realista para la humanidad. Por otro, se deberían comprender bien los procesos de

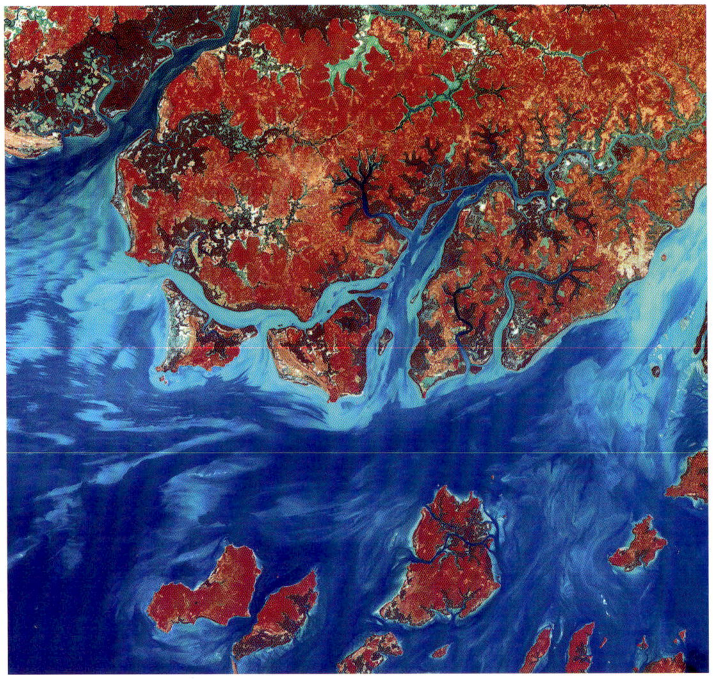

In der Küstenregion von Guinea-Bissau werden viele Sedimente ins Meer geschwemmt; siehe auch Seite 114 (Landsat)

Large amounts of sediment are flushed into the sea along the coast of Guinea-Bissau; compare with page 114 (Landsat)

Sur les côtes de la Guinée-Bissau de grandes quantités de sédiments se déversent dans la mer ; voir aussi page 114 (Landsat)

En la región costera de Guinea-Bissau se arrastran muchos sedimentos hasta el mar; véase también la página 114 (Landsat)

Das lehmige Wasser im Rio de la Plata wälzt sich in den Atlantik. Der weiße Fleck an der Südküste ist Buenos Aires (Envisat)

The Rio de la Plata carries silt far out into the Atlantic. The white area on the southern bank is Buenos Aires (Envisat)

Les eaux argileuses du Río de la Plata débouchent dans l'Atlantique. La tache blanche sur la côte sud représente la ville de Buenos Aires (Envisat)

El agua fangosa del Río de la Plata se descarga en el Atlántico. La mancha blanca de la costa meridional es Buenos Aires (Envisat)

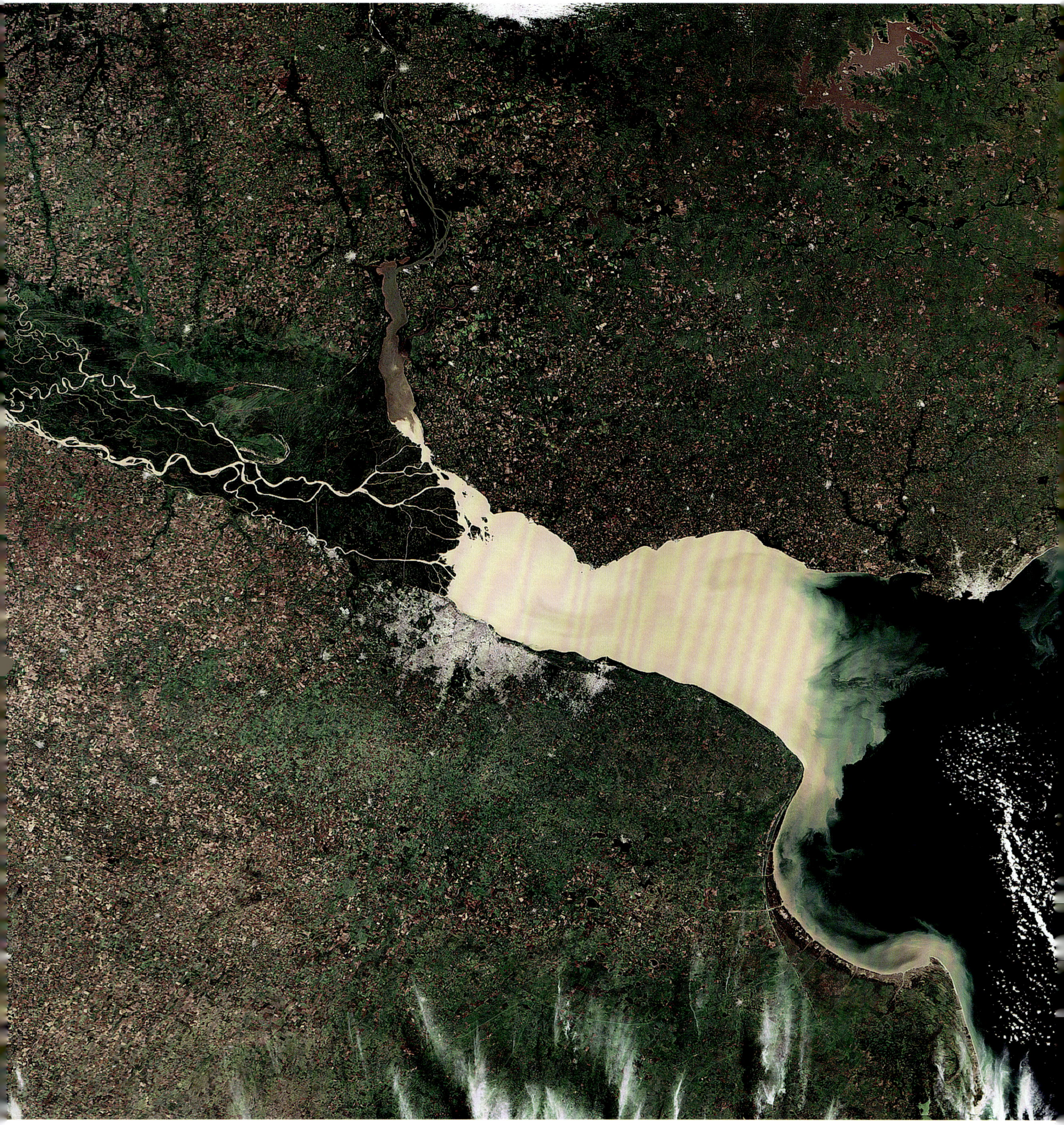

Unten: Die Bucht von Neapel mit der Insel Ischia und dem Vulkan Vesuv (rechts) in einer Radaraufnahme (Envisat)

Below: A radar image of the Bay of Naples, showing the island of Ischia and the volcano Vesuvius on the right (Envisat)

Ci-dessous : image de la baie de Naples, l'île Ischia et le volcan du Vésuve (à droite) acquise par radar (Envisat)

Abajo: El golfo de Nápoles con la isla Ischia y el volcán Vesubio (derecha) en una toma de radar (Envisat)

Rechte Seite: Verrücktes Klima – Wintereinbruch mit viel Schnee in der Türkei. Irak und Syrien im Süden (Envisat)

Right: A crazy climate—the advent of winter brings heavy snowfalls to Turkey, Iraq, and Syria to the south (Envisat)

Page de droite : surprenant climat. Offensive de l'hiver avec neiges abondantes en Turquie. Dans le sud, l'Irak et la Syrie (Envisat)

Página dcha.: El tiempo está loco: inicio del invierno con mucha nieve en Turquía. Al sur, Irak y Siria (Envisat)

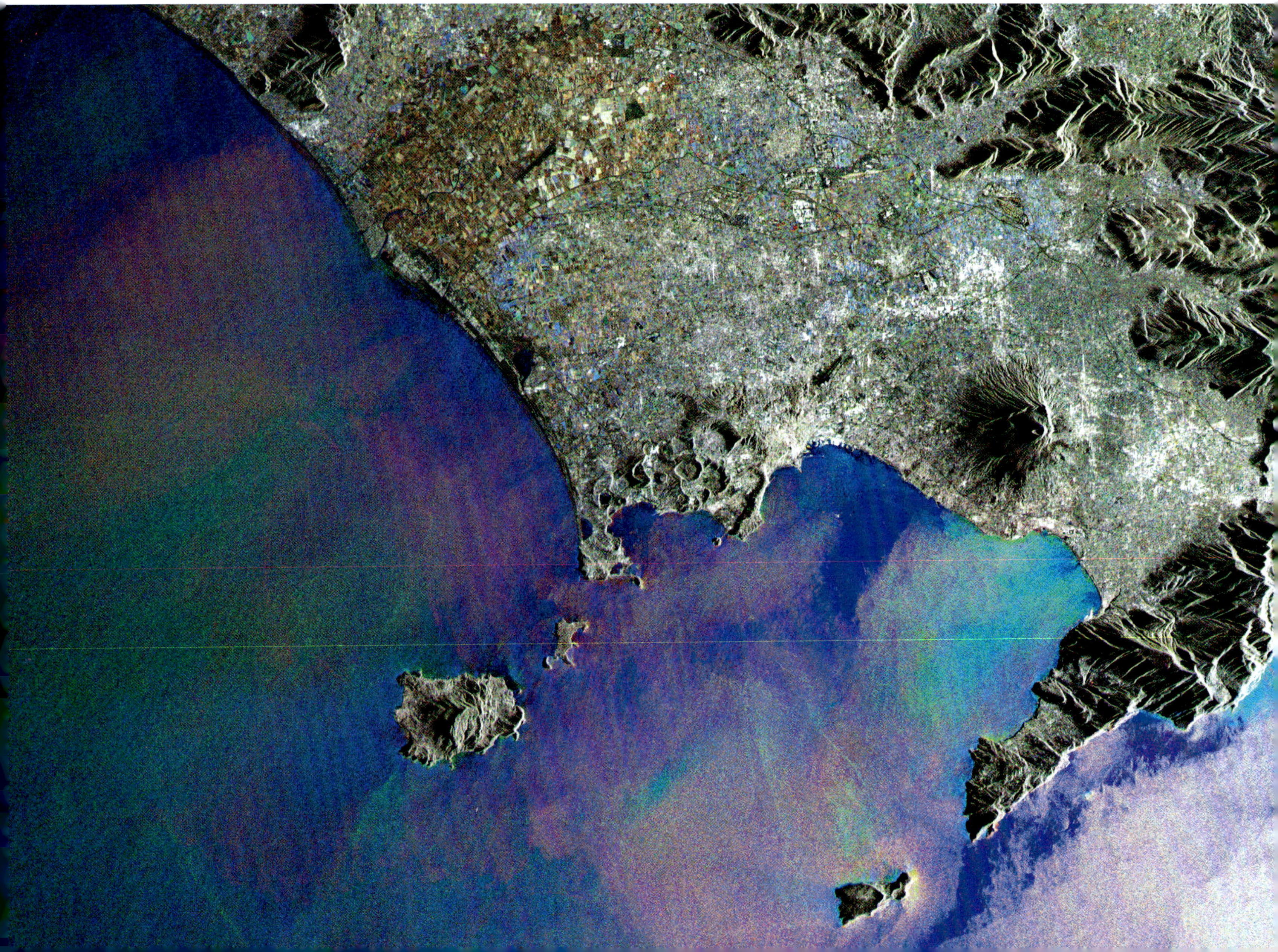

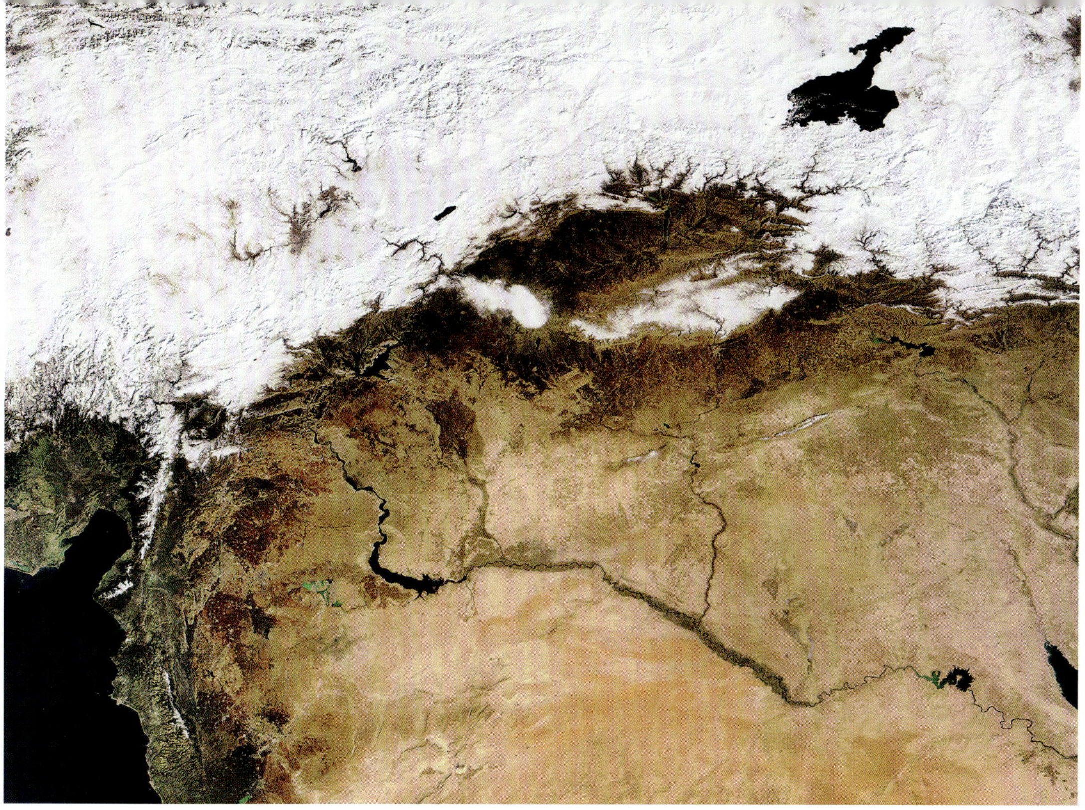

bleibt es auch. Zum einen würde so ein Umbau sehr lange dauern – viel zu lange, als dass Mars für uns Menschen eine realistische Ausweichmöglichkeit wäre. Zum anderen müsste man die Vorgänge auf dem Mars und in seiner Atmosphäre genauestens durchschauen, um sie gezielt zu verändern. Gerhard Schwehm schüttelt über solche Fantastereien nur den Kopf: *Wenn man nicht weiß, wie das gesamte System funktioniert, dann bastelt man an der einen und der anderen Ecke herum, aber man kann auf gut Deutsch manches verschlimmbessern. Das ist die Gefahr: Wir denken, wir wissen, was zu tun ist, aber dann läuft es völlig aus der Hand.* Selbst die Vorgänge in der Atmosphäre der Erde sind in vielem bis heute allenfalls in Grundzügen verstanden. Dies gilt erst recht für den Mars.

Hinzu kommt ein weiteres Problem. Unsere Erde hat einen Schutzschirm, den wir mit unseren Augen nie zu sehen bekommen: das Magnetfeld. Es lenkt die gefährlichen geladenen Teilchen aus dem All ab. Manchmal bekommen wir sie indirekt zu Gesicht, wenn sie verzweifelt als Polarlichter am Himmel leuchten, während Magnetfeld und Atmosphäre sie unschädlich machen. Diesen Schirm haben nicht alle Planeten. *Es ist zwar ganz spannend, von der Besiedlung*

like science fiction, and for the moment it is. For a start, such a change would take a long time—far too long for Mars to be a realistic asylum for humanity. In addition, the processes taking place on Mars and in its atmosphere would have to be understood in the greatest detail so that specific changes could be targeted. Gerhard Schwehm can only shake his head in wonderment at such notions. *If you don't understand how the whole system works, you start tinkering in one corner or another and with the best of intentions make things worse. That is the danger: we think we know what we are doing, but then it all runs away with us.* Many of the processes taking place in the Earth's atmosphere are understood only in broad strokes, and this is even more the case with Mars.

There is an additional problem. Our Earth has a protective screen, invisible to our eyes; its magnetic field deflects the many dangerous charged particles which arrive from space. We sometimes catch sight of these indirectly, as our magnetic field and atmosphere render them harmless and they light up the polar skies in a vain attempt to reach the surface. Not every planet has such protection. *It is exciting to read*

la tête devant de telles chimères : *Quand on ne sait pas comment fonctionne le système, on peut le modifier ici et là, mais on risque de faire pire en voulant faire mieux. Le réel danger consiste à croire savoir ce qu'il faut faire, puis perdre totalement le contrôle.* Même nos connaissances des processus atmosphériques terrestres sont encore rudimentaires. Que dire alors de nos connaissances de Mars.

À cela s'ajoute une autre difficulté. Notre Terre est protégée par un bouclier invisible : le champ magnétique. Il dévie les dangereuses particules chargées provenant de l'espace. Il nous arrive de les observer quand elles brillent désespérément dans le firmament sous forme d'aurores polaires, tandis que le champ magnétique et l'atmosphère les neutralisent. Toutes les planètes ne bénéficient pas d'une telle protection. *Les récits contant la colonisation de Mars sont certes passionnants, mais se heurtent à un problème de taille : le champ magnétique de Mars est quasi inexistant, la planète est donc entièrement exposée aux rayonnements,* déclare Albert Zaglauer, balayant ces idées hasardeuses d'un revers de main. Il a dirigé au sein d'Airbus la construction de la mission Swarm de

Marte y de su atmósfera para poder modificarlos selectivamente. Gerhard Schwehm mueve la cabeza, dubitativo, ante tales fantasías: *Si no se sabe cómo funciona todo el sistema, se pueden poner parches por aquí o por allá pero al final, puede ser peor el remedio que la enfermedad. Ahí reside el peligro: pensamos que sabemos lo que hay que hacer, pero después se nos va completamente de las manos.* Hasta hoy, incluso los procesos atmosféricos terrestres son entendidos, como mucho, únicamente en sus rasgos esenciales. En el caso de Marte, con más razón.

Además, hay otro problema. Nuestra Tierra tiene una pantalla protectora que no conseguimos ver con nuestros ojos: el campo magnético. Este desvía las peligrosas partículas cargadas del espacio. A veces llegamos a verlas indirectamente cuando lucen en forma de auroras boreales, mientras el campo magnético y la atmósfera las neutralizan. No todos los planetas tienen esta pantalla. *Es muy interesante leer sobre la colonización de Marte. Pero el mayor problema es que Marte no tiene prácticamente campo magnético, es decir, está expuesto a la radiación,* aclara Albert Zaglauer, descartando ideas temerarias. Dirigió como jefe de proyecto en la empresa Airbus, en

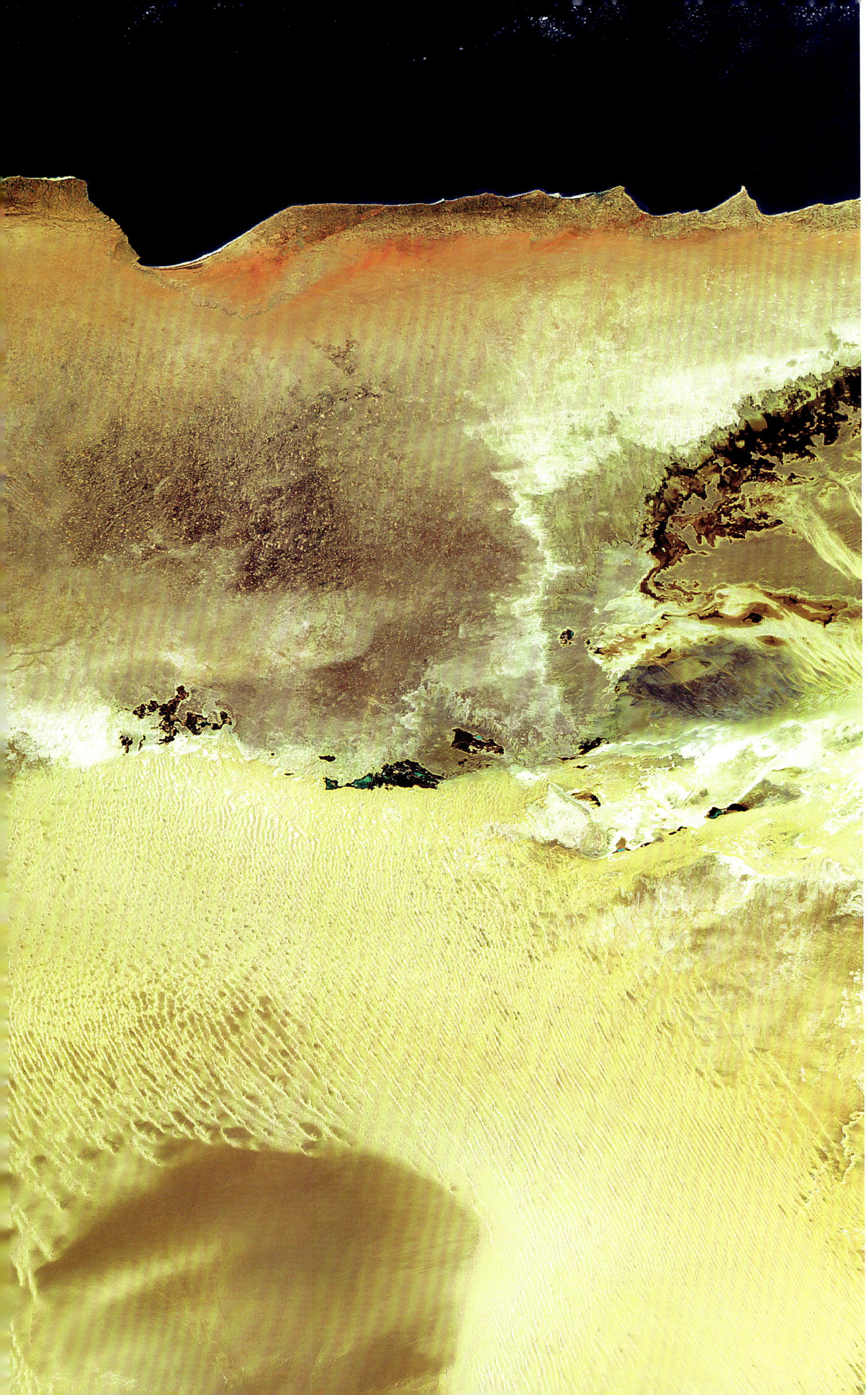

Die Sahara glänzt golden an der Grenze von Libyen und Ägypten. Der Ausschnitt zeigt 800 km von oben nach unten (Envisat)

The golden glint of the Sahara on the Libyan/Egyptian border. The image shows an area about 500 miles (800 km) across from top to bottom (Envisat)

Éclat doré du Sahara à la frontière entre la Libye et l'Égypte. Image couvrant 800 km de haut en bas (Envisat)

El Sahara resplandece dorado en la frontera de Libia con Egipto. El detalle muestra 800 km de norte a sur (Envisat)

Rechte Seite: Die Richat-Struktur in der Wüste Mauretaniens ist kein Einschlagkrater, sondern vulkanischen Ursprungs (Landsat)

Right: The Richat Structure in the Mauritanian desert resembles an impact crater but is of volcanic origin (Landsat)

Page de droite : la structure de Richat dans le désert de Mauritanie n'est pas un cratère d'impact, mais d'origine volcanique (Landsat)

Página dcha.: La estructura de Richat, en el desierto en Mauritania, no es un cráter de impacto, sino que tiene origen volcánico (Landsat)

des Mars zu lesen. Aber das große Problem ist, dass der Mars praktisch kein Magnetfeld hat, man dort also voll der Strahlung ausgesetzt ist, wischt Albert Zaglauer die kühnen Ideen beiseite. Er leitete als Projektleiter bei der Firma Airbus in Friedrichshafen den Bau der ESA-Mission Swarm. Sie soll das filigrane Magnetfeld unserer Erde untersuchen. Die drei Swarm-Satelliten funken auf den ersten Blick unspektakuläre Zahlenkolonnen zur Erde funken. Bunte Bilder wird es von Swarm nicht geben, erläutert Albert Zaglauer: *Es ist immer einfacher, der breiten Bevölkerung mit einem Bild zu vermitteln, was geleistet wird. Mit Messkurven wird kaum über die Expertenwelt hinaus wahrgenommen. Aber Magnetfelder sind unglaublich wichtig für unser Leben!* Vermeintlich langweilige Daten werden den Forschern helfen, eine unserer Lebensgrundlagen auf der Erde endlich genauer zu durchschauen. Denn die Satelliten im All haben Sinne für Vieles, was unseren Augen und Ohren entgeht – und was deswegen fälschlicherweise oft als eher unwichtig angesehen wird.

Umziehen unmöglich

Die Planetenwissenschaft steht bei Klima und Magnetfeldern erst ganz

about Mars colonies, but the great problem is that Mars has practically no magnetic field and we would be completely exposed to radiation there. Albert Zaglauer has little time for these bold ideas. The three Swarm satellites transmitted reams of seemingly unspectacular data back to Earth, and there won't be any color photography, as Zaglauer explains: *It is always easier to use an image to show the general public what we are doing. No one outside the scientific community will understand a Gauss curve, but magnetic fields are still incredibly important for life!* This rather dull-seeming data will help scientists toward a more complete understanding of the basis of life on Earth. Satellites in space can detect things which escape our eyes and ears and are thus often mistakenly disregarded.

Moving out is impossible

Interplanetary study of climate and magnetic fields is still in its infancy, and most phenomena are still completely unexplained. Before a planet can be rebuilt to a plan, exact knowledge of its atmosphere is required, including the air's

l'ESA, consacrée à l'étude du champ magnétique de la Terre. Les trois satellites Swarm transmettent des colonnes de chiffres à première vue très ordinaires. Swarm ne nous livrera pas de belles images, explique Albert Zaglauer : *Il est toujours plus facile de présenter nos résultats au grand public à l'aide de photographies. Les courbes de mesures passent quasi inaperçues en dehors des cercles d'experts. Pourtant, le champ magnétique est indispensable à notre vie !* Ces données présumées assommantes aideront ainsi les chercheurs à mieux comprendre les fondements de la vie sur Terre. Les satellites en orbite perçoivent en effet quantité de choses qui échappent à nos yeux et nos oreilles, et que nous croyons souvent, à tort, dénuées d'importance.

L'impossible exode

Dans l'étude du climat et du champ magnétique, la science planétaire n'en est qu'à ses balbutiements. La plupart des phénomènes sont encore largement inexpliqués. La transformation ciblée d'une planète exige une connaissance exacte de son atmosphère, des échanges surface-atmosphère, et du rôle des influences externes telles que les rayonnements

Friedrichshafen, la construcción de la misión Swarm de la ESA. Esta debe analizar los entresijos del campo magnético de la Tierra. Los tres satélites Swarm envían a la Tierra unas, a primera vista, anodinas columnas de números. Albert Zaglauer aclara que Swarm no producirá imágenes en color: *Siempre es más sencillo transmitir al público lo que se ha hecho mediante una imagen. Las gráficas de mediciones apenas se aprecian fuera del círculo de expertos. Pero los campos magnéticos son increíblemente importantes para nuestra vida.* Esos datos ayudarán a los investigadores a comprender finalmente con exactitud una de las razones de nuestra subsistencia en la Tierra. Los satélites espaciales son capaces de percibir muchas cosas que escapan a nuestros ojos y oídos y que, por eso mismo, consideramos erróneamente de poca importancia.

Traslado imposible

La planetología no ha hecho más que comenzar en lo que respecta al clima y a los campos magnéticos. La mayoría de los fenómenos continúan siendo enigmáticos para nosotros. Y antes de poder transformar selectivamente un planeta, se necesitan cono-

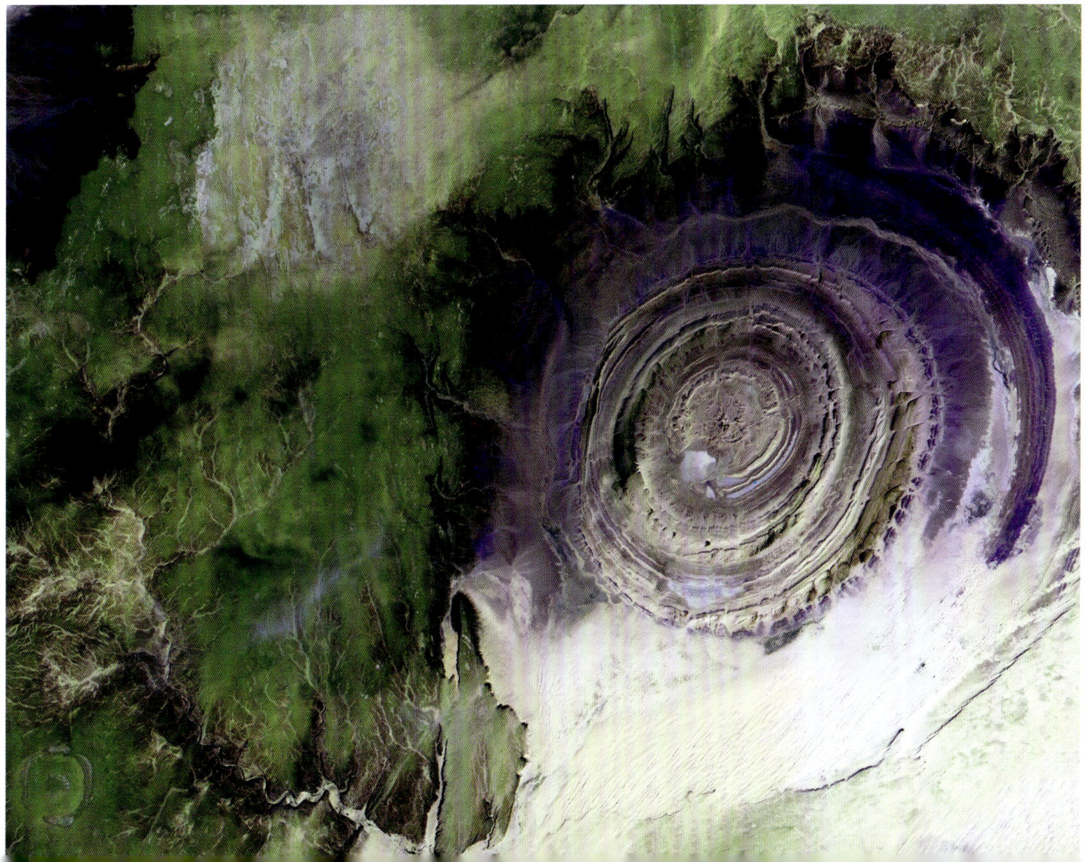

am Anfang. Die meisten Phänomene sind noch immer völlig rätselhaft. Bevor sich aber ein Planet gezielt umbauen lässt, bedarf es präziser Kenntnis seiner gesamten Atmosphäre, ihrer Wechselwirkung mit der Oberfläche und der Rolle der äußeren Einflüsse, etwa der Strahlungspartikel von der Sonne und aus den Tiefen des Weltraums. Da haben die Forscher noch einiges zu tun, sofern das Ziel überhaupt jemals erreichbar sein sollte.

So mahnt Gerhard Schwehm, man solle doch mit der Erde und den Zuständen hier zufrieden sein. Raumfahrt hilft uns, unseren Platz im All zu verstehen und unser Leben hier so einzurichten, dass es die Erde möglichst lange mit uns aushält. Aber Raumfahrt liefert uns keine Alternative zum irdischen Leben: *Es ist ein schöner Traum, mit Terraforming andere Planeten umzubauen. Aber vielleicht ist das vom Menschen auch viel zu egozentrisch gesehen. Wir sollten ein wenig mehr Demut zeigen und nicht überlegen, andere Planeten zu verändern, sondern erst einmal vor unserer eigenen Haustür sauber machen.*

Neben der Erforschung der Erde ist die Beobachtung der Sonne ein weiteres, derzeit geradezu boomendes Gebiet. In den vergangenen Jahren sind etliche Sonnensatelliten gestartet, andere sind in Bau oder werden gerade konzipiert. Die Sonne wird bereits rund um die Uhr überwacht. Denn der große gelbe Ball am Himmel ist bei weitem nicht so friedlich, wie es an einem klaren Sommertag scheinen mag. Sonne und Erde sind keineswegs getrennte Themen – sie sind aufs engste miteinander verknüpft. Das gilt zumindest für die kleine Erde, die den Launen der großen Sonne wenig entgegenzusetzen hat.

Die Erde befindet sich streng genommen noch in den Ausläufern der Sonnenatmosphäre. Fast alles, was in unserer Lufthülle geschieht, ist eine Reaktion auf die Strahlung der Sonne, die seit Jahrmilliarden auf die Erde trifft. Die Sonne treibt das Wetter an, ihre Hitze und ihre Ultraviolettstrahlung bestimmen die meisten

relationship with the surface and the role of external influences such as radiation particles from the Sun and from the depths of space. There is still much for scientists to do if this goal is ever to be achieved.

Gerhard Schwehm is of the opinion that we should be happy with the Earth and the conditions here. Space travel helps us to understand our place in the universe and to conduct our lives here such that the Earth will survive to support us for as long as possible, but it will not provide us with an alternative to life on Earth. *Using terraforming to reshape other planets is a beautiful dream, but perhaps it is also too egocentric as far as humanity is concerned. We should show a little humility and get our own house in order first before considering alterations on other planets.*

Alongside the exploration of the Earth, observation of the Sun is a field experiencing a current boom. A number of solar satellites have been launched in recent years, and more are under construction or in the planning phase. The Sun is now under surveillance 24 hours a day, and this great yellow ball has revealed itself to be far less peaceful than it might seem on a clear summer's day. Sun and Earth are not separate subjects at all—they are interdependent, and this is especially true of our tiny Earth, which has little to defend itself from the changing moods of the great Sun.

In the strictest sense, the Earth is located in the outermost reaches of the Sun's atmosphere. Almost everything occurring in the air that surrounds us is a reaction to the Sun's radiation, to which the Earth has been exposed for billions of years. The Sun determines the weather, and its heat and ultraviolet rays shape most of the chemical processes in our atmosphere. Its X-rays and clouds of charged particles continually bathe our planet, but to this day the Sun remains as enigmatic to astronomers as the Earth does to geologists. Only when we know precisely the nature and

solaire et cosmique. Dans ce domaine, il reste encore aux chercheurs beaucoup de travail pour atteindre l'objectif, s'il est effectivement atteignable.

C'est pourquoi Gerhard Schwehm nous engage à nous contenter de ce que nous offre la Terre. La recherche spatiale lève un coin du voile sur notre place dans l'univers pour nous permettre de mener ici-bas une vie encore longtemps supportable par la Terre, mais elle ne propose pas d'alternative à la vie sur Terre. *Transformer d'autres planètes grâce à la terraformation est un beau rêve. Mais les humains se laissent peut-être aller à un égocentrisme exacerbé. Nous devrions faire preuve d'un peu d'humilité, et cesser de réfléchir au moyen de transformer les planètes. Contentons-nous tout d'abord de balayer devant notre porte.*

Tout comme l'exploration de la Terre, l'observation du soleil est un domaine en plein essor. De nombreux satellites solaires ont été lancés ces dernières années. D'autres sont en construction ou en projet. Le soleil fait déjà l'objet d'une surveillance permanente. La grosse balle jaune suspendue dans le ciel est en effet loin d'être aussi paisible que ne le laisserait supposer une belle journée d'été. Soleil et Terre ne sont aucunement des matières indépendantes. Ces deux corps célestes sont étroitement liés. C'est en tout cas vrai pour notre petite Terre qui n'a pas grand-chose à opposer aux humeurs de ce soleil gigantesque.

La Terre se trouve à proprement parler dans les confins de l'atmosphère solaire. Pratiquement tous les processus observés dans notre atmosphère sont des réactions au rayonnement solaire qui frappe la Terre depuis des milliards d'années. Le soleil influence le climat terrestre. Sa chaleur et son rayonnement ultraviolet régissent la majorité des processus chimiques de notre atmosphère, tandis que ses rayons X et les particules chargées de ses vents sifflent constamment aux oreilles de la planète. Les mystères du soleil restent cependant aussi impénétrables pour les astronomes que les

cimientos precisos de toda su atmósfera, de sus interacciones con la superficie y del papel de los influjos externos, como las partículas irradiadas por el Sol y las procedentes de las profundidades del universo. Ahí todavía tienen algo que hacer los investigadores, en caso de que se pretenda alcanzar algún día el objetivo.

Gerhard Schwehm advierte que, a pesar de todo, debemos estar satisfechos con la Tierra y su situación. La astronáutica nos ayuda a entender nuestro lugar en el universo y a organizar la vida aquí de tal modo que la Tierra resista lo más posible con nosotros. *Transformar otros planetas con terraformación es un bonito sueño. Pero quizá la humanidad esté considerando esto desde un punto de vista demasiado egocéntrico. Deberíamos demostrar un poco más de humildad y, en lugar de pensar en cambiar otros planetas, podríamos empezar limpiando la puerta de nuestra propia casa.*

Junto a la investigación de la Tierra, la observación del Sol es otro campo que está experimentando un gran auge. En los últimos años se han lanzado algunos satélites solares y otros están en construcción o en fase de proyecto. El Sol ya está sometido a observación las 24 horas del día, porque esa gran pelota amarilla colgada del cielo no es, ni mucho menos, tan pacífica como podría parecer. El Sol y la Tierra no son, en absoluto, asuntos distintos. Muy al contrario, están íntimamente unidos. Esto es verdad, al menos en lo que respecta a la pequeña Tierra, que poco puede objetar ante los caprichos del gran Sol.

En rigor, la Tierra se encuentra aún en la periferia de la atmósfera solar. Casi todo lo que ocurre en nuestra atmósfera es una reacción a la radiación del Sol, que afecta a la Tierra desde hace miles de millones de años. El Sol es un motor meteorológico, su calor y su radiación ultravioleta determinan la mayoría de los procesos químicos que se producen en nuestra atmósfera. Sus rayos X y sus nubes de partículas cargadas barren permanentemente toda la Tierra. Pero hasta la actualidad, el Sol continúa siendo

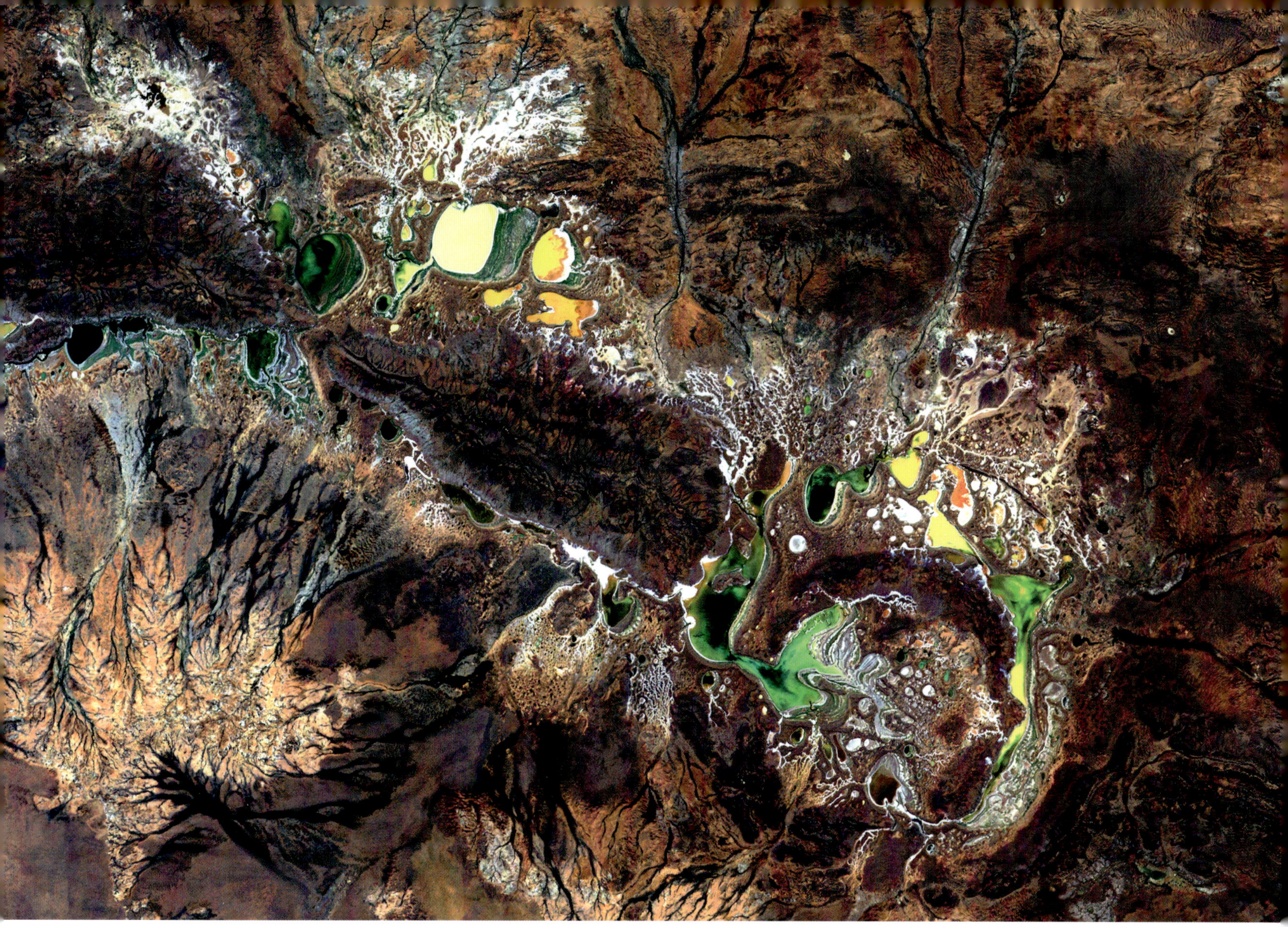

Spuren eines Besuchers: Die Shoemaker-Struktur in Australien entstand bei einem Einschlag vor 1,7 Milliarden Jahren (Landsat)

Traces of a visitor: the Shoemaker Crater in Australia was caused by an impact 1.7 billion years ago (Landsat)

Les traces d'un visiteur : la structure Shoemaker est apparue en Australie lors d'un impact il y a 1,7 milliard d'années (Landsat)

Trazas de un visitante: La estructura Shoemaker, en Australia, la produjo un impacto hace 1,7 millardos de años (Landsat)

Seite 214/215: Wüste überall? Nein. Ein Farmer widersetzt sich mit Bewässerung (roter Fleck) der Kalahari in Namibia (Landsat)

Pages 214–5: Just a desert? No. One farmer is fighting back against the Kalahari in Namibia using irrigation, visible as a red spot (Landsat)

Pages 214 et 215 : Une étendue désertique ? Un agriculteur namibien résiste par l'irrigation (tache rouge) dans le désert du Kalahari, en Namibie (Landsat)

Páginas 214/215: ¿Es todo desierto? No. Un granjero resiste con un regadío (mancha roja) en el Kalahari, en Namibia (Landsat)

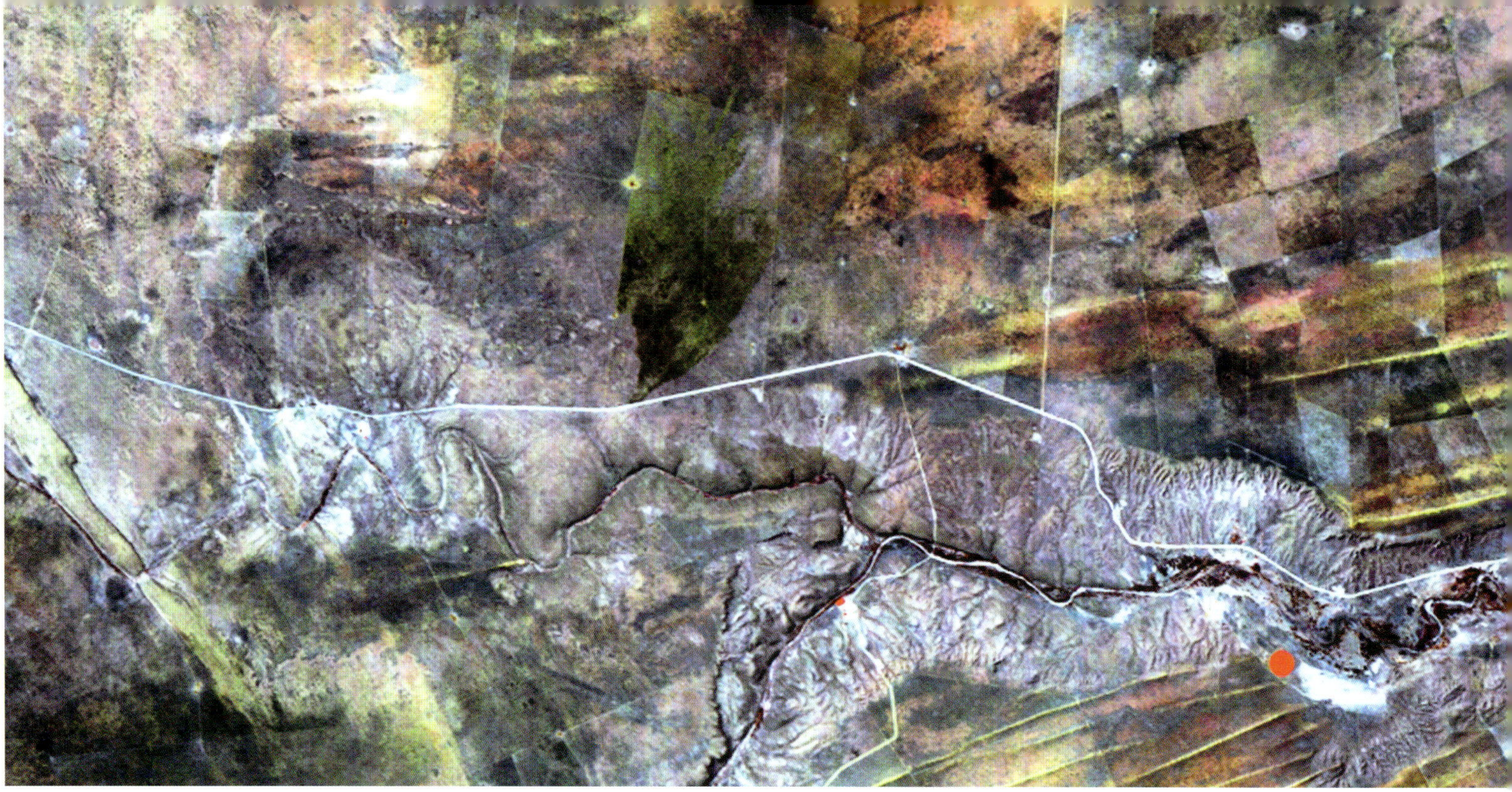

chemischen Vorgänge in unserer Atmosphäre. Ihre Röntgenstrahlung und Wolken aus geladenen Teilchen fegen der Erde ständig um die Ohren. Doch bis heute ist den Astronomen die Sonne mindestens so rätselhaft wie unsere Erde den Geologen. Nur wenn genau bekannt ist, welche Strahlung die Sonne *wie* ins Universum schickt, lässt sich auch verstehen, wie die Erde darauf reagiert. So ist auch Sonnenforschung eine zentrale Komponente der Erdbeobachtung.

Die lebensfreundlichen Bedingungen auf unserer Erde sind nicht selbstverständlich. Wir sind umgeben von einer Lufthülle, die nur scheinbar ein riesiges Luftmeer ist. Schon in fünf Kilometern Höhe kann kein Mensch auf Dauer leben. In hundert Kilometern Höhe fängt der Weltraum an. Sollte in einigen Lichtjahren Umgebung ein Stern als Supernova explodieren, würde die Schockwelle unsere Atmosphäre einfach wegpusten. Das Leben wäre sofort am Ende. Sollte die Sonne jetzt ausgehen, blieben noch acht Minuten Helligkeit. Dann würde die für uns unverzichtbare Licht- und Wärmequelle erlöschen. Binnen Tagen würde alles Leben im ewigen Eis erstarren. Krachte morgen ein zwei Kilometer großer Felsbrocken auf die Erde – und von denen gibt es im Sonnensystem

amount of all the radiation the Sun sends out into the universe will we be able to understand how the Earth reacts to it. Solar research is thus a central component of Earth observation.

The conditions on Earth so conducive to life should not be taken for granted. We are enveloped in an airy cloak with only the appearance of great depth. A human is unable to survive for very long at a height of just 3 miles (5 kilometers) and space itself begins at an altitude of just 60 miles (100 kilometers). The shock wave from a star exploding as a super-nova just a few light years away would be enough to blow the atmosphere away completely and life would be immediately at an end. Were the Sun to go out this second, there would be another eight minutes of light before our essential source of light and warmth was snuffed out. Within days, all life would be frozen in eternal ice. If a mile-long (1.5-kilometer) chunk of rock—and there are more of these in the solar system than we would care to imagine—were to crash into the Earth, it would wipe out an entire continent at the least. The explosion caused by the impact would probably extinguish all higher life on

secrets de la Terre pour les géologues. Quand nous connaîtrons avec exactitude les rayonnements émis par le soleil et leur origine, nous pourrons alors comprendre les réactions de la Terre. La recherche consacrée au soleil apparaît donc comme un élément clé de l'observation de la Terre.

Les conditions clémentes qui caractérisent notre Terre ne sont pas un fait acquis. Nous sommes entourés d'une atmosphère qui n'a que l'apparence d'une immense étendue d'air. Aucun être humain ne survivrait longtemps ne serait-ce qu'à cinq kilomètres d'altitude. L'univers commence à plusieurs centaines de kilomètres d'altitude. Si l'explosion d'une étoile à quelques années lumières donnait naissance à une supernova, son onde de choc aurait tôt fait de balayer notre atmosphère. La vie sur Terre s'évanouirait immédiatement. Si le soleil venait à périr, il nous resterait tout juste huit minutes de clarté, avant que ne s'éteigne notre indispensable source de lumière et de chaleur. En l'espace de quelques jours, toute vie se figerait dans la glace. Si demain une météorite de deux kilomètres de diamètre s'écrasait sur la Terre, et elles gravitent par centaines dans le système solaire, elle détruirait au moins un continent, et l'explosion causée par le choc

tan enigmático para los astrónomos como lo es la Tierra para los geólogos. Solo conociendo exactamente cuánta radiación emite el Sol al universo, podremos entender la consiguiente reacción de la Tierra. Así, la investigación solar es uno de los componentes centrales de la observación terrestre.

Las condiciones propicias para la vida de nuestra Tierra no son evidentes. Estamos rodeados de una cubierta gaseosa que, sólo aparentemente, es un gigantesco mar de aire. Ya a 5 km de altura ningún ser humano puede sobrevivir largo tiempo. A 100 km de altitud comienza el espacio sideral. Si una estrella estallase como supernova en un entorno de unos pocos años luz, la onda de choque se llevaría nuestra atmósfera. La vida se extinguiría inmediatamente. Si el Sol se apagara ahora mismo, aún nos quedarían ocho minutos de claridad. Después, desaparecería la fuente de luz y de calor que nos es indispensable. En unos pocos días, toda la vida se petrificaría en forma de hielo eterno. Si mañana chocase contra la Tierra una piedra de 2 km de diámetro, de las que en el Sistema Solar hay más de las que nos gustarían, destruiría al menos todo un continente. Probablemente, la explosión del impacto aniquilaría toda la escala

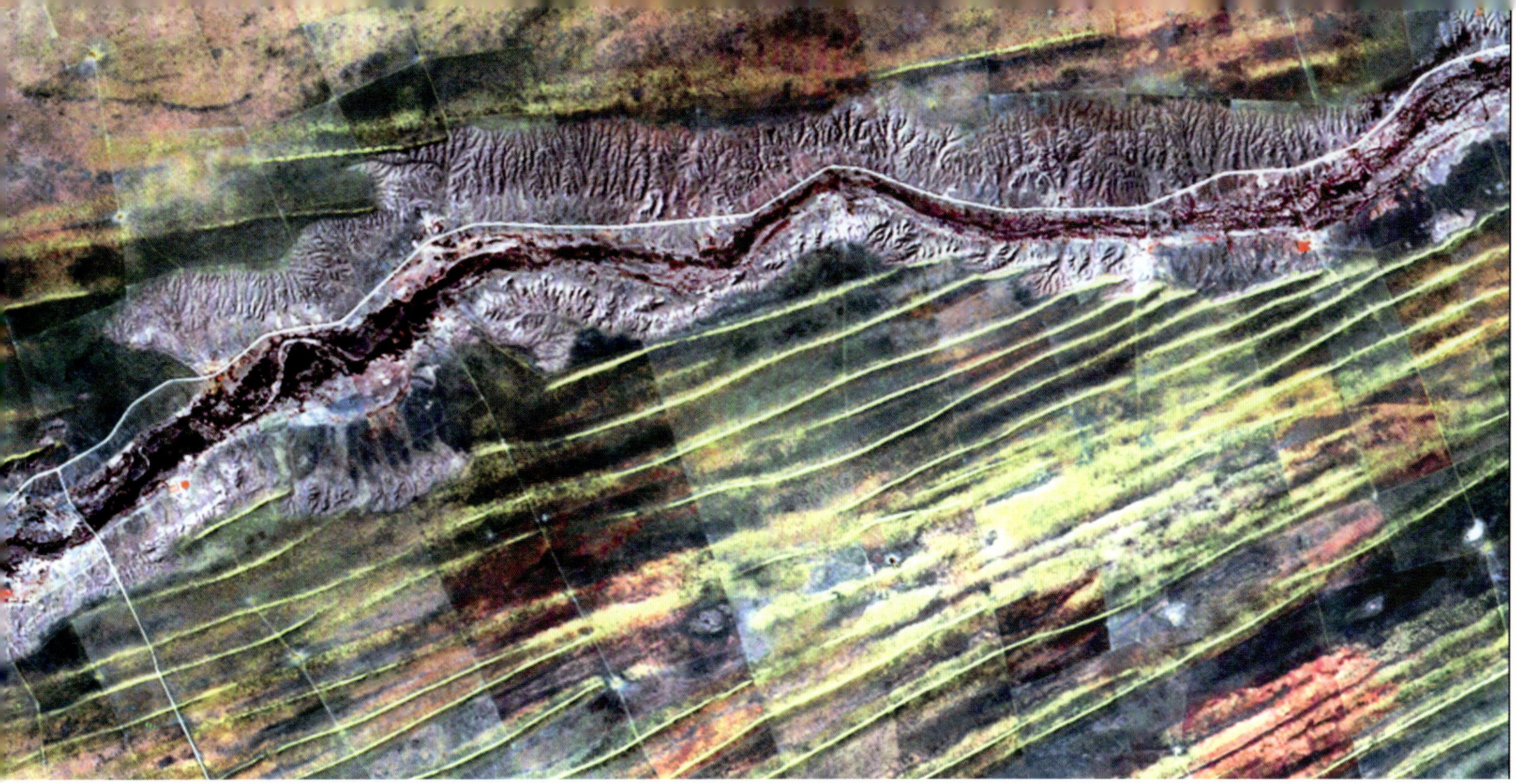

mehr, als uns lieb sein kann –, würde er mindestens einen gesamten Kontinent vernichten. Vermutlich würde die Explosion beim Aufprall praktisch alles hoch entwickelte Leben auslöschen. Sollte das Magnetfeld der Erde zusammenbrechen, kämen energiereiche Teilchen des Sonnenwinds und der kosmischen Strahlung ungehindert auf die Erdoberfläche. Katastrophale Mutationen im Erbgut wären die Folge. Leben, wie wir es kennen, wäre nicht mehr lange möglich.

Gefahr von innen

All diesen Gefahren sind wir schutzlos ausgeliefert. Sollten diese Ereignisse eintreten, würden sie unweigerlich das Ende der Menschheit bedeuten. Aber keine Sorge: So dramatisch, wie manche Hollywood-Streifen glauben machen wollen, läuft das Leben im All nicht ab. Einen Supernova-Kandidaten gibt es in der Nähe der Erde zurzeit nicht. Nach den Theorien der Astronomen wird unsere Sonne noch Milliarden Jahre gemächlich vor sich hin brennen – hoffen wir, dass die Modelle stimmen. Tatsächlich kommt die Erde immer wieder unter starken Beschuss aus dem All. Aber all zu oft scheint auch das nicht zu passieren, sonst hätte sich das Leben in den

Earth. If the Earth's magnetic field were to fail, charged particles from the solar wind and cosmic rays would reach the Earth's surface unhindered, resulting in catastrophic mutations to our DNA. Life as we know it would not be possible for long.

The danger from within

We are defenseless against all these dangers. Were such eventualities to occur, it would spell the end of the world as we know it. But don't worry—life in space is not as dramatic as many Hollywood movies make out. At the moment there are no candidates for a supernova near the Earth, and according to astronomers' theories, our Sun will continue to burn quietly for several billion years. Let's hope the models are accurate.

The Earth does indeed come under repeated attack from space, but this often seems to happen without our even noticing it, otherwise life would have been unable to develop so well in the past hundreds of millions of years. In short, there is little danger from outside, and the Earth seems to have dealt with cosmic perils pretty well during

réduirait probablement à néant toute forme de vie complexe. Si le champ magnétique terrestre s'effondrait, les particules ionisées du vent solaire et des rayonnements cosmiques bombarderaient librement la surface de la Terre, provoquant des mutations génétiques. La vie ne serait plus possible telle que nous la connaissons.

Les dangers intérieurs

Nous sommes à la merci de tous ces dangers. La survenue de tels événements signerait inéluctablement l'extinction de l'humanité. Il n'y a toutefois pas d'inquiétude à avoir : l'activité de l'univers n'est pas aussi alarmante que ne le laissent supposer les productions hollywoodiennes. Il n'existe à l'heure actuelle aucune supernova potentielle à proximité de la Terre. Selon les théories astronomiques, le soleil devrait continuer à se consumer lentement pendant plusieurs milliards d'années. Espérons que les calculs soient justes ! La Terre est, de fait, régulièrement soumise à de forts bombardements cosmiques. Mais cette fréquence ne semble pas trop élevée, sans quoi la vie n'aurait pu évoluer comme elle l'a fait pendant des centaines de millions d'années. En un mot, la Terre

superior de la vida. Si se colapsara el campo magnético de la Tierra, las partículas cargadas de alta energía del viento solar y la radiación cósmica llegarían sin obstáculos hasta la superficie terrestre. La consecuencia serían mutaciones catastróficas del patrimonio genético. La vida, tal como la conocemos, ya no sería posible durante mucho tiempo.

Peligros internos

Estamos expuestos, sin protección, a todos esos peligros. Si sobrevinieran estos acontecimientos, significarían inevitablemente el fin de la humanidad. Pero tampoco hay por qué preocuparse: la vida en el espacio no se desarrolla tan dramáticamente como algunas películas de Hollywood quieren hacernos creer. De momento, en las cercanías de la Tierra no hay ninguna candidata a supernova. Según las teorías de los astrónomos, nuestro Sol seguirá consumiéndose a sí mismo durante miles de millones de años... naturalmente, esperamos que los modelos estén en lo cierto. En realidad, la Tierra está sometida a un constante bombardeo desde el espacio. Pero no parece que todo esto pase muy frecuentemente ya que, de lo contrario, la vida no se hubiese

vergangenen hunderten Millionen von Jahren nicht so gut entwickeln können. Kurz: Von außen droht der Erde wenig Gefahr. Mit den kosmischen Fährnissen ist unser Planet in seiner langen Geschichte offenbar ganz gut zurechtgekommen. Doch nun droht Gefahr von innen. Das Leben hat sich so rasant entwickelt, dass erstmals die Bewohner auf der Erde in der Lage sind, sich selbst die Lebensgrundlage zu entziehen.

Ob sich diese Entwicklung aufhalten lässt, ist völlig ungewiss. Die Raumfahrtagenturen weltweit haben sich zumindest der Aufgabe verschrieben, unseren Planeten bestmöglich zu schützen, indem man ihn so gut wie möglich erkundet. Schutz durch Wissen: Nur wenn wir – vor allem dank der Satelliten, die die Erde permanent im Blick haben – verstehen, was auf unserem Planeten geschieht, welche Prozesse bedrohlich sind und wie man sie zum Besseren beeinflussen kann, werden wir uns hier dauerhaft aufhalten können.

Dass Erdbeobachtung und das Erkennen von Gefahren für das Leben auf der Erde keineswegs nur vergebliche Mühe ist, macht das Beispiel Ozon deutlich. Das Ozonloch und die Gefahren der ungehemmten Ultraviolettstrahlung waren viel diskutiertes Thema Ende der achtziger und Anfang der neunziger Jahre. Chemiker hatten schnell Fluorchlorkohlenwasserstoffe als Zerstörer des Ozons ausgemacht. Diese Stoffe fanden unter anderem als Treibgas in Spraydosen und als Kältemittel in Kühlschränken Verwendung. Die dramatischen Bilder vom Ozonschwund in der Atmosphäre heizten die Debatten an. Der Einsatz der gefährlichen Chemikalien wurde bald verboten. Gut ein Vierteljahrhundert nach seiner Entdeckung beginnt sich das Ozonloch langsam wieder zu schließen. Satellitendaten deuten an, dass sich der Ozongehalt in der oberen Atmosphäre wieder erholt. Das Ozonloch, einst klassisches Beispiel für die Bedeutung der Erdbeobachtung und wie Satelliten die Symptome der Erde diagnostizieren, kann nun zum Beispiel werden, wie Informationen aus dem All Entscheidungen auf der Erde positiv beeinflussen.

its long history. There is, however, now a danger within: life has developed so phenomenally quickly that the Earth's inhabitants are for the first time in a position to alter the very conditions which make life possible.

Whether this is a development that can be averted remains to be seen. Space agencies around the world have at least taken it upon themselves to protect the planet as best they can by exploring it as closely as possible. Such protection comes from knowledge gained principally from the satellites which now unceasingly observe the Earth—only when we understand what is happening on our planet, which processes are dangerous, and how we can influence them for the better, will we be able to maintain our existence here in the long term.

Proof that observation of the Earth and the recognition of the dangers presented to life here is not wasted effort is to be found with ozone. The hole in the ozone layer and the dangers of extensive exposure to ultraviolet rays were much-discussed topics at the end of the 1980s and the beginning of the 1990s. Chemists were quick to identify chlorofluorocarbons, substances employed as propellants for spray cans and coolants for refrigerators, as the agents of the ozone layer's destruction. Dramatic illustrations of falling levels of ozone in the atmosphere intensified debate, the use of these dangerous chemicals was soon banned, and a quarter of a century after its discovery, the hole in the ozone layer is slowly beginning to mend. Satellite data has revealed that ozone levels in the upper atmosphere are also recovering. The hole in the ozone layer, once a classic example of the importance of Earth observation and the use of satellites to diagnose the Earth's symptoms, now serves as an example of how information from space can have a positive influence on decisions made on the surface.

n'est pas menacée outre mesure par les dangers extérieurs. Notre planète semble s'être au cours de sa longue histoire assez bien accommodée des menaces cosmiques. Les véritables dangers proviennent désormais de l'intérieur. La vie a évolué à un tel rythme que les Terriens sont pour la première fois en mesure de détériorer eux-mêmes les conditions indispensables à leur vie.

Il n'est absolument pas certain que cette évolution puisse être enrayée. Les agences spatiales internationales se sont malgré tout fixé pour mission de protéger notre planète de leur mieux en l'explorant avec la plus grande précision possible. Étudier pour mieux protéger : si nous parvenons à comprendre, à l'aide des satellites qui surveillent la Terre en permanence, les phénomènes qui régissent la planète, les processus qui la menacent et la manière de les corriger, alors nous pourrons y séjourner durablement.

L'exemple de la couche d'ozone a clairement démontré les bienfaits de l'observation de la Terre et de l'identification des menaces. Le trou de la couche d'ozone et les dangers liés au rayonnement ultraviolet direct ont largement alimenté les discussions à la charnière des années 1980 et 1990. Les chimistes ont rapidement attribué la destruction de l'ozone aux chlorofluorocarbones, les fameux CFC. Ces gaz étaient notamment présents dans les bombes aérosols et les appareils réfrigérants. Les images dramatiques de la diminution de la couche d'ozone stratosphérique échauffaient les débats. L'emploi de ces produits toxiques n'a pas tardé à être interdit. Un bon quart de siècle après sa découverte, le trou d'ozone commence lentement à se refermer. Les données satellitaires indiquent par ailleurs que la concentration d'ozone dans la stratosphère se rétablit. Le trou dans la couche d'ozone, autrefois exemple classique de l'intérêt de l'observation de la Terre et de la faculté des satellites à diagnostiquer les maux de la planète, illustre désormais la capacité des informations spatiales à influencer les décisions prises sur Terre de manière positive.

podido desarrollar en los últimos cientos de millones de años. En definitiva, las amenazas exteriores tampoco parecen tan peligrosas. En toda su larga historia, nuestro planeta parece haber sabido arreglárselas con los peligros cósmicos. Pero ahora, el peligro amenaza desde dentro. La vida ha sufrido una evolución tan vertiginosa que, por primera vez, los habitantes de la Tierra están en condiciones de comprometer su propia subsistencia.

No se puede saber si se podrá detener esta evolución. Las agencias espaciales de todo el mundo se han entregado a la tarea de proteger nuestro planeta lo mejor posible, explorándolo hasta los límites disponibles. Protección mediante conocimientos: únicamente podremos permanecer aquí de forma indefinida si entendemos lo que sucede en el planeta, sobre todo gracias a los satélites que someten al planeta a una observación permanente para saber qué procesos son amenazadores y cómo se puede influir en obtener un beneficio de ellos.

La observación terrestre y el reconocimiento de los peligros para la vida en la Tierra no es en absoluto un esfuerzo inútil, como prueba el ejemplo del ozono. El agujero de ozono y los peligros de la falta de trabas a la radiación ultravioleta fueron asuntos muy debatidos a finales de la década de 1980 y a principios de la siguiente. Los químicos señalaron a los clorofluorocarbonos como los compuestos destructores del ozono. Estas sustancias se utilizaban como gases propelentes en aerosoles y refrigerantes en frigoríficos. Pronto se prohibió el uso de esas sustancias químicas peligrosas. Casi un cuarto de siglo después de su descubrimiento, el agujero de ozono comienza lentamente a cerrarse de nuevo. Los datos de los satélites dan a entender que se recupera de nuevo la concentración de ozono en la atmósfera superior. El agujero de ozono, que un día fue el ejemplo clásico del significado de la observación terrestre a través de los satélites, puede ser ahora el ejemplo de cómo esa información puede influir positivamente en las decisiones que se toman en el planeta azul.

Edrengiyn Nuruu ist die Übergangszone von den mongolischen Steppen im Norden zu den Wüsten in China (Landsat)

The Mongolian Steppe to the north merges with the deserts of China at Edrengiyn Nuruu (Landsat)

Edrengiyn Nuruu est la zone de transition entre les steppes mongoles au nord et le désert chinois (Landsat)

Edrengiyn Nuruu es la zona de transición entre las estepas mongolas del norte y los desiertos de China (Landsat)

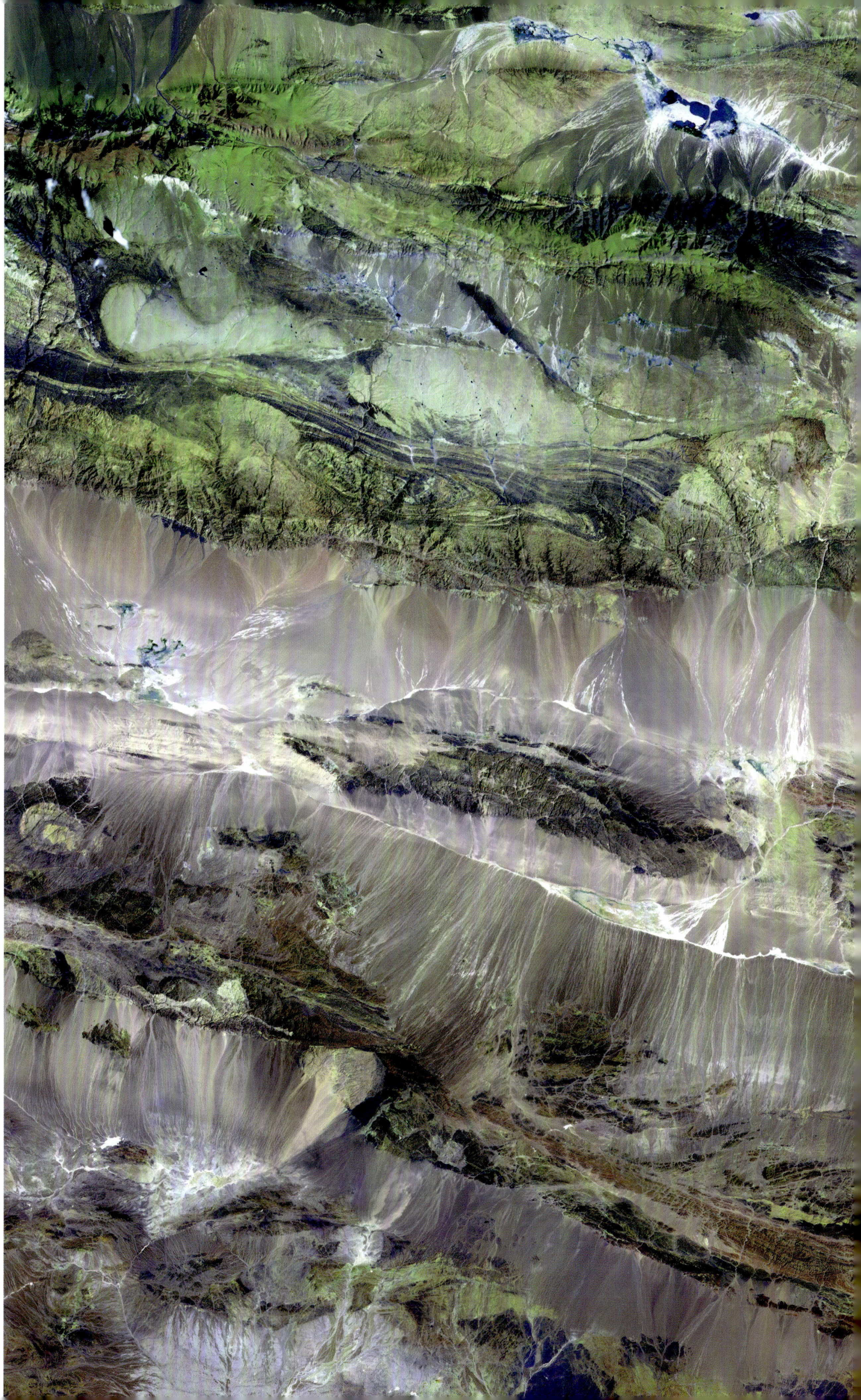

Seite 218: Great Sandy Desert in Australien. Die Wüste prägt eine sehr komplexe geologische Struktur (Landsat)

Page 218: The Great Sandy Desert in Australia. The wilderness is just part of an extremely complex geological structure (Landsat)

Page 218 : le Grand désert de sable en Australie. Le désert se caractérise par une structure géologique complexe (Landsat)

Página 218: Gran Desierto arenoso de Australia. El desierto se caracteriza por una compleja estructura geológica (Landsat)

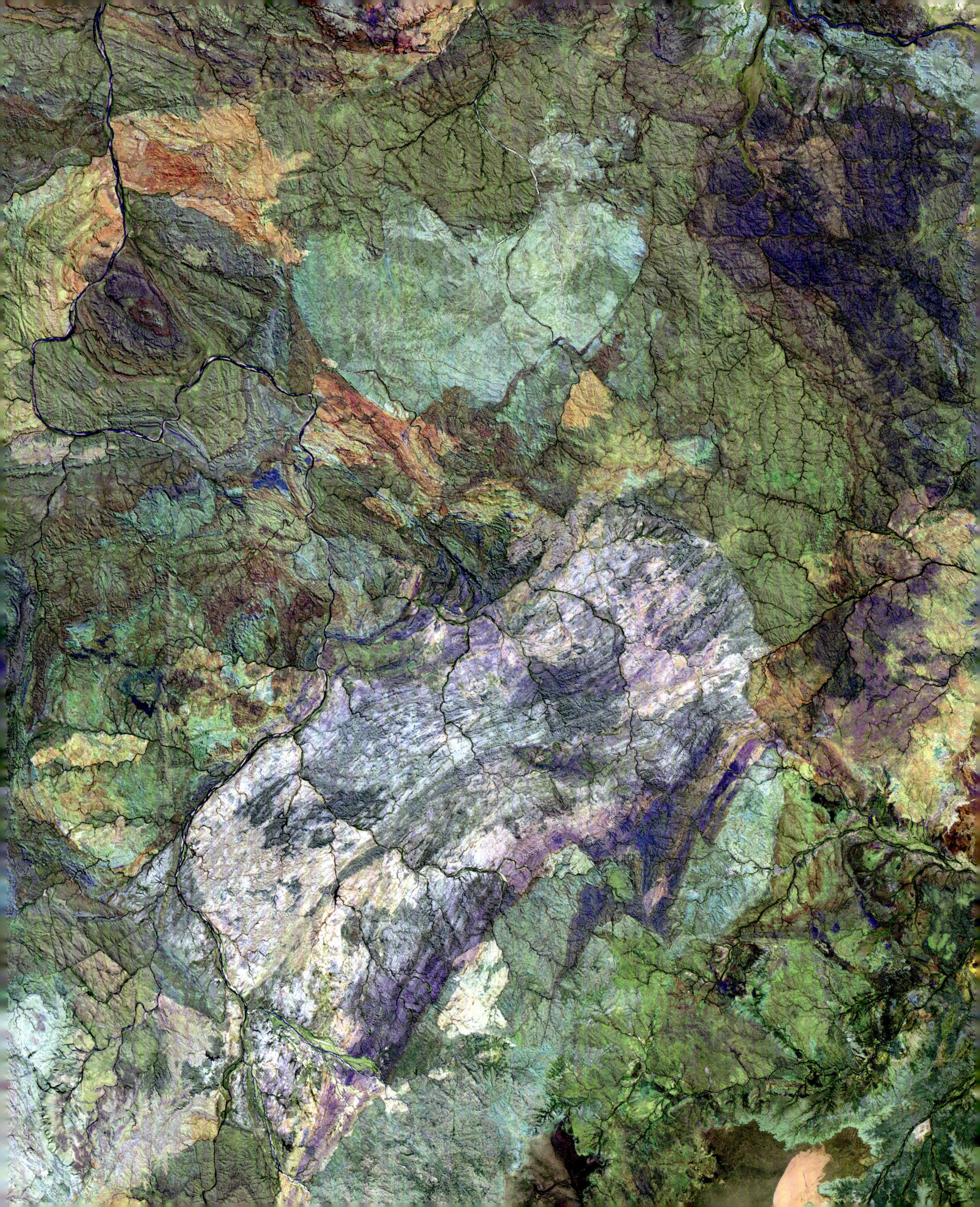

Erdbeobachter beobachten

Die Satelliten im All können uns nur informieren. Ändern können sie die Zustände auf der Erde nicht. Oder nur mittelbar: Denn manche Menschen verdanken ihre Arbeit mit dem Kosmos und der Erde dem Eindruck des wunderbaren Schauspiels, wenn ein Satellit am Firmament leuchtet. *Mein Interesse für das All wurde geweckt, als mein Vater mir nachts die Satelliten zeigte, die um die Erde kreisen,* erinnert sich Augustin Chicarro. Strahlend hell zieht häufig die Internationale Raumstation ISS von West nach Ost über den Himmel. Halten Sie in klarer Nacht Ausschau, ob Sie deutlich schwächere Lichtpunkte in etwa von Nord nach Süd oder Süd nach Nord zwischen den Sternen entlanglaufen sehen. Wenn Sie so ein Objekt entdecken, kann es ein Spionagesatellit sein, eine alte Raketenstufe oder eben einer der Erdbeobachtungssatelliten. Für Augustin Chicarro hat so ein Ereignis das Berufsleben bestimmt. Heute forscht er am Mars – und damit für die Erde. Wer weiß, was das Entdecken eines Satelliten mit Ihrem Leben macht.

Die Erdbeobachtung wird immer dringlicher. Dieser Wissenschaftszweig braucht alle klugen Köpfe, die sich für ihn interessieren. Die Erde zu untersuchen ist eine charmante Mischung aus Astrophysik, Geologie, Chemie, Biologie, Meteorologie, Meereskunde und vielen Disziplinen mehr. Die Aufgabe ist ebenso überwältigend wie das Vordringen in die Anfänge des Kosmos – und in gewisser Weise ebenso aussichtslos. Denn werden wir jemals das komplexe System Erde vollständig durchschauen? Das klingt genauso wahrscheinlich wie die Aussicht, künftig den Urknall komplett zu verstehen. Aber gerade die Beschäftigung mit dem fernen Kosmos zeigt, was für ein grandioser Ort unsere Erde in diesem unendlichen Universum ist.

Funkelnde Sternhaufen, Schwarze Löcher, ferne Galaxien und bunt schillernde Gaswolken mögen von betörender Schönheit sein. Sie mögen uns faszinieren und anspornen, den Horizont unserer Erkenntnis immer

Watching the Earth watchers

All satellites in space can do is provide information. They cannot change conditions on Earth. At least not directly, although there are many whose work with the cosmos and the Earth was inspired by the magnificent sight of a satellite rising into the night sky. *My interest in space was aroused in the evenings when my father would show me the satellites orbiting the Earth,* remembers Augustin Chicarro. The International Space Station (ISS) can be seen as a regular, brilliant streak across the sky from west to east. On a clear night, keep an eye out for much fainter points of light running between the stars from north to south or south to north. If you see such an object, it may be a spy satellite, an old rocket stage, or even an Earth observation satellite. Such an event determined Augustin Chicarro's choice of career, and he is now doing Mars research—for the sake of the Earth. Who knows what effect the sight of a satellite will have on your life?

Observing the Earth is of ever greater importance and this branch of science needs every smart brain that expresses an interest. Exploring the Earth is a beguiling mixture of astrophysics, geology, chemistry, biology, meteorology, oceanography, and many other disciplines. The task is as momentous as exploring the beginnings of the cosmos, and in some respects just as doomed—will we ever really understand the complex mechanisms of our Earth? It seems as unlikely as the hope of fully understanding the Big Bang, but it is just this study of the far cosmos which brings home to us what a fantastic place our Earth is in this boundless universe.

Twinkling star clusters, black holes, distant galaxies, and colorful gas clouds may be intoxicatingly beautiful and fascinating, inspiring us to push back the boundaries of our knowledge, but they are not a place where we can live. Galaxies merge and stars explode all the time in the universe, but our Earth

Les observateurs observés

Les satellites en orbite se contentent de fournir des informations. Ils sont incapables de modifier les conditions terrestres, si ce n'est indirectement : le merveilleux spectacle des satellites étincelant dans le firmament a en effet suscité des vocations pour l'étude du cosmos et de la Terre. *C'est en me montrant la nuit les satellites qui gravitaient autour de la Terre que mon père a éveillé mon intérêt pour l'espace*, se souvient Augustín Chicarro. Les lumières radieuses de la Station spatiale internationale (ISS) traversent souvent le ciel d'ouest en est. Observez le ciel, la nuit par temps dégagé, et voyez si vous distinguez des points lumineux sensiblement plus faibles qui zigzaguent entre les étoiles en direction du nord ou du sud. Si vous en repérez un, il peut s'agir d'un satellite espion, de l'ancien étage d'un lanceur, ou justement d'un satellite d'observation de la Terre. C'est ce genre d'expériences qui a influencé la carrière professionnelle d'Augustín Chicarro. Aujourd'hui il étudie Mars, au profit de la Terre. Qui sait l'influence qu'aura sur votre vie la découverte d'un satellite.

Il est devenu impératif d'observer la Terre. Cette discipline scientifique aura besoin de tous les esprits futés qui s'intéressent à elle. L'exploration de la Terre est un séduisant cocktail d'astrophysique, géologie, chimie, biologie, météorologie, océanographie, entre autres. Sa mission est tout aussi passionnante que l'étude de la naissance du cosmos, et d'une certaine manière tout aussi impossible. Serons-nous en effet jamais en mesure de déchiffrer entièrement ce système complexe qu'est la Terre ? Cela semble aussi probable que la perspective d'expliquer le Big-Bang dans son intégralité. Mais c'est justement l'étude du lointain cosmos qui révèle la splendeur de notre Terre dans l'immensité de l'univers.

Amas d'étoiles scintillants, trous noirs, galaxies lointaines et nuages gazeux multicolores sont peut-être d'une beauté envoûtante. Ils nous fascinent et nous incitent à repousser

Los observadores observados

Los satélites espaciales solo nos pueden informar. No son capaces de modificar la situación de la Tierra o, al menos, no directamente. *Mi interés en el espacio se despertó cuando mi padre me mostraba por las noches los satélites que giraban alrededor de la Tierra*, recuerda Agustín Chicarro. La estación espacial internacional ISS pasa frecuentemente por el cielo de oeste a este, con su claridad resplandeciente. En una noche clara, intente detectar a simple vista algún objeto moviéndose entre las estrellas, como un débil punto de luz que se desplaza de norte a sur o de sur a norte. Si descubre algún objeto, es posible que sea un satélite espía, una antigua fase de un cohete o incluso uno de los satélites de observación de la Tierra. Un acontecimiento así fue el que determinó la vida profesional de Agustín Chicarro. Hoy investiga en Marte y, por tanto, para la Tierra.

La observación terrestre es cada vez más apremiante. Esta rama de la ciencia necesita de todos los sabios que se interesan por ella. Investigar la Tierra es una encantadora mezcla de astrofísica, geología, química, biología, meteorología, oceanografía y muchas otras disciplinas. La tarea es tan imponente como penetrar en el origen del cosmos y, en cierto modo, tan poco prometedora. ¿Llegaremos algún día a comprender íntegramente el complejo sistema Tierra? Esto suena tan probable como la perspectiva de entender completamente la Gran Explosión. Pero precisamente es la reflexión sobre el remoto cosmos la que nos muestra la clase de lugar que es nuestra Tierra en este universo infinito.

Los centelleantes cúmulos estelares, agujeros negros, galaxias lejanas y nubes de gases de tornasolados colores, pueden llegar a ser de una hermosura fascinante. Pueden cautivarnos e incitarnos a ampliar el horizonte de nuestro conocimiento. Pero en ninguno de ellos hay sitio para la vida. En el cosmos, de forma permanente, las galaxias se fusionan y estallan estrellas. Pero nuestra

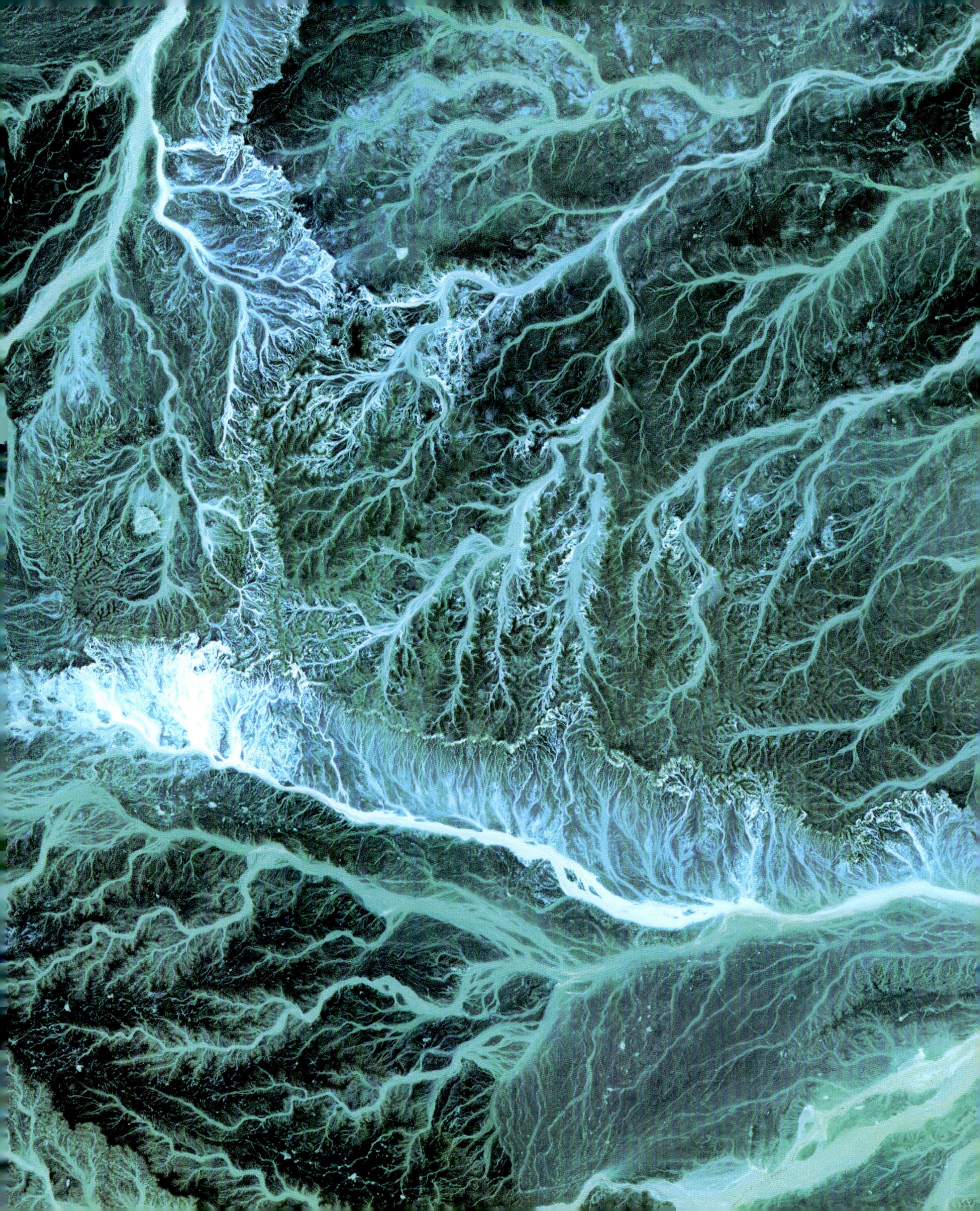

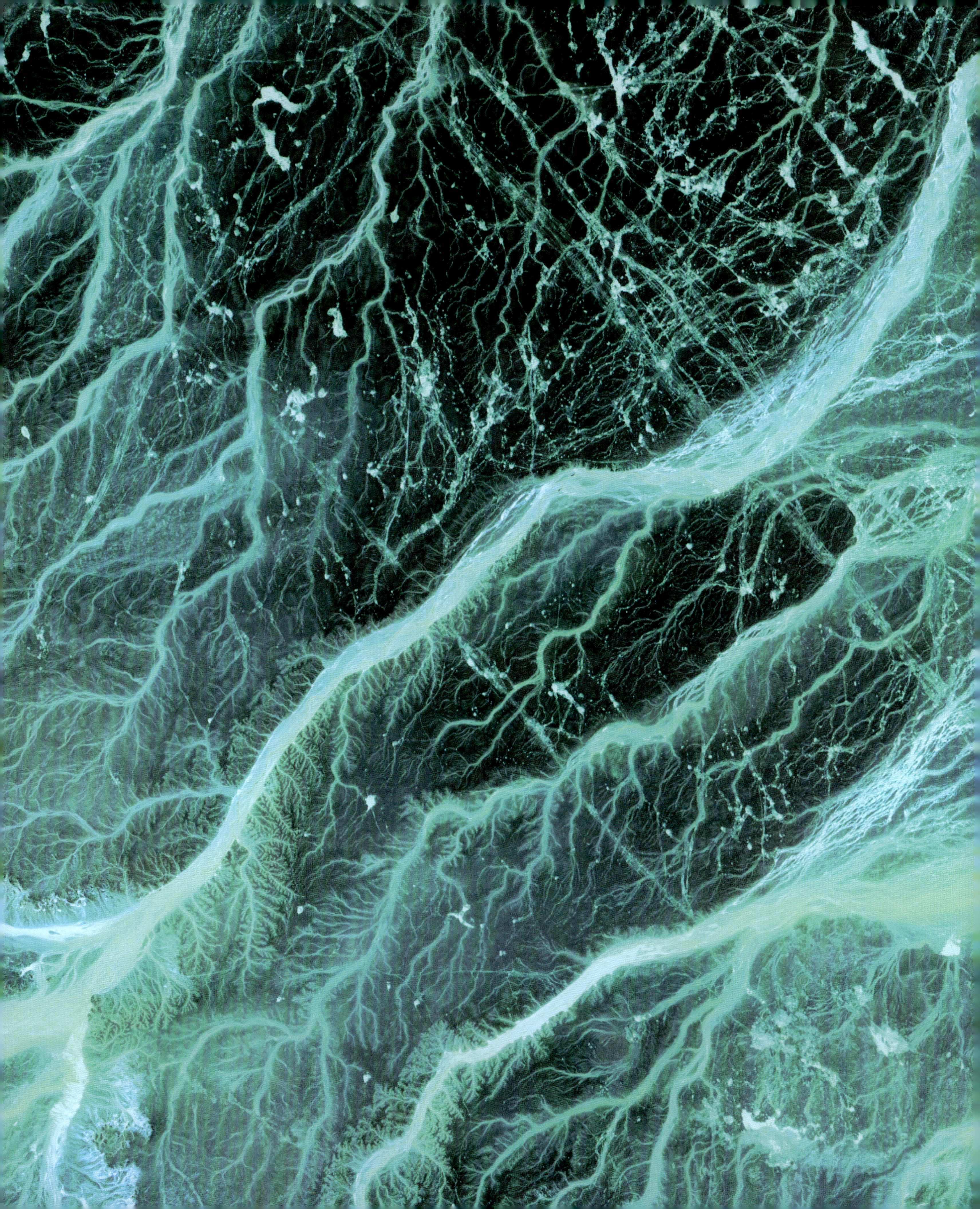

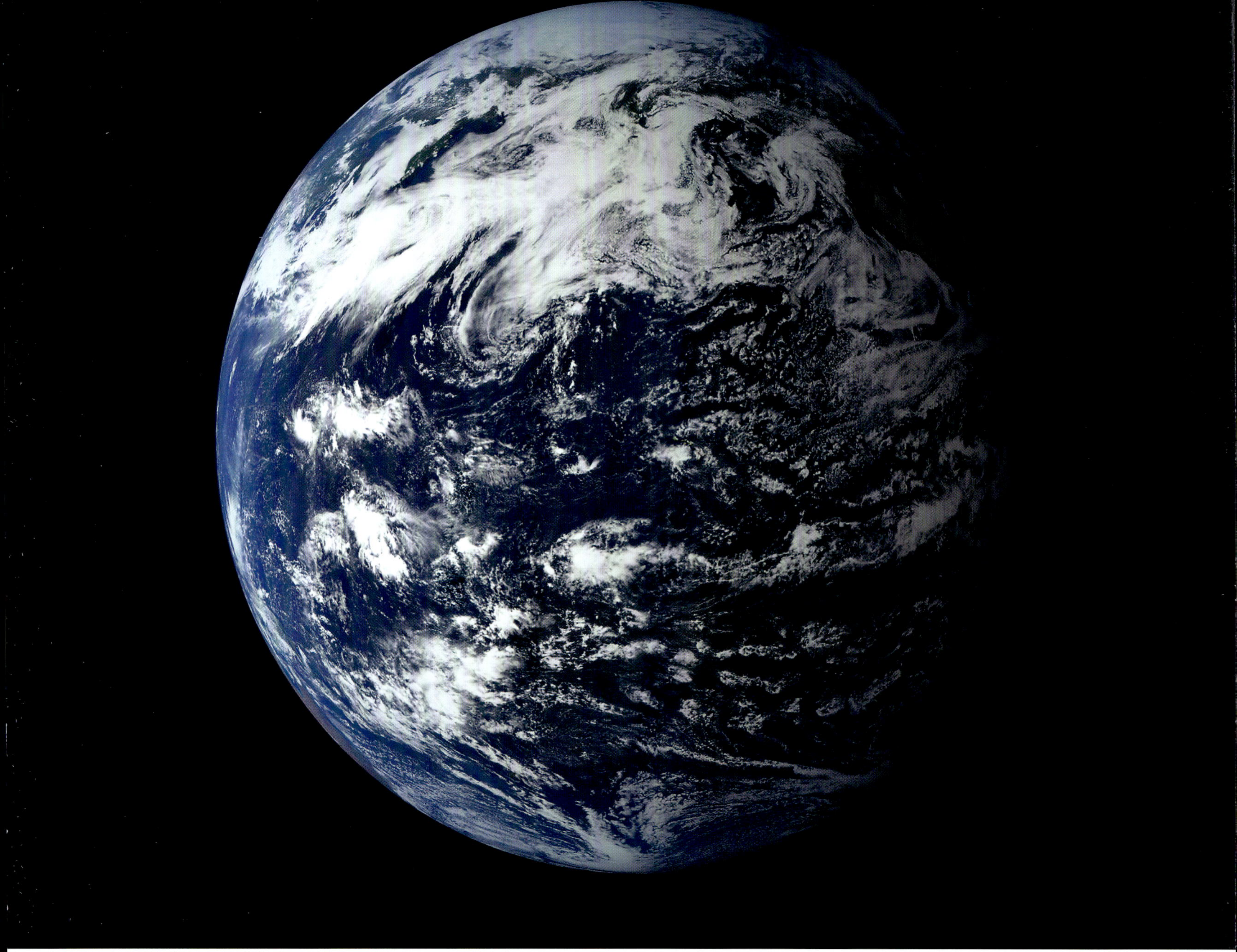

Seite 220/221: Durch die weit verzweigten Wadis des Jordantals fließt nur selten Wasser (Terra)

Pages 220–1: Water rarely flows through the dendritic pattern of wadis in the Jordan Valley (Terra)

Pages 220 et 221 : l'eau coule très rarement dans les vallées sèches du bassin du Jourdain (Terra)

Páginas 220/221: Raras veces el agua fluye por los muy ramificados wadis del valle del Jordán (Terra)

Oben: Blauer Planet – der pazifische Ozean dominiert eine Hälfte unserer Erde (Terra)

Above: The blue planet—the Pacific Ocean covers half of our Earth (Terra)

Ci-dessus : la planète bleue. L'océan Pacifique recouvre la moitié de notre Terre (Terra)

Arriba: Planeta azul: el océano Pacífico domina un hemisferio de nuestra Tierra (Terra)

Rechts: Erde und Mond im All. Foto der Messenger-Sonde aus 180 Millionen km Entfernung

Right: The Earth and the Moon in space. A photograph taken by the Messenger probe at a distance of 110 million miles (180 million km)

À droite : la Terre et la Lune dans l'espace. Photo acquise par la sonde Messenger à 180 millions de km de distance.

Dcha.: La Tierra y la Luna en el espacio. Foto de la sonda Messenger a 180 millones de kilómetros de distancia

weiter hinauszuschieben. Aber sie alle sind kein Platz zum Leben. Im Kosmos verschmelzen Galaxien, und ständig explodieren Sterne. Doch unsere Erde behauptet sich tapfer in den lebensfeindlichen Weiten des Weltalls. Davor kann man nur Respekt haben – und ein wenig Dankbarkeit könnte auch nicht schaden.

Die Erde zu erkunden geht nicht zum Nulltarif. Immerhin etwa eine Milliarde Euro pro Jahr lassen sich die ESA und ihre Mitgliedsstaaten die Erforschung und wissenschaftliche Überwachung unserer kosmischen Heimat kosten. Lohnt sich dieser finanzielle Einsatz? Gefragt, ob das legendäre Hubble-Weltraumteleskop die fast fünfzehn Milliarden US-Dollar wert sei, entgegnete einst der US-Astronom Mario Livio trocken, dass Hubble uns das Universum geschenkt habe – und dafür sei es sehr billig. Was Hubble für die Tiefen des Kosmos ist, sind die Sentinels, Swarm, GOCE, SMOS, Landsat & Co. für unseren Planeten: Sie öffnen uns die Augen. Sie geben uns die Erde zurück – und das ist jeden Euro wert

is holding its own in the hostile expanses of space. We can only respect this, and a little gratitude would not go amiss.

Exploring the Earth is not cheap. The ESA and its member states spend about a billion euros every year on scientific surveillance and research of our planetary home. Is this financial commitment paying off? When asked if the legendary Hubble telescope was worth the almost 15 billion US dollars it had cost, the US astronomer Marion Livio drily responded that it had given us the universe in return, and was therefore cheap at the price. What Hubble delivers from the depths of the cosmos is brought to us from our planet by the Sentinel, Swarm, GOCE, SMOS, Landsat, and the rest. They open our eyes and give us back the Earth, and that is worth every penny.

sans cesse les limites de nos connaissances. Mais aucun d'entre eux n'est propice à la vie. Le cosmos est le théâtre de fusions galactiques et d'explosions incessantes d'étoiles. Pourtant notre Terre se maintient vaillamment au sein de l'immensité hostile de l'univers. Un exploit qui mérite notre profond respect, et sans doute un minimum de gratitude.

L'exploration de la Terre n'est pas gratuite. L'ESA et ses États membres consacrent près d'un milliard d'euros par an à l'exploration scientifique et à la surveillance de notre berceau cosmique. De tels investissements sont-ils justifiés ? Alors qu'on lui demandait si le légendaire télescope spatial Hubble valait les quinze milliards de dollars qu'il avait coûté, l'astronome américain Mario Livio avait répondu : Hubble nous a offert l'univers, ce n'est donc pas cher payé. Hubble est au cosmos ce que les Sentinelles, Swarm, GOCE, SMOS, Landsat et leurs confrères sont à la Terre : ils nous ouvrent les yeux. Ils nous font prendre conscience de la singularité de la Terre – et cela n'a pas de prix.

Tierra se impone valiente en el hostil espacio para la vida que representa el universo. Ante ello sólo se puede manifestar respeto y quizá no estaría mal un poco de agradecimiento.

La exploración de la Tierra no es gratuita. La ESA y sus estados miembros gastan, al menos, la friolera de casi mil millones de euros anuales en investigación y control científico de nuestra patria cósmica. ¿Merece la pena esta apuesta financiera? A la pregunta de si el legendario telescopio espacial Hubble valía los casi quince mil millones de dólares que había costado, el astrónomo estadounidense Mario Livio respondía secamente que el Hubble nos había regalado el universo y que ya sólo por eso, había salido barato. Lo que el Hubble representa para las profundidades del cosmos, lo son para nuestro planeta los Sentinel, Swarm, GOCE, SMOS, Landsat y el resto de los satélites: ellos nos abren los ojos, nos devuelven la Tierra... y eso justifica hasta el último euro.

Picture Credits

Airbus: 72, 84, 85, 86–87, 144, 182

DLR: 6-7, 10–11, 33, 35, 38, 80, 81, 115, 117, 122, 127, 140, 177, 184, 185, 202–203

ESA: 5 near left, 5 centre, 8–9, 18, 25, 27 left, 28, 29, 30–31, 34, 36, 40, 41, 49 left, 49 right, 50, 53 top, 53 bottom, 55 top, 55 bottom, 56, 57, 58, 60, 62, 63 left, 63 right, 64 right, 67, 68, 69 right, 70, 71, 73, 77, 79 top, 82 all, 90, 91, 94, 95, 96 top, 96 bottom, 99, 100, 102 left, 102 right, 103, 106, 107, 108–109, 110, 111, 112, 114, 116, 121, 128, 129, 130, 131 left, 131 right, 133 top, 133 bottom, 135, 139, 141 left, 142 left, 142 right, 147, 163 bottom, 164, 168–169, 171, 174, 175, 176, 183, 191, 192, 193, 207, 208, 209, 210

ESA/Spot/CNES: 79 bottom, 97

Eumetsat: 5 far left, 22, 134

Infoterra: 64 left, 66, 118, 119, 120, 126, 180

JAXA/ESA: 47, 150–151, 153

NASA: front and back endpaper, 5 near right, 5 far right, 12–13, 21, 24 top, 24 bottom, 26, 27 right, 35, 37, 42, 46, 51, 65, 69 left, 76, 92, 93 left, 93 right, 98, 104–105, 136, 138, 141 right, 143 left, 143 right, 148 all, 149 all, 152, 154 top, 159 left, 159 right, 162, 163 left, 163 top, 165, 166 all, 167 all, 172–173, 181, 186, 188, 189, 190 all, 196, 197, 222, 223

NASA/DigitalGlobe: 145, 158 left, 158 right, 160 left, 160 right, 178 top, 178 bottom

NASA/GeoEye: 39, 154 bottom, 155, 156–157, 161 left, 161 right

Schulzendorff: 14

USGS/NASA: 44–45, 52, 59, 74–75, 88 top, 88 bottom, 125, 194–195, 198, 200, 201, 204, 205, 206, 211, 213, 214–215, 217, 218, 220–221

Special thanks to Mareike Döpke, Katrin Molch, Cordula Tegen, Robert Meisner, Siegfried Monser, Alexander Otto, and Andreas Schütz.

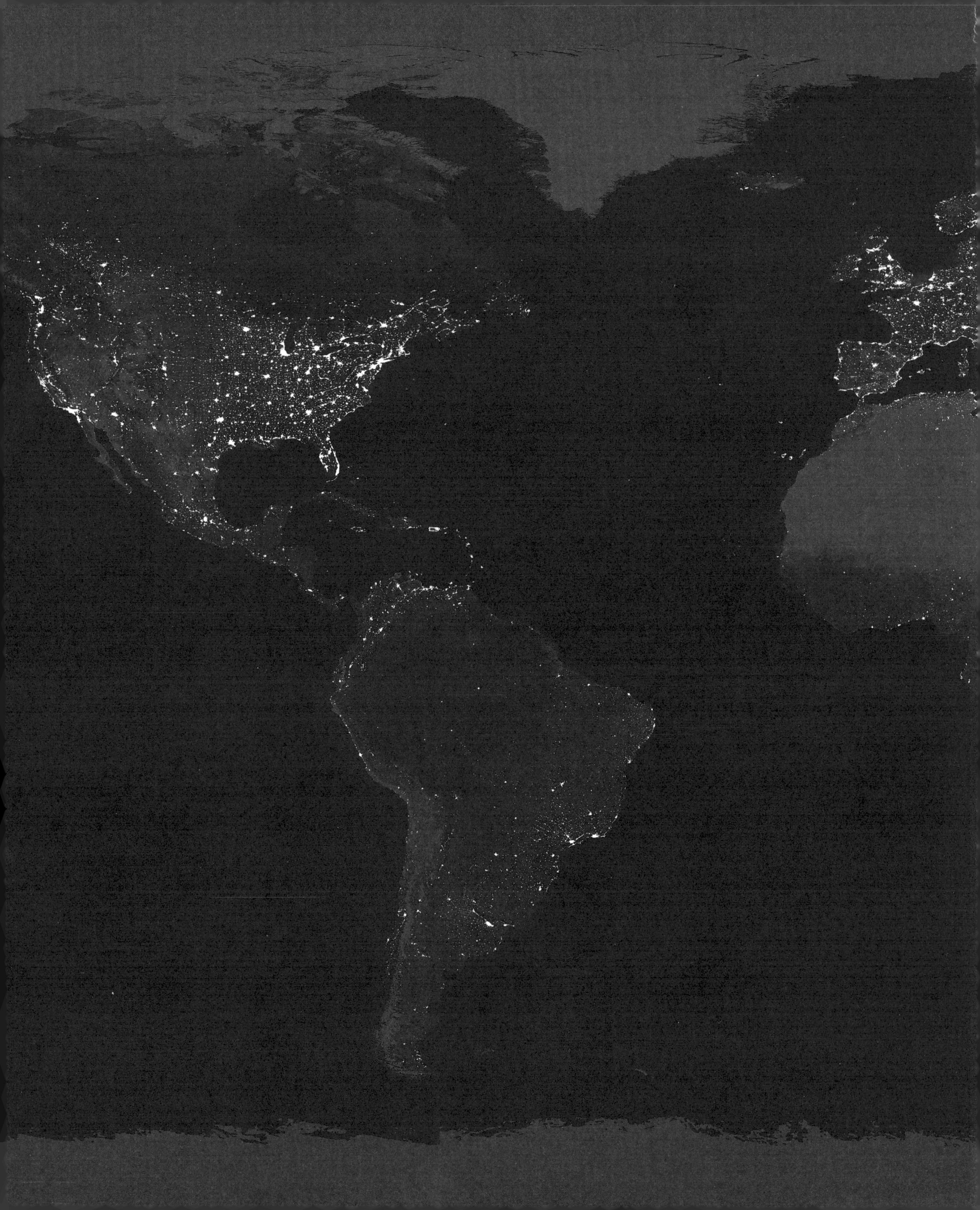